Multi-hazard Approaches to Civil Infrastructure Engineering

Paolo Gardoni • James M. LaFave
Editors

Multi-hazard Approaches to Civil Infrastructure Engineering

Springer

Editors
Paolo Gardoni
University of Illinois at Urbana-Champaign
Urbana, IL, USA

James M. LaFave
University of Illinois at Urbana-Champaign
Urbana, IL, USA

ISBN 978-3-319-29711-8 ISBN 978-3-319-29713-2 (eBook)
DOI 10.1007/978-3-319-29713-2

Library of Congress Control Number: 2016937406

Printed on acid-free paper

This Springer imprint is published by Springer Nature
The registered company is Springer International Publishing AG Switzerland

Preface

Natural and anthropogenic hazards cause billions of dollars of damage every year around the world, as well as several thousand fatalities and injuries and other societal impacts. The frequency of damaging natural events is expected to increase due to climate change, and their impact is also expected to increase due to growing populations in hazard prone regions. New engineering solutions are needed to help mitigate the effects of hazards and enhance the resilience of the built environment and communities. Past attempts at developing such engineering solutions have clearly shown that only integrated approaches, which incorporate knowledge beyond the traditional engineering disciplines, can be successful.

This volume focuses on the development of novel engineering approaches and interdisciplinary considerations needed to improve natural and anthropogenic hazard preparedness and mitigation. The volume addresses concerns related to overall safety, sustainability, and resilience of the built environment when subject to multiple hazards. While the primary attention is on engineering aspects of the problem, the volume has a strong interdisciplinary emphasis, with consideration also given to relevant aspects related to public policy, sociology, and law.

Most distinctively, this is the first book fully dedicated to the critical issue of multi-hazards. It examines a range of specific topics, including methodologies for vulnerability assessment of structures; new techniques to reduce structural demands through control systems; instrumentation, monitoring, and condition assessment of structures and foundations; new techniques for repairing structures that have suffered damage during past events or for structures that have been found in need of strengthening; development of new design and construction provisions that consider multiple hazards; novel considerations toward resilient infrastructure; as well as questions from law and the humanities pertaining to successful risk management for natural and anthropogenic hazards.

The edited volume collects contributions from some of the world's leading experts in each of the various fields covered. It builds upon the presentations, discussions, and related outcomes from the *International Conference on Multi-hazard Approaches to Civil Infrastructure Engineering* (ICMAE) that took place in Chicago, Illinois, on June 26–27, 2014, and it also includes a few additional contributions. Funding for the ICMAE was provided by the MAE Center: Creating a Multi-hazard Approach to Engineering.

Urbana, IL, USA Paolo Gardoni
Urbana, IL, USA James M. LaFave

Contents

Part VI Geo-Hazards

Part VII Societal Impact of Extreme Events

Part I
Introduction

Chapter 1
Multi-hazard Approaches to Civil Infrastructure Engineering: Mitigating Risks and Promoting Resilience

Paolo Gardoni and James M. LaFave

Abstract Natural and anthropogenic hazards cause billions of dollars of damage every year around the world, as well as several thousand fatalities or injuries and other societal impacts. While the frequency of damaging natural events is expected to increase due to climate change, new engineering solutions can help mitigate their impact and accelerate the recovery process. This volume is dedicated to the critical issue of creating successful solutions to multiple hazards. It examines a range of specific topics, including methodologies for vulnerability assessment of structures; new techniques to reduce structural demands through control systems; instrumentation, monitoring, and condition assessment of structures and foundations; new techniques for repairing structures that have suffered damage during past events, or for structures that have been found in need of strengthening; development of new design and construction provisions that consider multiple hazards; novel considerations toward resilient infrastructure; as well as questions from law and the humanities pertaining to successful management of natural and anthropogenic hazards. This book contains contributions from some of the world's leading experts in each of the fields covered by the edited volume.

The volume is organized into six main parts, after the introduction (Part I). Part II focuses on probabilistic methods and formulations needed for risk analysis. Part III starts the discussion on multiple hazards, by presenting recent advancements in earthquake engineering and then introducing concepts on disaster resilience and optimization. Part IV begins with a discussion of fires following earthquakes and then continues with other contributions specific to fire and blast, including both modeling and testing. Part V summarizes recent advances related to wind hazards, specifically considering tornadoes and hurricanes. Part VI focuses on geo-hazards, including the modeling of physical phenomena, condition assessment, and treatment of uncertainties. Finally, Part VII looks more generally at the impact of

P. Gardoni (✉) • J.M. LaFave
University of Illinois at Urbana–Champaign, Urbana, IL, USA
e-mail: gardoni@illinois.edu; jlafave@illinois.edu

P. Gardoni, J.M. LaFave (eds.), *Multi-hazard Approaches to Civil Infrastructure Engineering*, DOI 10.1007/978-3-319-29713-2_1

extreme events on society by discussing risk management, strategies for resilient communities, and new policy approaches. While each of Parts III–VI is primarily focused on a particular hazard, many of the parts include at least one chapter that looks beyond the specific hazard by bringing in considerations from a multi-hazard perspective and related to overall infrastructure resilience.

1.1 Introduction

Natural and anthropogenic hazards cause billions of dollars of damage every year, as well as several thousand fatalities or injuries and other societal impacts (Gardoni et al. 2016). While the frequency of damaging natural events is expected to increase due to climate change, new engineering solutions can help mitigate their impact and accelerate the recovery process. This edited volume focuses on the development of novel approaches and the interdisciplinary considerations needed to improve mitigation of natural and anthropogenic hazards. Work to date has focused on, and made considerable progress toward, the mitigation of individual hazards (earthquakes, wind, and so forth). The current volume addresses concerns related to overall safety, sustainability, and resilience of the built environment when subject to multiple hazards.

Multi-hazards can be classified as *concurrent* (e.g., wind and surge), *cascading* (e.g., fire following earthquake), or independent and likely to occur at different times (e.g., wind and earthquake). Two of the deadliest disasters due to *concurrent hazards* occurred in the Bay of Bengal in 1970 and Myanmar in 2008. A cyclone hit the coast of the Bay of Bengal and the storm surge killed approximately 500,000 people. In Myanmar, a storm surge killed more than 138,000 people. In 2005, the USA experienced one of the greatest recorded storm surges when Hurricane Katrina hit the coastline of the Gulf of Mexico. Another high storm surge hit the east coast of the USA when Hurricane Sandy landed in 2012. Fires following earthquakes are an example of *cascading hazards*. A number of seismic events around the world (including in Chile, Japan, and the USA) have been characterized by significant damage and loss of lives caused by fires following earthquakes. Finally, while some regions around the world might be at risk for a single hazard type, many parts of the world are likely to face multiple hazard types for which the occurrences are independent (*independent hazards*), and as a result they typically occur at different times.

When considering the annual probability of failure or of reaching a specified damage state for a structure/infrastructure, accounting for the most relevant hazards and their possible dependency is essential, instead of simply considering a single hazard type. Within the USA, for example, seismic, wind (tornado), and flood hazards have been among the most damaging. A number of regions in the Midwest are vulnerable to both seismic and wind hazards, and other regions to both wind and flood. Similarly, much of the coastline of Japan is at risk for both earthquakes and tropical storms.

1.2 Overall Goal and Contributions

The goal of this edited volume is to promote mitigation of the impact of natural and anthropogenic hazards on society. The volume has a strong interdisciplinary emphasis, focusing on issues related to reliability analysis, risk determination, risk evaluation, and risk management for various hazards, as well as on disaster response and recovery. While the primary attention is on engineering aspects of the problem, consideration is also given to relevant interdisciplinary aspects related to public policy, sociology, and law.

There has been a considerable amount of research on assessment of the seismic vulnerability of structures, development of new methods to control earthquake forces transmitted to a structure (through base isolation or energy dissipation devices), incorporation of newly acquired knowledge into design provisions, assessment of the condition of existing structures, and planning of retrofit and repair strategies using both conventional and new materials. However, there is now a need to put all these developments into proper perspective, to critically examine the many methodologies and techniques available, to recommend the most appropriate ones for each case, and to identify remaining research needs.

Furthermore, while the seismic hazard has in past years received significant attention, many other hazards (like those related to water and wind) have caused substantial damage and had a significant societal impact. Therefore, there is a clear need to make progress toward mitigation of multiple hazards and, when appropriate, to expand the methodologies developed for the seismic hazard to other hazards. In addition, while work has been done considering one hazard at a time, there has only been limited work toward a uniform reliability for infrastructure considering multiple hazards.

This volume is the first book fully dedicated to the critical issue of multi-hazards. It examines a range of specific topics, including methodologies for vulnerability assessment of structures; new techniques to reduce structural demands through control systems; instrumentation, monitoring, and condition assessment of structures and foundations; new techniques for repairing structures that have suffered damage during past events, or for structures that have been found in need of strengthening; development of new design and construction provisions that consider multiple hazards; novel considerations toward resilient infrastructure; as well as questions from law and the humanities pertaining to successful management of natural and anthropogenic hazards. The edited volume collects contributions from some of the world's leading experts in each of the various fields covered.

1.3 Structure and Overview of the Volume

This volume is organized into six main parts, after the introduciton (Part I). Because of the underlying uncertainties associated with natural hazards and their impact, Part II focuses on probabilistic methods and formulations needed for risk analysis. Part

III starts the discussion on multiple hazards, by presenting recent advancements in earthquake engineering and then introducing concepts on disaster resilience and optimization. An earthquake might be followed by fire, and so Part IV begins with a discussion of fires following earthquakes and then continues with other contributions specific to fire and blast, including both modeling and testing. Part V summarizes recent advances related to wind hazards, specifically considering tornadoes and hurricanes. Part VI focuses on geo-hazards, including the modeling of physical phenomena, condition assessment, and treatment of uncertainties. Finally, Part VII looks more generally at the impact of extreme events on society by discussing risk management, strategies for resilient communities, and new policy approaches. While each of Parts III–VI is primarily focused on a particular hazard, many of the parts include at least one chapter that looks beyond the specific hazard, bringing in considerations from a multi-hazard perspective and related to overall infrastructure resilience.

Part II focuses on probabilistic methods for risk analysis. Hazard maps are often used in risk analysis to express spatial variability in the likelihood of occurrence of hazards, but such maps have only been developed to provide the probability that an intensity measure of interest at a given location on the map will exceed a specified threshold (typically in 1 year). However, as Paolo Bocchini, Vasileios Christou, and Manuel Miranda, authors of the first chapter in this part (Chap. 2) titled "Correlated Maps for Regional Multi-Hazard Analysis: Ideas for a Novel Approach," point out, lifeline and regional risk assessment require knowledge of the spatial correlation between any two points in space. The chapter discusses a random field formulation to capture the spatial correlation between intensities of interest at any two points. The presented formulation is general and applicable to any hazard, and so it is therefore particularly well suited for multi-hazard analysis.

Aging and deterioration of structures and infrastructure can play an important role in their reliability and on the risk faced by communities. Chapter 3 titled "Supporting Life-Cycle Management of Bridges Through Multi-Hazard Reliability and Risk Assessment," by Jamie Padgett and Sabarethinam Kameshwar, studies the impact of multiple hazards on deteriorated structures (bridges in particular). The chapter uses meta-models to develop parameterized time-dependent bridge fragility estimates considering multiple hazards, including earthquakes and hurricanes. The chapter ends with applications showing the relative importance of earthquake and hurricane hazards when considering different bridge characteristics and states of deterioration.

Risk assessment often has high computational costs. Chapter 4 titled "Natural Hazard Probabilistic Risk Assessment Through Surrogate Modeling," by Alexandros Taflanidis, Gaofeng Jia, and Ioannis Gidaris, focuses on modeling techniques that can be used to alleviate such computational costs. Specifically, the authors discuss the use of kriging surrogate modeling, including its advantages over other approaches as well as implementation details. The effectiveness of the proposed kriging surrogate modeling is illustrated considering two hazards—in seismic risk assessment and hurricane risk assessment.

There is often a distinction between natural hazards (e.g., earthquakes, hurricanes, and tornadoes) and anthropogenic hazards (human errors and malevolent acts). However, some of the methods for risk analysis and decision-making can be common between the two types of hazard. Mark Stewart, in Chap. 5 titled "Risk and Decision-Making for Extreme Events: Climate Change and Terrorism," discusses how optimal decisions can be reached using risk-based approaches where not only the likelihood of the threat, the vulnerability to the threat, and the exposure to the threat are modeled probabilistically, but the effectiveness of protective strategies and their costs are also modeled accounting for their uncertainties. The chapter illustrates the presented methodology by considering risk-based assessment of counterterrorism and climate adaptation strategies.

Part III starts the discussion on multiple hazards by presenting recent advances in earthquake engineering and then introducing concepts on disaster resilience and optimization. Once hazard maps are defined (e.g., by using the procedure described in Chap. 2), there is a need to quantify the vulnerability of structures and infrastructure. In the case of seismic hazards, buildings that lack earthquake-resistant details and therefore are characterized by non-ductile behavior are particularly vulnerable. As a result, in case of a seismic event there is a high probability of partial or global collapse of such buildings. Khalid Mosalam and Selim Günay, in their chapter (Chap. 6) titled "Progressive Collapse Simulation of Vulnerable Reinforced Concrete Buildings," note that two of the current challenges are the identification of such vulnerable buildings and definition of the most effective and economical retrofitting strategies. The chapter discusses methods to model gravity load failure as well as the underlying uncertainties and research needs in this field.

Still on the subject of seismic assessment of existing buildings, Paolo Pinto and Paolo Franchin, in Chap. 7 titled "Probabilistic Seismic Assessment of Existing Buildings: The CNR-DT212 Italian Provisions," describe an overview of a probabilistic approach for seismic assessment of buildings developed by the Italian National Research Council (CNR). The approach is intended to provide a more rigorous theoretical base for revising the European code (Eurocode 8, Part 3) and is of particular value in cases where a more rigorous analysis is required. The chapter also focuses on the definition and modeling of the states of damage for both structural and nonstructural components.

Design optimization strategies have traditionally focused on considerations toward a single hazard, while Hussam Mahmoud and Akshat Chulahwat, in Chap. 8 titled "Multi-Hazard Multi-Objective Optimization of Building Systems with Isolated Floors Under Seismic and Wind Demands," extend concepts of optimization for design variables considering multiple hazards. The chapter optimizes a sliding slab system considering both seismic and wind hazards, indicating the effectiveness of the proposed sliding slab system to withstand seismic and wind loads, and of the optimization scheme in identifying the most suitable configuration of the system.

In addition to multi-hazards, consideration can also be given to other factors that might influence the design or retrofit of a structure. In their chapter (Chap. 9 titled "Energy Efficiency and Seismic Resilience: A Common Approach"), Gian Michele Calvi, Luis Sousa, and Cristiana Ruggeri note that seismic retrofit and enhancement for energy efficiency of buildings are currently subject to separate considerations as to their benefits and costs. This chapter presents a formulation for the assessment of integrated investment strategies targeted at improving both the seismic resilience and energy efficiency of buildings. The chapter shows how an optimal solution can be achieved through the proposed integrated approach, instead of considering seismic resilience and energy efficiency as separate individual aspects.

Part IV begins by shifting the focus of the discussion to fires following earthquakes and then continues with other treatment on fire and blast, including both modeling and testing. Chapter 10, by Negar Khorasani, Maria Garlock, and Paolo Gardoni, is titled "Probabilistic Evaluation Framework for Fire and Fire Following Earthquake," and it provides a probabilistic framework to evaluate the performance of a structure under fire and fire following earthquake, by studying the response of the structure considering several limit states and incorporating uncertainties in demand and capacity parameters. The multi-hazard framework is applied to a steel moment-resisting frame in order to evaluate structural performance under post-earthquake fires.

Recent events have shown that local damage to building frames is often followed by fire igniting near the location of damage, but in current practice little consideration is given to fire as a cascading hazard for progressive collapse-resistant design. With that in mind, Spencer Quiel, in his chapter (Chap. 11) titled "Progressive Collapse Resistance for Steel Building Frames: A Cascading Multi-Hazard Approach with Subsequent Fire," explores the effects of fire following an extreme event that causes failure of a column on the perimeter of a steel building frame, which is a representative damage scenario in the typical design case for assessing progressive collapse resistance. Results include estimates of time to collapse initiation and correlation between the level of remaining passive fire protection and collapse time, with overall guidance also provided in the chapter for the assessment of these hazards.

Still on the topic of fire performance of structures, Venkatesh Kodur, in Chap. 12 titled "Strategies for Enhancing Fire Performance of NSM FRP-Strengthened Reinforced Concrete Beams," discusses approaches for achieving required fire resistance in reinforced concrete (RC) beams that have been strengthened by the application of near-surface-mounted (NSM) fiber-reinforced polymers (FRP). NSM FRP strengthening is a common approach used for structural retrofit of RC members to enhance their flexure and/or shear capacity. Results from numerical and experimental studies are used to quantify the influence of various parameters on fire resistance of NSM FRP-strengthened RC beams, and guidelines are provided in the chapter for achieving optimum fire resistance in such cases.

Moving along to the blast hazard, Lauren Stewart and Bradley Durant note, in Chap. 13 titled "Experimental and Analysis Methods for Blast Mitigating Designs in Civil Infrastructure," that incorporation of blast and shock loading into a multi-hazard framework requires consideration of the mechanical/structural behavior of a component or system in the impulsive loading regime. For such high-rate loading, material and structural response must often be evaluated experimentally in order to produce basic mechanical properties, initial design validation, and final design acceptance for construction implementation into new or existing infrastructure. Using a case study of a curtain wall system for blast response, the chapter utilizes various analysis and experimental procedures to highlight the process of designing and validating systems for blast mitigation.

Part V includes recent advances related to wind hazards, specifically considering tornadoes and hurricanes. For example, in Chap. 14, titled "Woodframe Residential Buildings in Windstorms: Past Performance and New Directions," John van de Lindt and Thang Dao demonstrate that residential buildings in coastal areas are often at risk to hurricanes, which can result in both wind and storm surge damage, while tornadoes are one of the most devastating natural hazards that can occur all across the USA, both in terms of high wind loading and debris impact. Based in part on an examination of collected damage data, this chapter proposes a general procedure for performance-based wind engineering, as well as related research needs for its specific development and application to wood frame structures.

Typical tornado risk assessment methodologies currently used by various agencies utilize empirically derived loss models that rely on historical claim data for predicting future effects of tornadoes. Xinlai Peng, David Roueche, David Prevatt, and Kurtis Gurley, in their chapter (Chap. 15) titled "An Engineering-Based Approach to Predict Tornado-Induced Damage," note the potential limitations of this current approach and that a more rigorous strategy may be the development of engineering-based tornado damage assessment (ETDA) models, which can be made applicable to construction in any region and to any tornado size and strength. The chapter presents a framework for an ETDA of low-rise buildings, which is then shown to predict damage in good agreement with post-tornado damage observations for vulnerable non-engineered residential buildings.

As alluded to above, the landfall of a hurricane can involve different hazard sources (wind, wind-borne debris, flood, and rain) that interact to generate the overall hazard scenario for a given structure, and so hurricanes ought to be viewed as multi-hazard scenarios. With that in mind, Vipin Unnikrishnan and Michele Barbato, in Chap. 16 titled "Performance-Based Hurricane Engineering: A Multi-Hazard Approach," use a probabilistic performance-based hurricane engineering (PBHE) framework for risk assessment of pre-engineered and non-engineered residential structures subject to hurricane hazard. Results include annual probability of exceedance of various repair costs for the target residential buildings due to each hazard, as well as their combined effect, which highlights the importance of considering the interaction between different hazard sources.

Arindam Chowdhury, Mohammadtaghi Moravej, and Filmon Habte continue the treatment on hurricanes, in their chapter (Chap. 17) titled "Wall of Wind Research and Testing to Enhance Resilience of Civil Infrastructure to Hurricane Multi-Hazards." They report on two large-scale experimental wind studies, performed on low-rise building roof coverings, which included wind-induced roof surface pressure and roof panel deflection measurements. Observed failure modes under realistic wind loading conditions were different from what is typically observed using standard uniform pressure testing methods, and so the experiments reported in the chapter reveal new aspects of roof response to high wind speeds and highlight the importance of large-scale modeling of structures that can incorporate realistic component and connection details, as well as architectural features.

Part VI focuses on geo-hazards, including modeling of physical phenomena and condition assessment, as well as some related treatment of uncertainties. On that latter point, Robert Gilbert, Mahdi Habibi, and Farrokh Nadim, in their chapter (Chap. 18) titled "Accounting for Unknown Unknowns in Managing Multi-Hazard Risks," note that a significant challenge in managing multi-hazard risk pertains to accounting for the possibility of events beyond our range of experience. In such cases, classical statistical approaches and subjective assessments are of limited value, respectively, because there are no data to analyze and they are derived from within our range of experience. The chapter therefore proposes a new framework, called decision entropy theory, to assess probabilities and manage risks for possibilities in the face of limited information. From a practical perspective, the theory highlights the importance of considering how possibilities for natural hazards could impact the preferred alternatives for managing risks, underscoring the significance of developing adaptable approaches to manage multi-hazard risks.

In Chap. 19, titled "Bayesian Risk Assessment of a Tsunamigenic Rockslide at Åknes," Zenon Medina-Cetina, Unni Eidsvig, Vidar Kveldsvik, Sylfest Glimsdal, Carl Harbitz, and Frode Sandersen introduce the application of two methods for estimating the risk due to a potential tsunamigenic rockslide. The first method follows a classical approach for empirical relations between the risk components, while the second method follows a more recent approach based on Bayesian networks, which introduces the notion of causal effects. A key component in both approaches is the evidence assimilation of experts, who provide technical information and also their beliefs in terms of probability measures, which is a strategy that introduces a unique approach for incorporating fine engineering judgment into risk measures in a transparent and systematic manner. The chapter shows that rockslide risk estimates obtained from the methods yield significant qualitative differences in terms of inference capabilities, but their orders of magnitude for overall expected risk are relatively similar.

The next chapter (Chap. 20) investigates the relevant hydrologic and geotechnical processes triggering failure of steep hillside slopes under rainfall infiltration. This chapter, by Ronaldo Borja, Jinhyun Choo, and Joshua White, is titled "Rock Moisture Dynamics, Preferential Flow, and the Stability of Hillside Slopes." The work focuses on the triggering mechanisms of slope failure induced by rainfall events and highlights the multi-physical nature of the problem. Nonlinear finite element simulations of the failure of hypothetical hillside slopes, similar in configuration to two well-documented test slopes, are presented in the chapter, revealing the impacts of slope/bedrock topography, rainfall history, rock moisture dynamics, and preferential flow pattern on the failure of hillside slopes.

In Chap. 21, titled "Innovation in Instrumentation, Monitoring and Condition Assessment of Infrastructure," Kenichi Soga argues that the design, construction, maintenance, and upgrading of civil engineering infrastructure requires fresh thinking to minimize the use of materials, energy, and labor, which can only be achieved by understanding the performance of the infrastructure (both during its construction and throughout its design life) through innovative monitoring. It is hypothesized that the future of infrastructure relies on smarter information, such as the rich data obtained from embedded sensors, to act as a catalyst for new design, construction, operation, and maintenance processes for integrated infrastructure systems linked directly with user behavior patterns. The chapter also presents some examples of emerging sensor technologies.

Part VII focuses on the societal impact of extreme events—the four chapters in this part look more generally at the impact of extreme events on society by discussing risk management, strategies for resilient communities, and new policies. In the first chapter of the part (Chap. 22, titled "Theories of Risk Management and Multiple Hazards: Thoughts for Engineers from Regulatory Policy"), Arden Rowell looks at regulatory policy for managing multiple risks. After a discussion of the core challenges to multi-risk management faced by policymakers, the chapter then discusses three approaches for regulatory risk management: the precautionary principle, cost-benefit analysis, and the capabilities approach.

Chapter 23, titled "Disaster Risk Reduction Strategies in Earthquake-Prone Cities," by Lori Peek, Stacia Ryder, Justin Moresco, and Brian Tucker, describes specific risk reduction activities currently implemented in 11 earthquake-prone cities around the world. Understanding the tools and actual resources that practitioners and organizations have available can help in developing more effective strategies. The chapter is based on both survey results and in-depth interview data obtained from a variety of professionals in government, business, health care, and education, as well as from grassroots groups. The chapter ends with practical advice on how to develop effective hazard mitigation strategies.

A recent concept in risk reduction and recovery from extreme events is resilience. Chapter 24, titled "Community Resilience: The Role of the Built Environment," by Therese McAllister, brings in the notion of community resilience, namely, the ability of communities to prepare for a recovery from the occurrence of extreme events. Specifically, it looks at the roles of buildings and infrastructure systems in defining community resilience and argues that the dependencies/interdependencies between buildings and infrastructure systems are currently not properly modeled and accounted for. The chapter illustrates this limitation by considering the performance of the built environment in the cases of Hurricane Katrina and Superstorm Sandy. This chapter ends by putting forward recommendations to improve community resilience.

New technologies play an important role in mitigating risk and promoting resilience. In the final chapter of this volume (Chap. 25, titled "Digital Technologies, Complex Systems, and Extreme Events: Measuring Change in Policy Networks"), Louise Comfort examines methods of digital data collection and how data are analyzed, plus she describes models developed to evaluate the complex, dynamic interactions between extreme events and affected communities. The chapter presents specific applications considering three different types of hazard: a superstorm, an airport fuel leak, and a wildland fire. As a conclusion, the chapter presents a proposal to operationalize the monitoring and modeling of interactions between extreme events and the affected communities.

If only one point could be taken away from this volume, it would be that to promote mitigation of the impact of natural and anthropogenic hazards on society, there is a need to go beyond the traditional boundaries of hazard-specific research. Optimal strategies for the mitigation of risk and promotion of resilience should be developed in consideration of all the multiple, relevant hazards. Such strategies should consider the life cycle of a building or infrastructure (accounting for aging and deterioration.) Furthermore, technical engineering strategies can be successful only if they are well integrated with policymaking. This volume helps make a step toward the development of successful *multi-hazard approaches to civil engineering infrastructure*.

Acknowledgments This volume builds upon the presentations, discussions, and related outcomes from the *International Conference on Multi-Hazard Approaches to Civil Infrastructure Engineering* (ICMAE) that took place in Chicago, Illinois, on June 26–27, 2014. Funding for the ICMAE was provided by the MAE Center: Creating a Multi-Hazard Approach to Engineering. The opinions and findings presented in this volume are those of the authors of each chapter and do not necessarily reflect the views of the MAE Center or the editors.

Reference

Gardoni, P., Murphy, M., & Rowell, A. (2016). *Risk analysis of natural hazards*. Springer, International Publishing.

Part II
Probabilistic Methods for Risk Analysis

Chapter 2
Correlated Maps for Regional Multi-Hazard Analysis: Ideas for a Novel Approach

Paolo Bocchini, Vasileios Christou, and Manuel J. Miranda

Abstract The modeling of multiple hazards for regional risk assessment has attracted increasing interest during the last years, and several methods have been developed. Traditional hazard maps concisely present the probabilistic frequency and magnitude of natural hazards. However, these maps have been developed for the analysis of individual sites. Instead, the probabilistic risk assessment for spatially distributed systems is a much more complex problem, and it requires more information than what is provided by traditional hazard maps. Engineering problems dealing with interdependent systems, such as lifeline risk assessment or regional loss estimation, are highly coupled, and thus it is necessary to know the probability of having simultaneously certain values of the intensity measure (e.g., peak ground acceleration) at all locations of interest. Therefore, the problem of quantifying and modeling the spatial correlation between pairs of geographically distributed points has been addressed in several different ways. In this chapter, a brief review of some of these approaches is provided. Then, a new methodology is presented for the generation of an optimal set of maps, representing the intensity measure of a natural disaster over a region. This methodology treats the intensity measures as two-dimensional, non-homogeneous, non-Gaussian random fields. Thus, it can take advantage of probabilistic tools for the optimal sampling of multidimensional stochastic functions. Even with few sample maps generated in this way, it is possible to capture accurately the hazard at all individual sites, as well as the spatial correlation among the disaster intensities at the various locations. The framework of random field theory enables a very elegant formulation of the problem, which can be applied to all types of hazards with minimal adjustments. This makes the proposed methodology particularly appropriate for multi-hazard analysis.

P. Bocchini (✉) • V. Christou
Advanced Technology for Large Structural Systems (ATLSS) Engineering Research Center, Department of Civil and Environmental Engineering, Lehigh University, 117 ATLSS Drive, Imbt Labs, Bethlehem, PA 18015, USA
e-mail: paolo.bocchini@lehigh.edu; vac212@lehigh.edu

M.J. Miranda
Department of Engineering, 200D Weed Hall, Hofstra University, Hempstead, NY 11549-1330, USA
e-mail: manuel.j.miranda@hofstra.edu

P. Gardoni, J.M. LaFave (eds.), *Multi-hazard Approaches to Civil Infrastructure Engineering*, DOI 10.1007/978-3-319-29713-2_2

2.1 Introduction

The study of the effects of natural disasters on the built environment is an integral part of civil engineering, and a thorough data set that concisely presents the type and magnitude of the considered natural hazard was necessary. In particular, a probabilistic tool capable of providing the information on the magnitude of a hazard and the period of reoccurrence of the hazard event was important both for safe engineering design and for political and economical decision making. This led to the development of hazard maps, which have become an essential part of building code provisions and risk models used by the insurance industry.

Hazard maps are the most popular outcome of probabilistic hazard analysis. The information that hazard maps provide is the probability of a given intensity measure (IM) exceeding a certain threshold during a fixed period. The same information is sometimes provided as the value of the IM that is exceeded with a given probability over a fixed period. The IM of choice must be representative of the severity of an investigated natural extreme event effect at a certain location. In seismic engineering, for example, one metric to express the earthquake intensity at each site is the peak ground acceleration, whereas in hurricane hazard analysis, the magnitude of the wind speed is a popular representative metric for wind effects and the water elevation for storm surge effects.

Hazard maps have been successfully used for decades in structural design, assessment, and retrofit, but unfortunately this tool is only appropriate for the analysis of individual sites. For the analysis of distributed infrastructure systems, lifelines, and building stocks, hazard maps do not provide adequate information. For example, the total travel time of vehicles in a transportation network, a commonly used performance metric, depends on the combined effect of a large set of ground motion intensities (e.g., peak ground acceleration at the locations of all bridges of the network). The values of IM at various sites during the same extreme event show significant spatial correlation, which needs to be carefully modeled in order to accurately assess risk (Bocchini and Frangopol 2011a). Therefore, the information on the spatial correlation is very important for interdependent systems and highly coupled problems. Other practical cases where this issue arises include network reliability assessment, regional loss estimation, evacuation planning for storm surge, and lifeline resilience prediction (Bocchini and Frangopol 2012; Decò et al. 2013; Saydam et al. 2013) and all the cases where it is necessary to know the probability of having simultaneously certain values of the IM at all the locations of interest (Fig. 2.1).

In several recent studies, it has been shown that underestimating the importance of the spatial correlation may introduce significant inaccuracy. For example, Lee and Kiremidjian (2007) illustrated that seismic risk models that do not consider ground motion and damage correlation underestimate system risk, and, as a consequence, high-cost economic decisions may end up being nonconservative. Similarly, Bocchini and Frangopol (2011b) showed that the assumption of totally uncorrelated bridge damage states in a network reliability analysis leads to large

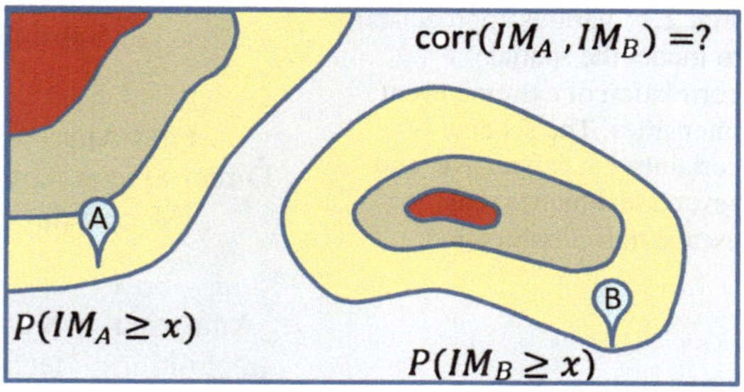

Fig. 2.1 Hazard maps provide the marginal hazard at all points, but not their correlation

nonconservative errors on the network performance. On the other hand, Crowley and Bommer (2006) demonstrated that assuming perfect spatial correlation leads to overestimate the loss exceedance curves when a probabilistic seismic hazard analysis (PSHA) is applied.

The scientific community has addressed the problem of incorporating the spatial correlation of the IM in different ways. In this chapter a brief review of these techniques is provided, and a recently developed perspective to the problem is presented. The first idea of this novel approach is to consider the IM of an extreme event over a region as a two-dimensional random field. Then, an effective technique called "functional quantization" (Luschgy and Pagès 2002; Miranda and Bocchini 2012, 2015) is used to generate a small set of maps that provides an optimal approximation (in the mean-square sense) of the desired random field. This new approach is characterized by a formulation of the problem that enables a truly multi-hazard paradigm, because it treats in the same way the intensities of all possible disasters. In fact, the methodology requires only to have an appropriate subroutine for the simulation of IM maps for the considered hazard, such as earthquake, flood, or hurricane.

Section 2.2 presents and discusses some recently developed methodologies aimed at modeling the spatial correlation of hazards for regional analysis. In Sect. 2.3 a new methodology called "hazard quantization" (HQ) for the optimal sampling of IM maps is showcased. Next, in Sect. 2.4 a brief discussion on the models for the simulation of IM maps for different hazards is provided, to illustrate how HQ can be applied to multiple disasters. In Sect. 2.5 a numerical example is provided for the case of seismic ground motion intensity maps to demonstrate the accuracy and potential of HQ. Finally, the most relevant conclusions are drawn in Sect. 2.6.

2.2 Review of Spatially Correlated Hazard Models

Hazard analysis is one part of the much broader framework of risk analysis. For example, the assessment of the probable losses in a highly seismic area for a single structure at a given site is usually performed using the analytical Pacific Earthquake

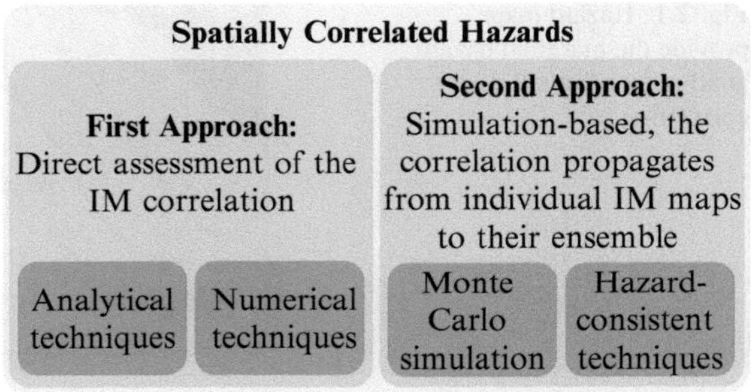

Fig. 2.2 Various approaches to model the spatial correlation of extreme event intensities. The scheme is certainly not exhaustive, and several techniques may escape this classification

Engineering Research Center (PEER) loss analysis framework (McGuire 2004). This framework breaks a complex problem into simpler tasks consisting in the computation of conditional probabilities, which are then combined using the total probability theorem. The first of these tasks deals with the probabilistic seismic hazard of the site, which is usually provided by the hazard maps.

Things become more complicated when spatially distributed systems are studied. In these cases, closed-form relationships or other computationally efficient solutions for the rest of the risk assessment tasks are less likely to be available than when dealing with an individual structure. Therefore, the hazard analysis should balance the conflicting objectives of keeping the computational cost limited while accounting for the spatial correlation.

To address these issues, researchers have proposed various techniques, which can be broadly classified based on two general approaches (Fig. 2.2). The first approach consists in the attempt to directly assess the correlation among the values of the IM at various locations. This can be achieved implicitly or explicitly with several analytical or computational techniques, but always using substantial simplifications and assumptions, which are convenient, but not necessarily adherent to reality (Bocchini and Frangopol 2011b; Gardoni et al. 2003; Moghtaderi-Zadeh and Kiureghian 1983).

The second approach exploits the idea that spatial correlation is naturally embedded in all real and realistic individual IM maps. Therefore, if these maps are used in a simulation-based method, the correlation present in each one of them propagates to their ensemble. A first distinction among the various techniques consists in the way a subset of representative IM maps is selected from a much larger suite of real or synthetic sample maps. For instance, Jayaram and Baker (2010) use k-mean clustering for the selection process, whereas Han and Davidson (2012) use an optimization formulation which minimizes errors in hazard curves over several return periods at all grid points. Then, the selected maps are usually weighted according to a certain criterion. The most popular subgroup of these techniques follows the so-called "hazard-consistent" weighting process. In this case, it is imposed that the marginal probability of exceedance of the reduced (and weighted) suite of IM maps matches the one obtained by a comprehensive probabilistic hazard analysis (e.g., from USGS hazard maps) for some selected discrete values and at some selected locations. The choice of the target values and the more or less regular grid of locations differentiates the various techniques of this class (Vaziri et al. 2012).

In the last years, the second approach (i.e., simulation-based techniques) has become more prominent; hence, it will be discussed in the following with more details. Monte Carlo simulation (MCS) is the forefather of all simulation techniques and is used as benchmark to assess the performance of other methodologies. In the literature, there are some variations on how MCS is conducted to generate IM maps. In general, MCS generates a random set of hazard scenarios and each one of them yields an IM map. For any given hazard and region of interest, MCS simulates the initial location of the disaster (e.g., epicenter of an earthquake or genesis point for a hurricane), magnitude, and other key parameters sampled from appropriate probability distributions, mostly based on historical data. Then the effects of the disaster are propagated, based on the (random) characteristics of the region, for instance, using seismic attenuation functions or integrating in time and space the fluid-dynamic status of the hurricane. When a large number of IM maps (i.e., at least in the order of 10^5) are considered, the full range of possible hazard scenarios that can occur in a region is captured and MCS is capable to represent both the marginal hazard at each considered location and the spatial correlation among sites.

As previously mentioned, the simulation of the IM maps is usually part of a much more complex analysis. It is followed by probabilistic damage assessment (e.g., through component or system fragility analyses), probabilistic estimation of direct and indirect losses and consequences, and probabilistic prediction of the short-, mid-, and long-term recovery, if resilience and sustainability are investigated (Bocchini et al. 2014). In a fully simulation-based approach, each one of these steps consists then in an additional nested simulation of other random variables or functions. When the number of samples required for each step is combined, the computational cost of the analysis grows very quickly, often beyond the practical computational capacity. For this reason, it is important that each one of the steps of the simulation captures all the relevant characteristics of the random quantities involved with the smallest possible number of samples.

Among the various techniques that have addressed this issue, some of the most effective and popular were reviewed and compared by Han and Davidson (2012). These techniques aim at selecting and weighting few representative IM maps, so that the resulting (weighted) ensemble matches the marginal hazard at some benchmark sites and captures the spatial correlation appropriately. The differentiation among the various techniques consists in the selection of the IM maps and in the weighting criteria.

The first relevant method of this family was introduced by Chang et al. (2000). This method selects a reduced set of earthquakes based on engineering judgment and estimates the hazard-consistent occurrence probability for a single site. In particular, it tries to match some points in the hazard curve produced from the reduced set of earthquakes with the hazard curve produced from the full set of all possible earthquakes. A similar approach to the selection of the earthquake scenarios (i.e., combination of source and magnitude) by the user was developed by Campbell and

Seligson (2003), but they estimate the hazard-consistent occurrence probabilities using optimization techniques to minimize the error in hazard curves over several return periods for all the locations of interest.

Although these analyst-driven methods are of easy application, it is not always clear which set of IM maps would yield optimal results, and thus more efficient and systematic techniques to select the hazard events have been developed. Kiremidjian et al. (2007) introduced the importance sampling technique to better control the selection of the earthquake events over a small range with strong magnitude. Similarly, Jayaram and Baker (2010) used importance sampling for the magnitude as well as the ground motion residuals and then used k-means clustering to combine "similar" IM maps and thus to further reduce their number for the subsequent loss analysis.

Before Jayaram and Baker's work, the proposed approaches favored the selection of (earthquake) scenarios defined by a magnitude and location, rather than IM maps. Due to the superposition of the residuals, which bundles all the intra-event uncertainty that is not explicitly modeled otherwise, each scenario could yield many different IM maps. The k-means clustering that Jayaram and Baker adopted gathers IM maps according to the Euclidean norm. Within each iteration, the centroid of each cluster is recalculated as the mean of all the maps in the cluster, and the scheme stops when no map reassignment takes place in one iteration. At the end one map is randomly selected from each cluster. Except for this last step, the technique presented by Jayaram and Baker (2010) strongly resembles the one that will be described in Sect. 2.3. However, the latter has been developed starting from a very different basis and is characterized by mean-square optimality of the results (Christou and Bocchini 2015; Miranda and Bocchini 2012, 2015). It will be shown that this yields a superior representation of the actual spatial correlation.

Another set of methodologies belonging to the hazard-consistent family is based explicitly on probabilistic optimization. This group of techniques minimizes the error between the marginal hazard yielded by a set of selected IM maps and the "exact" values (e.g., those provided by USGS) for a selected grid of points in a region. These optimization methodologies have been applied to different hazards. Vaziri et al. (2012) demonstrated the methodology for earthquakes, and Han and Davidson (2012) applied it to the area of Los Angeles as case study. Legg R. et al. (2010) introduced the constrained probabilistic optimization formulation to the case of hurricane wind hazard. Apivatanagul et al. (2011) extended this method so that the selected hurricanes and their annual occurrence probabilities match both wind and storm hazards.

Overall, these methods yield good results in terms of capturing the marginal hazard with a lower computational cost than MCS. However, they do not address explicitly the way in which the spatial correlation is modeled, which should be the main reason for the development of these methodologies. The implicit assumption is that the use of real or realistic synthetic realizations of the IM maps, which individually carry spatial correlation, will automatically imply a good representation of the correlation of the weighted ensemble. Instead, the technique proposed in the next section addresses this point explicitly, optimizing not only the representation of the marginal hazard but also of the spatial correlation.

2.3 Hazard Quantization Method

The HQ method diverges from the techniques previously described, approaching the regional hazard problem from another perspective. The key difference is that it embraces the nature of IM maps as random fields, which was only sporadically hinted at in the previous literature (Jayaram and Baker 2009). This approach yields several benefits. First, it allows to take direct advantage of several methodologies that have already been developed for the enhanced representation of generic random fields, which are supported by rigorous mathematical proofs of optimality. Second, it allows a more elegant treatment of the quantities involved in the problem. For example, in the case of earthquakes, there is no distinction between the random parameters that define the earthquake source (often called "scenario parameters") and those which model the inter- and intra-event variability (often referred to as "residuals"). HQ considers all parameters in the same way, without the need of a hierarchy and a specialized simulation technique for each group of them (which could be anyway included). This, in turn, yields the third advantage of HQ: its general perspective makes it a perfect paradigm for multi-hazard analysis. All the hazards can be addressed in the same way, with a consistent and uniform framework. The only subroutine that is hazard specific is the one for the generation of individual maps, as described in Sect. 2.4.

2.3.1 Theoretical Foundation

Functional quantization (FQ) is a technique for the optimal selection of representative samples of a random function, to be used in a simulation-based probabilistic analysis (Luschgy and Pagès 2002). There are several well-known variance reduction techniques that make the sampling process of random variables more effective, such as Latin hypercube sampling, importance sampling, and stratified sampling (Bucher 2009). Similarly, FQ is used for the optimal selection and weighting of random samples of a stochastic process or random field, so that even a small to moderate number of samples can capture the probabilistic characteristics of the simulated function.

The FQ approach consists of approximating a generic random function F with another random function F_N, which can be fully described in a probabilistic sense by a finite number of carefully selected samples $\{f_i\}_{i=1}^{N}$ and their associated weights $\{p_i\}_{i=1}^{N}$ such that

$$\sum_{n=1}^{N} p_i = 1 \tag{2.1}$$

In other words, FQ approximates a random function with a "simple function" (i.e., a collection of weighted realizations) rather than using the classical "parametric representation" such as Karhunen-Loéve, spectral representation, or polynomial chaos series (Ghanem and Spanos 2003; Stefanou 2009). This basic idea is shared by FQ with the stochastic reduced order model (SROM) technique, more known in the field of civil engineering (Grigoriu 2009), but the two methodologies differ both in their sampling and in the weighting scheme. The most defining characteristic of FQ is its optimality property, which results in the mean-square convergence of the approximate representation to the actual random function. This makes FQ particularly appropriate for hazard analysis in general, and regional IM maps in particular, where convergence is sought on both marginal distribution and correlation.

To take advantage of FQ, the IM map is considered as a stochastic function $F(\mathbf{x}, \omega)$, which is a bimeasurable random field defined in the probability space $(\Omega, \mathscr{F}, \mathbb{P})$ as:

$$F(\mathbf{x}, \omega) : \Xi \times \Omega \rightarrow \mathbb{R} \tag{2.2}$$

where Ξ is a two-dimensional space domain in $\mathbb{R}^2$, Ω is the sample space, $\mathbf{x}$ is a point in Ξ, and ω is an outcome in Ω. For the specific case of IM maps, the space domain Ξ represents the geographic region of interest, and the point $\mathbf{x}$ is a specific location in such region. A more convenient way to approach FQ is to interpret the random function defined above as a random variable $F(\omega)$ with values in the space of square integrable functions $L^2(\Xi)$:

$$F(\omega) : \Omega \rightarrow L^2(\Xi) \tag{2.3}$$

Every outcome ω of the sample space is mapped by the random function to a certain two-dimensional realization in the $L^2(\Xi)$ space. Each realization is a hypersurface, such as a certain IM map over the region of interest.

On the other hand, the random function F_N, which approximates F, is defined by the following equation:

$$F_N(\mathbf{x}, \omega) = \sum_{i=1}^{N} f_i(\mathbf{x}) \cdot 1_{\Omega_i}(\omega) \tag{2.4}$$

where the deterministic functions f_i are called "quanta" and 1_{Ω_i} is the indicator function associated with event $\Omega_i \subset \Omega$:

$$1_{\Omega_i}(\omega) = \begin{cases} 1, \text{ if } \omega \in \Omega_i \\ 0, \text{ otherwise} \end{cases} \tag{2.5}$$

Equation (2.4) can be interpreted assuming that the probability space Ω is partitioned into a mutually exclusive and almost collectively exhaustive set $\{\Omega_i\}_{i=1}^N$, and each event Ω_i has an associated probability $\mathbb{P}(\Omega_i)$ and a representative quantum f_i (Fig. 2.3). Quantum f_i represents all the sample functions associated with ω's

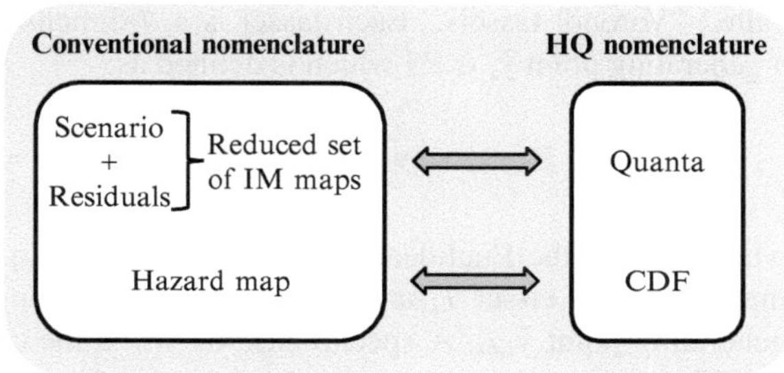

Fig. 2.3 Nomenclature of the hazard quantization technique and corresponding terms in the more traditional approaches

which belong to event Ω_i. The approximation F_N maps all the outcomes $\omega \in \Omega_i$ to the same quantum f_i, whereas random field F can map each outcome to a different realization. The quanta $\{f_i\}_{i=1}^N$ selected by FQ to represent the various $\{\Omega_i\}_{i=1}^N$ are always going to be optimal (in the mean-square sense) for a specific number N, called "quantizer size."

The same process of partitioning and approximation can be seen also in the space of square integrable functions $L^2(\Xi)$. From this perspective, the $L^2(\Xi)$ space is tasseled into $\{V_i\}_{i=1}^N$, where each tassel V_i collects all the realizations corresponding to $\omega \in \Omega_i$. Therefore, FQ induces a tessellation $\{V_i\}_{i=1}^N$ in the $L^2(\Xi)$ space and consequently a corresponding partition $\{\Omega_i\}_{i=1}^N$ in Ω. Based on this partition, the probability $\mathbb{P}(\Omega_i)$ associated with each event should still be computed. However, thanks to the mentioned relationship between the two spaces, it is possible to compute instead the probability $\mathbb{P}_F(V_i) = p_i$ of the corresponding tassel V_i, which is equal to the associated $\mathbb{P}(\Omega_i)$ (Miranda and Bocchini 2015).

In summary, a technique that performs FQ should accomplish the following tasks:

1. Obtain an optimal partition $\{\Omega_i\}_{i=1}^N$ or the associated tessellation $\{V_i\}_{i=1}^N$;
2. Determine the quanta $\{f_i\}_{i=1}^N$ that optimally represent the random function F with a fixed N according to Eq. (2.4);
3. Compute the probabilities $\mathbb{P}(\Omega_i) = \mathbb{P}_F(V_i) = p_i$ associated with the partition of the sample space $\{\Omega_i\}_{i=1}^N$ and the tessellation $\{V_i\}_{i=1}^N$.

From a practical point of view, there are several techniques to compute the quantizer of a certain random function. Some of the best known quantization techniques were presented by Luschgy and Pagès (2004) with names "quantizer designs I, II, III, and IV." However, these techniques cannot be applied to the case of non-Gaussian random fields, as required for the quantization of IM maps. Miranda and Bocchini (2012; 2013; 2015) developed another quantization technique, with the aim of being more versatile and extend FQ also to the case of multidimensional, non-Gaussian, and nonstationary random functions. This technique is called "Functional Quantization by Infinite-Dimensional Centroidal Voronoi Tessellation" (FQ-IDCVT) and is perfectly appropriate for the quantization of IM maps.

FQ-IDCVT exploits the idea of Voronoi tessellation (VT) which is a process of partitioning a finite-dimensional Euclidean space $\mathbb{R}^d$ into regions $\{T_i\}_{i=1}^N$, called "Voronoi tassels." Each tassel is a d-dimensional convex polyhedron with a generating point $\check{\mathbf{y}}_i \in \mathbb{R}^d$ which is defined as:

$$T_i = \left\{ \mathbf{y} \in \mathbb{R}^d \quad | \quad \|\mathbf{y} - \check{\mathbf{y}}_i\| < \|\mathbf{y} - \check{\mathbf{y}}_j\| \quad \text{for } j = 1, 2, \ldots, N; j \neq i \right\} \tag{2.6}$$

where $\| \cdot \|$ is the Euclidean norm. According to Eq. (2.6), all the points $\mathbf{y} \in \mathbb{R}^d$ that belong to tassel T_i are closer to the generating point $\check{\mathbf{y}}_i$ than to any other generating point $\check{\mathbf{y}}_{j \neq i}$. A special case of VT is the centroidal Voronoi tessellation (CVT), where each generating point $\check{\mathbf{y}}_i$ is also the centroid of tassel V_i. A CVT of a finite-dimensional Euclidean space can be computed using Lloyd's method (Ju et al. 2002).

FQ-IDCVT essentially enhances Lloyd's method and extends it to the infinite-dimensional Hilbert space of squared integrable functions $L^2(\Xi)$ (Miranda and Bocchini 2015). In this infinite-dimensional space, tassels are defined as follows:

$$\begin{aligned} T_i = \Big\{ F(\omega) \in L^2(\Xi) \quad | \quad & \|F(\omega) - \check{f}_i\|_{L^2(\Xi)} < \|F(\omega) - \check{f}_j\|_{L^2(\Xi)} \\ & \text{for } j = 1, 2, \ldots, N; j \neq i \Big\} \end{aligned} \tag{2.7}$$

where $\check{f}_i$ is the generating point of tassel T_i and $\| \cdot \|_{L^2(\Xi)}$ is the $L^2(\Xi)$ norm. Equation (2.7) denotes that all the realizations $F(\omega)$ closer to $\check{f}_i$ than to any other $\check{f}_{j \neq i}$ are clustered in tassel T_i. Note that the tassels generated by the CVT in Eq. (2.7) will be used as the tassels V_i's in the FQ sense. In other words, $T_i \equiv V_i$ and $\check{f}_i \equiv f_i$.

Tassels T_i's and their centroids $\check{f}_i$'s are generated by the iterative algorithm presented in Sect. 2.3.2. This algorithm minimizes an error metric called "distortion":

$$\Delta\left(\{V_i, f_i\}_{i=1}^N\right) = \sum_{i=1}^{N} \int_{V_i} \|F(\omega) - f_i\|^2_{L^2(\Xi)} \mathrm{d}\mathbb{P}_F \tag{2.8}$$

It has been proven (Miranda and Bocchini 2015) that the minimization of the distortion as defined in Eq. (2.8) ensures that a CVT of $L^2(\Xi)$ is obtained and it is optimal according to the mean-square criterion. The argument of the norm in Eq. (2.8) imposes convergence of the approximate representation to the random field, not focusing only on the first moment or the marginal distribution. Additionally, FQ-IDCVT has been shown to work particularly well against the curse of dimensionality that arises in these problems, and its applicability has been illustrated for Gaussian and non-Gaussian random fields (Bocchini et al. 2014; Christou et al. 2016).

Figure 2.4 shows how FQ can be used to simulate IM maps and, in turn, regional hazards in the context of a loss estimation analysis. This particular application of FQ is named HQ.

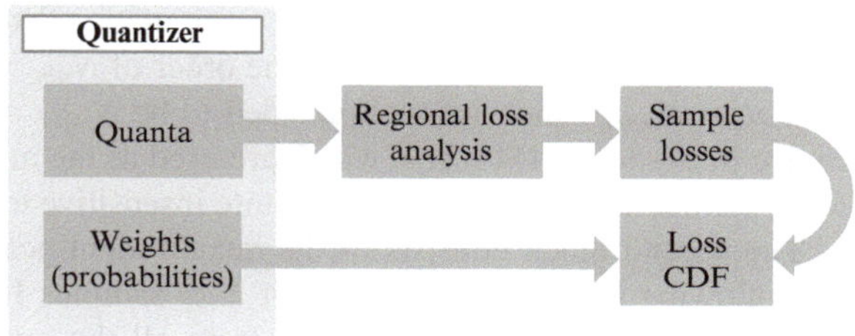

Fig. 2.4 Flowchart of the loss estimation using a quantizer to represent the regional hazard

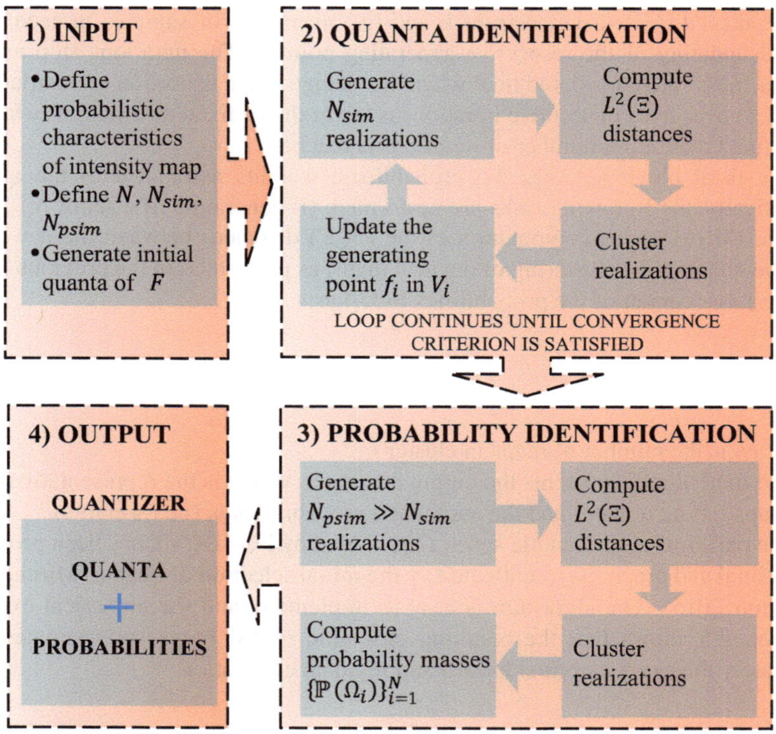

Fig. 2.5 Flowchart of the FQ-IDCVT algorithm

2.3.2 Computational Algorithm

Figure 2.5 shows the flowchart of the FQ-IDCVT algorithm, consisting of four blocks, which are briefly explained in the following. More details are provided by Miranda and Bocchini (2013).

The first block includes the required input data. These are (i) probabilistic characteristics of the stochastic parameters required to generate an IM map; (ii) "quantizer size" N that is the number of sample IM maps that will be used and essentially depends on the available computational resources; (iii) computational parameter N_{sim}, which is usually in the order of $N_{sim} = 100 \cdot N$; (iv) computational parameter N_{psim} which is recommended to be $N_{psim} = 10 \cdot N_{sim}$; and (v) an initial choice of N sample IM maps, which are used as the initial set of quanta (the quality of the output has been proven to be quite insensitive to this initial choice).

The second block consists of the quanta identification. Each iteration includes the following tasks: (i) generation of N_{sim} samples IM maps; (ii) computation of the $L^2(\Xi)$ distance of the IM maps from all the quanta $\{f_i\}_{i=1}^N$ in the 2D space; (iii) clustering of each realization j to the tassel m, where f_m is the quanta with the smallest $L^2(\Xi)$ distance from j; and (iv) averaging of samples in each tassel V_i and updating of the respective generating point f_i. The iterations stop and the quanta $\{f_i\}_{i=1}^N$ are considered final when the change in the distortion as computed by Eq. (2.8) over five consecutive iterations is lower than a fixed threshold or when the predefined maximum number of iterations is reached.

The third block assesses the probabilistic weights associated to the quanta. Specifically, the following tasks are performed: (i) generation of a set of N_{psim} new sample IM maps; (ii) computation of the $L^2(\Xi)$ distances between the new maps and the quanta; (iii) clustering of the new maps, as performed in the previous block; and (iv) assessment of the probabilities $\mathbb{P}(\Omega_i)$ as:

$$\mathbb{P}(\Omega_i) = p_i = \frac{N_i}{N_{psim}} \tag{2.9}$$

where N_i is the number of maps in cluster i.

The final block represents the output quantizer, which is the representative set of IM maps f_i (i.e., quanta) and the associated probabilities p_i (i.e., weights).

The partition of the sample space Ω provided by FQ-IDCVT has been proven to be optimal and practically unaffected by the initial selection of quanta (Miranda and Bocchini 2015). The algorithm is easy to implement, and the numerical example in Sect. 2.5 shows that the resulting ensemble of IM maps approximates very accurately the exact marginal hazard and regional correlation.

2.4 HQ for Multi-Hazard Analysis

As already mentioned, HQ is a very general and consistent framework that can be applied to all probabilistic hazard assessments, in a similar way to MCS (Fig. 2.6). For every hazard, the HQ approach needs a large set of historical IM maps and/or a hazard-specific model for the generation of synthetic IM maps. After the HQ is performed, the weights of the IM maps are used to obtain the full probability distribution of the marginal hazard. When multiple hazards are considered, the weights of the various hazard scenarios can be easily combined.

Even though the quantization process has a non-negligible computational cost, this is usually extremely smaller than all the subsequent probabilistic regional

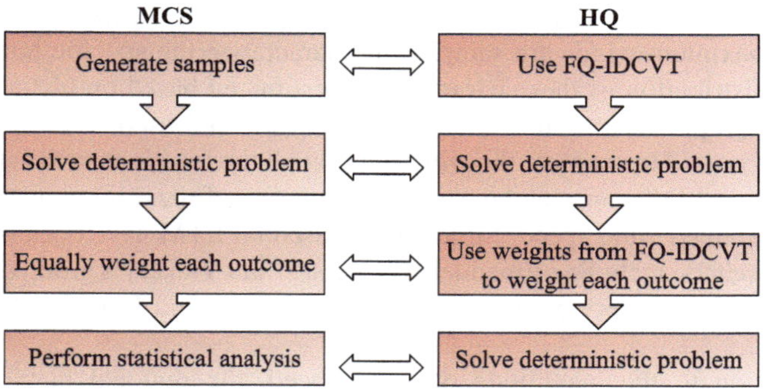

Fig. 2.6 Probabilistic analysis using HQ and corresponding tasks in a traditional Monte Carlo simulation

loss, risk, or resilience analyses. Therefore, the cost of quantization is usually overcompensated by the reduction in the number of samples that need to be considered for a complete probabilistic characterization of the hazard.

The following subsections provide few examples of hazard models that can be used within HQ to perform regional hazard analysis.

2.4.1 HQ for Seismic Hazard Analysis

In seismic hazard analysis, HQ requires only an existing ground-motion model, where the IM (e.g., spectral acceleration at a structural period of interest) is usually a function of the median spectral acceleration (provided by the ground-motion model) and of some other parameters whose effect is usually bundled in the intra-event and inter-event residuals (Crowley and Bommer 2006). The median spectral acceleration is usually a function of parameters such as earthquake magnitude, source-to-site distance, site classification, and fault rupture mechanism and can be obtained from any of the attenuation functions available in the literature, such as those developed by Abrahamson and Silva (1997) or Boore and Atkison (2008). In many practical applications, only the median spectral acceleration is considered and the residuals are ignored.

2.4.2 HQ for Hurricane Hazard Analysis

For hurricane wind hazard analysis, some of the most popular techniques are based on the empirical track model (ETM) introduced by Vickery and Twisdale (1995a). This model generates maps of the wind track at discrete time instants during the full evolution of a hurricane, from its formation on the ocean (genesis point) until its final dissipation. When HQ is utilized, parameters like the central pressure and the radius of maximum winds are sampled to characterize the specific hurricane. The statistical distribution of these parameters is estimated based on historical data on a broad geographic area. Once the hurricane track is assessed, at each step a wind field model is used to estimate the wind speed at all the locations in the study region. The wind speed IM maps can be expressed in terms of maximum gust wind speed or mean sustained wind speed. Simulations based on ETM have been presented by many researchers, including Emanuel et al. (2006) and Vickery's group (Vickery and Twisdale 1995b, Vickery et al. 2000b, 2000a, Vickery and Blanton 2008, Vickery et al. 2009).

In addition to the magnitude of wind speed, the regional hurricane hazard estimation requires the assessment of the surges that can occur throughout a region and the associated probabilities of occurrence. The storm surge assessment becomes particularly important when the region of interest is close to the coastline. In this case, a surge model is required to compute the inland values of the sea surface height above the mean sea level. For instance, Advanced Circulation (ADCIRC) is a high-resolution storm surge model that could be considered for this task (Westerink et al. 2008).

2.5 Numerical Application

As an example of application of the proposed methodology, a simplified representation of an earthquake ground motion is considered. Figure 2.7 shows the region of interest with two predefined faults. The moment magnitude of the earthquake and

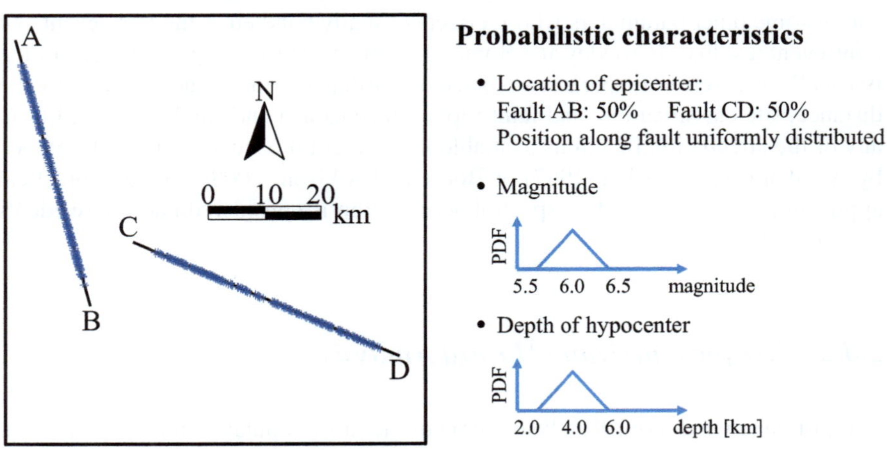

Fig. 2.7 Specified faults AB and CD in the region of interest. The *blue crosses* represent the sample epicenters

Table 2.1 Coordinates of the edge points of the faults in geographic coordinate system

Edge nodes	Latitude	Longitude
A	34.29	−118.39
B	34.26	−118.37
C	34.24	−118.36
D	34.27	−118.31

the hypocenter depth are assumed to have a triangular distribution with minimum, mode, and maximum equal to $[5.5, 6, 6.5]$ and $[2.0, 4.0, 6.0]km$, respectively. The fault type is considered to be a strike slip, and the surface fault rupture length is determined according to the model adopted by HAZUS-MH (DHS 2003):

$$\log_{10}(L) = a + b \cdot M \tag{2.10}$$

where L is the rupture length in km, M is the moment magnitude of the earthquake, and $a = -3.55$ and $b = 0.74$ are regression coefficients for a strike-slip fault type at the surface. The two line segments that represent the faults have edge coordinates listed in Table 2.1.

To model the seismic attenuation, the empirical regression model presented by Abrahamson and Silva (1997) is utilized for the generation of ground-shaking maps. The proposed functional form is in terms of median spectral acceleration at a structural period T and is given as follows:

$$\ln S_a = f(M, r_{rup}) \tag{2.11}$$

where S_a is the median spectral acceleration in g, M is the moment magnitude, and r_{rup} is the closest distance to the rupture plane in km. Function $f(M, r_{rup})$ is assumed to take the basic functional form of the attenuation for strike-slip events:

$$\begin{aligned} \text{for } (M, r_{rup}) \leq c_1 \quad f_1(M, r_{rup}) &= a_1 + a_2(M - c_1) + a_{12}(8.5 - M)^n \\ &+ [a_3 + a_{13}(M - c_1)] \ln(R) \\ \text{for } (M, r_{rup}) > c_1 \quad f_1(M, r_{rup}) &= a_1 + a_2(M - c_1) + a_{12}(8.5 - M)^n \\ &+ [a_3 + a_{13}(M - c_1)] \ln(R) \end{aligned} \tag{2.12}$$

where $R = \sqrt{r_{rup}^2 + c_4^2}$ and a_1, a_2, a_3, a_{12}, a_{13}, c_1, c_4, and n are coefficients for the average horizontal component of the median spectral acceleration listed in Table 2.2 for the case of a structural period $T = 0.10s$.

The example is analyzed using HQ, and the characteristics of the resulting quantizer are compared to the exact values of the autocorrelation and the marginal probability of exceeding a certain value of $S_a(T = 0.1s)$ computed by extensive MCS. The FQ-IDCVT parameters considered herein are $N = 500$, $N_{sim} = 100 \cdot N$, and $N_{psim} = 1{,}000 \cdot N$.

Table 2.2 Coefficients for the average horizontal component of the median spectral acceleration. Extracted from a table with data for all structural periods by Abrahamson and Silva (1997)

Period	c_4	a_1	a_2	a_3	a_{12}	a_{13}	c_1	n
0.10	5.50	2.160	0.512	−1.1450	0.0280	0.17	6.4	2

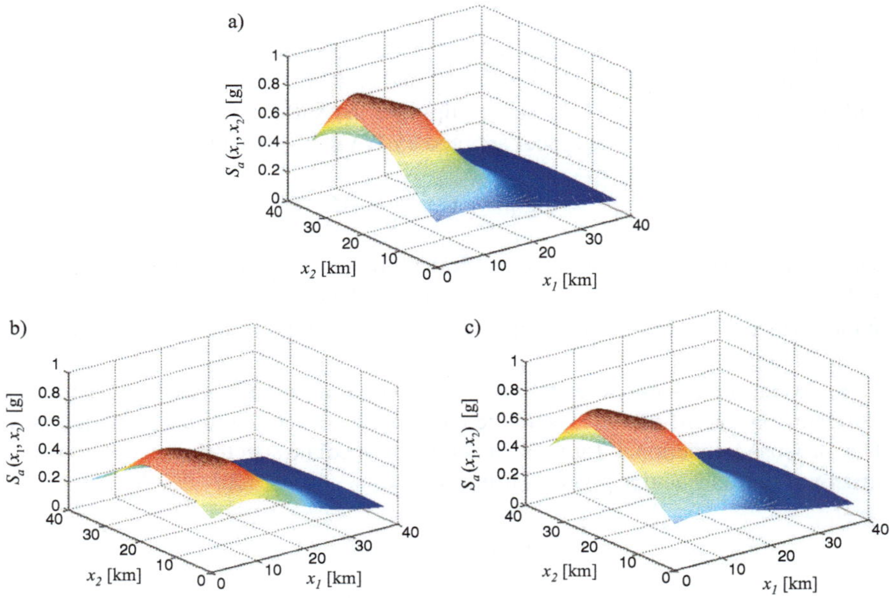

Fig. 2.8 (**a**) Random sample of the median spectral acceleration $S_a(x_1, x_2, T = 0.1s)$ and associated quanta obtained from HQ with (**b**) $N = 50$ and (**c**) $N = 500$

A random sample of the median spectral acceleration $S_a(x_1, x_2, T = 0.1s)$ and the associated quanta with $N = 50$ and $N = 500$ are presented in Fig. 2.8, over the spatial domain of points (x_1, x_2). Even though the quanta are not necessarily historical or synthetic realizations of the IM maps, they capture the overall trend in a realistic way. As expected, as the quantizer size increases, the quanta resemble more closely the samples of their cluster.

Figures 2.9, 2.10, and 2.11 compare the exact marginal hazard and the one obtained from HQ with $N = 500$ for different values of the median spectral acceleration. These results show that the marginal hazard obtained by HQ is in close agreement with the exact one, even at the tails of the distribution (Figs. 2.9a and 2.11c). Good results have been obtained also for other values of the structural period T.

To be able to plot the autocorrelation of the quantizer, it has been determined for different 1D stripes of the random field (the complete autocorrelation of a two-dimensional nonhomogeneous field is a 4D function). A comparison between the

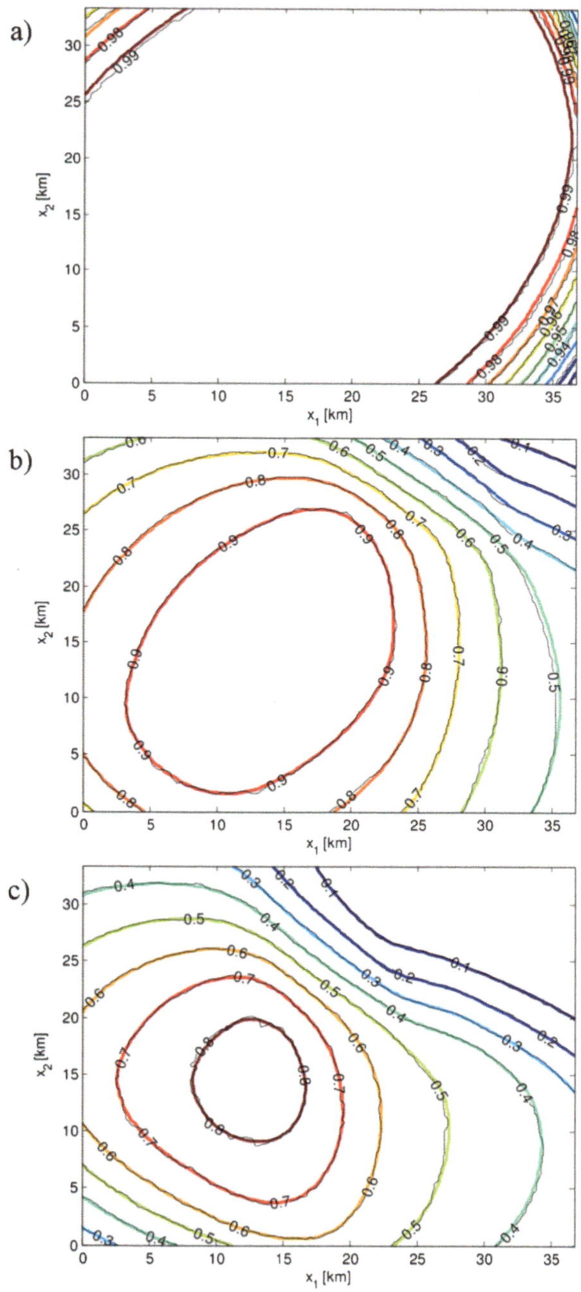

Fig. 2.9 Comparison between the exact probability of exceedance (*thick colored lines*) and the result obtained from HQ with $N = 500$ (*thin black lines*) for (**a**) $P[S_a(T = 0.1s) > 0.1\text{g}]$, (**b**) $P[S_a(T = 0.1s) > 0.2\text{g}]$, and (**c**) $P[S_a(T = 0.1s) > 0.3\text{g}]$

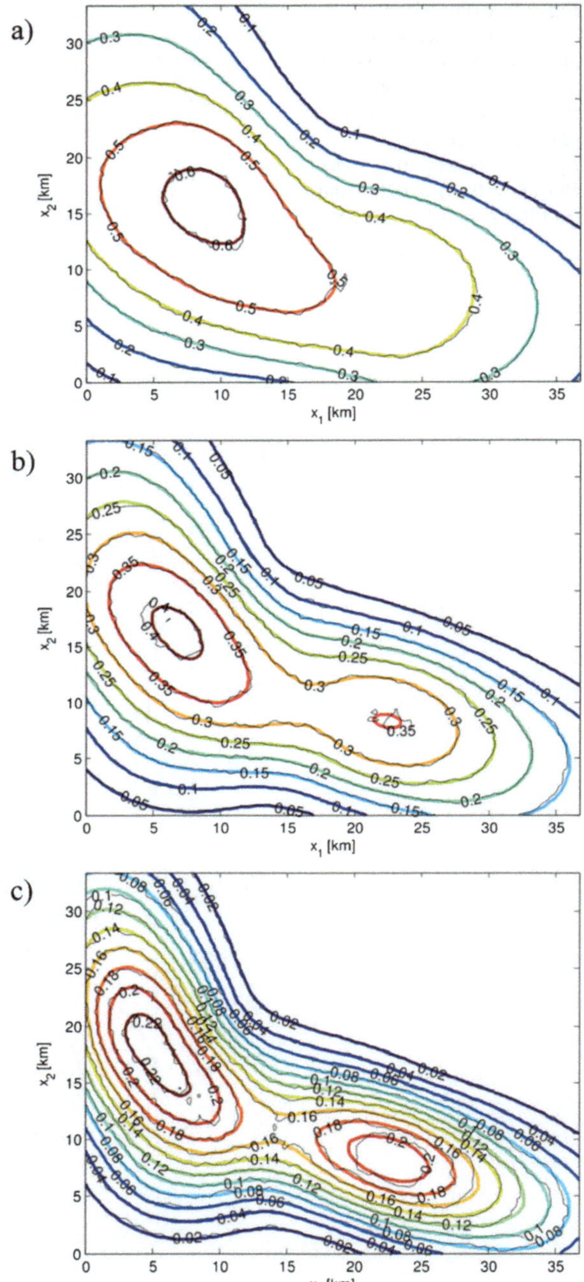

Fig. 2.10 Comparison between the exact probability of exceedance (*thick colored lines*) and the result obtained from HQ with $N = 500$ (*thin black lines*) for (**a**) $P[S_a(T = 0.1s) > 0.4\text{g}]$, (**b**) $P[S_a(T = 0.1s) > 0.5\text{g}]$, and (**c**) $P[S_a(T = 0.1s) > 0.6\text{g}]$

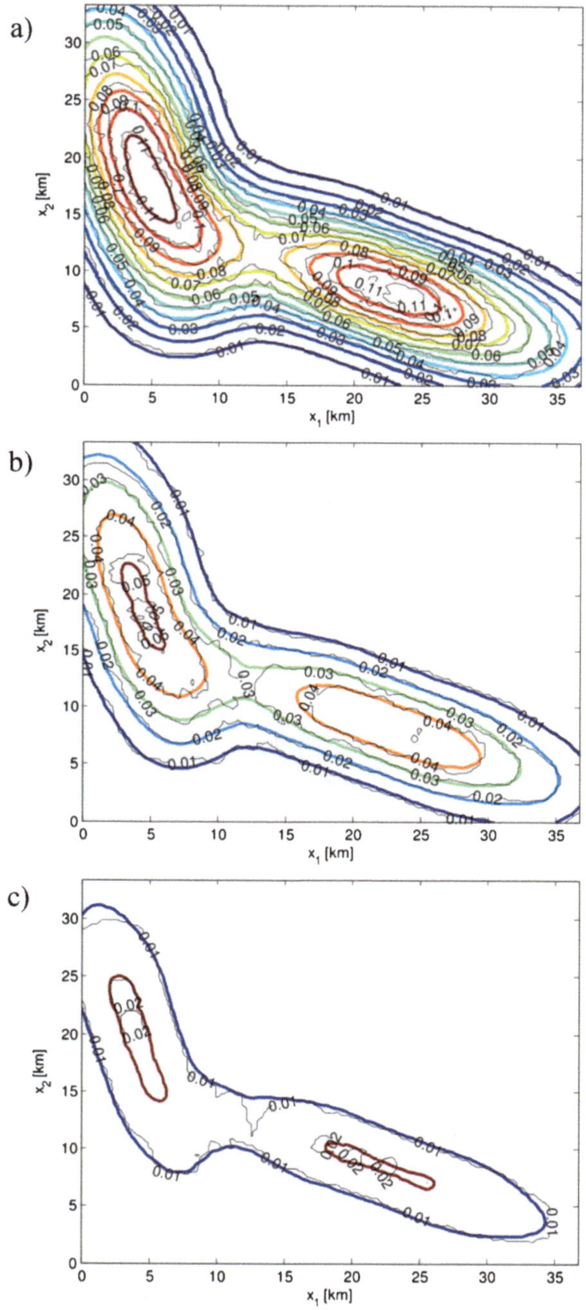

Fig. 2.11 Comparison between the exact probability of exceedance (*thick colored lines*) and the result obtained from HQ with $N = 500$ (*thin black lines*) for (**a**) $P[S_a(T = 0.1s) > 0.7\text{g}]$, (**b**) $P[S_a(T = 0.1s) > 0.8\text{g}]$, and (**c**) $P[S_a(T = 0.1s) > 0.9\text{g}]$

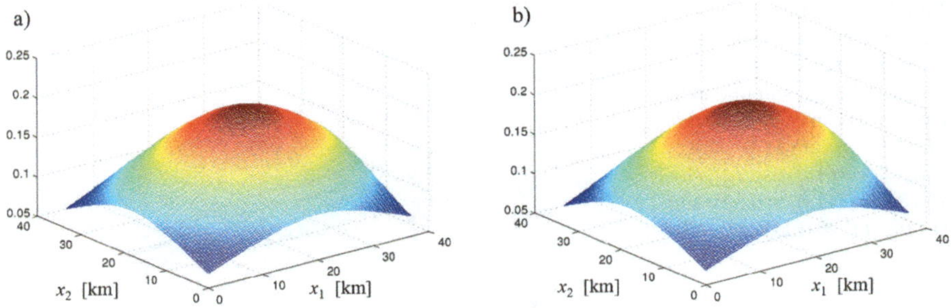

Fig. 2.12 Autocorrelation function of $S_a(T = 0.1s)$ determined for a stripe of the region. (**a**) Approximation obtained with a quantizer of size $N = 500$ and (**b**) exact

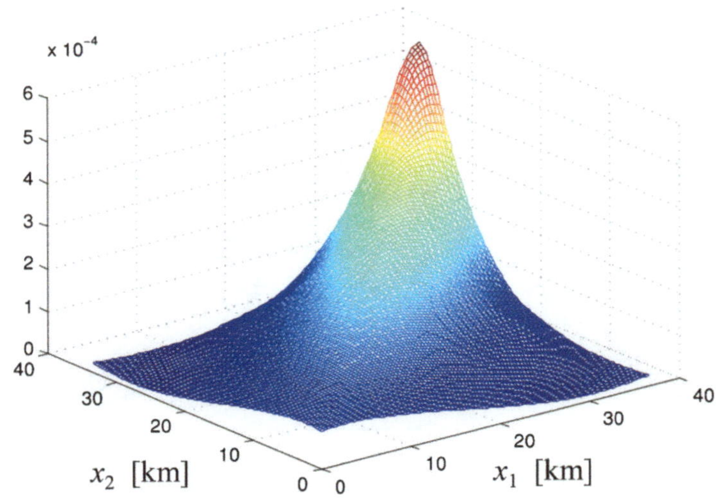

Fig. 2.13 Difference between approximated autocorrelation and exact obtained for a stripe of the region

exact autocorrelation and the one obtained from the quantizer is shown in Fig. 2.12. Looking at the figures, it is clear that the overall trend of the autocorrelation is captured very accurately. For a more quantitative comparison, Fig. 2.13 illustrates the difference between the ensemble autocorrelation of the quantizer and the exact. The error on this second-order statistic, notoriously difficult to capture, is considerably small, in the order of 0.1 %. Similar results have been obtained for all other regions of the field and for median spectral accelerations at other structural periods. Han and Davidson (2012) provided a similar comparison for the best hazard-consistent techniques available and showed that the error was in the order of 2 % (considering also the residuals). Even though a more direct comparison on the same test bed and with the same type of residuals is needed, this example suggests that HQ yields an error that is practically negligible, at least on the median spectral acceleration.

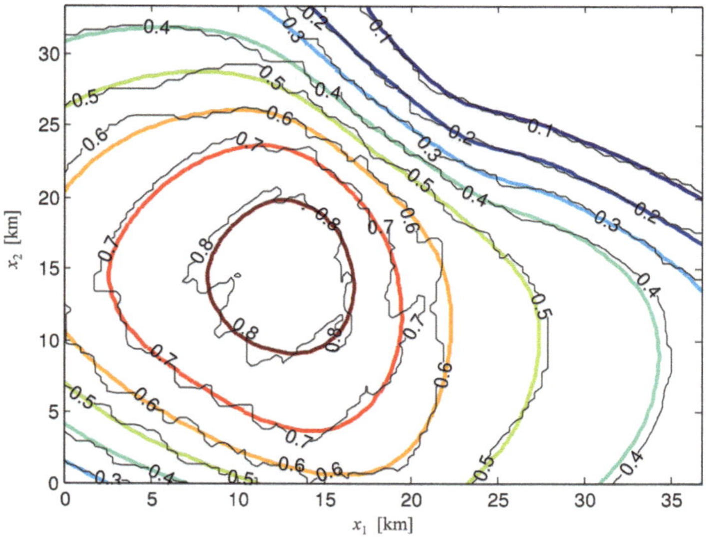

Fig. 2.14 Comparison on the probability of exceedance between the exact (*thick colored lines*) and the result obtained from HQ with $N = 50$ (*thin black lines*) for $P[S_a(T = 0.1s) > 0.30\text{g}]$

Figure 2.14 compares the exact marginal hazard and the one obtained by HQ with $N = 50$. It is evident that HQ performs well even for such a small number of samples (and, in turn, of loss analyses), which makes the method very appealing for regional hazard analysis.

2.6 Conclusions

A new methodology called "hazard quantization" is presented for the generation of an optimal set of maps representing the intensity of a natural disaster over a region. The proposed approach is rooted in the idea of considering explicitly the IM maps of any hazard as a two-dimensional, non-Gaussian, nonhomogeneous random field. Adopting this perspective, an advanced tool called "FQ-IDCVT" is used for the optimal sampling of these random functions. For highly correlated random fields, such as IM maps for any type of hazard, FQ-IDCVT ensures that the weighted ensemble of the samples tends to match particularly well all the properties of the field, including the marginal hazard and the spatial correlation.

Compared to similar techniques available in the literature, HQ presents important advantages. First and foremost, its outcome has been proven to be the best possible representation of a random field (i.e., of an IM map) for a given number of samples N, in the mean-square sense. This results in the error in both marginal hazard and spatial correlation to be practically negligible, even with small sample sizes.

Other advantages are that the implementation of the technique is very simple and straightforward and that its application to any possible hazard requires only a subroutine capable of generating appropriate samples of the IM map. Based on these promising advantages, HQ will be further studied, validated, and compared against the other available methodologies.

Nomenclature

Notation	Definition
ADCIRC	Advanced Circulation
CVT	Centroidal Voronoi tessellation
F, $F(\mathbf{x}, \omega)$	Random function
F_N	Random function used to approximate F
FQ	Functional quantization
FQ-IDCVT	Functional Quantization by Infinite-Dimensional Centroidal Voronoi Tessellation
f_i	Deterministic function representative of F_N over Ω_i (i.e., quantum)
$\check{f}_i$	Generating point of tassel T_i
$\{f_i\}_{i=1}^N \cup \{p_i\}_{i=1}^N$	Quantizer
HQ	Hazard quantization
IM	Intensity measure
$L^2(\Xi)$	Hilbert space of square integrable functions
MCS	Monte Carlo simulation
N	Quantizer size, equal to the number of IM maps
N_i	Number of samples f_i that belong to tassel V_i
N_{sim}	Computational parameter (recommended $N_{sim} = 100 \cdot N$)
N_{psim}	Computational parameter (recommended $N_{psim} = 10 \cdot N_{sim}$)
PEER	Pacific Earthquake Engineering Research Center
PSHA	Probabilistic seismic hazard analysis
$\mathbb{P}_F(V_i)$	Probability mass of tassel V_i
$\mathbb{P}(\Omega_i)$	Probability associated with event Ω_i
p_i	Probability mass
$\mathbb{R}^d$	Euclidean d-dimensional space
S_a	Median spectral acceleration
SROM	Stochastic reduced order models
$\{T_i\}_{i=1}^N$	Set of regions of the finite-dimensional Euclidean space $\mathbb{R}^d$ (i.e., "Voronoi tassels")
USGS	United States Geological Survey
VT	Voronoi tessellation

Notation	Definition
$\{V_i\}_{i=1}^N$	Set of tassels corresponding to events $\{\Omega\}_{i=1}^N$
$\mathbf{x}$	Point in Ξ
y_i	Point that belongs to tassel T_i (finite-dimensional case)
$\check{y}_i$	Generating point of tassel T_i (finite-dimensional case)
Δ	Distortion functional
Ξ	Spatial domain of interest in $\mathbb{R}^2$
Ω	Sample space
$(\Omega, \mathscr{F}, \mathbb{P})$	Probability space
1_{Ω_i}	Indicator function of Ω_i

References

Abrahamson, N. A., & Silva, W. J. (1997). "Empirical response spectral attenuation relations for shallow crustal earthquakes." *Seismological Research Letters, 68*(1), 94–127.

Apivatanagul, P., Davidson, R., Blanton, B., & Nozick, L. (2011). "Long-term regional hurricane hazard analysis for wind and storm surge." *Coastal Engineering, 58*(6), 499–509.

Bocchini, P., Frangopol, D., Ummenhofer, T., & Zinke, T. (2014). "Resilience and sustainability of civil infrastructure: Toward a unified approach." *ASCE Journal of Infrastructure Systems, 20*(2), 04014004.

Bocchini, P., & Frangopol, D. M. (2011a). "Generalized bridge network performance analysis with correlation and time-variant reliability." *Structural Safety, 33*(2), 155–164.

Bocchini, P., & Frangopol, D. M. (2011b). "A stochastic computational framework for the joint transportation network fragility analysis and traffic flow distribution under extreme events." *Probabilistic Engineering Mechanics, 26*(2), 182–193.

Bocchini, P., & Frangopol, D. M. (2012). "Restoration of bridge networks after an earthquake: multi-criteria intervention optimization." *Earthquake Spectra, 28*(2), 426–455.

Bocchini, P., Miranda, M. J., & Christou, V. (2014). "Functional quantization of probabilistic life-cycle performance models." In H. Furuta, D. M. Frangopol, & M. Akiyama (Eds.), *Life-Cycle of Structural Systems: Design, Assessment, Maintenance and Management* (pp. 816–823). Tokyo, Japan: Taylor and Francis.

Boore, M. D., & Atkison, M., G. (2008). "Ground-motion prediction equations for the average horizontal component of PGA, PGV, and 5%-damped PSA at spectral periods between 0.01s and 10.0s." *Earthquake Spectra, 24*(1), 99–138.

Bucher, C. (2009). *Computational Analysis of Randomness in Structural Mechanics*. Balkema: CRC Press. Leiden, The Netherlands: Taylor & Francis.

Campbell, K., & Seligson, H. (2003). "Quantitative method for developing hazard-consistent earthquake scenarios." *Advancing Mitigation Technologies and Disaster Response for Lifeline Systems*, 829–838.

Chang, E. S., Shinozuka, M., & Moore, E., J. (2000). "Probabilistic earthquake scenarios: Extending risk analysis methodologies to spatially distributed systems." *Earthquake Spectra, 16*(3), 557–572.

Christou, V., & Bocchini, P. (2015). "Efficient computational models for the optimal representation of correlated regional hazard." In *Proceedings of the 12th International Conference on Applications of Statistics and Probability in Civil Engineering* (pp. 1–8), Vancouver, Canada, July 12–15.

Christou, V., Bocchini, P., & Miranda, M. J. (2016). "Optimal representation of multi-dimensional random fields with a moderate number of samples: application to stochastic mechanics." *Probabilistic Engineering Mechanics, Elsevier*, in press. DOI: 10.1016/j.probengmech.2015.09.016.

Crowley, H., & Bommer, J. J. (2006). "Modelling seismic hazard in earthquake loss models with spatially distributed exposure." *Bulletin of Earthquake Engineering, 4*, 249–273.

Decò, A., Bocchini, P., & Frangopol, D. M. (2013). "A probabilistic approach for the prediction of seismic resilience of bridges." *Earthquake Engineering & Structural Dynamics, 42*(10), 1469–1487.

DHS. (2003). In *HAZUS-MH MR4 Earthquake Model Technical Manual*. Department of Homeland Security; Emergency Preparedness and Response Directorate; Whashington, D.C.: Federal Emergency Management Agency; Mitigation Division.

Emanuel, K., Ravela, S., Vivant, E., & Risi, C. (2006). "A statistical deterministic approach to hurricane risk assessment." *Bulletin of the American Meteorological Society, 87*(3), 299–314.

Gardoni, P., Mosalam, K., & Der Kiureghian, A. (2003). "Probabilistic seismic demand models and fragility estimates for rc bridges." *Journal of Earthquake Engineering, 7*, 79–106.

Ghanem, R., & Spanos, P. D. (2003). *Stochastic finite elements: a spectral approach*, revised edition. New York: Dover.

Grigoriu, M. (2009). "Reduced order models for random functions. Application to stochastic problems." *Applied Mathematical Modelling, 33*(1), 161–175.

Han, Y., & Davidson, R. A. (2012). "Probabilistic seismic hazard analysis for spatially distributed infrastructure." *Earthquake Engineering & Structural Dynamics, 41*, 2141–2158.

Jayaram, N., & Baker, J. W. (2009). "Correlation model for spatially distributed ground-motion intensities." *Earthquake Engineering & Structural Dynamics, 38*, 1687–1708.

Jayaram, N., & Baker, J. W. (2010). "Efficient sampling and data reduction techniques for probabilistic seismic lifeline risk assessment." *Earthquake Engineering & Structural Dynamics, 39*, 1109–1131.

Ju, L., Du, Q., & Gunzburger, M. (2002). "Probabilistic methods for centroidal Voronoi tessellations and their parallel implementations." *Parallel Computing, 28*(10), 1477–1500.

Kiremidjian, S. A., Stergiou, E., & Lee, R. (2007). "Issues in seismic risk assessment of transportation networks." In K. D. Pitilakis (Ed.), *Earthquake Geotechnical Engineering* (pp. 461–480). Netherlands: Springer.

Lee, R. G., & Kiremidjian, A. S. (2007). "Uncertainty and correlation in seismic risk assessment of transportation systems." *Technical Report 2007/05*, Pacific Earthquake Engineering Research Center (PEER).

Legg, R. M., Nozick k. L., & Davidson, A. R. (2010). "Optimizing the selection of hazard-consistent probabilistic scenarios for long-term regional hurricane loss estimation." *Structural Safety, 32*(1), 90–100.

Luschgy, H., & Pagès, G. (2002). "Functional quantization of Gaussian processes." *Journal of Functional Analysis, 196*(2), 486–531.

Luschgy, H., & Pagès, G. (2004). "Sharp asymptotics of the functional quantization problem for Gaussian processes." *The Annals of Probability, 32*(2), 1574–1599.

McGuire, K. R. (2004). *Seismic hazard and risk analysis*. Oakland, CA: Earthquake Engineering Research Institute.

Miranda, M., & Bocchini, P. (2012). "Mean square optimal approximation of random processes using functional quantization." In *Proceedings of the 2012 Joint Conference of the Engineering Mechanics Institute and the 11th ASCE Joint Specialty Conference on Probabilistic Mechanics and Structural Reliability*. Notre Dame, IN, June 17–20, 2012.

Miranda, M. J., & Bocchini, P. (2013). "Functional quantization of stationary gaussian and non-gaussian random processes." In G. Deodatis, B. R. Ellingwood, & D. M. Frangopol (Eds.), *Safety, Reliability, Risk and Life-Cycle Performance of Structures and Infrastructures* (pp. 2785–2792). New York, NY: Columbia University, CRC Press, Taylor and Francis Group.

Miranda, M. J., & Bocchini, P. (2015). "A versatile technique for the optimal approximation of random processes by functional quantization." *Applied Mathematics and Computation, Elsevier, 271*, 935–958.

Moghtaderi-Zadeh, M., & Kiureghian, A. D. (1983). "Reliability upgrading of lifeline networks for post-earthquake serviceability." *Earthquake Engineering & Structural Dynamics, 11*, 557–566.

Saydam, D., Bocchini, P., & Frangopol, D. M. (2013). "Time-dependent risk associated with deterioration of highway bridge networks." *Engineering Structures, 54*, 221–233.

Stefanou, G. (2009). "The stochastic finite element method: past, present and future." *Computer Methods in Applied Mechanics and Engineering, 198*(9–12), 1031–1051.

Vaziri, P., Davidson, R., Apivatanagul, P., & Nozick, L. (2012). "Identification of optimization-based probabilistic earthquake scenarios for regional loss estimation." *Journal of Earthquake Engineering, 16*, 296–315.

Vickery, P., Masters, F., Powell, M., & Wadhera, D. (2009). "Hurricane hazard modeling: The past, present, and future." *Journal of Wind Engineering and Industrial Aerodynamics, 97*, 392–405.

Vickery, P., & Twisdale, L. (1995a). "Prediction of hurricane wind speeds in the united states." *Journal of Structural Engineering, 121*(11), 1691–1699.

Vickery, P., & Twisdale, L. (1995b). "Wind-field and filling models for hurricane wind-speed predictions." *Journal of Structural Engineering, 121*(11), 1700–1709.

Vickery, P. J., & Blanton, B. O. (2008). "North carolina coastal flood analysis system hurricane parameter development." *Technical report*, Renaissance Computing Technical Report TR-08-06. University of North Carolina at Chapel Hill, North Carolina.

Vickery, P. J., Skerlj, P. F., Steckley, A. C., & Twisdale, L. A. (2000). "Hurricane wind field model for use in hurricane simulations." *Journal of Structural Engineering, 126*(10), 1203–1221.

Vickery, P. J., Skerlj, P. F., & Twisdale, L. A. (2000). "Simulation of hurricane risk in the U.S. using empirical track model." *Journal of Structural Engineering, 126*(10), 1222–1237.

Westerink, J. J., Luettich, R. A., Feyen, J. C., Atkinson, J. H., Dawson, C., Roberts, H. J., et al. (2008) "A basin- to channel-scale unstructured grid hurricane storm surge model applied to southern louisiana." *Monthly Weather Review, 136*(10), 833–864.

Chapter 3
Supporting Life Cycle Management of Bridges Through Multi-Hazard Reliability and Risk Assessment

Jamie E. Padgett and Sabarethinam Kameshwar

Abstract Bridge infrastructure is susceptible to damage from a large host of threats including natural hazards, aging and deterioration, and demands that increase with population growth and urbanization. Life cycle management of bridge infrastructure requires an understanding of the relative contribution of these threats to the risk of damage or impending consequences, such as life cycle costs. Traditionally, limited attention has been given to understanding the hazard risk profile to bridge infrastructure, defined as the relative risks posed by multiple hazards and the synergies or trade-offs in protecting for different hazards. Furthermore, effective strategies are needed to jointly consider cumulative damage (e.g., from aging) and punctuated damage (e.g., from natural hazards) when assessing the influence of design or upgrade decisions that may mitigate risks from multiple potentially competing hazards. This chapter utilizes metamodels as an efficient strategy for developing parameterized time-dependent bridge fragilities for multiple hazards, thereby facilitating multi-hazard risk assessment and life cycle management. Threats considered in the case studies include earthquakes, hurricanes, aging and deterioration, and live loads. The applications illustrate the relative contribution of earthquake and hurricane hazards to the risk of losses given variation in bridge parameters, the influence of considering aging when assessing the hazard risk profile, and the impact of concurrent threats (e.g., truck and earthquake) on the life cycle risk.

3.1 Introduction

Bridge infrastructure in the United States is susceptible to multiple hazards such as earthquakes, hurricanes, floods, and collisions. Even individual bridges within a regional portfolio of bridges may be subjected to multiple hazards during their

J.E. Padgett (✉) • S. Kameshwar
Department of Civil and Environmental Engineering, Rice University, 6100 Main Street, MS-318, Houston, TX 77005, USA
e-mail: jamie.padgett@rice.edu; sk56@rice.edu

P. Gardoni, J.M. LaFave (eds.), *Multi-hazard Approaches to Civil Infrastructure Engineering*, DOI 10.1007/978-3-319-29713-2_3

life span. For example, bridges in Charleston, South Carolina may be susceptible to earthquakes and hurricanes; while bridges in the Houston, Texas ship channel region may be susceptible to hurricane surge and vessel collision. The issue of multiple hazards has been acknowledged by bridge design engineers in several states and they also consider multiple hazards in the design process (Lee et al. 2011). However, in order to optimally design bridges subjected to multiple hazards, the risk profile of hazards in consideration, i.e., trade-offs and synergies in risk, should be understood well. Several studies have comprehensively studied the risk to bridges due to individual hazards; for example, seismic reliability and risk has been extensively studied in the literature (Gardoni et al. 2002, 2003; Mackie et al. 2008; Nielson 2005; Ghosh and Padgett 2011). However, only recently studies have started focusing on multi-hazard risk assessment. In contrast, several studies exist on multi-hazard risk assessment for building structures. For example, risk assessment of residential wood buildings considering earthquake, hurricane wind, snow, and similar extreme loads has been the focus of many studies (Li and Ellingwood 2009; Ellingwood et al. 2004; Li and van de Lindt 2012; Yin and Li 2011). McCullough and Kareem (2011) have proposed a general performance-based design framework for designing coastal structures susceptible to multiple hazards. Even though these studies significantly improve the existing multi-hazard risk assessment procedures, they cannot be directly applied to bridges due the unique complexities of bridge behavior under hazard loading as well as the advances required to efficiently apply these concepts for comparative analysis across a range of design parameters. However, a few studies have recently focused on multi-hazard risk assessment of bridge structures. Decò and Frangopol (2011) assess risk due to several hazards including earthquakes, pier scour, and live loads. Effects of aging were also included in seismic and live load performance. In seismic performance assessment, fragility parameters were modified with time as per Ghosh and Padgett (2010), and for live load reliability, the load carrying capacity of girders was decreased with age. Kameshwar and Padgett (2014) have proposed a parameterized multi-hazard risk assessment framework for a portfolio of bridges and showed its application for earthquake and hurricane hazards. Liang and Lee (2013a, b) assess load effects and estimate bridge failure probabilities for concurrent occurrence of scour, earthquake, and truck collision. Wang et al. (2014) consider combined effects of scour and earthquake to evaluate load factors for concrete bridges. Furthermore, several studies have also studied the effect of aging on seismic performance of bridges (Choe et al. 2009; Ghosh and Padgett 2010).

Most of the abovementioned studies solely focus on risk assessment due to several nonconcurrent hazards or concurrent hazards. However, decision making under multiple hazards should acknowledge and consider the effect of multi-hazard combinations on the performance of bridges. For example, seismic risk may be exacerbated by the presence of additional loads due to a concurrent hazard, or two nonconcurrent hazards may have competing influence on selection of a specific design parameter. Moreover, joint consideration of hazards with aging of structures and its implications on the life cycle risks should also be considered.

Kameshwar and Padgett (2014) consider the effect of nonconcurrent hazards, earthquakes and hurricanes, on the selection of optimal column height. However, current literature lacks studies that explore and categorize different types of multi-hazard combinations and their effects on bridge reliability, risk, and design parameter selection while considering the effects of aging. Therefore, to address these gaps, this chapter will define categories of multi-hazard combinations based on occurrence, influence on fragility and risk, and influence on design parameter selection. Furthermore, several examples of multi-hazard combinations will be categorized into the abovementioned groups based on risk assessment of a case study bridge situated in Charleston, South Carolina. Since the bridge is located in coastal Charleston, the bridge is subjected to hurricane wave and surge loads in addition to earthquake and truck loads while considering the effects of aging.

The following section will define the categories of multi-hazard combinations described above. Section 3.3 will characterize the hazards considered in this study, i.e., the probabilities of hazard occurrence and load patterns for each hazard are established. In Sect. 3.4, the demands imposed by the hazards for different combinations of design parameters are evaluated using metamodels. The demands are used along with component capacities to evaluate bridge fragility in Sect. 3.5 which is further used to evaluate risk, quantified herein as the annual failure probability of the bridge. The results from the risk assessment procedure are discussed in Sect. 3.6 where the multi-hazard combinations are categories into different groups. Finally, the conclusions of this study are presented in Sect. 3.7.

3.2 Categorization of Multi-Hazard Combinations

Categorization of multi-hazard combinations based on criteria such as occurrence, effect on reliability, and bridge design is an important first step in understanding the effect of different hazards for multi-hazard decision making. Occurrence-based classification would help in determining the load combinations that the bridge would have to resist during extreme events. This classification of hazards based on occurrence is relatively straightforward and is discussed in the following subsection. Understanding of the demands imposed by multiple hazards and their subsequent classification is crucial for multi-hazard design and decision making, since in a multi-hazard scenario, a remedial action may be potentially detrimental to the performance of the bridge during another hazard. Even though classification of multiple hazards based on their effects on bridge reliability and design parameter selection is important for multi-hazard decision making, current literature lacks guidance or efficient methods to support classifying multiple hazards based on this criteria. Therefore, this study aims to classify the hazards based on occurrence, on effect on bridge reliability and risk, and on influence on design parameter selection, which are discussed in Sects. 3.2.1, 3.2.2, and 3.2.3, respectively.

3.2.1 *Based on Hazard Occurrence*

The first approach for classifying multi-hazard combinations requires assessment of the hazard occurrence potential. Based on occurrence of the hazard events, multiple hazards can be broadly classified in to the following categories:

3.2.1.1 Nonconcurrent Hazards

This category includes hazard combinations whose probability of occurring simultaneously is very low. For example, earthquakes and hurricanes have very low joint probability of occurrence. Similarly, hurricane loads and truck loads have low chance of simultaneous occurrence since people either evacuate before the hurricane or take shelter during a hurricane. Disjoint occurrence of hazards allows independent modeling of load effects on the bridge. However, independence of load effects does not necessarily imply uncorrelated influence on design parameter selection.

3.2.1.2 Concurrent Hazards

As the name suggests, this category of hazards includes combination of hazards which either always act simultaneously or have appreciable probability of joint occurrence. For such hazards, modeling of load effects must consider the joint load effects due to the multiple hazards. For example, during hurricanes coastal bridges may be subjected to combined wave and surge forces; similarly, earthquakes may happen while trucks are passing over the bridges, as observed in past events. Furthermore, the probability of joint occurrence of hazards must be obtained to evaluate risk.

3.2.1.3 Cascading Hazards

Occurrence of a hazard may trigger other hazards; for example, earthquakes or vessel collisions may cause fire. The main hazard and the subsequent hazards can be considered collectively as cascading hazards. Cascading hazards may also be considered as a special case of nonconcurrent hazards where the main hazard and the subsequent hazards occur within very short duration of time. This category of multiple hazards is one of the most challenging and least studied categories of multiple hazards. Evaluation of reliability and risk under such multiple hazards involves accumulation of damage due to the main extreme event and the following cascading events. To add to the complexity of the problem, probabilities of occurrence of the cascading events also have to be evaluated which may depend on the damage caused due to the main event. This study will focus on nonconcurrent and concurrent hazards for reliability and risk assessment; cascading hazards will be addressed in future research.

3.2.2 Based on Influence on Fragility and Risk

The second category for classifying combinations of hazards entails evaluating the fragility and risk to the structure under multi-hazard exposure. In multi-hazard design and decision-making situations, a better understanding of the risk portfolio of the bridge may help in choosing optimal retrofit options or design parameters. Categorization of hazard combinations based on their effect on reliability and risk can improve the understanding of the risk portfolio. Therefore, in this category, this chapter classifies hazard combinations as amplifying or diminishing.

3.2.2.1 Amplifying Hazards

Hazard combinations where the presence of one hazard increases the vulnerability (decreases the reliability) of the bridge during the occurrence of other hazards can be classified as amplifying hazards. For example, in some cases pier scour has been shown to be detrimental to seismic performance of bridges, so scour and earthquakes can be considered as amplifying hazards. Identification of amplifying hazards is important because the overall risk to the bridge increases due to such hazard combinations.

3.2.2.2 Diminishing Hazards

Bridge performance may improve during a hazard, i.e., increase in reliability may be observed, due to the presence of other hazards or additional loads due to other hazards. Such combinations of hazards can be included into the category of diminishing hazards. For example, in some cases the presence of trucks on the bridge deck may actually improve the seismic reliability of bridges, due to vehicle bridge interaction or in some cases due to a favorable shift in the natural period of the system.

3.2.3 Based on Influence on Design Parameter Selection

The final category for classifying hazard combinations includes exploration of the design parameter space and its influence on reliability and risk. Understanding this influence of design parameter variation is important in multi-hazard decision making since improving the performance of the bridge to one hazard may inadvertently worsen its performance during the other hazard. Further, identification of design parameters which can improve the bridge performance for several hazards or optimize the ultimate design parameter selection is also important for economical design. However, literature lacks classification of hazards into such categories that may shed light on practical design consideration. Therefore, this study will classify the combination of hazards into groups based on their influence on selection of design parameters as competing or complementary hazards.

3.2.3.1 Competing Hazards

Hazard combinations that have opposing or competing influence on bridge design parameter selection may be categorized as competing hazards. For example, earthquakes and hurricanes may have competing influence on column height selection. Increase in column height has been shown to improve the reliability of bridges subjected to wave and surge loads; however, increase in column height alone may increase the seismic risk (Kameshwar and Padgett 2014).

3.2.3.2 Complementary Hazards

The group of hazards where mitigation of one of the hazards serves as a remedial action for the bridge during other hazards, or a combination of hazards for which change in a design parameter improves the performance of the bridge for all the hazards in the combination, may be categorized as complementary hazards. For example, improving the ductility of bridge columns may improve bridge performance during seismic events and collision events involving trucks or vessels with bridge columns. Identification of this type of hazard combination is also important since cognizance of these hazards and their associated preferable design parameters may reduce the overall cost of reaching a target system reliability or risk level.

3.3 Characterization of Hazards

In order to accomplish the goals of this study, i.e., to classify the hazards into various categories, the risk must be evaluated for the case study bridge due to earthquakes, hurricane, and combined seismic and truck loads. The first step in the risk assessment is to categorize the multiple hazards based on occurrence and the second step involves estimating probabilities of occurrence and the related load effects. Among the hazards, earthquakes and hurricanes are treated as nonconcurrent hazards since the probability of their joint occurrence is extremely low. Therefore, their probabilities of occurrence and load effects can be evaluated independently. On the other hand, earthquake and truck loads are considered to be concurrent hazards. This implies that that their joint probability of occurrence and joint load effects must be determined. In addition to these hazards, threats due to deterioration of the case study bridge due to aging are also taken into consideration.

Table 3.1 Bridge parameter values

Variable	Range
Column height (H_c)	3.60–9.40 m
Column diameter (D_c)	0.76–1.52 m
Longitudinal reinforcement ratio (ρ_l)	0.02–0.04
Transverse reinforcement ratio (ρ_t)	5.00×10^{-3} –1.10×10^{-3}

The case study bridge, situated in coastal Charleston, South Carolina, is assumed to be a simply supported concrete girder bridge. The bridge has three 22.3 m-long equal spans, each 7.6 m wide, while other bridge parameters such as column height, diameter, and reinforcement ratios in transverse and longitudinal direction are varied; the range of the variables is shown in Table 3.1. The bridge is modeled in OpenSees (Mazzoni et al. 2006) following the general modeling recommendations outlined by Nielson (2005). For this case study bridge, the following section elaborates the procedure used to evaluate the probability of hazard occurrence at the bridge site and their load effects on the case study bridge.

3.3.1 Earthquakes and Truck Loads

Seismic response of the case study bridge is studied by simulating the response of the bridge for a suite of ground motions. Since recorded ground motions are not available for the Charleston region, suites of synthetic ground motions developed for the Central and Southeastern United States are used. The suite of ground motions developed by Fernandez and Rix (2008), consisting of 288 ground motions, is used along with a second suite consisting of 60 ground motions which was developed by Wen and Wu (2001). Next, seismic hazard occurrence data for the Charleston region is obtained from the US Geological Survey (USGS) (Petersen et al. 2008). The seismic hazard data is fit to a hyperbolic expression proposed by Bradley et al. (2007) to evaluate the risk in Sect. 3.5.

Joint occurrence of earthquakes and truck loads is modeled by placing a truck on the bridge and simultaneously exciting the bridge with ground motions. The truck loads are applied to the truck by placing a WB-20 truck at the centerline of the bridge at various locations along the length of the bridge. The truck weight is assumed to follow the bimodal distribution obtained by Ghosh et al. (2014). The probability of a truck being present on the bridge as per Ghosh et al. (2014) is

$$P\,(\text{one truck}) \approx (L-18)\,Q \times 10^{-5} \tag{3.1}$$

where Q is the flow rate in trucks per hour and L is the length of the bridge (66.9 m). The effect of joint occurrence of trucks and earthquakes on the risk estimates is discussed with the risk assessment procedure in Sect. 3.5.

3.3.2 *Hurricanes*

Maximum wave and surge load estimates on the bridge are obtained from the coastal guideline specification by the American Association of State Highway and Transportation Officials (AASHTO) (2008). The maximum forces obtained from the AASHTO guidelines are distributed in a phenomenological model of the wave load time series following Ataei et al. (2010). The maximum wave and surge forces are functions of hazard intensity parameters, wave height and surge height, and random variables such as wave period and wave length. In order to assess the risk to the bridge, the probability of occurrence of hurricane and the joint probability distribution of the hazard intensity parameters must be assessed. Hurricane occurrence in the Charleston region is assumed to follow a Poisson process with a mean annual rate of 0.23 (Scheffner and Carson 2001). The joint probability distribution of the hazard intensity parameters for an assumed water depth of 3.0 m and fetch length of 5.0 km is obtained using the procedure outlined in Kameshwar and Padgett (2014).

3.3.3 *Aging*

Deterioration due to aging may not be considered as a hazard; however, it poses significant threat to extreme event performance of bridges, such as seismic performance. Therefore, this study considers the effects of aging by modeling the reduction in the diameter of the steel reinforcement bars, which decreases the reinforcement ratio and confinement of core concrete, and oxidation of elastomeric bearing pads, which increases the stiffness of the elastomeric bearing pads. For tidal exposure conditions with 40 mm cover, rebar corrosion initiation time is modeled using DuraCrete (2000) and corrosion propagation follows the model proposed by Choe et al. (2008); while for modeling oxidation of the elastomeric bearing pads, formulation proposed by Itoh and Gu (2009) is used. The deck is assumed to be simply placed on the bearings over the substructure without any vertical connection such as dowels. Therefore, aging is assumed to have no effect on the hurricane response of the bridge.

3.4 Demand Assessment

Loads from the concurrent and nonconcurrent hazard combinations, described above, are applied to the case study bridge to estimate the demands on bridge components. However, several parameters of the bridge, listed in Table 3.1, are varied to study the effect of these parameters on bridge reliability and risk. Furthermore, parameters such as concrete strength, steel strength, friction coefficients at bearings, and gap between abutments and deck are considered to be random variables. Each combination of the parameters leads to a new bridge sample and a large number of such combinations may exist. However, simulating all the possible combinations is practically infeasible; therefore, metamodels are used in this study to estimate demands on the bridge with limited number of simulations. Metamodels are efficient mathematical tools which detect underlying relation between input parameters, i.e., hazards and bridge parameters in this study, and the output, i.e., component response. In order to model the component response, a set of design parameters, listed in Table 3.1, and random variables is generated that represents the entire space of variables. For this purpose, Latin hypercube sampling (LHS) (McKay et al. 1979) is used. For each hazard, the set of parameters generated by LHS is randomly paired intensity measures and an age value, where deterioration is considered. Therefore, age of the bridge also becomes a variable which is used to predict the response of the bridge. In case of earthquake and truck loads, the parameters are randomly paired with a ground motion and a truck weighing between 0.0 and 60.0 tons. While for hurricanes, the parameters are randomly paired with a set of wave height and surge height values, which are generated on an evenly spaced grid with surge height ranging between 0.0 and 6.0 m and wave height varying from 0.0 to 3.5 m.

Under seismic excitation and joint truck and seismic excitation, the response of bridge components such as columns, abutments, and bearings is modeled using different metamodels. In this study, the component responses are modeled using response surfaces with higher-order polynomials, Adaptive Basis Function Construction (ABFC) (Jekabsons 2010) which is also a polynomial-based metamodel, multivariate adaptive regression splines (MARS) (Friedman 1991), and radial basis functions (RBF) (Hardy 1971). Each of the aforementioned metamodels has certain advantages and disadvantages. Polynomial-based methods are transparent, but they are not suitable for extrapolation; MARS is a very quick method, but it may overfit the data; and RBF achieve very good accuracy, but the method requires scaling of the input data. Since the metamodels may have different performances in predicting the response of a component, performance measures are used to assess the fit of the selected metamodels. The performance of the metamodels was compared based on goodness of fit measures such as R^2 value, root mean square error (RMSE), and mean R^2 value in 5-fold cross validations. Based on these performance metrics, the fourth-order polynomial response surface generated using the sequential forward selection (SFS) method is found to perform best in predicting bearing deformation. While third-order polynomial, obtained using SFS, is observed to perform best in predicting column drift and abutment displacement. Using SFS for generating polynomial response surface ensures that only most significant polynomial terms are introduced in the polynomial equation. A normally distributed model error term with zero mean and standard deviation equal to the RMSE of the model is also added to each of the metamodels. For brevity, the response surface models are not included herein.

In this study, the bridge is assumed to be safe after the hurricane if the bridge deck is not displaced; however, failure is assumed if the deck is displaced due to the hurricane wave and surge forces. Categorization of the response of the bridge into the two categories leads to a classification problem. Therefore, response prediction of bridges subjected to hurricanes is performed using a different type of metamodels called binary classifiers. This category of metamodels can easily predict failure or survival of the bridge as a binary variable. Random forest (Pavlov 2000) and Support Vector Machines (SVM) (Cristianini and Shawe-Taylor 2000) are used for hurricane response prediction of bridges. Since the response of the classifiers is a binary variable, the performance metrics used for component response prediction under seismic and truck loads cannot be used for these metamodels. So, the performance of these metamodels is assessed using a confusion matrix (Kohavi and Provost 1998) which counts the number of true positives, true negatives, false positives, and false negatives, which can be further used to measure the accuracy of prediction. Among random forest and SVM, random forest was observed to perform better, and therefore random forest is selected for hurricane response prediction in this study.

3.5 Reliability and Risk Assessment

The demands imposed by the hazards on the bridge components are compared with their capacities to assess component reliability. For seismic response such as column drifts, bearing deformation, and abutment displacements, Table 3.2 shows the component capacities for the complete damage limit state and their corresponding distribution. In the expression for mean drift capacity of columns in Table 3.2, ALR is the axial load ratio, L is half the column height, and $\alpha = (1 - s/d)^2$, where s is the spacing between transverse reinforcement and d is the effective depth of the column cross section. As seen from Table 3.2, the limit states for bearings and abutments are invariant to the presence of trucks. However, the presence of trucks is indirectly accounted for in the column drift capacity limit states by including the axial load ratio in the capacity limit state. Moreover, the effect of aging on the capacity of the columns is also included by reducing the reinforcement ratio corresponding to decrease in rebar diameter. Using these demand and capacity estimates, the reliability of the bridge components can be estimated when subjected to seismic loads or to joint seismic and truck loads. In order to assess the reliability, first, a

Table 3.2 Component capacity for complete damage limit state

Component	Median/mean	Coefficient of variation (%)	Distribution
Bearing (Ramanathan et al. 2012)	255.0 mm (median)	47.0	Lognormal
Abutment (Ramanathan et al. 2012)	55.0 mm (median)	47.0	Lognormal
Column (Panagiotakos and Fardis 2001)	$0.9\left(0.2^{\text{ALR}}\right)(6.89 f_c)^{0.275}\left(\frac{L}{D}\right)^{0.45}\left(1.1^{\frac{100\alpha\rho_t f_y}{f_c}}\right)$ (mean)	47.0	Lognormal

set of 30×10^{-3} design parameters, described in Table 3.1, and random variables such as concrete and steel strength is generated using LHS. Next, the demands on the bridge components are assessed using the metamodels, described above, and the capacities of the components are obtained from Table 3.2. Demands and capacities are compared with each other, and the outcome is represented by a binary variable; 1 represents failure and 0 represents a survival. For the complete damage limit state considered in this study, the bridge is considered to be a series system where failure of a component leads to system failure. The component binary output is used to evaluate the binary system output which is further used in logistic regression to evaluate fragility of the system. In case of hurricanes, the response from the classifiers is already in binary form; therefore, it is directly used in logistic regression. Failure probability is obtained using logistic regression as

$$P\left(\text{Fail}\,|X,\text{IM},t\right) = \frac{e^{g(X,\text{IM},t)}}{1+e^{g(X,\text{IM},t)}} \tag{3.2}$$

In the above equation, $P(\text{Fail} \mid X, \text{IM}, t)$ is the failure probability conditioned on parameters X, described in Table 3.1, intensity measures IM, and age t. The function $g(X)$ is the logit function which predicts the logarithm of odds in favor of failure. In this study, the logit function is a polynomial in X, IM, and t.

Risk, i.e., the annual failure probability, is assessed for each of the multi-hazard combinations by convolving the corresponding fragility with hazard occurrence. Since the annual failure probabilities (p_f) are small, seismic risk to the bridge can be written as (Der Kiureghian 2005)

$$p_f = \int_{\text{pga}} P\left[\text{Fail}\,|X,\text{pga},t\right] \left|\frac{\mathrm{d}\widehat{\nu}}{\mathrm{d}\left(\text{pga}\right)}\right| \mathrm{d}\left(\text{pga}\right) \tag{3.3}$$

In the above equation, $P[\text{Fail}\,|X,\text{pga},t]$ is the seismic fragility where X represents bridge parameters; pga refers to peak ground acceleration, the intensity measure; t is age of the bridge; and $\widehat{\nu}$ is the seismic hazard curve obtained from the USGS. Similar to Eq. (3.3), in the case of joint earthquake and truck presence, the annual probability of failure can be estimated as

$$p_f = \int_{\text{pga}} P_{LL+EQ}\left[\text{Fail}\,|X,\text{pga},t\right] \left|\frac{\mathrm{d}\widehat{\nu}}{\mathrm{d}\left(\text{pga}\right)}\right| \mathrm{d}\left(\text{pga}\right) \tag{3.4}$$

The term $P_{LL+EQ}\left[\text{Fail}\,|X,\text{pga},t\right]$ is the joint seismic and truck load fragility which is estimated using the total probability theorem as

$$\begin{aligned} P_{LL+EQ}\left[\text{Fail}\,|X,\text{pga},t\right] &= \left[1-P\left(\text{truck}\right)\right] * P\left[\text{Fail}\,|X,\text{pga},t\right] \\ &+ P\left(\text{truck}\right) * P\left[\text{Fail}\,|X,\text{pga},t,\text{truck}\right] \end{aligned} \tag{3.5}$$

where $P[\text{Fail}\,|X, \text{pga}, t]$ is the seismic fragility without the effect of trucks, as in Eq. (3.3), $P(\text{truck})$ represents the probability of truck presence, and $P[\text{Fail}\,|X, \text{pga}, t, \text{truck}]$ is the bridge fragility function which is also conditional on truck presence and is given as

$$P\left[\text{Fail}\,|X, \text{pga}, t, \text{truck}\right] = \int_{W} P\left[\text{Fail}\,|X, \text{pga}, t, \text{truck}, w\right] f_W(w)\, \mathrm{d}\, w \tag{3.6}$$

In Eq. (3.6), w is the truck weight, $f_W(w)$ is the probability distribution of truck weights obtained from Ghosh et al. (2014), and $P[\text{Fail}\,|X, \text{pga}, t, \text{truck}, w]$ is the seismic fragility conditioned on the truck loads, in addition to other parameters. Different truck locations were also considered; however, the truck location was found to have insignificant effect on the response of bridge components. Therefore only truck weight is considered in the joint fragility function in Eq. (3.6). In the joint live load and seismic risk assessment, it is assumed that probability of multiple truck presence is negligible. So, probability of truck absence is calculated as the complement of presence of one truck. However, depending upon the route on which the bridge falls and size of the bridge, presence of multiple trucks may have significant probability and its influence on reliability and risk to the bridge may be studied in future work.

Hurricane risk is also evaluated using a procedure similar to the seismic risk assessment procedure described above. In case of hurricanes, the deck uplift fragility does not depend on age due to lack of any physical tie-down between the deck and the bent; therefore, the risk is independent of age. The annual probability of failure due to hurricanes can be obtained using

$$p_f = \lambda \int_{S}\int_{H} P\left[\text{Fail}\,|X, s, h\right] f_{s,h}\,(h, s)\, \mathrm{d}\, h\, \mathrm{d}\, s \tag{3.7}$$

$P[\text{Fail}\,|X, s, h]$ is the uplift fragility of bridge deck which is conditioned on the bridge parameters (X) and the intensity measures, wave height (H) and surge height (S), $f_{S,H}(h, s)$ is the joint probability distribution of wave height and surge height, and λ is the annual rate of hurricane occurrence. Using the above equations, risk to the bridge is evaluated as various bridge parameters are varied. The trends in risk for variation in parameters are discussed in the following section.

3.6 Results and Discussions

One of the primary aims of this study is to understand the nature of multiple hazard combinations and categorize them based on their effect on reliability and design parameter selection. First, the effect of two nonconcurrent hazards, i.e., earthquakes and hurricanes, is studied. Figure 3.1 shows the effect of varying column height

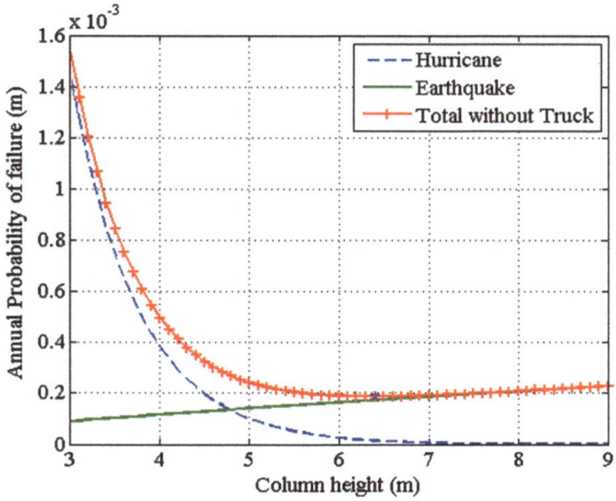

Fig. 3.1 Variation of hurricane, seismic, and total annual failure probability with column height

while keeping other parameters constant ($D = 1.0$ m, $\rho_l = 3.0\,\%$ and $\rho_t = 0.5\,\%$), on the hurricane risk and seismic risk, without aging or truck loads. It can be observed that as the column height increases, the hurricane risk decreases rapidly; however, increase in the column height also leads to a slow increase in earthquake risk. Since the two hazards are independent, the total risk shown as "Total without Truck" in Fig. 3.1 can be obtained as the sum of seismic and hurricane risk. With initial increase in column height, the total risk decreases sharply due to the rapid decrease in hurricane risk, which dominates at lower column heights. However, as the column height increases further, the total risk starts to increase since seismic risk dominates at larger column heights. These observations show that earthquakes and hurricanes have competing requirements for the column height. Therefore, these two hazards can be categorized as competing hazards with respect to their influence on column height.

Bridges are often exposed to harsh environments leading to deterioration due to aging, which significantly affects the risk over the lifetime of the bridge. In order to assess the effects of aging on the bridge during its life cycle, the bridge is assumed to be exposed to tidal exposure conditions. Figure 3.2 shows the total annual probability of failure along the life span of the bridge as the column height varies. The hurricane risk is assumed to remain constant along the life cycle of the bridge and variation in the total risk is due to seismic hazard only. Qualitatively, at each value of age, the variation in risk with changing column height is similar to that in Fig. 3.1. However, with increase in age, the hazard risk changes and increases up to 10 % in comparison to a pristine bridge. Thus, deterioration due to aging has an amplifying effect on the seismic risk. This result suggests although aging considerations are currently are not included in the modern design and retrofit codes, future research should support the development of design guidelines where time-evolving hazard risks, which may be significant as the bridge ages, are accounted for when designing for extreme events.

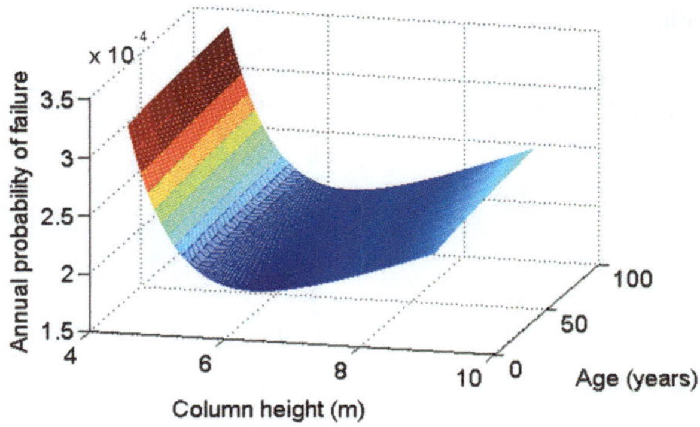

Fig. 3.2 Variation in total risk with age and change in column height

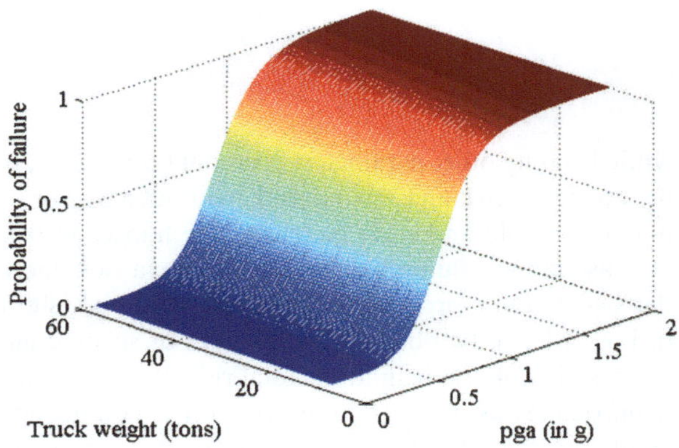

Fig. 3.3 Joint seismic and truck load fragility

Performance of the bridge ($H = 7.0$ m, $D = 1.0$ m, $\rho_l = 3.0\,\%$ and $\rho_t = 0.5\,\%$) under the two concurrent hazards, earthquakes and truck loads, is shown in Fig. 3.3, which shows the joint fragility of the bridge due to truck and seismic loads. From the figure, the effect of pga can be clearly seen on the failure probability. As expected, with increase in pga, the probability of failure increases. However, the effect of truck weight is not apparent from the figure. So, Fig. 3.4 shows the change in the median pga for failure as the truck weight increases. It can be seen that the median pga increases as the truck weight increases, implying that the fragility decreases due to the presence of the truck. The decrease in fragility can be attributed to two reasons: firstly, due to increase in the drift capacity of the column because of higher axial load ratio with increased truck loads and, secondly, due to the ground motions used in this study. The mean response spectra of all the ground motions used in this study show that the spectral acceleration decreases after natural period of 0.25 s, and all the bridges used in this study have periods larger than 0.25 s. Therefore presence of a truck, which increases the period of the bridge, decreases the spectral acceleration demand on the bridge. The effect of truck presence and age on the seismic risk is shown in Fig. 3.5. As the flow rate of trucks, Q, increases, the annual probability of

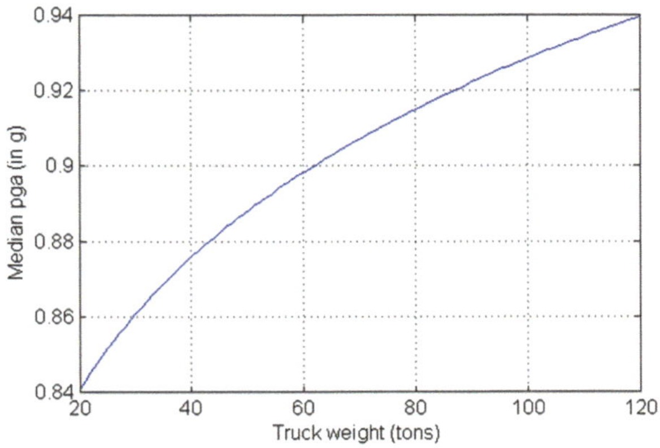

Fig. 3.4 Variation in median pga with truck weight

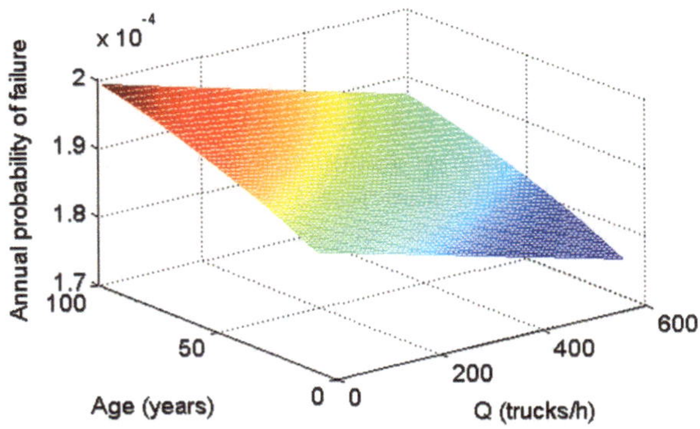

Fig. 3.5 Effect of truck presence on seismic risk

failure decreases. The probability of truck presence described in Eq. (3.1) increases with Q and presence of a truck decreases the fragility; therefore, increased truck flow decreases the risk. With age, the variation is similar to Fig. 3.2, with about 10 % change in risk between a pristine and 100-year-old bridge. Therefore, the multi-hazard combination of truck and seismic loads can be considered as diminishing in nature for this case study since presence of trucks actually helps decrease the seismic risk. However, it is acknowledged that these results are case study specific; variation in design details may lead to different results.

3.7 Conclusions

This study offers potential nomenclature and taxonomy for categorizing multi-hazard cases of interest for life cycle management of bridges. Furthermore, it explores various multi-hazard cases offering categorization of the class of multi-hazards and insights from the multi-hazard risk assessment to a case study bridge.

Such an assessment is achieved through application of a proposed parameterized multi-hazard risk assessment framework which also includes the effects of deterioration due to aging. Multi-hazard combinations are categorized in this chapter into several groups based on their occurrence, i.e., concurrent, nonconcurrent, and cascading, based on their influence on reliability and risk as amplifying or diminishing, and based on their effects on design parameter selection as competing or complementary.

The categorization of hazards is performed on the basis of applying the parameterized risk assessment framework for the case study bridge in Charleston, South Carolina. While considering the effects of aging, the bridge is subjected to earthquakes, hurricanes, and joint seismic and truck loads. For each of these hazards or hazard combinations, this study employs metamodel-based demand assessment to assist in exploration of the parameter space without additional simulations. The results from the application of the multi-hazard risk assessment procedure on the case study bridge provide important insight to the risk portfolio. The results highlight the competing influence of the two nonconcurrent hazards, earthquakes and hurricanes, on column height. This competing nature of the two hazards shows the importance of risk assessment considering nonconcurrent multiple hazards and categorization of multiple hazards according to their influence on selection of bridge design parameters. Reliability assessment of the bridge for concurrent occurrence of seismic and truck loads show that truck presence decreases the seismic fragility of the case study bridge. Consequently, for this particular case study, the risk, i.e., the annual probability of failure, decreases due to presence of a truck whose magnitude depends on the flow rate of the trucks. Thus, the results uncover the diminishing nature of the seismic and truck load combination. Future work will focus on classifying additional multi-hazard combinations and extending the proposed framework to a portfolio of bridges. In addition to this, effects of cascading hazards on reliability and selection of design parameters should also be explored.

Acknowledgments The authors would like to gratefully acknowledge the support of this research by the National Science Foundation (NSF) under Grant No. CMMI-1055301. Any opinions, findings, and conclusions or recommendations expressed in this material are those of the authors and do not necessarily reflect the views of the National Science Foundation. The authors would also like to acknowledge computational facilities provided by the Data Analysis and Visualization Cyberinfrastructure (NSF grant OCI-0959097).

References

AASHTO. (2008). *Guide specifications for bridges vulnerable to coastal storms*, Washington, DC, American Association of State Highway Transportation Officials.

Ataei, N., Stearns, M., & Padgett, J. E. (2010). Response sensitivity for probabilistic damage assessment of coastal bridges under surge and wave loading. *Transportation Research Record: Journal of the Transportation Research Board, 2202*(1), 93–101.

Bradley, B. A., Dhakal, R. P., Cubrinovski, M., Mander, J. B., & MacRae, G. A. (2007). Improved seismic hazard model with application to probabilistic seismic demand analysis. *Earthquake Engineering and Structural Dynamics, 36*(14), 2211–2225.

Choe, D.-E., Gardoni, P., Rosowsky, D., & Haukaas, T. (2008). Probabilistic capacity models and seismic fragility estimates for RC columns subject to corrosion. *Reliability Engineering & System Safety, 93*(3), 383–393. http://dx.doi.org/10.1016/j.ress.2006.12.015.

Choe, D.-E., Gardoni, P., Rosowsky, D., & Haukaas, T. (2009). Seismic fragility estimates for reinforced concrete bridges subject to corrosion. *Structural Safety, 31*(4), 275–283.

Cristianini, N., & Shawe-Taylor, J. (2000). *An introduction to support vector machines and other kernel-based learning methods*. Cambridge University Press, Cambridge.

Decò, A., & Frangopol, D. M. (2011). Risk assessment of highway bridges under multiple hazards. *Journal of Risk Research, 14*(9), 1057–1089. doi:10.1080/13669877.2011.571789.

Der Kiureghian, A. (2005). Non-ergodicity and PEER's framework formula. *Earthquake Engineering and Structural Dynamics, 34*(13), 1643–1652. doi:10.1002/eqe.504.

DuraCrete. (2000). Statistical quantification of the variables in the limit state function. Report No. BE95-1347/R9.

Ellingwood, B., Rosowsky, D., Li, Y., & Kim, J. (2004). Fragility assessment of light-frame wood construction subjected to wind and earthquake hazards. *Journal of Structural Engineering, 130*(12), 1921–1930. doi:10.1061/(ASCE)0733-9445(2004)130:12(1921).

Fernandez, J. A., & Rix, G. J. (2008). Seismic hazard analysis and probabilistic ground motions in the upper Mississippi embayment. In *Geotechnical earthquake engineering and soil dynamics* (Vol. IV, pp. 1–10). doi:10.1061/40975(318)8.

Friedman, J. H. (1991). Multivariate adaptive regression splines. *Annals of Statistics, 19*(1), 1–67. doi:10.2307/2241837.

Gardoni, P., Der Kiureghian, A., & Mosalam, K. M. (2002). Probabilistic capacity models and fragility estimates for reinforced concrete columns based on experimental observations. *Journal of Engineering Mechanics, 128*(10), 1024–1038.

Gardoni, P., Mosalam, K. M., & der Kiureghian, A. (2003). Probabilistic seismic demand models and fragility estimates for RC bridges. *Journal of Earthquake Engineering, 7*(spec01), 79–106.

Ghosh, J., Caprani, C. C., & Padgett, J. E. (2014). Influence of traffic loading on the seismic reliability assessment of highway bridge structures. *Journal of Bridge Engineering,* 10.1061/(ASCE)BE.1943-5592.0000535, 04013009.

Ghosh, J., & Padgett, J. E. (2010). Aging considerations in the development of time-dependent seismic fragility curves. *Journal of Structural Engineering, 136*(12), 1497–1511.

Ghosh, J., & Padgett, J. E. (2011). Probabilistic seismic loss assessment of aging bridges using a component-level cost estimation approach. *Earthquake Engineering and Structural Dynamics, 40*(15), 1743–1761. doi:10.1002/eqe.1114.

Hardy, R. L. (1971). Multiquadric equations of topography and other irregular surfaces. *Journal of Geophysical Research, 76*(8), 1905–1915. doi:10.1029/JB076i008p01905.

Itoh, Y., & Gu, H. (2009). Prediction of aging characteristics in natural rubber bearings used in bridges. *Journal of Bridge Engineering, 14*(2), 122–128.

Jekabsons, G. (2010). Adaptive basis function construction: An approach for adaptive building of sparse polynomial regression models. In Y. Zhang (Ed.), *Machine learning* (pp. 127–155), INTECH Open Access Publisher.

Kameshwar, S., & Padgett, J. E. (2014). Multi-hazard risk assessment of highway bridges subjected to earthquake and hurricane hazards. *Engineering Structures, 78*, 154–166.

Kohavi, R., & Provost, F. (1998). Glossary of terms. *Machine Learning, 30*(2–3), 271–274.

Lee, G. C., Liang, Z., Shen, J. J., & O'Connor, J. S. (2011). *Extreme load combinations: A survey of state bridge engineers*. MCEER, Buffalo.

Li, Y., & Ellingwood, B. (2009). Framework for multihazard risk assessment and mitigation for wood-frame residential construction. *Journal of Structural Engineering, 135*(2), 159–168. doi:10.1061/(ASCE)0733-9445.

Li, Y., & van de Lindt, J. W. (2012). Loss-based formulation for multiple hazards with application to residential buildings. *Engineering Structures, 38*, 123–133.

Liang, Z., & Lee, G. C. (2013a). Bridge pier failure probabilities under combined hazard effects of scour, truck and earthquake. Part I: Occurrence probabilities. *Earthquake Engineering and Engineering Vibration, 12*(2), 229–240.

Liang, Z., & Lee, G. C. (2013b). Bridge pier failure probabilities under combined hazard effects of scour, truck and earthquake. Part II: Failure probabilities. *Earthquake Engineering and Engineering Vibration, 12*(2), 241–250.

Mackie, K. R., Wong, J.-M., & Stojadinović, B. (2008). *Integrated probabilistic performance-based evaluation of benchmark reinforced concrete bridges*. Pacific Earthquake Engineering Research Center, University of California, Berkeley.

Mazzoni, S., McKenna, F., Scott, M. H., & Fenves, G. L. (2006). *OpenSees command language manual*. Mazzoni, Silvia, et al. OpenSees command language manual. Pacific Earthquake Engineering Research (PEER) Center, Berkeley.

McCullough, M., & Kareem, A. (2011). A framework for performance-based engineering in multi-hazard coastal environments. In *Structures Congress 2011*, April 14–16 2011 Las Vegas, Nevada. (pp. 1961–1972). ASCE.

McKay, M. D., Beckman, R. J., & Conover, W. J. (1979). Comparison of three methods for selecting values of input variables in the analysis of output from a computer code. *Technometrics, 21*(2), 239–245.

Nielson, B. G. (2005). *Analytical fragility curves for highway bridges in moderate seismic zones*. PhD thesis, Georgia Institute of Technology, Atlanta.

Panagiotakos, T. B., & Fardis, M. N. (2001). Deformations of reinforced concrete members at yielding and ultimate. *ACI Structural Journal, 98*(2), 135–148.

Pavlov, Y. L. (2000). *Random forests*. VSP, Utrecht.

Petersen, M. D., Frankel, A. D., Harmsen, S. C., Mueller, C. S., Haller, K. M., Wheeler, R. L., et al. (2008). *Documentation for the 2008 update of the United States national seismic hazard maps*. US Geological Survey.

Ramanathan, K., DesRoches, R., & Padgett, J. E. (2012). A comparison of pre- and post-seismic design considerations in moderate seismic zones through the fragility assessment of multispan bridge classes. *Engineering Structures, 45*, 559–573.

Scheffner, N. W., & Carson, F. C. (2001). *Coast of South Carolina storm surge study*. Vicksburg, MS: U.S. Army Engineer Research and Development Center, Coastal and Hydraulics Laboratory.

Wang, Z., Padgett, J. E., & Dueñas-Osorio, L. (2014). Risk-consistent calibration of load factors for the design of reinforced concrete bridges under the combined effects of earthquake and scour hazards. *Engineering Structures, 79*, 86–95.

Wen, Y. K., & Wu, C. L. (2001). Uniform hazard ground motions for mid-America cities. *Earthquake Spectra, 17*(2), 359–384. doi:10.1193/1.1586179.

Yin, Y.-J., & Li, Y. (2011). Probabilistic loss assessment of light-frame wood construction subjected to combined seismic and snow loads. *Engineering Structures, 33*(2), 380–390. doi:10.1016/j.engstruct.2010.10.018.

Chapter 4
Natural Hazard Probabilistic Risk Assessment Through Surrogate Modeling

Alexandros A. Taflanidis, Gaofeng Jia, and Ioannis Gidaris

Abstract Assessment of risk under natural hazards is associated with a significant computational burden when comprehensive numerical models and simulation-based methodologies are involved. Despite recent advances in computer and computational science that have contributed in reducing this burden and have undoubtedly increased the popularity of simulation-based frameworks for quantifying/estimating risk in such settings, in many instances, such as for real-time risk estimation, this burden is still considered as prohibitive. This chapter discusses the use of kriging surrogate modeling for addressing this challenge. Kriging establishes a computationally inexpensive input/output relationship based on a database of observations obtained through the initial (expensive) simulation model. The up-front cost for obtaining this database is of course high, but once the surrogate model is established, all future evaluations require small computational effort. For illustration, two different applications are considered, involving two different hazards: seismic risk assessment utilizing stochastic ground motion modeling and real-time hurricane risk estimation. Various implementation issues are discussed, such as (a) advantages of kriging over other surrogate models, (b) approaches for obtaining high efficiency when the output under consideration is high dimensional through integration of principal component analysis, and (c) the incorporation of the prediction error associated with the metamodel into the risk assessment.

4.1 Introduction

Prediction of the performance of civil infrastructure systems exposed to natural hazards is associated with significant uncertainties, pertaining to the description of the hazard characteristics as well as to the properties of the system under consideration (Ellingwood 2001; Vickery et al. 2006; Resio et al. 2012). This is especially true when life-cycle analysis is considered (Wen and Kang 2001; Taflanidis and Beck 2009) since the aforementioned description needs to address

---

A.A. Taflanidis (✉) • G. Jia • I. Gidaris
Department of Civil and Environmental Engineering and Earth Sciences,
University of Notre Dame, Notre Dame, IN, USA
e-mail: a.taflanidis@nd.edu

P. Gardoni, J.M. LaFave (eds.), *Multi-hazard Approaches to Civil Infrastructure Engineering*, DOI 10.1007/978-3-319-29713-2_4

the anticipated exposure (Kumar et al. 2015) and system behavior over a large time period. A probabilistic approach provides a rational and consistent framework for addressing such uncertainties (Jaynes 2003) using probability models to describe the relative likelihood of different properties of the natural hazard and of the system itself. This then facilitates the description of the performance through the *natural hazard risk*, quantified by the probabilistic integral that corresponds to the expected value of some risk consequence measure over the established probability models for the system and its excitation (hazard).

Assessment of this risk entails ultimately evaluation of the probabilistic integral quantifying it. Analytical approximations and specialized approaches (Rackwitz 2001; Der Kiureghian 1996; Taflanidis 2010) can be adopted for this purpose but include an unknown, unavoidable error and can further impose restrictions on the complexity of the models adopted to characterize and analyze risk. On the other hand, approaches relying on stochastic (i.e., Monte Carlo) simulation offer a high-accuracy solution and more importantly impose no constraints on the complexity of the assumed numerical and probability models (Au and Beck 2003; Taflanidis and Beck 2009). They involve, however, higher computational cost, a feature which prohibited for some time their widespread adoption. Advances over the last decade in computer hardware and simulation algorithms, in particular the wide use of distributed/parallel computing (Fujimoto 2001), have contributed in reducing this computational burden traditionally associated with stochastic simulation approaches, and have facilitated the detailed modeling and solution of problems that were until recently considered as computationally intractable (Resio and Westerink 2008; Pellissetti 2008; Hardyniec and Charney 2015), increasing the popularity of simulation-based frameworks for quantifying/estimating natural hazard risk. Still in many instances, for example, for real-time risk estimation or in applications with complex nonlinear dynamical models, the computational burden associated with simulation-based approaches is still prohibitive.

The use of surrogate models (also frequently referenced as metamodels) is a popular approach for addressing this challenge and for further alleviating the computational cost associated with such simulation-based frameworks for natural hazard risk assessment (Gavin and Yau 2007; Tsompanakis et al. 2009; Gidaris et al. 2014). Surrogate models offer a computationally inexpensive input/output relationship based on a database of observations obtained through the initial (expensive) simulation model. The up-front cost for obtaining this database is of course high, but once the surrogate model is established, all future evaluations require small computational burden. This chapter discusses the adoption of kriging metamodels within this context. For illustration, two different applications are considered, involving two different hazards: seismic risk assessment utilizing stochastic ground motion modeling and real-time hurricane risk estimation. Various implementation issues are discussed, such as (a) the advantages of kriging metamodeling approach over other surrogate models, (b) the approaches for obtaining high efficiency when the output under consideration is high dimensional (over 10,000) through integration of principal component analysis, and (c) the explicit incorporation of the prediction error associated with the kriging metamodel into the risk formulation.

The discussions demonstrate the great benefits that the combination of kriging and stochastic simulation provides for natural hazard risk assessment. This combination can foster a comprehensive and detailed characterization of risk, in terms of the models selected for the system and the natural hazard and for the uncertainty quantification, and at the same time an efficient estimation for it. Initially, the general framework for simulation-based risk quantification and assessment is presented, and then the discussion focuses on kriging implementation and the two specific applications examined.

4.2 Risk Quantification and Assessment

Risk Quantification Evaluation of response/performance under natural hazards requires adoption of appropriate numerical models for (1) the natural hazard (excitation), (2) the system of interest, and (3) the system performance (Fig. 4.1). The combination of the first two models provides the system response vector, denoted $\mathbf{z} \in \mathbb{R}^{n_z}$ herein with individual response quantities denoted as z_k. The performance evaluation model assesses, then, the favorability of this response, based on the chosen criteria.

The characteristics of these models are not known with absolute certainty. Uncertainties may pertain to: (1) the variability of primary characteristics of excitation events, for example, intensity or occurrence rates, or of secondary properties typically given by predictive relationships based on these primary characteristics (Holland 1980; Rezaeian and Der Kiureghian 2010; Mavroeidis and Papageorgiou 2003), such as duration of excitation or pressure distribution along domains of

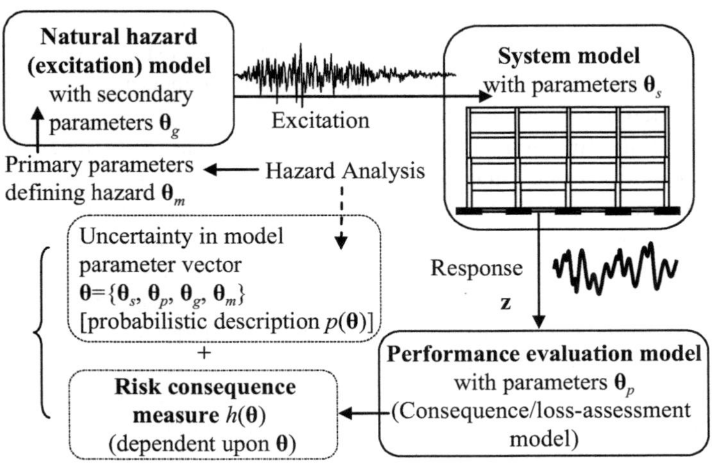

Fig. 4.1 Model for natural hazard risk description. Illustration example corresponds to seismic risk

interest; (2) the properties of the system itself, for example, related to parameters influencing restoring forces or to damping characteristics for structural systems (Liel et al. 2009); and (3) the parameters related to the performance of the system, for example, thresholds defining fragility (i.e., acceptable performance) of system components (Gardoni et al. 2002). Characterizing these uncertainties through a probabilistic description leads then to a versatile quantification of natural hazard risk (Taflanidis and Beck 2009; Taflanidis et al. 2013a).

To formalize these ideas, let $\boldsymbol{\theta} \in \Theta \subset \mathbb{R}^{n\theta}$ denote the augmented n_θ-dimensional vector of model parameters where Θ represents the space of possible model parameter values. As illustrated in Fig. 4.1, vector $\boldsymbol{\theta}$ is composed of *all* the model parameters for the individual system, $\boldsymbol{\theta}_s$; excitation, $\boldsymbol{\theta}_m$ (primary) and $\boldsymbol{\theta}_g$ (secondary); and performance evaluation, $\boldsymbol{\theta}_p$, models. For addressing the uncertainty in $\boldsymbol{\theta}$, a probability density function (PDF) $p(\boldsymbol{\theta})$ is assigned to it that quantifies our available knowledge in the context of probability logic (knowledge on hazard characteristics or properties of system under consideration). For given values for the model parameters $\boldsymbol{\theta}$ the risk consequence measure, representing the utility of the response from a decision-theoretic point of view, is given by $h(\boldsymbol{\theta}) : \mathbb{R}^{n\theta} \rightarrow \mathbb{R}^+$. This measure is related to the performance/consequences that can be calculated based on the estimated response $\mathbf{z}$ (performance given that an excitation event has occurred), whereas it can be additionally dependent, for example within life-cycle analysis studies, to assumptions made about the rate of occurrence of excitation events (incorporation of the probability of such events occurring). *Natural hazard risk*, R, is finally described by the *probabilistic integral* that corresponds to the expected value of $h(\boldsymbol{\theta})$ over the probability models:

$$R = \int_{\Theta} h(\boldsymbol{\theta}) p(\boldsymbol{\theta}) d\boldsymbol{\theta}. \tag{4.1}$$

Through different selections of the risk consequence measure, different risk quantifications can be addressed, ranging from life-cycle cost to reliability (Taflanidis and Beck 2009; Jia et al. 2014). A specific consequence measure utilized in a variety of different risk applications, for example, within system reliability analysis or life-cycle cost estimation (Ellingwood 2001; Goulet et al. 2007; Jia and Taflanidis 2013), is the probability that some response quantity z_k (e.g., peak interstory drift for a structure) will exceed some threshold β_k that determines acceptable performance. For certain applications, for example, within seismic risk assessment, where this concept can be used to represent the fragility of system components, it is common to incorporate a prediction error in this definition (Porter et al. 2006; Taflanidis et al. 2013b); this can be equivalently considered as the aforementioned threshold corresponding to an uncertain quantity with some chosen distribution (this distribution ultimately determines the cumulative distribution function for the component fragilities). A common choice for the latter (Porter et al. 2006) is lognormal distribution. The equivalent representation is then that the threshold that determines acceptable performance is given by $(\beta_k \bullet \varepsilon_{\beta k})$ with $\varepsilon_{\beta k}$ having a lognormal distribution with median equal to one and logarithmic standard deviation $\sigma_{\beta k}$. This then leads to the following risk consequence measure

$$h(\boldsymbol{\theta}) = P\left[z_k(\boldsymbol{\theta}) \geq \beta_k \varepsilon_{\beta k} \middle| \boldsymbol{\theta}\right] = P\left[\ln\left(\varepsilon_{\beta k}\right) \leq \ln\left(z_k(\boldsymbol{\theta})\right) - \ln\left(\beta_k\right) \middle| \boldsymbol{\theta}\right]$$
$$= \Phi\left(\frac{\ln\left(z_k(\boldsymbol{\theta})\right) - \ln\left(\beta_k\right)}{\sigma_{\beta k}}\right), \tag{4.2}$$

where $\Phi(.)$ denotes the standard normal cumulative distribution (CDF). For $\sigma_{\beta k} = 0$, representing the case that no uncertainty is considered in the description of β_k, this measure simplifies to an indicator function, being one if $z_k > \beta_k$ and zero if not.

Coupled with stochastic simulation (i.e., Monte Carlo) approaches for estimating the probabilistic integral in Eq. (4.1), as will be discussed next, the framework illustrated in Fig. 4.1 for risk quantification imposes no restriction on the complexity of the adopted numerical or probability modes and ultimately facilitates a generalized, versatile description of natural hazard risk and has been implemented successfully for studies considering a variety of hazards (wind, surge, waves, earthquakes) and structural systems (Taflanidis et al. 2011, 2013a, b; Gidaris and Taflanidis 2015).

Risk Assessment The estimation of risk given by Eq. (4.1) requires calculation of a multidimensional probabilistic integral. To support adoption of probability and numerical models with higher complexity, this calculation can be established through a stochastic (i.e., Monte Carlo) simulation. Using a finite number, N, of samples of $\boldsymbol{\theta}$ drawn from proposal density $q(\boldsymbol{\theta})$, an estimate for R and the coefficient of variation for that estimate (quantifying its accuracy), δ, are given by

$$\widehat{R} = \frac{1}{N}\sum_{j=1}^{N} h\left(\boldsymbol{\theta}^j\right)\frac{p\left(\boldsymbol{\theta}^j\right)}{q\left(\boldsymbol{\theta}^j\right)} \text{ and } \delta \approx \frac{1}{\sqrt{N}}\sqrt{\frac{1/N\sum_{j=1}^{N}\left(h\left(\boldsymbol{\theta}^j\right)p\left(\boldsymbol{\theta}^j\right)/q\left(\boldsymbol{\theta}^j\right)\right)^2}{\widehat{R}^2} - 1}, \tag{4.3}$$

where $\boldsymbol{\theta}^j$ denotes the jth sample. The proposal densities may be used to improve the accuracy of this estimation, i.e., reduce the coefficient of variation, by focusing the computational effort on regions of the Θ space that contribute more to the integrand of the probabilistic integral of Eq. (4.1)—this corresponds to the concept of importance sampling (IS). For problems with a large number of model parameters, choosing efficient importance sampling densities for all components of $\boldsymbol{\theta}$ is challenging (Taflanidis and Beck 2008) and can lead to convergence problems for the estimator in Eq. (4.3); thus, it is preferable to formulate IS densities only for the important components of $\boldsymbol{\theta}$, i.e., the ones that have the biggest influence on the seismic risk, and use $q(\boldsymbol{\theta}) = p(\boldsymbol{\theta})$ for the rest (Taflanidis and Beck 2008). For natural hazard risk applications, the primary parameters related to the hazard ($\boldsymbol{\theta}_m$ in Fig. 4.1) are generally expected to have the strongest impact on the calculated risk (Taflanidis and Beck 2009), so selection of IS densities may focus only on them.

Evaluating, now, the computational efficiency of this simulation-based estimation, the most demanding task in most practical applications is the calculation of the model response **z**. The computational burden for generating the required samples, for estimating performance/consequences given that response, or for calculating the sample average is typically very small. Thus, the formulation allows to seamlessly integrate recent advances in high-performance computing (parallel/distributed computing) to perform the required N evaluations of the system performance independently, in parallel mode. This significantly reduces the computational barriers that have been traditionally associated with approaches based on stochastic simulation. It also forms the foundation of an efficient assessment of risk for different seismic risk quantifications as well as the efficient estimation of risk under different design scenarios corresponding to different assumptions for $p(\boldsymbol{\theta})$ through some appropriate selection of the proposal densities $q(\boldsymbol{\theta})$ [more details may be found in (Gidaris and Taflanidis 2015)]. For applications, though, involving computationally intensive, high-fidelity numerical models, the burden for this analysis can be still prohibitive, especially for real-time applications. Surrogate modeling can be adopted in these cases to improve efficiency.

4.3 Kriging Metamodeling for Natural Hazard Risk Assessment

Surrogate models (metamodels) provide a simplified representation of the input/output relationship of complex processes, requiring large computational cost for their evaluation. Various such models have been proposed in the literature, such as neural networks (NNs) (Hajela and Berke 1992), response surface approximations (RSAs) (Gavin and Yau 2007), moving least squares RSA (MLS RSA) (Breitkopf et al. 2005), or kriging (Sacks et al. 1989), sharing the same principle; they generate the approximate, surrogate model based on information from a sufficient (typically small) number of intelligently selected evaluations of the exact model (typically referenced as support points or training set) or even a combination of model evaluations and experimental data (Gardoni et al. 2002, 2003). That surrogate model is then adopted as an approximation to the input/output relationship for the exact, complex model.

Preliminary Considerations for Surrogate Modeling Within Risk Assessment Within natural hazard risk assessment, metamodels can be implemented to approximate high-fidelity numerical models utilized for providing the response vector **z**. The underlying assumption for this selection is that the performance evaluation model in Fig. 4.1 is for most applications numerically simple, so establishing the surrogate model for the response **z**, rather than directly for the risk consequence measure $h(\boldsymbol{\theta})$, is necessary and also more advantageous since it additionally removes one level of approximation which provides a higher accuracy in the surrogate modeling approach (Jin et al. 2001). This also means that the input vector $\mathbf{x} \in \mathbb{R}^{n_x}$ that needs to be considered for the surrogate model is composed of only $\boldsymbol{\theta}_m$, $\boldsymbol{\theta}_g$, and $\boldsymbol{\theta}_s$ in the context of Fig. 4.1 representation, that is, excluding any parameters related to the performance evaluation model $\boldsymbol{\theta}_p$. Furthermore, the metamodel can be built to approximate a function of the response and not the response directly, for example, if risk consequence measure of Eq. (4.2) is utilized then the surrogate model should be built for approximating the logarithm of each response quantity $\ln(z_k)$ rather than the response quantity itself z_k.

Within this context, let $\mathbf{x} \in \mathbb{R}^{n_x}$ and $\mathbf{y} \in \mathbb{R}^{n_y}$ denote, respectively, the input and output vectors considered for the surrogate model implementation. For forming the metamodel initially, a database with n_m observations is obtained that provides information for the **x**-**y** pair. This process is also known as the design of experiments (DoE). For this purpose, n_m samples for $\{\mathbf{x}^j\ j = 1, \ldots, n_m\}$, also known as support points, are created within some domain X. Preliminary selection of the samples can be based on some space-filling approach (Latin hypercube sampling), with adaptive refinements also an option (Dubourg et al. 2011). The domain X should cover the expected range of values possible for each x_i (informed by the range of possible values within Θ) that will be needed in the evaluation of the risk integral. It should be stressed that this does not require a firm definition for $p(\boldsymbol{\theta})$, simply knowledge of the range for which the kriging metamodel will be used so that the support points extend over this range. Using this dataset the metamodel can be formulated and a kriging metamodel is considered here for this purpose.

Kriging Formulation A quick overview of kriging implementation is presented next. More details on the fundamental principles and computational details behind this implementation may be found in Sacks et al. (1989) and Lophaven et al. (2002).

The kriging predictor for each component y_i of $\mathbf{y}$ corresponds to a Gaussian variable $N(\hat{y}_i(\mathbf{x}), \sigma_i^2(\mathbf{x}))$ with mean $\hat{y}_i(\mathbf{x})$ and variance $\sigma_i^2(\mathbf{x})$ (Sacks et al. 1989). The response output can be approximated through this predictor, leading to

$$y_i(\mathbf{x}) = \widehat{y}_i(\mathbf{x}) + \varepsilon_i^g \sigma_i(\mathbf{x}), \tag{4.4}$$

where ε_i^g is a standard Gaussian variable. Here, we will present the case that a single surrogate model is developed for the entire output vector $\mathbf{y}$. This approach significantly reduces computational complexity but could potentially reduce accuracy (if the optimal surrogate models corresponding to different components y_i are expected to be drastically different).

The fundamental building blocks of kriging are the n_p dimensional basis vector, $\mathbf{f}(\mathbf{x})$, and the correlation function, $R(\mathbf{x}^l,\mathbf{x}^m)$, defined through hyper-parameter vector $\mathbf{s}$. The former provides a "global" model in the X space [and is ultimately similar to the global prediction provided by RSA], while the latter creates a "localized" deviation/correction weighting the points in the training set based on their closeness to the target point $\mathbf{x}$. The general concept is similar to the moving character of

MLS RSA (Breitkopf et al. 2005). Typical selections for these functions are a full quadratic basis and a generalized exponential correlation, respectively, leading to

$$\mathbf{f}(\mathbf{x}) = \begin{bmatrix} 1 & x_1 & \cdots & x_{n_x} & x_1^2 & x_1 x_2 & \cdots & x_{n_x}^2 \end{bmatrix}^T \; ; \; n_p = (n_x + 1)(n_x + 2)/2$$
$$R\left(\mathbf{x}^l, \mathbf{x}^m\right) = \prod_{i=1}^{n_x} \exp\left[-s_i \left|\mathbf{x}_i^l - \mathbf{x}_i^m\right|^{s_{n_x+1}}\right]; \; \boldsymbol{s} = \left[s_1 \; \cdots \; s_{n_x+1}\right]. \tag{4.5}$$

Then, for the set of n_m observations (training set) with input matrix $\mathbf{X} = \left[\mathbf{x}^1 \cdots \mathbf{x}^{n_m}\right]^T \in \mathbb{R}^{n_m \times n_x}$ and corresponding output matrix $\mathbf{Y} = \left[\mathbf{y}\left(\mathbf{x}^1\right) \cdots \mathbf{y}\left(\mathbf{x}^{n_m}\right)\right]^T \in \mathbb{R}^{n_m \times n_y}$, we define the basis matrix $\mathbf{F} = \left[f\left(\mathbf{x}^1\right) \ldots f\left(\mathbf{x}^{n_m}\right)\right]^T \in \mathbb{R}^{n_m \times n_p}$ and the correlation matrix $\mathbf{R} \in \mathbb{R}^{n_m \times n_m}$ with the lm element defined as $R(\mathbf{x}^l, \mathbf{x}^m)$, l, m=1,…,n_m. Also for every new input $\mathbf{x}$, we define the correlation vector $\mathbf{r}(\mathbf{x}) = \left[R\left(\mathbf{x}, \mathbf{x}^l\right) \ldots R\left(\mathbf{x}, \mathbf{x}^{n_m}\right)\right]^T$ between the input and each of the elements of $\mathbf{X}$. The mean kriging prediction (given as row vector) is then

$$\widehat{\mathbf{y}}(\mathbf{x}) = \mathbf{f}(\mathbf{x})^T \boldsymbol{\alpha}^* + \mathbf{r}(\mathbf{x})^T \boldsymbol{\beta}^*, \tag{4.6}$$

where matrices $\boldsymbol{\alpha}^* \in \mathbb{R}^{n_p \times n_y}$ and $\boldsymbol{\beta}^* \in \mathbb{R}^{n_m \times n_y}$ are given by

$$\boldsymbol{\alpha}^* = \left(\mathbf{F}^T \mathbf{R}^{-1} \mathbf{F}\right)^{-1} \mathbf{F}^T \mathbf{R}^{-1} \mathbf{Y}; \; \boldsymbol{\beta}^* = \mathbf{R}^{-1}\left(\mathbf{Y} - \mathbf{F}\boldsymbol{\alpha}^*\right). \tag{4.7}$$

Through the proper tuning of the hyper-parameters $\mathbf{s}$ of the correlation function, kriging can efficiently approximate very complex functions. The optimal selection of $\mathbf{s}$ is typically based on the maximum likelihood estimation (MLE) principle, where the likelihood is defined as the probability of the n_m observations, and maximizing this likelihood with respect to $\mathbf{s}$ ultimately corresponds to optimization

$$\mathbf{s}^* = \arg\min_{\mathbf{s}} \left[|\mathbf{R}|^{\frac{1}{n_m}} \sum_{i=1}^{n_y} \tilde{\sigma}_i^2 / \gamma_i \right], \tag{4.8}$$

where $|.|$ stands for determinant of a matrix; γ_i is a weight for each output quantity, typically chosen as the variance over the observations $\mathbf{Y}$; and $\tilde{\sigma}_i^2$ corresponds to the process variance (mean square error of the metamodel), given by the diagonal elements of the matrix $(\mathbf{Y} - \mathbf{F}\boldsymbol{\alpha}^*)^T \mathbf{R}^{-1} (\mathbf{Y} - \mathbf{F}\boldsymbol{\alpha}^*)/n_m$. Standard approaches for solving this optimization are given in (Lophaven et al. 2002).

An estimate for $\sigma_i^2(\mathbf{x})$, which can be equivalently considered as the variance for the prediction error between the real process y_i and the kriging prediction $\hat{y}_i$, is also provided through the kriging metamodel. This is a local estimate, meaning that it is a function of the input $\mathbf{x}$ and not constant over the entire domain X, and for output y_i is given by

$$\sigma_i^2(\mathbf{x}) = \tilde{\sigma}_i^2 \left[1 + \mathbf{u}(\mathbf{x})^T \left(\mathbf{F}^T \mathbf{R}^{-1} \mathbf{F}\right)^{-1} \mathbf{u}(\mathbf{x}) - \mathbf{r}(\mathbf{x})^T \mathbf{R}^{-1} \mathbf{r}(\mathbf{x})\right], \tag{4.9}$$

where $\mathbf{u}(\mathbf{x}) = \mathbf{F}^T \mathbf{R}^{-1} \mathbf{r}(\mathbf{x}) - \mathbf{f}(\mathbf{x})$.

Derivative information can be also easily obtained by noting that vectors $\boldsymbol{\alpha}^*$ and $\boldsymbol{\beta}^*$ are independent of $\mathbf{x}$. Denoting by $\mathbf{J}_\mathrm{f}$ and $\mathbf{J}_\mathrm{r}$ the Jacobian matrices with respect to $\mathbf{x}$ of $\mathbf{f}$ and $\mathbf{r}$, respectively, the gradients for the median predictions and the error variance are

$$\nabla\hat{\mathbf{y}}(\mathbf{x}) = \mathbf{J}_\mathrm{f}(\mathbf{x})^T\boldsymbol{\alpha}^* + \mathbf{J}_\mathrm{r}(\mathbf{x})^T\boldsymbol{\beta}^*$$
$$\nabla\sigma_\mathrm{i}^2(\mathbf{x}) = 2\tilde{\sigma}_\mathrm{i}^2\left[\mathbf{u}(\mathbf{x})^T\left(\mathbf{F}^T\mathbf{R}^{-1}\mathbf{F}\right)^{-1}\left(\mathbf{F}^T\mathbf{R}^{-1}\mathbf{J}_\mathrm{r}(\mathbf{x}) - \mathbf{J}_\mathrm{f}(\mathbf{x})\right) - \mathbf{r}(\mathbf{x})^T\mathbf{R}^{-1}\mathbf{J}_\mathrm{r}(\mathbf{x})\right]. \tag{4.10}$$

This information can be used, for example, for calculating the design points for the integrand of Eq. (4.1) for forming IS densities or when the application of interest corresponds to a design optimization problem and kriging is simultaneously developed for risk assessment as well as for performing the optimization with respect to the system design variables (Gidaris et al. 2014).

Validation of Metamodel The performance of the metamodel can be validated directly by the process variance $\tilde{\sigma}_\mathrm{i}^2$ or by calculating different error statistics for each one of the components of the output vector $\mathbf{y}$, such as the coefficient of determination RD_i^2 or the mean percent error ME_i, using a leave-one-out cross-validation approach (Kohavi 1995). This approach is established by removing sequentially each of the observations from the database, using the remaining support points to predict the output for that one and then evaluating the error between the predicted and real responses. The validation statistics are subsequently obtained by averaging the errors established over all observations. For the considered implementation for risk assessment, where one is concerned about providing adequate accuracy over the ensemble of scenarios considered (rather than for each separate scenario), high values for the coefficient of determination are of particular importance since they indicate that the kriging model can describe very well the variability within the initial database. The performance of the metamodel can be improved primarily by increasing the number of support points n_m or by their proper selection (Picheny et al. 2010). Other potential strategies for such performance improvement could be the change of the correlation function or the basis functions (Jia and Taflanidis 2013).

Advantages of Kriging Compared to other surrogate modeling approaches, especially approaches that entail matrix manipulations only, such as RSA and MLS RSA, kriging offers some distinct advantages:

- It corresponds to an interpolation metamodel, meaning that the predictions for any input $\mathbf{x}$ that belongs in the initial dataset $\mathbf{X}$ will match the exact corresponding output. The same is not necessarily true for many other metamodels (like RSA) that establish a local averaging (regression metamodels).
- It provides a variance for the prediction error which is also a function of the location $\mathbf{x}$. In other surrogate modeling approaches, this variance is typically treated as constant over the entire input domain (Taflanidis et al. 2013a).

- It involves only matrix manipulations for its implementation with matrix inversions that need to be performed only once, in the definition of $\boldsymbol{\alpha}^*$ and $\boldsymbol{\beta}^*$ in Eq. (4.7). This ultimately should be attributed to the fact that the correlation matrix **R** is dependent only on the training set **X**. In MLS RSA, that can provide similar accuracy as kriging (Simpson et al. 2001), the equivalent matrix of weights (establishing the local correction aspects of the methodology and having a similar role as **R**) is explicitly dependent on the input **x**, meaning that the inversions involved are different for each different input **x**. The implications of this property are significant. Kriging implementation requires keeping in memory only matrices $\boldsymbol{\alpha}^*$ and $\boldsymbol{\beta}^*$ (rather than the high-dimensional matrices **Y**, **F**, and **R**). Also, evaluations over a large number of different inputs, as required within stochastic simulation setting, can be efficiently established through proper matrix manipulations, simply by augmenting vectors **f**(**x**) and **r**(**x**) over all these inputs. It should be noted, though, that for evaluation over a single point, the complexity of kriging is higher than the one for MLS RSA (Simpson et al. 2001).
- The optimization in Eq. (4.8) for the parameters of the correlation function can be performed highly efficiently (at least for identifying local minima). The established approaches for optimization of the parameters related to the weight matrix in RSA are more computationally intensive (Loweth et al. 2010) requiring some cross-validation approach over the training set.

Risk Assessment Through Kriging Modeling Once the metamodel has been established, it can be directly used to approximate the response **z** and subsequently the consequence measure within the stochastic simulation-based evaluation in Eq. (4.3). Additionally, the prediction error of the metamodel can be directly incorporated in this estimation, altering ultimately the consequence measure. For example, for the measure given by Eq. (4.2) and assuming that the kriging prediction is developed for $\ln(z_k)$, giving $\ln z_k = \ln \widehat{z}_k(\mathbf{x}) + \varepsilon_k^g \sigma_k(\mathbf{x})$, we have

$$\begin{aligned} h(\boldsymbol{\theta}) &= P\left[\ln\left(\varepsilon_{\beta k}\right) \leq \ln\left(z_k(\boldsymbol{\theta})\right) - \ln\left(\beta_k\right) \left|\boldsymbol{\theta}\right.\right] \\ &= P\left[\ln\left(\varepsilon_{\beta k}\right) \leq \ln\left(\widehat{z}_k(\mathbf{x})\right) + \varepsilon_k^g \sigma_k(\mathbf{x}) - \ln\left(\beta_k\right) \left|\boldsymbol{\theta}\right.\right] \\ &= P\left[\ln\left(\varepsilon_{\beta k}\right) - \varepsilon_k^g \sigma_k(\mathbf{x}) \leq \ln\left(\widehat{z}_k(\mathbf{x})\right) - \ln\left(\beta_k\right) \left|\boldsymbol{\theta}\right.\right] = \Phi\left(\frac{\ln(\widehat{z}_k(\mathbf{x})) - \ln(\beta_k)}{\sqrt{\sigma_{\beta k}^2 + \sigma_k^2(\mathbf{x})}}\right), \end{aligned} \tag{4.11}$$

where the last equality is based on the fact that since $\ln(\varepsilon_{\beta k})$ and $\varepsilon_k^g \sigma_k(\mathbf{x})$ are zero mean Gaussian variables, their difference will also be a normal variable with zero mean and standard deviation the quantity in the denominator within the Gaussian CDF in Eq. (4.11).

The kriging implementation is demonstrated next in two examples, where additional characteristics for it are showcased.

4.4 Seismic Risk Assessment Through Stochastic Ground Motion Modeling

In the last decades, significant advances have been established in seismic risk decision management through the development of assessment methodologies based on detailed socioeconomic metrics quantifying life-cycle performance (Goulet et al. 2007). Powerful frameworks, widely acknowledged to provide the basis for these advances, have been the consequence-based engineering (CBE) (Abrams et al. 2002) of the Mid-America Earthquake (MAE) Center and the performance-based earthquake engineering (PBEE) (Moehle and Deierlein 2004; Bozorgnia and Bertero 2004) of the Pacific Earthquake Engineering Research (PEER) center, which represent two of the most important advances for probabilistic description of seismic risk. Within this setting, comprehensive risk quantification can be established by evaluating the structural performance through nonlinear dynamic response analysis (Goulet et al. 2007), rather than through simplified approaches such as pushover analysis. This has implications both for the system model and for the natural hazard model; for the latter, an excitation model needs to be considered that can provide a description for the entire ground acceleration time history. The framework in Fig. 4.1 is consistent with the aforementioned approaches though it is founded upon a system-theoretic formulation for the problem: consideration of exposure/vulnerability/consequence modules and quantification of the parametric modeling uncertainty within the description of each of these models. Note that in this case, each response quantity z_k corresponds to a different engineering demand parameter (EDP) that is utilized to describe the performance of the structural system (Goulet et al. 2007).

The implementation of surrogate modeling in this setting can greatly enhance the efficiency of seismic risk assessment (Zhang and Foschi 2004; Buratti et al. 2010), especially for studies examining design optimization (Möller et al. 2009; Gidaris et al. 2014). This implementation is discussed here considering applications where stochastic ground motion models are utilized to describe the excitation. The discussion starts with a quick overview of this hazard modeling approach.

Seismic Risk Modeling Through Stochastic Ground Motion Models Stochastic ground motion models (Boore 2003; Rezaeian and Der Kiureghian 2010; Vetter et al. 2016) have been gaining increased attention within the structural engineering community for description of seismic hazard (Jensen and Kusanovic 2014; Gidaris et al. 2014). They are based on modulation of a stochastic sequence (typically white noise), $\mathbf{w} \in W$, through functions that address the frequency and time-domain characteristics of the excitation. The parameters of these functions (corresponding to the secondary parameters $\boldsymbol{\theta}_g$ in Fig. 4.1) represent characteristics such as the duration of excitation or average frequency content and can be related to seismological parameters (corresponding to primary parameters $\boldsymbol{\theta}_m$ in Fig. 4.1), such as the moment magnitude, M, or rupture distance, r_{rup}, by appropriate predictive relationships. Description of the uncertainty in the seismological parameters and these predictive relationships facilitates then the comprehensive description of the seismic hazard. For the former, this is established through a probabilistic hazard analysis (Kramer 1996), whereas for the latter, this is established through the process of developing the predictive relationships for the ground motion model itself (Rezaeian and Der Kiureghian 2010).

One additional attractive feature of this ground motion modeling is the fact that potential near-fault effects can be easily incorporated within it. This is facilitated through the addition to the broadband component of the excitation, described through a stochastic ground motion model, of a directivity pulse (Mavroeidis and Papageorgiou 2003) addressing the longer-period component of the excitation. This pulse has its own parameters (included in $\boldsymbol{\theta}_g$ in Fig. 4.1) that are dependent upon $\boldsymbol{\theta}_m$. Therefore, the probabilistic foundation for describing the pulse characteristics is the same as for the broadband component. The fact that not all near-fault excitations include directivity pulses can be addressed through adoption of a model describing the probability of occurrence of such pulses (Shahi and Baker 2011), also dependent upon $\boldsymbol{\theta}_m$. When pulses are not included, the ground motion is described only through its broadband component. This approach ultimately supports a complete probabilistic description of the seismic hazard in close proximity to active faults (Gidaris and Taflanidis 2015).

Implications to Kriging Implementation The use of stochastic ground motion models to describe the hazard leads ultimately to system response that is a function of not only $\boldsymbol{\theta}$ but $\mathbf{w}$ as well. Due to the high dimensionality of $\mathbf{w}$ (stemming from partitioning of the entire duration of the excitation to appropriate time intervals), the development of a surrogate model for the entire input vector composed of both $\boldsymbol{\theta}$ and $\mathbf{w}$ is impractical. To address this challenge, an alternative formulation can be considered (Zhang and Foschi 2004; Schotanus et al. 2004; Gidaris et al. 2015) by separating the input space into two vectors: the first corresponding to the stochastic sequence and the second to the remaining model parameters. The impact of the first one (stochastic sequence) is addressed by assuming that under its influence, each response quantity z_k follows a lognormal distribution with median $\bar{z}_k$ and logarithmic standard deviation σ_{zk}, which corresponds to a common assumption within earthquake engineering (Zhang and Foschi 2004; Jalayer and Cornell 2009; Aslani and Miranda 2005; Shome 1999). This means that

$$\ln(z_k) = \varepsilon_k \sigma_{zk} + \ln(\bar{z}_k), \tag{4.12}$$

with ε_k corresponding to a standard Gaussian variable. The metamodel needs to be developed with respect to only the low-dimensional $\boldsymbol{\theta}$ vector, to provide predictions for these two statistical quantities, corresponding to the statistics for the EDPs of interest due to the influence of the white noise. Therefore, the output vector $\mathbf{y}$ is in this case composed of both $\ln(\bar{z}_k)$ and σ_{zk} for all the EDPs of interest. Once the metamodel has been established, it can be directly used to estimate the risk consequence measure, which then needs to be appropriately modified to take into account the approximation of Eq. (4.12). Assuming that the kriging prediction

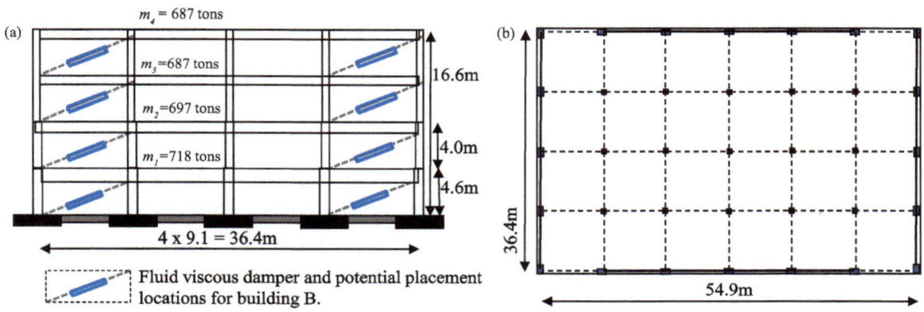

Fig. 4.2 Four-story reinforced benchmark building, (**a**) elevation and (**b**) plan views

error for σ_{zk} is small, this leads to the following modification for the consequence measure (Gidaris et al. 2015):

$$h(\boldsymbol{\theta}) = \Phi\left(\frac{\ln\left(\widehat{\overline{z}}_{\mathrm{k}}(\mathbf{x})\right) - \ln(\beta_{\mathrm{k}})}{\sqrt{\sigma_{\beta\mathrm{k}}^2 + \sigma_{\mathrm{k}}^2(\mathbf{x}) + \widehat{\sigma}_{\mathrm{zk}}^2(\mathbf{x})}}\right), \tag{4.13}$$

where $\widehat{\overline{z}}_{\mathrm{k}}(\mathbf{x})$ and $\widehat{\sigma}_{\mathrm{zk}}^2(\mathbf{x})$ correspond to the median kriging predictions for the $\ln(\overline{z}_{\mathrm{k}})$ and σ_{zk}, respectively, and $\sigma_{\mathrm{k}}^2(\mathbf{x})$ is the kriging prediction error variance for $\ln(\overline{z}_{\mathrm{k}})$.

Illustrative Implementation This approach is demonstrated next considering the structure and hazard description in Gidaris et al. (2015). The structure, shown in Fig. 4.2, corresponds to design *A* in the benchmark study presented in Goulet et al. (2007). The total masses per floor are also shown in the figure. To demonstrate the versatility of the framework and its ability to assess risk for structures equipped with seismic protective devices, an additional case study is considered through incorporation of fluid viscous dampers. The structure without dampers is referenced as building *A*, whereas the structure with the dampers, building *B*.

Structural Model The lateral system consists of two exterior moment-resisting frames in each direction, with interior intermediate gravity frames. The resultant structural model corresponds to a two-dimensional four-bay frame modeled in OpenSees (McKenna 2011). The nonlinear hysteretic behavior of the structure is taken into account through lumped plasticity beam-column elements, modeled by using the modified Ibarra-Medina-Krawinkler nonlinear hinge model (Ibarra et al. 2005) with degrading strength and stiffness characteristics. To reduce the number of random variables, the approach proposed in Liel et al. (2009) is adopted here; perfect correlation is assumed for strength/stiffness and ductility characteristics for each one of the ten different potential plastic hinges. Under this assumption, 20 independent variables need to be considered, namely, column strength/stiffness ($c_{\mathrm{s,nc}}$) and column ductility ($c_{\mathrm{d,nc}}$) for six different columns and beam strength/stiffness ($b_{\mathrm{s,nb}}$) and beam ductility ($b_{\mathrm{d,nb}}$) for four different beams. Finally, the structural model is assumed to have Rayleigh damping with damping ratio ζ associated with the first and third modes. The structural model parameter vector is ultimately $\boldsymbol{\theta}_s = \{c_{s,nc}\ c_{d,nc}\ b_{s,nb}\ b_{d,nb}\ \zeta;\ nb = 1,\ldots,4,\ nc = 1,\ldots,6\}$. Building *B* is upgraded with fluid viscous dampers. A velocity exponent equal to 0.5 is considered for all dampers, representing a common value for seismic applications, whereas the dampers are placed in the exterior bays of the moment-resisting frame as indicated in Fig. 4.2. The damping coefficients are chosen as [9370, 4370, 2620, 2050] (kN(s/m)$^{0.5}$) for the dampers within each story following the design process in Gidaris and Taflanidis (2015). These dampers are modeled in OpenSees utilizing the ViscousDamper material, corresponding to a Maxwell model implementation (Christopoulos and Filiatrault 2006), whereas the axial stiffness is taken as 250,000 kN/m for all dampers.

Seismic Hazard Model For describing the seismic hazard, the same excitation model as in Jia et al. (2014) is adopted. The broadband component for the excitation is represented through a point source model (Boore 2003; Taflanidis and Beck 2009) based on a parametric description of the temporal envelope and radiation spectrum of the ground motion, both given as function of M and r_{rup}. Near-fault characteristics are incorporated through the velocity pulse model proposed by Mavroeidis and Papageorgiou (2003) that has as input parameters the pulse period amplitude T_p, a parameter that controls its amplitude A_p, the oscillatory character (number of half cycles) γ_p, and its phase ν_p. Ultimately, the excitation model parameter vector is $\boldsymbol{\theta}_g = [M, r, A_p, T_p, \gamma_p, \nu_p]$ when considering excitations with directivity pulses and $\boldsymbol{\theta}_g = [M, r]$ when considering excitations without such pulses. Metamodels need to be separately developed for each of the excitation models and will be abbreviated as *P* and *NP*, respectively.

Metamodel Formulation Two different structural models (buildings *A* and *B*) and two different excitation models (*P* and *NP*) are considered for the surrogate model development leading to four different cases. The response quantities approximated are the peak interstory drifts δ_k and absolute peak floor accelerations $\ddot{a}_k$ for all floors $k = 1,..,4$. The model parameter vector $\mathbf{x}$ has 23 components for the *NP* excitation and 27 components for the *P* excitation, whereas the domains defining X are chosen (Gidaris et al. 2015) based on the anticipated value range for each parameter. For example, for the *P* excitation, the selection is based on the characteristics of ground motions exhibiting near-fault components and the observed properties of those components (Shahi and Baker 2011; Mavroeidis and Papageorgiou 2003).

An adaptive refinement strategy is established to select the number of support points. Table 4.1 reports the number of support points n_m used for each metamodel, as well as validation metrics such as the average coefficient of determination and average mean error over δ_k and $\ddot{a}_k$ denoted as ARD^2_{δ}, $\text{ARD}^2_{\ddot{a}}$, AME_{δ}, and $\text{AME}_{\ddot{a}}$, respectively, calculated through a cross-validation approach. It is evident that challenges are encountered in developing the metamodel for building *A* and *P* excitation, leading to a larger value for the total number of support points. This stems from resonance conditions created by the directivity pulse which contribute to significantly higher variability in the response and ultimately to greater difficulty

Table 4.1 Validation metrics of the different surrogate models developed

Building	Excitation	Statistic	n_m	ARD^2_{δ}	$ARD^2_{\ddot{a}}$	$AME_{\delta}(\%)$	$AME_{\ddot{a}}(\%)$
A	*P*	$\ln(\bar{z}_k)$	10,000	0.96	0.97	15.0	5.10
A	*P*	σ_{zk}	10,000	0.91	0.79	13.3	7.32
A	*NP*	$\ln(\bar{z}_k)$	2000	1.00	1.00	1.10	1.75
A	*NP*	σ_{zk}	2000	0.94	0.88	2.50	3.53
B	*P*	$\ln(\bar{z}_k)$	1000	0.96	0.98	13.2	5.65
B	*P*	σ_{zk}	1000	0.94	0.88	8.45	4.83
B	*NP*	$\ln(\bar{z}_k)$	1000	1.00	1.00	1.20	1.85
B	*NP*	σ_{zk}	1000	0.95	0.91	2.10	2.75

in the developed model to accurately capture this variability. The results indicate that the accuracy established is high, with average coefficients of determination higher than 96 % for $\ln(\bar{z}_k)$ (79 % for σ_{zk}) and average mean errors lower than 15 % (lower than 13.3 % for σ_{zk}). As discussed previously, the high values for the coefficient of determination are especially important for the considered risk assessment implementation. The metamodel performance for the *NP* excitation is exceptionally good, showing that adequate accuracy would be possible with even lower values for the number of support points, but even for the *P* excitation the metamodel performance is more than adequate, especially for the acceleration responses. The lower overall accuracy for the drift responses should be attributed to the stronger impact from the nonlinear hysteretic structural behavior that results in larger variability for the drift responses. For the building equipped with dampers, the protection against such large inelastic responses offered by the dampers results in reduction of that variability and ultimately in higher accuracy of the established surrogate model.

Risk Assessment Next, the optimized kriging metamodel is utilized for estimating seismic risk. For the basic comparisons in this section, the risk is defined as the probability that the response will exceed acceptable threshold β_k (i.e., a range will be considered for β_k demonstrating the efficiency of the approach for different risk levels) with $\sigma_{\beta k}$ taken equal to 0.2. For the seismic hazard, two different cases are considered. In the first case, denoted as *NP* hazard, it is assumed that no excitations include near-fault effects (so only the *NP* excitation model is utilized). In the second case, denoted as *PP* hazard, the possibility of including a near-fault pulse is considered through the probability model developed by Shahi and Baker (2011) that quantifies the probability of an excitation to include a pulse dependent upon other seismicity characteristics (distance to fault rupture, moment magnitude). A detailed description of the seismic hazard characterization for this case is provided in Jia et al. (2014) and Gidaris and Taflanidis (2015). This leads to risk quantification as

$$R = \int_{\Theta} \sum_{\varepsilon_p=\{yes,no\}} h(\boldsymbol{\theta}|\varepsilon_p)P(\varepsilon_p|M,r)p(\boldsymbol{\theta})d\boldsymbol{\theta} \tag{4.14}$$

where ε_p is a binary (outcomes {yes, no}) random variable describing the probability of pulse existence, $P(\varepsilon_p|M, r)$ is the probability model for it, and the risk consequence measure $h(\boldsymbol{\theta}|\varepsilon_p)$ is estimated based on the *P* excitation (or surrogate model) if $\varepsilon_p = $ yes and the *NP* excitation (or surrogate model) if $\varepsilon_p = $ no.

Details for the chosen probability models $p(\boldsymbol{\theta})$ are included in Gidaris et al. (2015). Stochastic simulation with $N = 10{,}000$ samples is utilized for the estimation of the seismic risk. Importance sampling is established for *M* with density chosen [based on prior experience (Taflanidis and Beck 2009)] as truncated Gaussian with mean 6.8 and standard deviation 1.0. For the *PP* hazard, importance sampling is also formulated for ε_p based on observed sensitivity in Jia et al. (2014), with 80 % of the excitations taken to include a pulse. The estimates from the surrogate model, denoted as SE, are compared against the estimates calculated through the high-fidelity model, denoted as HF. Indicative results are presented in Fig. 4.3, in all cases as plots of the probability of failure (risk) against the threshold β_k. The coefficient of variation (for the stochastic simulation) for the risk estimated through the surrogate model and all considered cases is not higher than 9.0 and 17.0 % for probabilities of failure as low as 10^{-2} and 10^{-3}, respectively, demonstrating the relatively high accuracy that can be established even for rare-event simulations.

The comparison between the risk estimated from the high-fidelity model and the kriging metamodel indicates that the accuracy achieved is very high. Even for building *A* and the *PP* hazard, utilizing primarily a surrogate model (excitation model *P*) that encountered greater challenges to provide satisfactory accuracy, the agreement achieved is high. As anticipated from the accuracy characteristics of the kriging metamodel, the agreement is closer for the *NP* hazard (compared to the *PP* hazard) and for building *B* (compared to building *A*). Comparison between buildings *A* and *B* shows that retrofitting with fluid viscous dampers greatly contributes to mitigating risk for drift responses.

The total CPU time (computational burden) for the seismic risk assessment utilizing the surrogate model for building *A* is only 530 s for the *PP* case and 108 s for the *NP* case. The larger value for the former stems from the larger number of support points utilized for the *P* excitation model. These numbers represent a significant reduction of computational burden when compared to the high-fidelity model which requires 210.9 h for the *PP* case and 187.4 h for the *NP* case. The corresponding CPU times for building *B* are 64 s for the *PP* case and 39 s for the *NP* case when surrogate model is used, whereas the high-fidelity model requires 180.5 h for the *PP* case and 165.7 h for the *NP* case (smaller values because the nonlinear dampers contribute to less severe inelastic structural response).

It is evident, therefore, that the surrogate modeling implementation facilitates a highly efficient and accurate seismic risk assessment for structures equipped or not with seismic protective devices; a large number of samples can be used with a negligible computational effort, whereas the provided risk estimates are close to the ones obtained from the high-fidelity approach. Of course, some initial computational effort is required for the development of the database to inform the surrogate model.

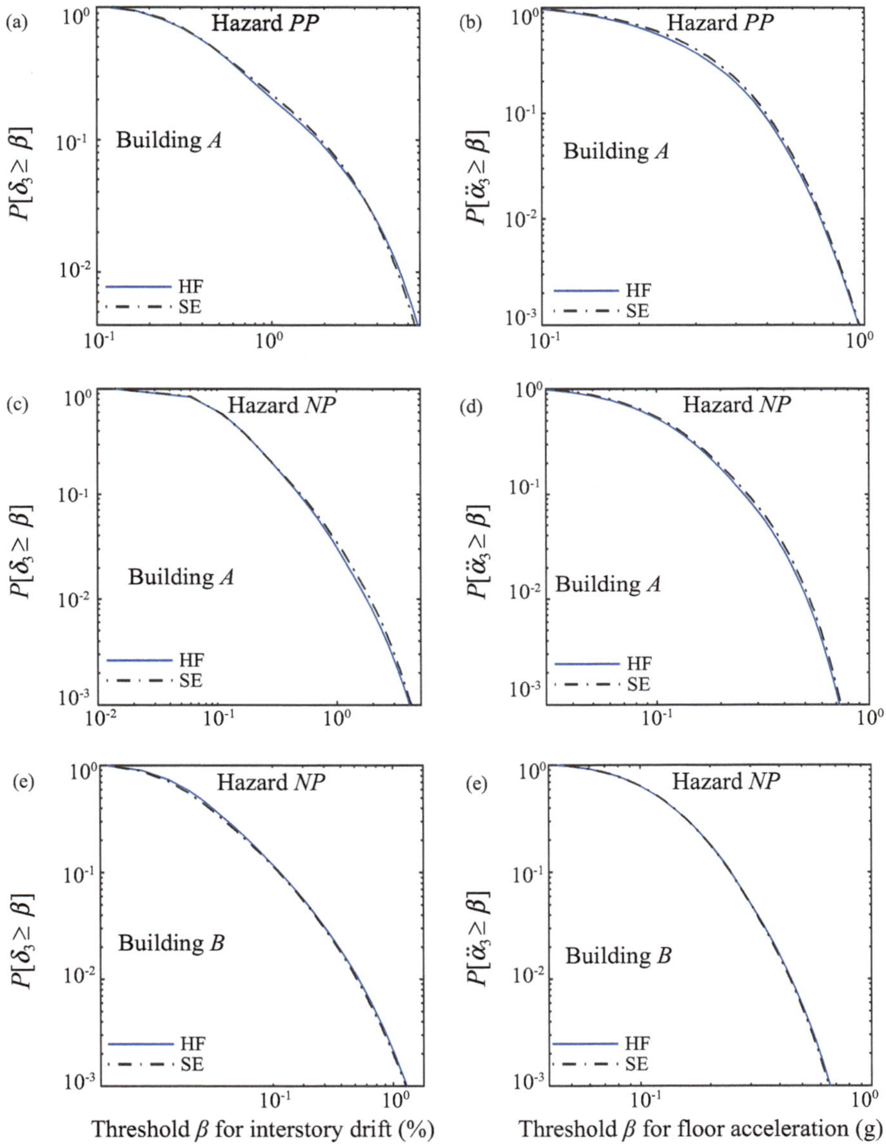

Fig. 4.3 Probability $P[z_k \geq \beta]$ of exceeding a specific threshold based on high-fidelity model (OpenSees) and kriging metamodel for peak interstory drift [parts (**a**), (**c**), (**e**)] and peak floor acceleration [parts (**b**), (**d**), (**f**)] of the third floor of buildings *A* and *B*. Top [parts (**a**), (**b**)] and middle [parts (**c**), (**d**)] rows correspond to *PP* and *NP* hazards, respectively, for building *A* and bottom row to building *B* for *NP* hazard

Table 4.2 Expected repair cost per seismic event for partition walls and ceiling for buildings *A* and *B* for seismic hazard *PP*

	Building *A*				Building *B*			
	Partition cost ($\$10^3$)		Ceiling cost ($\$10^3$)		Partition cost ($\$10^3$)		Ceiling cost ($\$10^3$)	
Floor	HF	SE	HF	SE	HF	SE	HF	SE
1	16.48	17.10	6.66	6.62	6.23	5.97	10.73	10.19
2	21.84	22.43	5.14	5.22	11.64	11.08	6.02	5.54
3	26.86	27.39	2.64	2.88	14.29	13.82	3.06	2.88
4	24.57	25.01	4.36	4.42	10.04	9.51	4.18	3.62

Once this model is established, it can be used, though, for any risk quantification desired.

For demonstration, the risk quantified as the expected repair cost per seismic event for the ceiling or the partition walls is further calculated, utilizing the subassembly approach to estimate repair cost (Porter et al. 2001). The vulnerability information from FEMA-P-58 2012 is utilized; three different damage states are considered for both subassemblies, with the probability of exceeding each damage described by a fragility function as in Eq. (4.2) and approximated through the kriging surrogate model through Eq. (4.13). This fragility information is then coupled with the repair cost for each damage state to provide the total repair cost. The repair costs per seismic event are reported in Table 4.2 for both buildings for seismic hazard *PP*. The high accuracy achieved using the kriging approximation is again evident by comparing the results obtained from the HF and SE approaches.

4.5 Real-Time Hurricane Risk Assessment

Hurricane risk assessment has received a lot of attention in the past decade, in response to the 2005 and 2008 devastating hurricane seasons. Of special interest in this case is the development of real-time tools that can provide efficient assessment during landfalling events to guide decisions of emergency response managers, whereas one of the greater advances in this field has been the development and adoption of high-fidelity numerical simulation models (corresponding to the system model for the description in Fig. 4.1) for reliable and accurate prediction of surge/wave responses for a specific hurricane event (Resio and Westerink 2008). These models permit a detailed representation of the hydrodynamic processes, albeit at the cost of greatly increased computational effort (more than a few thousand CPU hours for analyzing each hurricane event). They are based on a high-resolution grid description of the entire coastal region of interest (including more than a few million nodes), using detailed bathymetric data, and with the wind pressure time history of the hurricane (excitation model for Fig. 4.1 description) as input (Vickery et al. 2009) can simulate the surge and wave responses. The adoption of such models increases though, significantly, the computational cost for estimating hurricane risk. This is intensified by the fact that for appropriately assessing the hurricane impact, the simulation needs to extend a few (4–5) days prior to landfall. This is essential for both numerical convergence and for capturing all changes in the wave and surge environment that can be of significant importance (Dietrich et al. 2011).

To address this challenge, surrogate modeling concepts have been considered by various researchers in the past decade (Irish et al. 2009; Taflanidis et al. 2012; Das et al. 2010) with kriging (Jia and Taflanidis 2013) facilitating a computationally efficient implementation especially for real-time risk assessment. This is discussed in this section, starting with a review of hurricane modeling to fit within the description provided in Fig. 4.1.

Hurricane Modeling Among the various methodologies for hurricane risk assessment, a probabilistic approach, frequently referenced as the joint probability method (JPM), has been gaining popularity within the engineering community (Toro et al. 2010; Resio et al. 2007). The approach relies on a simplified description of each hurricane/storm scenarios through a small number of model parameters, corresponding to the characteristics close to landfall. These primary parameters, representing vector $\boldsymbol{\theta}_{\mathrm{m}}$ in Fig. 4.1, are the location of landfall x_{o}, the angle of approach at landfall α, the central pressure c_{p}, the forward speed during final approach to shore v_{f}, and the radius of maximum winds R_{m} leading to $\boldsymbol{\theta}_{\mathrm{m}} = [x_{\mathrm{o}}\ \alpha\ c_{\mathrm{p}}\ v_{\mathrm{f}}\ R_{\mathrm{m}}]^T$. The variability of the hurricane track and characteristics prior to landfall is also important, but directly incorporating this variability in the hurricane description would increase significantly the number of model parameters and so it is avoided. Instead, this variability is approximately addressed by appropriate selection of the hurricane track history prior to landfall, so that important anticipated variations, based on regional historical data, are described.

This modeling approach leads to characterization of hurricane risk through the probabilistic integral in Eq. (4.1) with main source of uncertainty $\boldsymbol{\theta} = \boldsymbol{\theta}_{\mathrm{m}}$. This quantification can be adopted for describing the long-term risk in a region (Resio et al. 2007, 2012) as well as for real-time applications (Smith et al. 2011). In the former case, $p(\boldsymbol{\theta})$ is chosen based on anticipated regional trends and climatological models, whereas in the latter case, it is provided by the national weather service prior to landfall in the form of a most probable hurricane track/intensity prediction along with statistical errors associated with this prediction. Part (b) of Fig. 4.4 demonstrates an example for the latter.

An important challenge in this application is the fact that the number of output response quantities of interest, representing the impact of the hurricane over a large coastal region, is typically large. Examples of such responses include (1) the storm surge (ζ), i.e., still-water level, defined as the average sea level over a several-minute period; (2) the significant wave height (H_{s}) (possibly along with the corresponding peak period T_{p}); (3) the wave run-up level, defined as the sea level including run-up of wind waves on the shore; and (4) the time that normally dry locations are inundated. Temporal and spatial variation is very important. With respect to

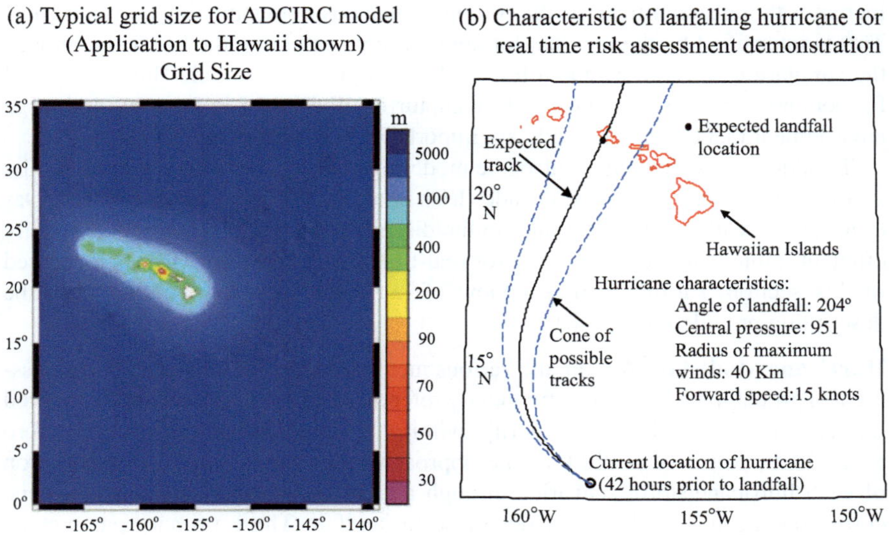

Fig. 4.4 (**a**) Typical grid size for ADCIRC high-fidelity model and (**b**) details of landfalling to Oahu hurricane considered in the demonstration example (results shown in Fig. 4.5)

the first aspect, the response variables may refer to maximum responses over the entire hurricane history or to responses at specific time instances prior to landfall. With respect to the second aspect, the response will be typically estimated in a large number of locations, expressed in some grid format [either corresponding to the initial grid for the high-fidelity model or to some lower resolution interpolated version (Taflanidis et al. 2013a)], which is the main characteristic of the analysis contributing to the large dimension of the output vector. Each component of $\mathbf{z}$ corresponds ultimately to a specific response variable [e.g., any of the (1)–(4) described above] for a specific coastal location and specific time. The dimension of n_z can easily exceed 10^6, depending on the type of application.

Dimension Reduction Through Principal Component Analysis This large dimension of the response output imposes significant challenges in terms of both computational speed and, perhaps more importantly, memory requirements (Jia and Taflanidis 2013). The latter are particularly important for supporting the development of cyber-enabled platforms (Kijewski-Correa et al. 2014) that can be deployed in real time, allowing emergency response managers to simultaneously perform different types of analyses.

To address this challenge, the adoption of principal component analysis (PCA) as dimensional reduction technique was proposed in Jia and Taflanidis (2013) to reduce the dimensionality of the output vector by extracting a smaller number of latent outputs to represent the initial high-dimensional output. Considering the strong potential correlation between responses at different times or locations in the same coastal region (which are the main attributes contributing to the high dimension of the output vector), this approach can significantly improve computational efficiency without compromising accuracy.

PCA starts by converting each of the output components into zero mean and unit variance under the statistics of the observation set (composed of n_{m} observations) through the linear transformation

$$\underline{z}_{\mathrm{k}} = \frac{z_{\mathrm{k}} - \mu_{\mathrm{k}}^{z}}{\sigma_{\mathrm{k}}^{z}}, \text{ with } \mu_{\mathrm{k}}^{z} = \frac{1}{n_{\mathrm{m}}}\sum_{h=1}^{n_{\mathrm{m}}} z_{\mathrm{k}}^{h}, \sigma_{\mathrm{k}}^{z} = \sqrt{\frac{1}{n_{\mathrm{m}}}\sum_{h=1}^{n_{\mathrm{m}}} \left(z_{\mathrm{k}}^{h} - \mu_{\mathrm{k}}^{z}\right)^{2}}. \tag{4.15}$$

The corresponding (normalized) vector for the output is denoted by $\underline{\mathbf{z}} \in \mathbb{R}^{n_z}$ and the observation matrix by $\underline{\mathbf{Z}} \in \mathbb{R}^{n_{\mathrm{m}} \times n_z}$. The idea of PCA is to project the normalized $\underline{\mathbf{Z}}$ into a lower dimensional space by considering the eigenvalue problem for the associated covariance matrix $\underline{\mathbf{Z}}^T\underline{\mathbf{Z}}$ and retaining only the m_{c} largest eigenvalues. Then $\underline{\mathbf{Z}}^T = \mathbf{P}\mathbf{Y}^T + \boldsymbol{\tau}$ where $\mathbf{P}$ is the $n_z \times m_{\mathrm{c}}$ dimension projection matrix containing the eigenvectors corresponding to the m_c largest eigenvalues, $\mathbf{Y}$ is the corresponding $n_{\mathrm{m}} \times m_{\mathrm{c}}$ observation matrix for the principal components (latent outputs), and $\boldsymbol{\tau}$ is the error introduced by not considering all the eigenvalues (Tipping and Bishop 1999). If λ_{i} is the ith largest eigenvalue, m_{c} can be selected so that the ratio

$$\sum\nolimits_{j=1}^{m_c} \lambda_j / \sum\nolimits_{j=1}^{n_z} \lambda_j \tag{4.16}$$

is greater than some chosen threshold r_0 [typically chosen as 99 %]. This then means that the selected latent outputs can account for at least r_0 of the total variance of the data (Tipping and Bishop 1999). It is then $m_{\mathrm{c}} < \min(n_{\mathrm{m}}, n_{\mathrm{z}})$ with m_{c} being usually a small fraction of $\min(n_{\mathrm{m}}, n_{\mathrm{z}})$. For $n_{\mathrm{m}} << n_{\mathrm{z}}$, obviously, $m_{\mathrm{c}} << n_{\mathrm{z}}$, leading to a significant reduction of the dimension of the output.

The latent outputs, denoted by $\mathbf{y}_i$, $i = 1, \ldots, m_{\mathrm{c}}$ are the outputs with observations that correspond to the ith column of $\mathbf{Y}$ and the outputs for which the kriging metamodel is ultimately developed (in other words, $m_{\mathrm{c}} = n_{\mathrm{y}}$ in the terminology established in Sect. 4.3). The relationship between the initial output vector and the vector of the latent outputs is $\underline{\mathbf{z}} = \mathbf{P}\mathbf{y}$. Kriging is then formulated for the output vector $\mathbf{y}$. Of particular importance is the fact that in this case the elements of $\mathbf{y}$ have an associated relevance, represented by its variance, which is proportional to λ_j, the portion of the variability within the initial database represented from this latent output. This means that, contrary to common approaches for normalizing vector $\mathbf{y}$ within the surrogate model optimization of Eq. (4.8) through the introduction of weights γ_i, in this case, no normalization should be established, i.e., $\gamma_i = 1$. This equivalently corresponds to latent outputs with larger values of λ_j being given higher priority in the surrogate model optimization. The alternative approach is to develop a separate surrogate model for each output separately (Jia and Taflanidis 2013). This does not increase memory requirements but has an impact though on the computational time for developing and implementing the surrogate mode.

The linear relationships between **y** and $\underline{\mathbf{z}}$ finally allow for a direct transformation of the probability models for the predictions for these two quantities [a Gaussian variable under linear transformation still follows a Gaussian distribution]. Considering additionally the inverse of transformation in Eq. (4.15) to transform these predictions back to the original space, we have (Jia and Taflanidis 2013) that the kriging-based predictor for **z** follows a Gaussian distribution with mean

$$\widehat{\mathbf{z}}(\mathbf{x}) = \boldsymbol{\Sigma}^z\left(\mathbf{P}\widehat{\mathbf{y}}(\mathbf{x})\right) + \boldsymbol{\mu}^z \tag{4.17}$$

and variance $\underset{\sim zk}{\sigma}^2(\mathbf{x})$ for $z_k(\mathbf{x})$ corresponding to the diagonal elements of the matrix

$$\boldsymbol{\Sigma}^z\left[\mathbf{P}\boldsymbol{\Sigma}(\mathbf{x})\mathbf{P}^T + \upsilon^2\mathbf{I}\right]\boldsymbol{\Sigma}^z, \tag{4.18}$$

where $\boldsymbol{\Sigma}^z$ is the diagonal matrix with elements σ_k^z, $\boldsymbol{\mu}^z$ is the vector with elements μ_k^z $k = 1,\ldots,n_z$, $\boldsymbol{\Sigma}(\mathbf{x})$ is the diagonal matrix with elements $\sigma_k^2(\mathbf{x})$, $k = 1,\ldots,m_c$, and $\upsilon^2\mathbf{I}$ stems from error $\boldsymbol{\tau}$. An estimate for the latter is given by

$$\upsilon^2 = \sum\nolimits_{j=m_c+1}^{n_z} \lambda_j/(n_z - m_c) \tag{4.19}$$

corresponding to the average variance of the discarded dimensions when formulating the latent output space. This Gaussian predictor for **z** can be then used in risk assessment with the error, characterized through variance $\underset{\sim zk}{\sigma}^2(\mathbf{x})$ used to provide an appropriate modification of the risk consequence measure (Jia and Taflanidis 2013), similar to the approach discussed earlier in Eq. (4.11) [$\underset{\sim zk}{\sigma}^2(\mathbf{x})$ needs to replace $\sigma_k^2(\mathbf{x})$ in this case].

Illustrative Implementation This approach has been implemented to develop efficient tools for real-time hurricane risk assessment for the Hawaiian islands (Taflanidis et al. 2013a; Kijewski-Correa et al. 2014; Jia and Taflanidis 2013) and recently for New Orleans (Taflanidis et al. 2014). Results from the former are discussed briefly here. The database utilized in this case comes from a regional flood study (Kennedy et al. 2012) consisting of 603 storms with $\boldsymbol{\theta} = \boldsymbol{\theta}_m$ representing the only source of uncertainty in the risk characterization. The high-fidelity model chosen to accurately predict the surge and wave response is a combination of the ADCIRC and SWAN numerical models (Bunya et al. 2010; Kennedy et al. 2012) and consists of 1,590,637 nodes and 3,155,738 triangular elements. Each simulation utilizing this model requires over 1500 CPU hours to complete. Grid characteristics are also shown in part (a) of Fig. 4.4.

The response output considered here corresponds to the maximum (over the hurricane duration) significant wave heights H_s in the region extending from 157.392°W to 158.584°W and 21.11°N to 21.90°N, close to Oahu island, and storm surge ζ for near-shore/inland locations around the coast of the island of Oahu with average distance of 300 m and up to the 4 m contour. The former has dimension

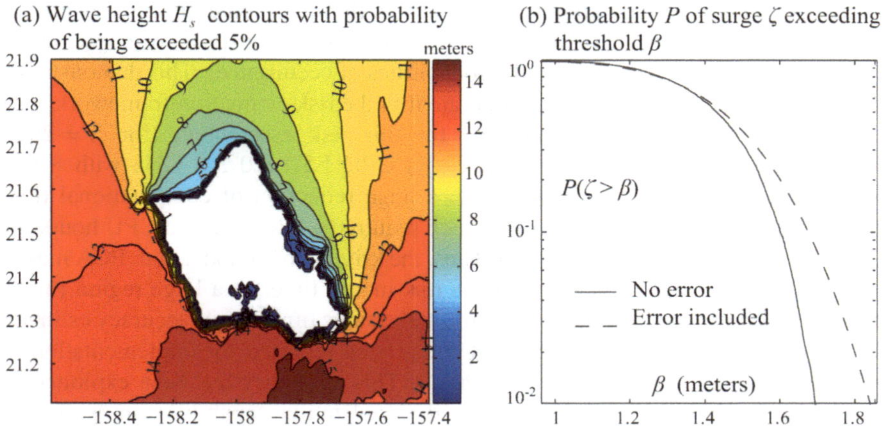

Fig. 4.5 Risk assessment results for scenario illustrated in Fig. 4.4b. (**a**) Wave height with 5 % probability of exceedance close to Oahu and (**b**) probability of surge exceeding thresholds β at location with coordinates 21.3769°N, 157.9666°W

$n_z = 12{,}800$ and the latter $n_z = 77{,}175$ leading to high-dimensional application. Based on the database, a single kriging metamodel with PCA is implemented with $m_c = 40$, which account for 99 % of the total variability in the corresponding initial outputs. This selection reduces the sizes of the matrices that need to be stored in memory by over 90 %. For the optimized metamodel, the average coefficient of determination and average mean error over all nodal points in the initial response space are 0.948 and 4.51 % for significant wave height and 0.930 and 5.33 % for storm surge, respectively. The probability of misclassification for the surge (i.e., identifying a location as inundated when it is not and vice versa) is 2.6 %. These error statistics show that the kriging metamodel provides high-accuracy approximations to the hurricane response (small errors).

This computational efficiency of the established metamodel can then be utilized to support the development of stand-alone tools (Taflanidis et al. 2013a) or, perhaps more importantly, cyber-enabled portals supporting wide online dissemination and collaborative environments (Kijewski-Correa et al. 2014). As indicated previously, these tools can be used to estimate the regional long-term risk or provide real-time predictions during landfalling events. The latter is demonstrated in Figs. 4.4 and 4.5. Part (b) of Fig. 4.4 shows the hurricane scenario considered; in this case, the risk assessment is performed 42 h prior to landfall. Details on the quantification of the uncertainty on $\boldsymbol{\theta}$ based on standard meteorological prediction errors are included in Taflanidis et al. (2013a). Figure 4.5 then shows results for the wave height contours with probability of being exceeded 5 % [part (a)] and the probability that surge will exceed threshold β [part (b)] for location with coordinates 21.3769°N, 157.9666°W, which is near the shoreline of East Loch inside Pearl Harbor. The predictions with and without considering the kriging prediction error (i.e., taking $\underset{\sim zk}{\sigma}^2(\mathbf{x}) = 0$) are included in this plot. The comparison indicates that the prediction error can have a significant impact on the calculated risk, and it will lead to more conservative estimates for rare events (with small probabilities of occurrence). This demonstrates that it is important to explicitly incorporate it in the risk estimation framework.

The total evaluation time required for this risk assessment (for $N = 2000$ samples) is only 25 s on Intel(R) Xeon(R) CPU E5-1620 3.6 GHz with 8 GB of memory. These results correspond to a huge reduction of computational time compared to the high-fidelity model, which required a few thousand CPU hours for analyzing a single hurricane scenario. Thus, the kriging metamodel with PCA makes it possible to efficiently assess hurricane risk in real time for a large region (high-dimensional correlated outputs) providing at the same time a high-accuracy estimate for the calculated risk. Similar efficiency has been reported for implementation to New Orleans region (Taflanidis et al. 2014). This efficiency has been exploited to develop cyber portals that offer enhanced visualization capabilities as well as a versatile online collaborative environment (Kijewski-Correa et al. 2014).

4.6 Conclusions

Simulation-based modeling or risk estimation approaches facilitate a comprehensive and detailed characterization of natural hazard risk, with advances in computer and computational science dramatically reducing the computational burden associated with these approaches. This chapter examined the integration of kriging surrogate modeling in this context for further reduction of this burden. Kriging establishes a computationally inexpensive input/output relationship based on a database of observations obtained through the initial (expensive) simulation model. It enjoys a straightforward optimization (to improve its accuracy) and relies only on matrix manipulations (with no matrix inversions needed for its implementation), which supports a highly efficient calculation of the output for multiple inputs, as required within a stochastic simulation setting. Additionally, it provides a prediction error that is also a function of the input (and not constant over the examined domain), whereas the incorporation of that error in the risk assessment can significantly impact the risk estimates and improve their accuracy, especially when analyzing rare events. For applications with high-dimensional output, kriging can be integrated with principal component analysis to improve computational efficiency and reduce the memory requirements for the metamodel deployment. These characteristics are particularly useful for the development of automated risk assessment tools and were demonstrated in this chapter considering implementation for real-time hurricane risk estimation. In the other example considered, the seismic risk assessment when stochastic ground motion modeling is considered for the hazard description, it was shown that despite the high dimensionality of the input, stemming from the stochastic sequence involved in the ground motion model, through proper assumptions, approximately addressing the impact of the stochastic sequence in this case, kriging can still facilitate an efficient and accurate risk assessment. Overall, the chapter demonstrated the potential that kriging offers within a simulation-based setting for describing natural hazard risk.

References

Abrams, D. P., Elnashai, A. S., & Beavers, J. E. (2002). *A new engineering paradigm: Consequence-based engineering*. Linbeck Lecture Series in Earthquake Engineering: Challenges of the New Millennium, University of Notre Dame, Linbeck Distinguished Lecture Series, Notre Dame, IN.

Aslani, H., & Miranda, E. (2005). Probability-based seismic response analysis. *Engineering Structures, 27*(8), 1151–1163.

Au, S. K., & Beck, J. L. (2003). Subset simulation and its applications to seismic risk based on dynamic analysis. *Journal of Engineering Mechanics, ASCE, 129*(8), 901–917.

Boore, D. M. (2003). Simulation of ground motion using the stochastic method. *Pure and Applied Geophysics, 160*, 635–676.

Bozorgnia, Y., & Bertero, V. (2004). *Earthquake engineering: From engineering seismology to performance-based engineering*. Boca Raton, FL: CRC Press.

Breitkopf, P., Naceur, H., Rassineux, A., & Villon, P. (2005). Moving least squares response surface approximation: Formulation and metal forming applications. *Computers & Structures, 83*(17–18), 1411–1428.

Bunya, S., Dietrich, J. C., Westerink, J. J., Ebersole, B. A., Smith, J. M., Atkinson, J. H., et al. (2010). A high resolution coupled riverine flow, tide, wind, wind wave and storm surge model for Southern Louisiana and Mississippi. Part I: Model development and validation. *Monthly Weather Review, 138*(2), 345–377.

Buratti, N., Ferracuti, B., & Savoia, M. (2010). Response surface with random factors for seismic fragility of reinforced concrete frames. *Structural Safety, 32*(1), 42–51.

Christopoulos, C., & Filiatrault, A. (2006). *Principles of passive supplemental damping and seismic isolation*. Pavia, Italy: IUSS Press.

Das, H. S., Jung, H., Ebersole, B., Wamsley, T., & Whalin, R. W. (2010). *An efficient storm surge forecasting tool for coastal Mississippi*. Paper presented at the 32nd International Coastal Engineering Conference, Shanghai, China.

Der Kiureghian, A. (1996). Structural reliability methods for seismic safety assessment: A review. *Engineering Structures, 18*(6), 412–424.

Dietrich, J. C., Zijlema, M., Westerink, J. J., Holthuijsen, L. H., Dawson, C., Luettich, R. A., et al. (2011). Modeling hurricane waves and storm surge using integrally-coupled, scalable computations. *Coastal Engineering, 58*, 45–65.

Dubourg, V., Sudret, B., & Bourinet, J.-M. (2011). Reliability-based design optimization using kriging surrogates and subset simulation. *Structural Multidisciplinary Optimization, 44*(5), 673–690.

Ellingwood, B. R. (2001). Earthquake risk assessment of building structures. *Reliability Engineering & System Safety, 74*(3), 251–262.

FEMA-P-58. (2012). *Seismic performance assessment of buildings*. Redwood City, CA: American Technology Council.

Fujimoto, R. M. (2001). Parallel simulation: Parallel and distributed simulation systems. In: *Proceedings of the 33rd Winter Simulation Conference* (pp. 147–157). Arlington, Virginia.

Gardoni, P., Der Kiureghian, A., & Mosalam, K. H. (2002). Probabilistic capacity models and fragility estimates for reinforced concrete columns based on experimental observations. *Journal of Engineering Mechanics, 128*(10), 1024–1038.

Gardoni, P., Mosalam, K. M., & der Kiureghian, A. (2003). Probabilistic seismic demand models and fragility estimates for RC bridges. *Journal of Earthquake Engineering, 7*, 79–106.
Gavin, H. P., & Yau, S. C. (2007). High-order limit state functions in the response surface method for structural reliability analysis. *Structural Safety, 30*(2), 162–179.
Gidaris, I., & Taflanidis, A. A. (2015). Performance assessment and optimization of fluid viscous dampers through life-cycle cost criteria and comparison to alternative design approaches. *Bulletin of Earthquake Engineering, 13*(4), 1003–1028.
Gidaris, I., Taflanidis, A. A., & Mavroeidis, G. M. (2014). *Multiobjective formulation for the life-cycle cost based design of fluid viscous dampers*. Paper presented at the IX International Conference on Structural Dynamics (EURODYN 2014), Porto, Portugal, June 30–July 2.
Gidaris, I., Taflanidis, A. A., & Mavroeidis, G. P. (2015). Kriging metamodeling in seismic risk assessment based on stochastic ground motion models. *Earthquake Engineering and Structural Dynamics. 44*(14), 2377–2399.
Goulet, C. A., Haselton, C. B., Mitrani-Reiser, J., Beck, J. L., Deierlein, G., Porter, K. A., et al. (2007). Evaluation of the seismic performance of code-conforming reinforced-concrete frame building-From seismic hazard to collapse safety and economic losses. *Earthquake Engineering and Structural Dynamics, 36*(13), 1973–1997.
Hajela, P., & Berke, L. (1992). Neural networks in engineering analysis and design: An overview. *Computing Systems in Engineering, 31*(1–4), 525–538.
Hardyniec, A., & Charney, F. (2015). A new efficient method for determining the collapse margin ratio using parallel computing. *Computers & Structures, 148*, 14–25.
Holland, G. J. (1980). An analytic model of the wind and pressure profiles in hurricanes. *Monthly Weather Review, 108*(8), 1212–1218.
Ibarra, L. F., Medina, R. A., & Krawinkler, H. (2005). Hysteretic models that incorporate strength and stiffness deterioration. *Earthquake Engineering and Structural Dynamics, 34*(12), 1489–1511.
Irish, J., Resio, D., & Cialone, M. (2009). A surge response function approach to coastal hazard assessment. Part 2: Quantification of spatial attributes of response functions. *Natural Hazards, 51*(1), 183–205.
Jalayer, F., & Cornell, C. (2009). Alternative non-linear demand estimation methods for probability-based seismic assessments. *Earthquake Engineering and Structural Dynamics, 38*(8), 951–972.
Jaynes, E. T. (2003). *Probability theory: The logic of science*. Cambridge, UK: Cambridge University Press.
Jensen, H. A., & Kusanovic, D. S. (2014). On the effect of near-field excitations on the reliability-based performance and design of base-isolated structures. *Probabilistic Engineering Mechanics, 36*, 28–44.
Jia, G., Gidaris, I., Taflanidis, A. A., & Mavroeidis, G. P. (2014). Reliability-based assessment/design of floor isolation systems. *Engineering Structures, 78*, 41–56.
Jia, G., & Taflanidis, A. A. (2013). Kriging metamodeling for approximation of high-dimensional wave and surge responses in real-time storm/hurricane risk assessment. *Computer Methods in Applied Mechanics and Engineering, 261–262*, 24–38.
Jin, R., Chen, W., & Simpson, T. W. (2001). Comparative studies of metamodelling techniques under multiple modelling criteria. *Structural and Multidisciplinary Optimization, 23*(1), 1–13.
Kennedy, A. B., Westerink, J. J., Smith, J., Taflanidis, A. A., Hope, M., Hartman, M., et al. (2012). Tropical cyclone inundation potential on the Hawaiian islands of Oahu and Kauai. *Ocean Modelling, 52–53*, 54–68.
Kijewski-Correa, T., Smith, N., Taflanidis, A. A., Kennedy, A., Liu, C., Krusche, M., et al. (2014). CyberEye: Development of integrated cyber-infrastructure to support rapid hurricane risk assessment. *Journal of Wind Engineering and Industrial Aerodynamics, 133*(211–224).
Kohavi, R. (1995). A study of cross-validation and bootstrap for accuracy estimation and model selection. In: *Proceedings of the International Joint Conference on Artificial Intelligence* (pp. 1137–1145). Montreal, Canada.

Kramer, S. L. (1996). *Geotechnical earthquake engineering*. Upper Saddle River, NJ: Prentice Hall.

Kumar, R., Cline, D. B. H., & Gardoni, P. (2015). A stochastic framework to model deterioration in engineering systems. *Structural Safety, 53*, 36–43.

Liel, A. B., Haselton, C. B., Deierlein, G. G., & Baker, J. W. (2009). Incorporating modeling uncertainties in the assessment of seismic collapse risk of buildings. *Structural Safety, 31*(2), 197–211.

Lophaven, S. N., Nielsen, H. B., & Sondergaard, J. (2002). *DACE-A MATLAB kriging toolbox*. Technical University of Denmark.

Loweth, E. L., De Boer, G. N., & Toropov, V. V. (2010). *Practical recommendations on the use of moving least squares metamodel building*. Paper presented at the Thirteenth International Conference on Civil, Structural and Environmental Engineering Computing, Crete, Greece.

Mavroeidis, G. P., & Papageorgiou, A. S. (2003). A mathematical representation of near-fault ground motions. *Bulletin of the Seismological Society of America, 93*(3), 1099–1131.

McKenna, F. (2011). OpenSees: A framework for earthquake engineering simulation. *Computing in Science & Engineering, 13*(4), 58–66.

Moehle, J., & Deierlein, G. (2004). A framework methodology for performance-based earthquake engineering. In: *Proceedings of the 13th World Conference on Earthquake Engineering*, Vancouver, Canada, August 1–6, 2004.

Möller, O., Foschi, R. O., Quiroz, L. M., & Rubinstein, M. (2009). Structural optimization for performance-based design in earthquake engineering: Applications of neural networks. *Structural Safety, 31*(6), 490–499.

Pellissetti, M. (2008). Parallel processing in structural reliability. In: *Proceedings of the 4th International Conference on Advances in Structural Engineering and Mechanics (ASEM)*.

Picheny, V., Ginsbourger, D., Roustant, O., Haftka, R. T., & Kim, N. H. (2010). Adaptive designs of experiments for accurate approximation of a target region. *Journal of Mechanical Design, 132*(7), 071008.

Porter, K. A., Kennedy, R. P., & Bachman, R. E. (2006). *Developing fragility functions for building components* (Report to ATC-58). Applied Technology Council, Redwood City, CA.

Porter, K. A., Kiremidjian, A. S., & LeGrue, J. S. (2001). Assembly-based vulnerability of buildings and its use in performance evaluation. *Earthquake Spectra, 18*(2), 291–312.

Rackwitz, R. (2001). Reliability analysis—A review and some perspectives. *Structural Safety, 23*, 365–395.

Resio, D. T., Boc, S. J., Borgman, L., Cardone, V., Cox, A., Dally, W. R., et al. (2007). *White paper on estimating hurricane inundation probabilities*. Consulting Report prepared by USACE for FEMA.

Resio, D. T., Irish, J. L., Westering, J. J., & Powell, N. J. (2012). The effect of uncertainty on estimates of hurricane surge hazards. *Natural Hazards, 66*(3), 1443–1459.

Resio, D. T., & Westerink, J. J. (2008). Modeling of the physics of storm surges. *Physics Today, 61*(9), 33–38.

Rezaeian, S., & Der Kiureghian, A. (2010). Simulation of synthetic ground motions for specified earthquake and site characteristics. *Earthquake Engineering and Structural Dynamics, 39*(10), 1155–1180.

Sacks, J., Welch, W. J., Mitchell, T. J., & Wynn, H. P. (1989). Design and analysis of computer experiments. *Statistical Science, 4*(4), 409–435.

Schotanus, M., Franchin, P., Lupoi, A., & Pinto, P. (2004). Seismic fragility analysis of 3D structures. *Structural Safety, 26*(4), 421–441.

Shahi, S. K., & Baker, J. W. (2011). An empirically calibrated framework for including the effects of near-fault directivity in probabilistic seismic hazard analysis. *Bulletin of Seismological Society of America, 101*(2), 742–755.

Shome, N. (1999). Probabilistic seismic demand analysis of nonlinear structures. *Ph.D Thesis*. Stanford University, Stanford, CA.

Simpson, T. W., Peplinski, J. D., Koch, P. N., & Allen, J. K. (2001). Metamodels for computer-based engineering design: Survey and recommendations. *Engineering with Computers, 17*, 129–150.

Smith, J. M., Westerink, J. J., Kennedy, A. B., Taflanidis, A. A., & Smith, T. D. (2011). SWIMS Hawaii hurricane wave, surge, and runup inundation fast forecasting tool. In: *Proceedings of the 2011 Solutions to Coastal Disasters Conference*, Anchorage, Alaska, June 26–29, 2011.

Taflanidis, A. A. (2010). Reliability-based optimal design of linear dynamical systems under stochastic stationary excitation and model uncertainty. *Engineering Structures, 32*(5), 1446–1458.

Taflanidis, A. A., & Beck, J. L. (2008). An efficient framework for optimal robust stochastic system design using stochastic simulation. *Computer Methods in Applied Mechanics and Engineering, 198*(1), 88–101.

Taflanidis, A. A., & Beck, J. L. (2009). Life-cycle cost optimal design of passive dissipative devices. *Structural Safety, 31*(6), 508–522.

Taflanidis, A. A., Jia, G., Kennedy, A. B., & Smith, J. (2012). Implementation/Optimization of moving least squares response surfaces for approximation of hurricane/storm surge and wave responses. *Natural Hazards, 66*(2), 955–983.

Taflanidis, A. A., Jia, G., Norberto, N.-C., Kennedy, A. B., Melby, J., & Smith, J. M. (2014). *Development of real-time tools for hurricane risk assessment.* Paper presented at the Second International Conference on Vulnerability and Risk Analysis and Management/Sixth International Symposium on Uncertainty Modeling and Analysis, Liverpool, England, July 13–16.

Taflanidis, A. A., Kennedy, A. B., Westerink, J. J., Smith, J., Cheung, K. F., Hope, M., et al. (2013a). Rapid assessment of wave and surge risk during landfalling hurricanes; probabilistic approach. *Journal of Waterway, Port, Coastal, and Ocean Engineering, 139*(3), 171–182.

Taflanidis, A. A., Loukogeorgaki, E., & Angelides, D. A. (2011). *Risk assessment and sensitivity analysis for offshore wind turbines.* Paper presented at the 21st International Offshore (Ocean) and Polar Engineering Conference, Maui, Hawaii, June 19–24.

Taflanidis, A. A., Vetter, C., & Loukogeorgaki, E. (2013b). Impact of modeling and excitation uncertainties on operational and structural reliability of Tension Leg Platforms. *Applied Ocean Research, 43*, 131–147.

Tipping, M. E., & Bishop, C. M. (1999). Probabilistic principal component analysis. *Journal of the Royal Statistical Society B, 61*, 611–622.

Toro, G. R., Resio, D. T., Divoky, D., Niedoroda, A., & Reed, C. (2010). Efficient joint-probability methods for hurricane surge frequency analysis. *Ocean Engineering, 37*, 125–134.

Tsompanakis, Y., Lagaros, N. D., Psarropoulos, P. N., & Georgopoulos, E. C. (2009). Simulating the seismic response of embankments via artificial neural networks. *Advances in Engineering Software, 40*(8), 640–651.

Vetter, C. R., Taflanidis, A. A., & Mavroeidis, G. P. (2016). Tuning of stochastic ground motion models for compatibility with ground motion prediction equations. *Earthquake Engineering and Structural Dynamics, 45*(6), 893–912.

Vickery, P. J., Skerlj, P. F., Lin, J., & Twisdale, L. A. (2006). HAZUS-MH hurricane model methodology. II: Damage and loss estimation. *Natural Hazards Review, 7*(2), 94–103.

Vickery, P. J., Wadhera, D., Powell, M. D., & Chen, Y. (2009). A hurricane boundary layer and wind field model for use in engineering applications. *Journal of Applied Meteorology and Climatology, 48*(2), 381–405.

Wen, Y. K., & Kang, Y. J. (2001). Minimum building life-cycle cost design criteria. I: Methodology. *Journal of Structural Engineering, 127*(3), 330–337.

Zhang, J., & Foschi, R. O. (2004). Performance-based design and seismic reliability analysis using designed experiments and neural networks. *Probabilistic Engineering Mechanics, 19*(3), 259–267.

Chapter 5
Risk and Decision-Making for Extreme Events: Climate Change and Terrorism

Mark G. Stewart

Abstract Terrorism and climate change are extreme events that frighten and alarm. This makes decision-making for these hazards or threats all the more difficult, particularly when decision-makers are risk averse. This chapter will describe how risk-based approaches are well suited to optimising decisions related to these extreme events. Stochastic methods are used to model threat likelihood, vulnerability, effectiveness of protective strategies, exposure and costs. The concepts will be illustrated with current research of risk-based assessment of counterterrorism and climate adaptation strategies. The case studies consider (1) protection of new bridges against terrorist attack and (2) climate change and cost-effectiveness of designing new port facilities to be less vulnerable to severe corrosion.

5.1 Introduction

Cyclones, earthquakes, tsunamis and floods are natural hazards that cause significant loss of life and economic and social losses. Added to this are 'man-made' hazards such as climate change and terrorism. These hazards are low-probability—high-consequence—events which in recent times are more commonly referred to as 'extreme events'. Extreme events illicit extreme reactions—risk aversion, probability neglect, cost neglect, and worst-case thinking—that may distort the decision-making process in an effort by policymakers to be seen to be 'doing something' irrespective of the actual risks involved. Policymaking in these circumstances becomes a 'risky business' (Hardaker et al. 2009). If rational approaches to public policymaking are not utilised, then politically driven processes 'may lead to raising unnecessary fears, wasting scarce resources, or ignoring important problems' (Paté-Cornell 2002).

Terrorism and climate change are extreme events of much interest. They can engender fear in the community, and predictions of impending doom are often overstated. Many terrorism and climate change 'risk' and 'risk management' reports

M.G. Stewart (✉)
Centre for Infrastructure Performance and Reliability, The University of Newcastle, Newcastle, NSW 2308, Australia
e-mail: mark.stewart@newcastle.edu.au

P. Gardoni, J.M. LaFave (eds.), *Multi-hazard Approaches to Civil Infrastructure Engineering*, DOI 10.1007/978-3-319-29713-2_5

dwell on lists of vulnerabilities and consequences. There is seldom mention of probabilities, quantitative measures of vulnerability or the likelihood of losses. While useful for initial risk screening, intuitive and judgement-based risk assessments are of limited utility to complex decision-making since there are often a number of climate or threat scenarios, adaptation or counterterrorism options, limited funds and doubts about the cost-effectiveness of protective measures. In this case, the decision-maker may still be uncertain about the best course of action. For this reason, there is a need for sound system and probabilistic modelling that integrates the performance of infrastructure systems with the latest developments in stochastic modelling, structural reliability and decision theory.

There is increasing research that takes into account the changing climate risks and life cycle costs in engineering to reduce the vulnerability or increase the resiliency of infrastructure—we refer to this as 'climate adaptation engineering'. Climate adaptation engineering is defined as measures taken to reduce the vulnerability or increase the resiliency of built infrastructure to a changing climate; this may include, for example, enhancement of design standards (higher design loads or flood levels), retrofitting or strengthening of existing structures, utilisation of new materials and changes to inspection and maintenance regimes (Stewart et al. 2014; Stewart and Deng 2015). The IPCC (2012) reports that 'vulnerability is a key factor in disaster losses, yet it is not well accounted for'. Probabilistic terrorism risk assessment methods have been developed to assess the risks of terrorism and effectiveness of risk-reducing measures (Mueller and Stewart 2011a, b, 2016). While the jargon differs, the decision support approaches to counterterrorism and climate adaptation measures have much in common, as do the challenges. The chapter aims to draw out these issues in more detail.

This chapter will describe how risk-based approaches are well suited to optimising decisions related to extreme events, in this case, climate adaptation strategies and counterterrorism measures. An important aspect is assessing when protective measures become economically viable, if protection can be deferred, and decision preferences for future costs and benefits (many of them intergenerational). Stochastic methods are used to model threat likelihood, vulnerability, effectiveness of protective strategies, exposure and costs. The concepts will be illustrated with current research of risk-based assessment of counterterrorism and climate adaptation strategies. The case studies consider (1) protection of new bridges against terrorist attack and (2) climate change and cost-effectiveness of designing new port facilities to be less vulnerable to severe corrosion caused by an increase in seawater temperature.

5.2 Key Issues

There are a number of issues and questions related to controversial and emotive issues such as terrorism, climate change and other extreme events. These contribute to risk aversion and are discussed as follows.

5.2.1 Worst-Case Thinking

Worst-case thinking, or hyperbole, tends to dominate the thinking of many climate change and terrorism experts. In 2008, Department of Homeland Security (DHS) Secretary Michael Chertoff proclaimed the 'struggle' against terrorism to be a 'significant existential' one (Mueller and Stewart 2011a). And in 2014, Mayor Bill de Blasio of New York at a UN summit proclaimed that 'We know humanity is facing an existential threat' from climate change (Grynbaum 2014). The notion that a threat short of all-out nuclear war could be existential to humanity is hard to fathom. If business-as-usual predictions are biased towards impending doom, then this justifies any response no matter the cost in loss of civil liberties, quality of life and treasure.

5.2.2 Cost Neglect

While it is not difficult to list threats and vulnerabilities, what is more challenging is to ascertain the cost to reduce these threats and vulnerabilities and to decide who pays and when. There is a notion that safety is infinitely good, and no cost is too high. There is no attempt to compare costs against benefits.

5.2.3 Probability Neglect

Many analysts base their findings on threats or scenarios that they assume will occur. There is no consideration of the likelihood of a terrorist attack, that a specific CO_2 emission scenario will occur or that adaptation will be effective. For example, a US 2014 climate risk assessment report predicts trillions in dollars of damage due to climate change for the business-as-usual scenario—i.e. the USA continues in its current path (Risky Business 2014). There is no attempt to quantify the likelihood that CO_2 emissions will continue unabated for the next 85 years, that CO_2 mitigation measures will be implemented, that adaptation measures are implemented, or of the impact of improved or game-changing technologies. Sunstein (2003) terms this as 'probability neglect' that 'people's attention is focused on the bad outcome itself, and they are inattentive to the fact that it is unlikely to occur'. There is no certainty with predictions, nicely summed up by physicist Niels Bohr: 'Prediction is very difficult, especially if it's about the future'.

5.2.4 Opportunity Costs

Policymakers that act before they carefully consider the implications of their actions can result in undesirable outcomes which are often referred to as 'opportunity costs'. For example, increased delays and added costs at US airports due to new security procedures provide incentive for many short-haul passengers to drive to their destination rather than flying, and, since driving is far riskier than air travel, the extra automobile traffic generated has been estimated to result in 500 or more extra road fatalities per year (Blalock et al. 2007). Using a DHS-mandated value of statistical life of $7.5 million (Robinson et al. 2010), this equates to a loss of $3.75 billion per year or nearly $50 billion over the period 2002–2014. A CO_2 mitigation strategy that reduces economic growth, particularly in developing countries, may reduce their ability to adapt. Weather- and climate-related fatality rates and economic losses are also 3–10 times higher in developing countries (IPCC 2012). Clearly then, if people are wealthier in the future, their well-being will be higher (Goklany 2008).

5.2.5 *Acceptable Risk*

The notion of acceptable risk is rarely raised in public discussions. The world is not risk-free. The generally accepted level of annual fatality risk (AFR) is one in a million (e.g. Stewart and Melchers 1997); see, for example, Murphy and Gardoni (2008) and Gardoni and Murphy (2014) for a fuller discussion on risk acceptability. The probability that an American will be killed by a terrorist in the USA, with the events of 2001 included in the count, stands at about one in four million per year (Mueller and Stewart 2016), or the probability an airline passenger will be killed by a terrorist act is a low one in 90 million per year (Mueller and Stewart 2016). By comparison, an American's chance of being killed in an automobile crash is about one in 8000, the chance of becoming a victim of homicide is about one in 22,000, and the chance of being killed by lightning is one in seven million per year. How much should we be willing to reduce a risk that is already very low, and is the risk reduction worth the cost?

5.3 Risk-Based Decision Support

Decision criteria for extreme events are typically based on (1) AFR and (2) cost-effectiveness of protective measures. Risk for a system exposed to a threat is

$$E(L) = \sum \Pr(T) \Pr\left(H \middle| T\right) \Pr\left(D \middle| H\right) \Pr\left(L \middle| D\right) L \tag{5.1}$$

where $\Pr(T)$ is the annual probability that a specific threat will occur (a terrorist attack, an emission scenario), $\Pr(H|T)$ is the annual probability of a hazard (wind, heat, explosion) conditional on the threat, $\Pr(D|H)$ is the probability of damage or other undesired effects conditional on the hazard (also known as vulnerability or fragility) for the baseline case of no extra protection (i.e. 'business as usual'), $\Pr(L|D)$ is the conditional probability of a loss (economic loss, loss of life, etc.) given occurrence of the damage (resilience) and L is the loss or consequence if full damage occurs. In some cases, 'damage' may equate to 'loss' and so a vulnerability function may be expressed as $\Pr(L|H)$ which is equal to the product $\Pr(D|H)\Pr(L|D)$. The summation sign in Eq. (5.1) refers to the number of possible threats, hazards, damage levels and losses. If the loss refers to a monetary loss, then $E(L)$ represents an economic risk.

If the loss refers to fatalities, then $E(L)$ represents an AFR. Stewart and Melchers (1997) and Mueller and Stewart (2011a) reviewed the quantitative safety goals used by the US Nuclear Regulatory Commission, UK Health and Safety Executive, Australian and European hazardous industrial development regulators, US environmental carcinogenic exposure regulators and others. These government regulators are concerned with low-probability—high-consequence—failures. The consensus risk acceptance criteria obtained for involuntary fatality risk to an individual are:

- AFRs higher than 1×10^{-3}–1×10^{-4} are deemed unacceptably high.
- AFRs in the range of 1×10^{-4}–1×10^{-6} are generally tolerable if the benefits outweigh the risks to provide an economic or social justification of the risk.
- AFRs smaller than 1×10^{-6} are deemed as negligible and further regulation is not warranted. Risk is broadly acceptable (or tolerable) as long as precautions are maintained, and further improvements are not required if these involve high costs.

If we modify Eq. (5.1) where ΔR is the reduction in risk caused by protective measures (e.g. climate adaptation or counterterrorism), then expected loss after protection is

$$E_{\text{protect}}(L) = \sum (1 - \Delta R)\, E(L) - \Delta B \tag{5.2}$$

where ΔR is the reduction in risk caused by the protective measure, $E(L)$ is the 'business-as-usual' expected loss (risk) given by Eq. (5.1) and ΔB is the co-benefit such as reduced losses to other hazards, increased energy efficiency of new materials, etc. If there is an opportunity cost associated with a new measure, then ΔB becomes a negative value. Protective measures should result in risk reduction (ΔR) that may arise from a combination of reduced likelihood of the hazard, damage states, safety hazards and people exposed to the safety hazard.

The challenging aspect of risk-based decision theory is predicting values of $\Pr(T)$, $\Pr(H|T)$, $\Pr(D|H)$, $\Pr(L|D)$ and ΔR. This information may be inferred from expert opinions, scenario analysis and statistical analysis of prior performance data, as well as system and reliability modelling. Since there is uncertainty associated with such predictions, the use of probability distributions to describe mean, variance and distribution type is recommended.

If the AFR lies in the generally tolerable region (e.g. 1×10^{-4} to 1×10^{-6}), then three criteria may be used to assess if the benefits of protective measures exceed their cost:

1. Net present value (NPV)
2. Probability of cost-effectiveness or Pr(NPV > 0)
3. Benefit-to-cost ratio (BCR)

The 'benefit' of a protective measure is the reduction in damages or losses associated with the protective strategy, and the 'cost' is the cost of the protective strategy. The net benefit or NPV is equal to benefit minus the cost. The decision problem is to maximise the NPV

$$\mathrm{NPV} = \sum E(L)\Delta R + \Delta B - C_{\mathrm{protect}} \tag{5.3}$$

where C_{protect} is the protection cost including opportunity costs that reduces risk by ΔR. Figure 5.1 shows how protective costs increase with risk reduction, while benefits increase. The optimal protection occurs when the NPV is a maximum, leading to optimal risk reduction. Relevant is what level of expenditure and risk reduction gives the greatest benefit and when does the law of diminishing returns kick in. The first dollars spent on protective measures are likely to be worthwhile, even if the last is not.

The BCR is

$$\mathrm{BCR} = \frac{\sum E(L)\Delta R + \Delta B}{C_{\mathrm{protect}}} \tag{5.4}$$

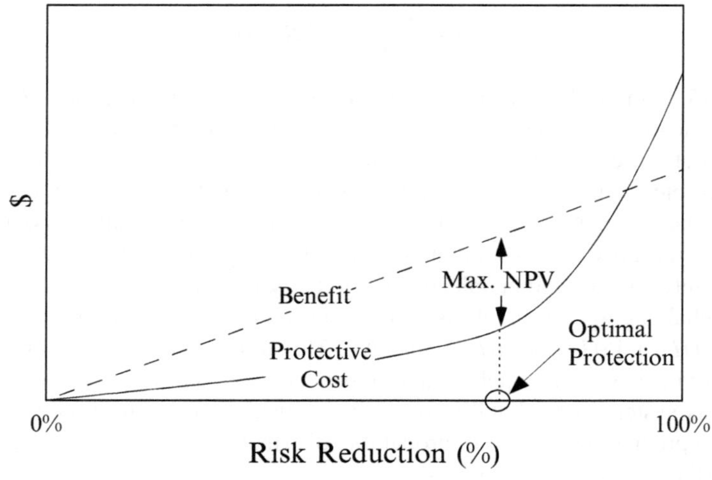

Fig. 5.1 Schematic of the NPV showing optimal protection

If parameters Pr(T), Pr($H|T$), Pr($D|H$), Pr($L|D$) L, ΔR, ΔB and/or C_{protect} are random variables, then the output of the analysis (NPV or BCR) is also variable. This allows confidence bounds of the NPV or BCR to be calculated, as well as the probability that an adaptation measure is cost-effective denoted herein as Pr(NPV > 0). If the NPV > 0 or BCR > 1, then there is a net benefit and so the protective measure is cost-effective. Other notations and formulae can be used to provide optimal protection, but ultimately these also mostly rely on maximising the NPV.

If the probability that a specific threat will occur or Pr(T) is too unreliable, then a decision analysis based on scenario analysis where threat probability is decoupled from Eq. (5.1) provides an alternative decision-making criteria based on expected costs. The above equations can be generalised for any time period, discounting of future costs and more detailed time-dependent cost and damage consequences.

Threat, vulnerability, loss and protective costs are subject to considerable uncertainty due to lack of available data and models. For this reason, calculations of risks, costs and benefits will be imprecise. Hence, a 'breakeven' analysis may be useful where minimum threat probability, minimum risk reduction or maximum protective cost necessary for protective measures to be cost-effective is selected such that there is 50 % probability that benefits equal cost—i.e. mean(NPV) = 0. For example, if the actual cost of protection exceeds the predicted breakeven value, then protection is not cost-effective. Decision-makers can then judge whether a protective strategy meets these breakeven values.

Governments and their regulatory agencies normally exhibit risk-neutral attitudes in their decision-making as described by Eqs. (5.3) and (5.4) above. This is confirmed by the US Office of Management and Budget (OMB) which specifically states that 'the standard criterion for deciding whether a government program can be justified on economic principles is NPV—the discounted monetized value of expected net benefits (i.e. benefits minus costs)'—and that 'expected values (an unbiased estimate) is the appropriate estimate for use' (OMB 1992). This entails using mean or average estimates for risk and cost-benefit calculations and not worst-case or pessimistic estimates. Probability neglect is a form of risk aversion as decision-makers are clearly averse to events of large magnitude irrespective of the probability of it actually occurring. Utility theory can be used if the decision-maker wishes to explicitly factor risk aversion or proneness into the decision process (e.g. Stewart et al. 2011).

It is important to note that the issue of risk aversion is not a new one, but has been well researched and documented for politically sensitive and controversial decisions associated with nuclear power safety, aviation safety, pharmaceutical benefits scheme, environmental pollution and other extreme events. In these cases, risk acceptance criteria have been developed based on AFRs and net benefit analysis using expected (mean) values. In principle, decisions related to terrorism, climate change and other extreme events should be made with similar risk-based methodologies.

5.4 Terrorism Case Study: Design of New Bridges Against Terrorist Attack

Highway bridges are often seen as an attractive target for terrorists. There are 600,000 highway bridges in the USA and bridges seem to be especially vulnerable. As Chairman Bennie Thompson of the House of Representatives' Committee on Homeland Security insists, 'The U.S. highway system is particularly vulnerable to potential terrorist attacks because of its openness—vehicles and their operators can move freely and with almost no restrictions, and some bridge and tunnel elements are easily accessible and located in isolated areas making them more challenging to secure' (GAO 2009). However, a bridge is very difficult to damage severely because its concrete and steel construction already makes it something of a hardened structure. Building facades (glass, masonry, cladding) are far more vulnerable (Norville et al. 1999). The Global Terrorism Database shows that of the 14 bridges attacked by insurgents in the war zones of Iraq and Afghanistan between 1998 and 2007, the total number of fatalities was relatively few at 59, and no more than 10 were killed in any single attack.

The preferred method of attack is improvised explosive devices (IEDs). An IED is relatively simple to design and manufacture if done by well-trained personnel, resulting in reliabilities in excess of 90 % (Grant and Stewart 2012). However, the probability of an IED creating a damaging effect (damage in excess of $1 million or attack resulting in casualties) reduces to 23 % for terrorists in Western countries where there is less opportunity for IED operational skills to be acquired (Grant and Stewart 2015). In the USA this figure drops to 15 %. This was clearly evident from the second attack on the London Underground on 21 July 2005 where four IEDs failed to initiate and in Glasgow International Airport in 2007 and Times Square in 2010 where vehicle-borne improvised explosive devices (VBIEDs) failed to initiate. The probability of successful attacks using IEDs increases to 65 % for terrorists or insurgents in the Middle East (Grant and Stewart 2012).

An explosive blast will not blow up a bridge, but will more likely damage and weaken supporting elements, causing only partial collapse. Even if a bridge collapses, however, not all vehicle occupants on it will be killed. For example, the collapse of the ten-lane, 14-span, 580 m I35W bridge in Minneapolis in 2007 killed 13 people, but 111 vehicles were on the bridge at the time of collapse (NTSB 2008). A bridge collapse over the Arkansas River in 2002 killed 14 people when 11 vehicles, of the many that were on the bridge, plunged into the river (Bai et al. 2006). The unexpectedly high survival rates arise not only because the bridge only partially collapses but also because a car is designed to crumple on impact and thus absorb energy.

The replacement cost for a typical interstate highway bridge is set at $20 million. In addition to the economic cost of traffic diversion, there are other social and economic costs to a community. These are harder to quantify but may be in the order of tens to hundreds of millions of dollars because, although the loss of one bridge will not isolate a community, it will generally cause considerable inconvenience and disruption. It is assumed that the replacement cost and social and economic costs to the community sum to \$100 million. The expected number of fatalities is assumed as 20, at a cost of \$150 million based on the value of statistical life of \$7.5 million (Robinson et al. 2010). The total losses for a damaged bridge including both the loss of life and economic considerations is $L = \$250$ million.

Measures to enhance security for new bridges typically focus on strengthening columns and girders, additional steel reinforcement, minimum dimensions, adding lateral bracing and increasing standoff by bollards, security fences and vehicle barriers. Although there is much information available about design and retrofitting bridges to mitigate the effects of blast damage, there is little information about their cost. It is assumed that substantial mitigation of blast effects can be achieved for a new bridge at a cost of 5 % of a bridge's replacement value. If the bridge replacement value is \$20 million, the cost of enhancing its design is then \$1 million. Annualised over a design life of 75 years at 4 % and 7 %, discount rates result in security costs of \$44,000 and \$70,000, respectively. A middle value for strengthening results in a security cost of $C_{\text{protect}} = \$50{,}000$ per year.

It is generously assumed that protective measures reduce the risk by $\Delta R = 95\,\%$. We also include in these calculations that hazard likelihood (IED or VBIED detonating and causing a damaging effect) is rounded up to $\Pr(H|T) = 20\,\%$ as obtained from the GTD (Grant and Stewart 2015). It is then assumed there is 50 % likelihood that the VBIED will completely destroy the bridge killing 20 people ($\Pr(L|H) = 50\,\%$).

Table 5.1 shows the breakeven annual threat (attack) probabilities $\Pr(T)$ required at a minimum for security expenditures on protecting a bridge to be cost-effective. This breakeven analysis shows that protective measures that cost \$50,000 per year and that successfully protect against an attack that would otherwise inflict \$250 million in damage would be cost-effective only if the probability of a successful terrorist attack without them exceeds 0.2 % or one in 500 per bridge per year. If we assume risk is reduced only by 50 %, the minimum attack probability per year required for bridge protective measures to be considered cost-effective increases to 0.4 % per bridge. If the average cost of construction is halved to only \$10 million per bridge, then C_{protect} is halved to \$25,000, but if losses remain at \$250 million, then Table 5.1 shows that the annual attack probability needs to exceed 0.1 % per bridge per year for counterterrorism protective measures to be cost-effective.

As a conservative estimate, it is now assumed in these calculations that bridges are 100 % vulnerable to attack—i.e. a VBIED will always detonate ($\Pr(H|T) = 100\,\%$), then destroying the bridge every time and always killing 20 people ($\Pr(L|H) = 100\,\%$). This is unlikely to be the case since there is not 100 % surety that an IED will initiate successfully, and that the blast will then cause bridge collapse and maximum consequences. In other words, the calculations assume that every attack will achieve 100 % success. In this unrealistic case, the breakeven attack probabilities shown in Table 5.1 will decrease tenfold. The evidence to date suggests that such a high attack probability is not being observed.

On the other hand, the co-benefit of counterterrorism protective measures may be considerable if strengthening a bridge to be more blast resistant has the co-benefit

Table 5.1 Probability of an otherwise successful terrorist attack, in percentage per year, required for protective security expenditures to be cost-effective, assuming the expenditures reduce the risk by 95 %

Cost of protective measures $C_{protect}$ (per year)	Losses from a successful terrorist attack (L)					
	$100 million	$250 million	$1 billion	$2 billion	$10 billion	$100 billion
$25,000	0.3	0.1	0.03	0.01	0.00	0.00
$50,000	0.5	0.2	0.05	0.03	0.01	0.00
$100,000	1.1	0.4	0.11	0.05	0.01	0.00
$250,000	2.6	1.1	0.26	0.13	0.03	0.00
$500,000	5.3	2.1	0.53	0.26	0.05	0.01
$1 million	10.5	4.2	1.1	0.53	0.11	0.01
$5 million	52.6	21.1	5.3	2.6	0.53	0.05
$10 million	105.3	42.1	10.5	5.3	1.1	0.11
$100 million	1052.6	421.1	105.3	52.6	10.5	1.1

Note: Probability of 100 % denotes one attack per bridge per year

of reducing the risks from seismic, flood or other hazards. In this case, breakeven attack probabilities would reduce.

If there were one attack on a highway bridge every year in the USA, the attack probability would be only one in 600,000 per bridge per year (0.0002 %) because there are 600,000 bridges in the country. This probability is nowhere near the one in 500 likelihood of a successful attack required for bridge protective measures to be cost-effective. If the attack probability is a high 0.01 % per bridge per year then the BCR is only 0.05—i.e. $1 of cost buys only 5 cents of benefits. In fact, the only threat against a US highway bridge in the USA since 9/11 was a terrorist plot to target the four-lane Brecksville-Northfield High Level Bridge near Cleveland, Ohio, in 2012.

If $\Pr(T)$ is taken as one in 600,000 or 0.0002 % per bridge per year, $\Pr(H|T) = 20\,\%$ and $\Pr(L|H) = 50\,\%$, the AFR (without protective measures) is 1.7×10^{-7} fatalities per year. This is less than the risk acceptance criteria of 1×10^{-6} fatalities per year, and so further protection is not warranted.

If there is a specific threat such that the likelihood of attack is massively increasing or if a bridge is deemed an iconic structure such that its perceived value is massively inflated, bridge protective measures may begin to become cost-effective. Thus, San Francisco's Golden Gate Bridge or New York's Brooklyn Bridge might be a more tempting target for terrorists than a more typical highway bridge.

Finally, it may seem prudent to provide counterterrorism protective measures for new bridges as the additional cost for a single new bridge may seem modest at approximately $50,000 per bridge per year or a 5 % increase in construction costs and higher costs to retrofit existing bridges. The ASCE 2013 Infrastructure Report Card recommends that $20.5 billion is needed annually to replace or repair existing bridges in the USA (ASCE 2013). Up to an additional $2 billion per year in funding would then be needed to provide counterterrorism protective measures for these new bridges. This is a significant sum of money and could be better spent elsewhere if the aim is to reduce risk and save lives, such as flood levee banks, tornado shelters or other infrastructure to reduce risks from natural hazards. See Mueller and Stewart (2011a) and Stewart and Mueller (2014c) for further details.

For assessments of risks, costs and benefits of building and airport protection, aviation security and policing, see Stewart and Mueller (2011, 2013, 2014a, b) and Mueller and Stewart (2011a, 2014, 2016).

5.5 Climate Adaptation Case Study: Deterioration of Port Infrastructure

The 2014 Intergovernmental Panel on Climate Change Fifth Assessment (AR5) Synthesis Report concluded that the 'Warming of the climate system is unequivocal, and since the 1950s, many of the observed changes are unprecedented over decades to millennia. The atmosphere and ocean have warmed, the amounts of snow and ice have diminished, sea level has risen, and the concentrations of greenhouse gases have increased'. What is less certain is the impact that rising temperatures will have on rainfall, wind patterns, sea level rise and other phenomena.

Steel sheet piling is commonly used in many ports and harbours worldwide. However, corrosion of steel sheet piling can result in metal loss and reduced structural capacity, which can then lead to failure (see Fig. 5.2). Corrosion results from a chemical reaction, so an increase in seawater temperature can accelerate the corrosion process. The Fourth Assessment Report of the Intergovernmental Panel

Fig. 5.2 Example of failure of a sheet pile retaining wall (photo courtesy of R Jeffrey)

on Climate Change predicts that average seawater surface temperature is 'likely' to increase by 6 °C over the next 100 years (IPCC 2007).

The corrosion of concern for this type of coastal infrastructure is a phenomenon known as accelerated low water corrosion (ALWC) (Melchers and Jeffrey 2013). The vulnerability $Pr(D|H)$ of sheet piling to ALWC is obtained from a time-dependent structural reliability analysis. It is assumed, as is reasonable in practice, that the piles are unprotected, having no protective paint coatings or cathodic protection. Damage to the retaining wall will halt all dock works and associated services. Damage is defined as excessive deformation of the wall such as by visually noticeable deformation of pavements and dock areas.

Current design practice results in the installation of AU 25 U-profile sheet piles. However, if corrosion loss is expected to accelerate due to a changing climate, then a climate adaptation measure may be to select a stronger sheet pile with a larger thickness. In this case, an AU 26 sheet pile is 0.3–0.5 mm thicker and 3 % stronger than the AU 25 sheet pile.

The structural reliability analysis includes the stochastic variability of loads, soil properties, steel material properties, dimensions and corrosion processes. The vulnerability $Pr(D|H)$ for the existing AU 25 and proposed AU 26 sheet piles allowing for a 6 °C seawater temperature increase over the next 100 years is shown in Fig. 5.3. The risk reduction arising from using the higher-capacity AU 26 steel pile is shown in Fig. 5.4. Clearly, even though the adaptation measure is the installation of slightly larger (3 %) piles, the risk reduction reaches 20 % early in the service life of the sheet piles.

It can be assumed that damage shown in Fig. 5.2 will lead to 100 % likelihood of loss, hence, $Pr(L|D) = 100\,\%$. The economic loss (L) from damage of sheet piling can be considerable. The cost to repair damage is likely to be at least $1 million,

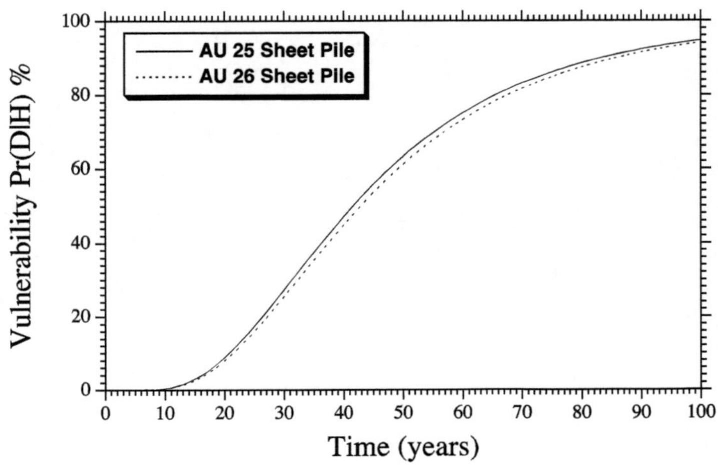

Fig. 5.3 Time-dependent vulnerability for sheet piles

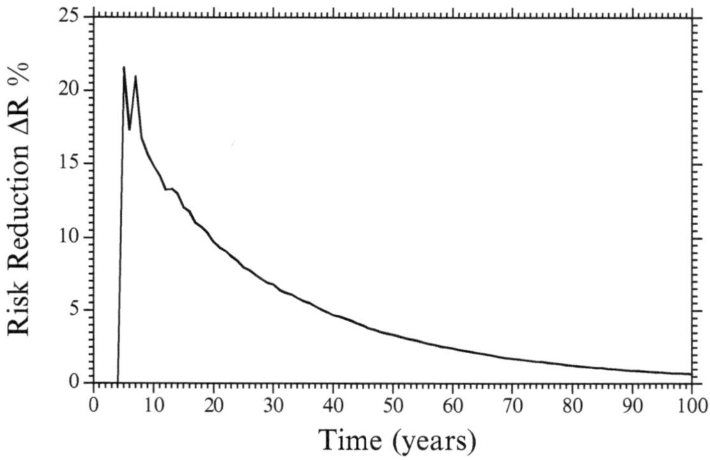

Fig. 5.4 Risk reduction from adaptation

and repair time is at least one month. To assess the indirect loss to the owner of a port, the economics of the Port Botany container terminal in Sydney is used as an illustrative example. The economic activity of the 12 shipping container berths runs to over \$2 billion per year (Sydney Ports 2008). This includes costs to the asset owner, trucking costs, worker wages and economic gains from the efficient import and export of goods in Australia. If one of these 12 berths is unavailable due to sheet piling damage, then shipping may be diverted to other berths. However, if all berths are busy, then delays can be expected at a pro-rata cost of \$14 million for loss of one berth for one month. An upper bound of economic loss, when also considering direct repair costs, is \$15 million. A lower bound is \$1 million assuming loss of one berth for one month does not disrupt normal shipping. A mid-estimate of $L = \$8$ million is thus reasonable.

The adaptation cost (C_{protect}) is based on the additional cost of purchasing larger AU 26 sheet piles. The AU 26 sheet piles are 2.5 % heavier than AU 25 piles. The additional material cost for a 200 m-long dock using 30 m-deep piles is approximately \$10,000.

The existing present value risk calculated from Eq. (5.1) for a scenario-based analysis ($\Pr(T) = \Pr(H|T) = 100\,\%$) of a 6 °C increase in seawater temperature in 100 years and 7 % discount rate is $E(L) = \$745{,}830$. The average risk reduction over 100 years is 5.4 %. Assuming no co-benefits, the NPV (or net benefit) of this adaptation measure is NPV = \$30,500. The benefit-to-cost ratio is BCR = 4.05. The use of larger AU 26 sheet piles is cost-effective for this climate scenario. This adaptation measure remains cost-effective even if adaptation costs double or economic losses are halved. The NPV will increase for discount rates lower than 7 %. Figure 5.5 shows the NPV as a function of time. The payback period (when benefit exceeds cost) is only 12 years.

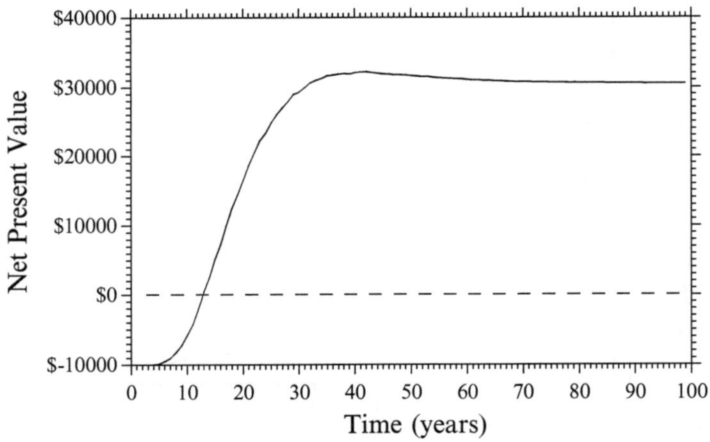

Fig. 5.5 Net present value for adaptation

This illustrative example shows that an adaptation measure that is low cost with a low risk reduction can still be cost-effective, particularly if the losses from infrastructure damage are relatively high. In other words, modest (or small) reductions in infrastructure vulnerability can be very cost-effective.

Finally, there is no certainty that existing design and construction practices are optimal. Design standards often are based on past experience, as well as new knowledge. However, they are seldom subject to a cost-benefit analysis due to modelling complexity and, more often than not, scarce resources to undertake work of this nature. Hence, it is desirable to assess the costs and benefits of existing designs. Moreover, there is likely to be uncertainty about climate scenarios. As such, it is useful to conduct a risk-based cost-benefit assessment for infrastructure assuming the current climatic conditions. The analysis reveals that for no change in seawater temperature, the NPV is $24,900. Hence, even if there is no change in seawater temperature, it is cost-effective to increase the size of sheet piling, in the present case, to AU 26. Hence, even if climate projections are overly conservative, adaptation measures still satisfies a ‘no regrets’ or ‘win-win’ policy (Susskind 2010).

Other case studies consider climate change and cost-effectiveness of designing new houses in Australia to be less vulnerable to severe storms (Stewart et al. 2014; Stewart 2014). To be sure, there are other case studies of assessing the efficiency and cost-effectiveness of climate adaptation strategies for built infrastructure, for example, floods and sea level rise (e.g. Hinkel et al. 2010; Hall et al. 2012; Botzen et al. 2013; Kundzewicz et al. 2013; Holden et al. 2013; Val et al. 2013), cyclones and severe storms (Bjarnadottir et al. 2011, 2013, 2014; Nishijima et al. 2012) and corrosion-reinforced concrete (Stewart and Peng 2010; Peng and Stewart 2014; Bastidas-Arteaga and Stewart 2015, 2016). For a general review, see Stewart et al. (2014).

5.6 Conclusions

Terrorism and climate change are extreme events that engender fear and anxiety in the community. Policymakers are also susceptible to these emotions. Risk-based approaches are suitable to assess the acceptability of risks and the cost-effectiveness of measures to reduce terrorism and climate impact risks. The concepts were illustrated with state-of-the-art applications of risk-based assessment for (1) the protection of new bridges against terrorist attack and (2) climate change and cost-effectiveness of designing new port facilities to be less vulnerable to severe corrosion caused by an increase in seawater temperature.

Acknowledgements The author thanks Dr Lizhengli Peng for generating the data for Fig. 5.3. The author also appreciates the financial support of the Australian Research Council and the Commonwealth Scientific and Industrial Research Organisation (CSIRO) Flagship Cluster Fund through the project Climate Adaptation Engineering for Extreme Events in collaboration with the Sustainable Cities and Coasts Theme, the CSIRO Climate Adaptation Flagship.

References

ASCE. (2013, March). *2013 Infrastructure Report Card*. Reston, VA: American Society of Civil Engineers.

Bai, Y., Burkett, W., & Nash, P. (2006). Lessons learnt from the emergency bridge replacement project. *Journal of Construction Engineering and Management, 132*(4), 338–344.

Bastidas-Arteaga, E., & Stewart, M. G. (2015). Damage risks and economic assessment of climate adaptation strategies for design of new concrete structures subject to chloride-induced corrosion. *Structural Safety, 52*(A, January), 40–53.

Bastidas-Arteaga, E., & Stewart, M. G. (2016). Economic assessment of climate adaptation strategies for existing RC structures subjected to chloride-induced corrosion. *Structure and Infrastructure Engineering, 12*(4), 432–449.

Bjarnadottir, S., Li, Y., & Stewart, M. G. (2011). A probabilistic-based framework for impact and adaptation assessment of climate change on hurricane damage risks and costs. *Structural Safety, 33*(3), 173–185.

Bjarnadottir, S., Li, Y., & Stewart, M. G. (2013). Hurricane risk assessment of power distribution poles considering impacts of a changing climate. *Journal of Infrastructure Systems, 19*(1), 12–24.

Bjarnadottir, S., Li, Y., & Stewart, M. G. (2014). Risk-based economic assessment of mitigation strategies for power distribution poles subjected to hurricanes. *Structure and Infrastructure Engineering, 10*(6), 740–752.

Blalock, G., Kadiyali, V., & Simon, D. H. (2007, November). The impact of post-9/11 airport security measures on the demand for air travel. *Journal of Law and Economics, 50*(4), 731–755.

Botzen, W. J. W., Alerts, J. C. J. H., & van den Bergh, J. C. J. M. (2013). Individual preferences for reducing flood risk to near zero through elevation. *Mitigation and Adaptation Strategies for Global Change, 18*(2), 229–244.

GAO. (2009, January). *Federal efforts to strengthen security should be better coordinated and targeted on the nation's most critical highway infrastructure*. Washington, DC: United States Government Accountability Office.

Gardoni, P., & Murphy, C. (2014). A scale of risk. *Risk Analysis, 34*(7), 1208–1227.

Goklany, I. M. (2008, February 5). *What to do about climate change*. Policy Analysis, No. 609. Washington, DC: Cato Institute.
Grant, M., & Stewart, M. G. (2012). A systems model for probabilistic risk assessment of improvised explosive device attack. *International Journal of Intelligent Defence Support Systems, 5*(1), 75–93.
Grant, M., & Stewart, M. G. (2015). Probabilistic risk assessment for improvised explosive device attacks causing significant building damage. *Journal of Performance of Constructed Facilities, 29*(5), B4014009.
Grynbaum, M. M. (2014, September 23). At U.N., de Blasio Warns of 'existential threat' from climate change. *New York Times*.
Hall, J. W., Brown, S., Nicholls, R. J., Pidgeon, N. F., & Watson, R. T. (2012). Proportionate adaptation. *Nature Climate Change, 2*, 833–834.
Hardaker, J. B., Fleming, E., & Lien, G. (2009). How should governments make risky policy decisions? *Australian Journal of Public Administration, 68*(3), 256–271.
Hinkel, J., Nicholls, R. J., Vafeidis, A. T., Tol, R. S. J., & Avagianou, T. (2010). Assessing risk of and adaptation to sea-level rise in the European Union: An application of DIVA. *Mitigation and Adaptation Strategies for Global Change, 15*(7), 703–719.
Holden, R., Val, D. V., Burkhard, R., & Nodwell, S. (2013). A network flow model for interdependent infrastructures at the local scale. *Safety Science, 53*(3), 51–60.
IPCC. (2007). Contribution of working groups I, II and III to the fourth assessment report on intergovernmental panel on climate change. In R. K. Pachauari & A. Reisinger (Eds.) (Core writing team), *Climate change 2007: Synthesis report*. Geneva, Switzerland: IPCC.
IPCC. (2012). A special report of working groups I and II of the intergovernmental panel on climate change. In C. B. Field et al. (Eds.), *Managing the risks of extreme events and disasters to advance climate change adaptation*. Cambridge: Cambridge University Press.
Kundzewicz, Z. W., Lugeri, N., Dankers, R., Hirabayashi Doll, P., Pinskwar, I., Dysarz, T., et al. (2013). Assessing river flood risk and adaptation in Europe—Review of projections for the future. *Mitigation and Adaptation Strategies for Global Change, 15*(7), 641–656.
Melchers, R. E., & Jeffrey, R. (2013). Accelerated low water corrosion of steel pilling in harbours. *Corrosion Engineering Science and Technology, 48*, 496–505.
Mueller, J., & Stewart, M. G. (2011a). *Terror, security, and money: Balancing the risks, benefits, and costs of homeland security*. Oxford: Oxford University Press.
Mueller, J., & Stewart, M. G. (2011b). The price is not right: The U.S. spends too much money to fight terrorism. *Playboy, 58*(10), 149–150.
Mueller, J., & Stewart, M. G. (2014). Evaluating counterterrorism spending. *Journal of Economic Perspectives, 28*(3), 237–248.
Mueller, J., & Stewart, M. G. (2016). *Chasing ghosts: The policing of terrorism*. Oxford: Oxford University Press.
Murphy, C., & Gardoni, P. (2008). The acceptability and the tolerability of societal risks: A capabilities-based approach. *Science and Engineering Ethics, 14*(1), 77–92.
Nishijima, K., Maruyama, T., & Graf, M. (2012). A preliminary impact assessment of typhoon wind risk of residential buildings in Japan under future climate change. *Hydrological Research Letters, 6*(1), 23–28.
Norville, H. S., Harvill, N., Conrath, E. J., Shariat, S., & Mallonee, S. (1999). Glass-related injuries in Oklahoma city bombing. *Journal of Performance of Constructed Facilities, 13*(2), 50–56.
NTSB. (2008, November 14). Highway accident report: Collapse of I-35W Highway Bridge, Minneapolis, Minnesota, August 1, 2007. Accident Report NTSB/HAR-08/03. Washington, DC: National Transportation Safety Board.
OMB. (1992). *Guidelines and discount rates for benefit-cost analysis of federal programs (revised)*. Circular No. A-94, October 29, 1992. Washington, DC: Office of Management and Budget.
Paté-Cornell, E. (2002). Risk and uncertainty analysis in government safety decisions. *Risk Analysis, 22*(3), 633–646.

Peng, L., & Stewart, M. G. (2014). Spatial time-dependent reliability analysis of corrosion damage to concrete structures under a changing climate. *Magazine of Concrete Research, 66*(22), 1154–1169.

Risky Business. (2014, June). Risky business: The economic risks of climate change in the United States. RiskyBussiness.org.

Robinson, L. A., Hammitt, J. K., Aldy, J. E., Krupnick, A., & Baxter, J. (2010). Valuing the risk of death from terrorist attacks. *Journal of Homeland Security and Emergency Management, 7*(1).

Stewart, M. G. (2014). Risk and economic viability of housing climate adaptation strategies for wind hazards in southeast Australia. *Mitigation and Adaptation Strategies for Global Change, 20*(4), 601–622.

Stewart, M. G., & Deng, X. (2015). Climate impact risks and climate adaptation engineering for built infrastructure. *ASCE-ASME Journal of Risk and Uncertainty in Engineering Systems, Part A: Civil Engineering, 1*(1), 04014001.

Stewart, M. G., Ellingwood, B. R., & Mueller, J. (2011). Homeland security: A case study in risk aversion for public decision-making. *International Journal of Risk Assessment and Management, 15*(5/6), 367–386.

Stewart, M. G., & Melchers, R. E. (1997). *Probabilistic risk assessment of engineering systems*. London: Chapman & Hall.

Stewart, M. G., & Mueller, J. (2013). Terrorism risks and cost-benefit analysis of aviation security. *Risk Analysis, 33*(5), 893–908.

Stewart, M. G., & Mueller, J. (2014a). Cost-benefit analysis of airport security: Are airports too safe? *Journal of Air Transport Management, 35*(March), 19–28.

Stewart, M. G., & Mueller, J. (2014b). Risk and cost-benefit analysis of police counter-terrorism operations at Australian airports. *Journal of Policing, Intelligence and Counter Terrorism, 9*(2), 98–116.

Stewart, M. G., & Mueller, J. (2014c). Terrorism risks for bridges in a multi-hazard environment. *International Journal of Protective Structures, 5*(3), 275–289.

Stewart, M. G., & Peng, J. (2010). Life cycle cost assessment of climate change adaptation measures to minimise carbonation-induced corrosion risks. *International Journal of Engineering under Uncertainty: Hazards, Assessment and Mitigation, 2*(1–2), 35–46.

Stewart, M. G., Val, D., Bastidas-Arteaga, E., O'Connor, A., & Wang, X. (2014). Climate adaptation engineering and risk-based design and management of infrastructure. In D. M. Frangopol & Y. Tsompanakis (Eds.), *Maintenance and safety of aging infrastructure* (pp. 641–684). Leiden: CRC Press.

Stewart, M. G., Wang, X., & Willgoose, G. R. (2014). Direct and indirect cost and benefit assessment of climate adaptation strategies for housing for extreme wind events in Queensland. *Natural Hazards Review, 15*(4), 04014008(12).

Stewart, M. G., & Mueller, J. (2011). Cost-benefit analysis of advanced imaging technology full body scanners for airline passenger security screening. *Journal of Homeland Security and Emergency Management, 8*(1), Article 30.

Sunstein, C. R. (2003). Terrorism and probability neglect. *Journal of Risk and Uncertainty, 26*(2–3), 121–136.

Susskind, L. (2010). Responding to the risks posed by climate change: Cities have no choice but to adapt. *Town Planning Review, 81*(10), 217–235.

Sydney Ports. (2008, March). *Port botany container terminal expansion overview*.

Val, D. V., Holden, R., & Nodwell, S. (2013). Probabilistic assessment of failures of interdependent infrastructures due to weather related hazards. In G. Deodatis, B. R. Ellingwood, & D. M. Frangopol (Eds.), *Safety, reliability, risk and life-cycle performance of structures and infrastructure* (pp. 1551–1557). London: Taylor & Francis Group.

Part III
Earthquakes

Chapter 6
Progressive Collapse Simulation of Vulnerable Reinforced Concrete Buildings

Khalid M. Mosalam and Selim Günay

Abstract There are many vulnerable reinforced concrete (RC) buildings located in earthquake-prone areas around the world. These buildings are characterized by the lack of seismic details and corresponding non-ductile behavior and significant potential of partial and global collapse. One of the current challenges of the earthquake engineering profession and research communities is the identification of such buildings and determination of effective and economical retrofit methods for response enhancement. Identification of these buildings is not a trivial task due to the various sources of non-ductile behavior and the large number of involved sources of uncertainty. Furthermore, accurate determination of collapse-prone buildings is important from an economical perspective. Unfortunately, there are not enough economical resources to retrofit all the non-ductile buildings that have the symptoms for collapse potential. In order to use the available monetary resources in an effective manner, these buildings should be accurately and reliably ranked to identify those that are most vulnerable to collapse. This chapter intends to provide a contribution to the accurate determination of the most collapse-vulnerable non-ductile RC buildings by discussing the methods from existing literature and exploring the research needs related to (a) gravity load failure modeling and (b) consideration of different uncertainty sources in an efficient manner.

6.1 Introduction

It is a well-known fact that there are many vulnerable reinforced concrete (RC) buildings located in earthquake-prone areas around the world. These buildings are characterized by the lack of seismic details (such as lack of confinement at the beam and column ends and the beam-column joints, strong beam-weak column proportions, and presence of shear-critical columns) and the corresponding non-ductile behavior and significant potential of partial and global collapse, posing threats to human life. One of the current challenges of the earthquake engineering-related

K.M. Mosalam (✉) • S. Günay
Department of Civil and Environmental Engineering, University of California, Berkeley, CA 94720, USA
e-mail: mosalam@berkeley.edu

P. Gardoni, J.M. LaFave (eds.), *Multi-hazard Approaches to Civil Infrastructure Engineering*, DOI 10.1007/978-3-319-29713-2_6

research is the identification of such buildings and the determination of effective and economical retrofit methods for enhancing their seismic response. Identification of these buildings is not an easy task due to the various sources of non-ductile behavior and a significant amount of uncertainty involved in material characteristics, reinforcement ratio, geometry, etc. Furthermore, accurate determination of collapse-prone buildings is important from an economical perspective. Unfortunately, there are not sufficient economical resources to retrofit all the non-ductile buildings that have the symptoms for collapse potential. In order to effectively use these limited monetary resources, these buildings should be accurately and reliably ranked to identify those that are most vulnerable to collapse. This chapter intends to provide a contribution to the accurate determination of the most collapse-vulnerable non-ductile buildings by discussing relevant methods from existing literature and exploring the research needs related to (a) gravity load failure modeling and (b) consideration of uncertainty in an efficient manner.

The current state of knowledge and practice in nonlinear static and dynamic analyses has the ability to determine side-sway collapse that occurs due to lateral dynamic instability at excessive lateral displacements, when the lateral strength and stiffness of the structure degrades significantly. However, non-ductile RC buildings mostly collapse by losing gravity load-carrying capacity much before reaching these excessive displacements, as demonstrated in Fig. 6.1. At point A of this figure, the lateral load resistance starts to degrade because of various events such as in-plane or out-of-plane (OOP) failure of infill walls or shear failure of columns. This is followed by the loss of the gravity load-carrying capacity of the lateral and gravity load-resisting components, starting with points B and C, respectively. The first part of this paper explores this relatively neglected, but significantly important, issue of gravity load-carrying capacity by discussing the methods of gravity load failure modeling in collapse simulations and pointing out further research needs. It is to be noted that the term "gravity load failure" is used in the rest of the paper to refer to the loss of gravity load-carrying capacity.

There are two alternative options that can be considered for gravity load failure modeling of the elements of a structure: (a) explicit modeling and (b) implicit modeling, i.e., non-simulated failure. Explicit modeling of gravity load failure consists of two stages: (1) detection of gravity load failure and (2) post-failure modeling. In the next section of the paper, the advantages and disadvantages of the explicit and implicit modeling options are presented. Moreover, three different approaches are discussed for the purpose of employing them in the second stage of explicit modeling. These approaches are (1) element removal, (2) assigning low stiffness to a failed element, and (3) representation of the post-failure response of failed elements with degradation. A subsequent section of the paper explores the models that can be used for failure detection of various structural members in explicit modeling. The final section related to gravity failure modeling is comprised of the modeling of the structural elements primarily designed to resist the gravity loads with insignificant contribution to the lateral load resistance. While the gravity load-resisting system can be approximately considered for the case of side-sway collapse, its explicit modeling significantly complements the explicit gravity load failure modeling of the primary lateral resisting system, as shown in Fig. 6.1.

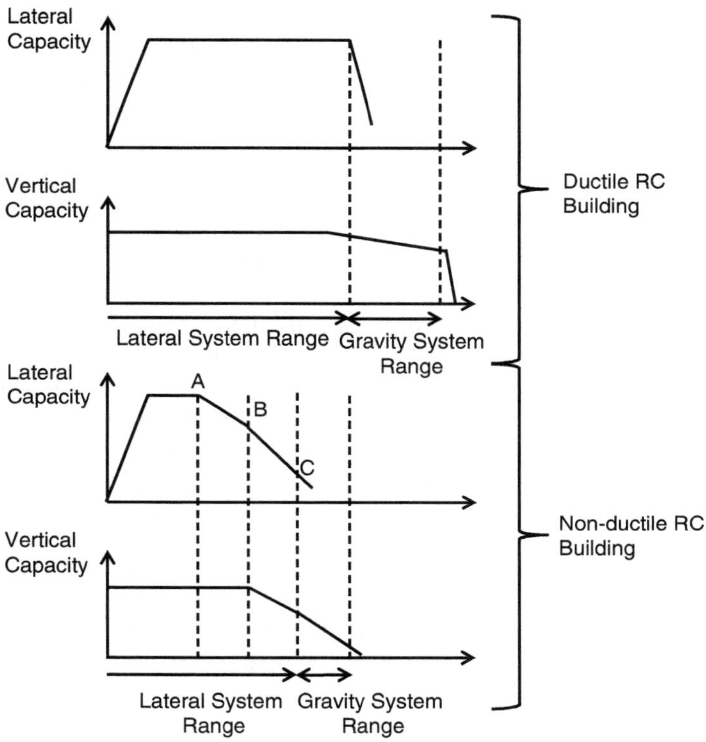

Fig. 6.1 Lateral and vertical responses of RC buildings (modified from Holmes 2000)

As mentioned above, non-ductile RC buildings generally involve a significant amount of uncertainties. Therefore, a substantial number of collapse simulations may need to be conducted for accurate identification of buildings which are most vulnerable to collapse. The final part of the paper presents a deterministic sensitivity analysis method from the literature, the so-called tornado diagram, as a method to identify the sources of uncertainties which are most influential on the seismic response. Accordingly, this practical method can handle the effect of uncertainty more efficiently in collapse simulations of non-ductile RC buildings by only incorporating these influential sources of uncertainty as probabilistic, while those that are less influential can be incorporated as deterministic.

6.2 Gravity Load Failure Modeling of Lateral Load-Resisting Components

In order to model the gravity load failure of the lateral load-resisting components of non-ductile RC buildings, there are two alternative options that can be considered. The first option is the explicit modeling of gravity load failure, while the second one is the determination of gravity failure of the components implicitly through post-processing without modeling of gravity failure (non-simulated failure). In the second approach, which has been the commonly utilized approach up to date, engineering demand parameters (EDPs), such as drifts or accelerations, obtained as a result of the analyses, can be used to determine collapse by comparing these EDPs with the limits provided by available gravity loss models (e.g., Elwood and Moehle 2005).

Implicit modeling of gravity load failure (non-simulated failure) can only be feasible in some cases where the first element failure, for example, a column axial failure, is sufficient to define global collapse. Such a case may occur when all the columns at a story have similar properties and failure of all columns is likely to take place almost simultaneously. In this case, there is no need to explicitly consider the consequences of an axially failed column and first column axial failure can be sufficient to define global collapse. In all other cases, explicit modeling of gravity load failure is essential for accurate determination of collapse. This distinction can be further supported by considering one of the collapse indicators discussed in the NIST report for collapse assessment and mitigation strategies for existing RC buildings (2010). The considered collapse indicator is the "maximum fraction of columns at a story experiencing axial failures," which requires the identification of the number of columns experiencing gravity load failures. Such identification would potentially be inaccurate if conducted by post-processing of results without an explicit consideration of gravity load failure, because the gravity load failure of a column (or a beam-column joint) is likely to affect the response and failure potential of the other columns and the overall system.

Explicit modeling of gravity load failure consists of two stages. The first stage is the detection of gravity load failure. Available models in literature for columns and beam-column joints that can be used for this purpose are presented in the next section. The second stage is the post-failure modeling, possible options of which are: (1) element removal, (2) assigning low stiffness to a collapsed element, and (3) representing the post-failure response with degradation. The first approach consists of the direct removal of the element from the structural model upon its failure (e.g., Talaat and Mosalam 2007, 2009). The second approach consists of reducing the stiffness of the collapsed element using a small multiplier (e.g., Grierson et al. 2005) in order to eliminate its contribution to the global structural stiffness matrix and restoring force vector. In the third approach, the post-failure response is represented with a degraded force-displacement relationship (e.g., Elwood and Moehle 2005). Advantages and disadvantages of these three approaches are summarized in Table 6.1 along with those of the implicit gravity load failure approach (non-simulated failure) mentioned above.

The element removal approach of Talaat and Mosalam (2007) is based on dynamic equilibrium and the resulting transient change in system kinematics and it constitutes the basis of a corresponding progressive collapse algorithm. This algorithm is implemented in OpenSees for automatic removal of collapsed elements during an ongoing simulation (Fig. 6.2). The implementation is carried out as a new OpenSees module, designed to be called by the main analysis module after

Table 6.1 Advantages and disadvantages of different gravity load failure modeling methods

Component failure method	Advantages	Disadvantages
Explicit modeling Option 1: element removal	1. Numerical problems associated with ill-conditioned stiffness matrices are eliminated 2. Enforcing dynamic equilibrium enables: (a) Computation of the resulting increase in nodal accelerations (b) Inclusion of the system's complete kinematic state at time of element collapse to determine if it can survive to a new equilibrium state 3. Motion of the collapsed element can be tracked relative to the damaged system to estimate the time and kinetics of a subsequent collision with the intact structural part 4. Elimination of the numerical convergence problems related to the iterative formulation of some element and material types by removing them (refer to Item 1 in the disadvantages of degraded post-failure response)	1. Requirement of additional bookkeeping operations to update the nodal masses and to check nodal forces, constraints, restraints, dangling nodes, floating elements, etc. 2. An additional computational burden introduced by the redefinition of degrees of freedom of a structural model and the corresponding connectivity, upon removal of an element or several elements 3. Convergence problems, not on the element or material levels, but on the numerical integration level, as a result of the sudden updating of mass, stiffness and damping matrices, and the local vibrations triggered as a consequence of the resulting transient effect (refer to discussions on the methods to overcome these convergence problems)
Explicit modeling Option 2: assigning low stiffness to a failed element	Additional tasks (Items 1 and 2 in disadvantages of element removal) related to the element removal process are avoided	1. Numerical problems associated with ill-conditioned stiffness matrices 2. Not possible to explicitly consider the consequences of component failure (Items 2 and 3 in advantages of element removal)
Explicit modeling Option 3: degraded post-failure response	Additional tasks (Items 1 and 2 in disadvantages of element removal) related to the element removal process are avoided	1. Numerical convergence problems related to the iterative formulation of some types of elements, e.g., force-based beam-column, and materials, e.g., Bouc-Wen, since the failure state generally occurs at a negatively sloped portion of the constitutive relationship 2. Not possible to explicitly consider the consequences of component failure (Items 2 and 3 in advantages of element removal)
Implicit modeling (non-simulated failure)	Suitable for fast and simplified analyses and in some special cases, e.g., having similar columns in one story	Inaccurate results due to the lack of realistic representation of the post-failure response

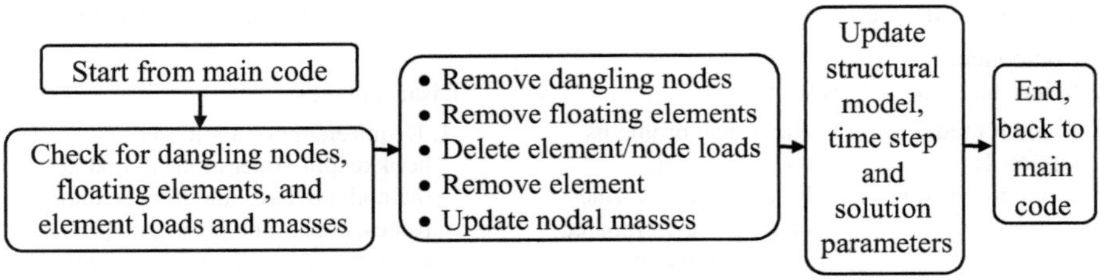

Fig. 6.2 Element removal algorithm of Talaat and Mosalam (2007)

each converged integration time step to check each element for possible violation of its respective removal criteria, where the relevant models presented in the next section can be used for defining the removal criteria. A violation of a predefined removal criterion triggers the activation of the algorithm on the violating element before returning to the main analysis module. Activation of the element removal algorithm includes updating nodal masses, checking if the removal of the collapsed element results in leaving behind dangling nodes or floating elements, which must be removed, and removing all associated element and nodal forces, imposed displacements, and constraints. It is noted that the gravity loads at the node of a column, which is common with the other elements, is not removed. Accordingly, the gravity loads on the structure are not reduced upon removal of a column, allowing for the analysis model to capture the redistribution of the gravity loads to the other intact columns.

Since the structural elements lose their ability to support gravity loads after gravity load failure, the removal of a failed element is the most representative approach to model gravity load failure. Hence, the discussion in the following two paragraphs are based on the comparison of the element removal approach (Option 1) with Options 2 and 3. It should be noted that Option 1 approach assumes that the gravity load support is lost instantaneously. As mentioned previously, the first stage of gravity load failure in explicit modeling is the detection of this failure. Such detection is based on equations derived from tests where the loss of the gravity load support of the test specimen was defined by a single point and there is no data obtained from the tests beyond this point. Accordingly, the assumption of instantaneous gravity load failure is dictated by the gravity failure detection models. As an alternative, the detection equations can be constructed with a probability distribution, e.g., in the form of a set of equations for the median and median plus/minus a dispersion. However, such equations require further experimental research.

The removal of a collapsed element requires several bookkeeping operations to update the nodal masses and to check nodal forces, constraints, restraints, dangling nodes, floating elements, etc. Also, there is an additional computational burden introduced by the redefinition of degrees of freedom of a structural model and the corresponding connectivity upon removal of one or more elements. Such additional tasks are avoided in Option 2, which consists of assigning low stiffness to failed elements. However, there are three important advantages of Option 1 compared to Option 2. First, it avoids numerical problems due to ill-conditioned stiffness matrices. Second, enforcing the dynamic equilibrium enables: (1) the computation of the resulting increase in nodal accelerations and (2) the inclusion of the system's complete kinematic state at time of element collapse to determine if the structure can successfully redistribute the forces from the removed element and survive to a new equilibrium state. Third, the motion of the collapsed element can be tracked relative to the damaged system to estimate the time and kinetics of a subsequent collision with the intact structural part.

Although representing the post-failure response with a degraded force-displacement relationship in Option 3 is realistic for most of the failed components, it may introduce numerical problems. The failure state generally corresponds to a negatively sloped portion of the constitutive relationship, where the iterative formulation of some types of elements, e.g., force-based beam-column, and materials, e.g., Bouc-Wen, is likely to experience convergence problems. The removal of such elements automatically eliminates the associated numerical problems. Analyses conducted to estimate the responses obtained from shaking table tests of a non-ductile RC frame showed that the analyses considering and not considering the element removal (ER) were both successful in predicting the observed collapse of the non-ductile members (Mosalam et al. 2009). On the other hand, the response after collapse was rather jagged and close to being unstable for the case without element removal, whereas the analysis with element removal provided a more reasonable response (Fig. 6.3). It should be noted that the analyses without the element removal used the degraded post-failure approach (Option 3).

Element removal may introduce convergence problems, not on the element or material levels but on the numerical integration level, as a result of the sudden updating of mass, stiffness and damping matrices, and the triggered local vibrations as a consequence of the resulting transient effect due to the sudden changes in the matrices. A possible solution to such convergence problems is adaptive switching of solver type and convergence criteria and reduction of the integration time step (Talaat and Mosalam 2007). It is to be noted that this strategy has been used for the analyses of the non-ductile RC frame mentioned in the above paragraph. Another effective solution is the use of transient integrators which do not require iterations, e.g., operator-splitting methods (Hughes et al. 1979). Analyses conducted on bridge systems (Mosalam et al. 2013) showed that the operator-splitting method results in exactly the same solution as the commonly used implicit Newmark integration even for cases with highly nonlinear response.

It is noted that the objective of this paper is not the recommendation of a specific explicit modeling option; rather, it is to present the available options of collapse determination to the interested reader. Furthermore, the most suitable option may change from one building to another. Hence, the presented advantages and disadvantages are expected to provide guidance to the readers in choosing the best starting point and possibly switching between different options.

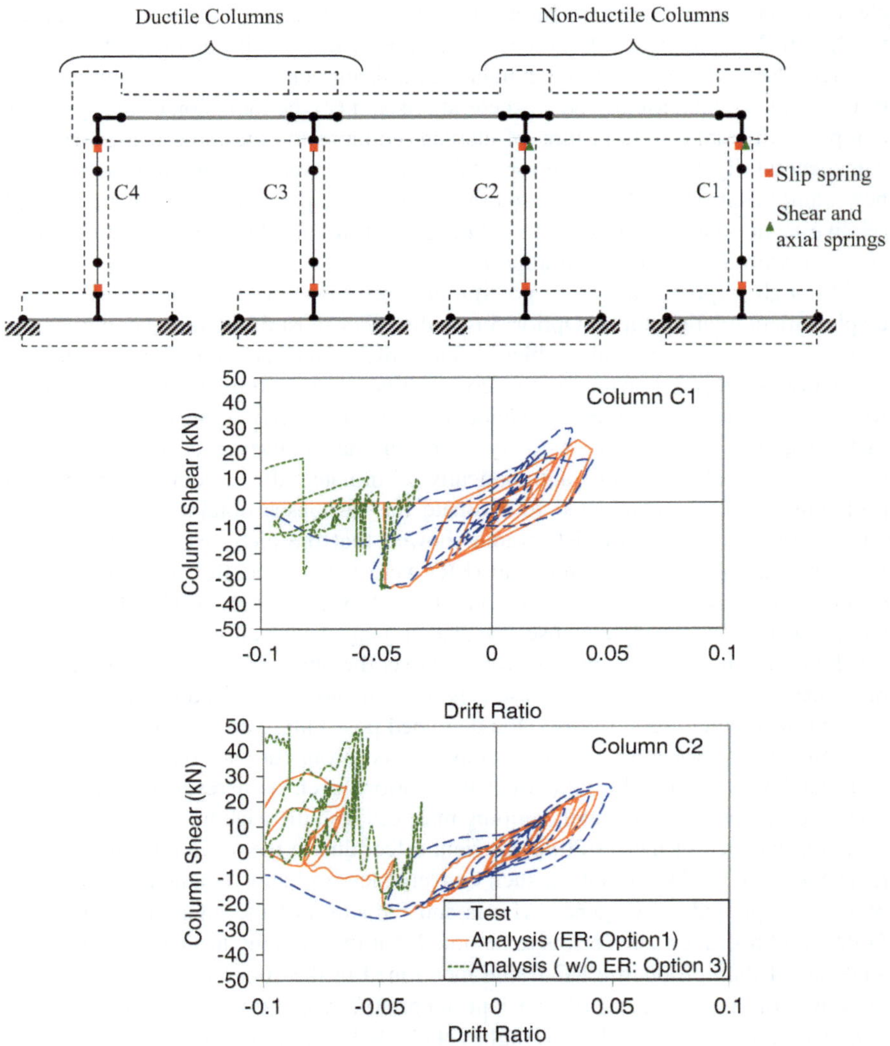

Fig. 6.3 Response of a system with non-ductile columns experiencing axial failure from shaking table tests and analyses with and without element removal (Mosalam et al. 2009)

6.3 Detection of Gravity Load Failure

As mentioned previously, the first stage of the explicit modeling of gravity load failure is the detection of gravity load failure. Models that can be used for the detection of gravity load failure of columns, beam-column joints, slab-column joints, and infill walls are described in the following subsections. Collapse of buildings with shear walls, due to loss of gravity load carrying capacity, has rarely been observed in the last 50 years (Wallace et al. 2008). Therefore, detection of gravity load failure of shear walls is not covered in this paper.

6.3.1 Columns

One of the models that can be used to detect the gravity load failure of columns is proposed by Elwood and Moehle (2005), where the drift at axial failure of a shear-damaged column is represented as follows:

$$\left(\frac{\Delta}{L}\right)_{\text{axial}} = \frac{4}{100}\frac{1 + (\tan\theta)^2}{\tan\theta + P\left(s/\left(A_{\text{st}} f_{\text{yt}} d_{\text{c}} \tan\theta\right)\right)} \tag{6.1}$$

where $(\Delta/L)_{\text{axial}}$ is the drift ratio at axial failure; P is the column axial force; A_{st}, f_{yt}, and s are respectively area, yield strength, and spacing of the transverse reinforcement; d_{c} is the column core depth (center to center of tie); and θ is the critical crack angle from the horizontal (assumed 65°).

Elwood and Moehle (2005) stated that this axial failure model is based on data from 12 columns, where all columns were constructed from normal strength concrete, had the same height-to-width ratio, were designed for longitudinal reinforcement yielding prior to shear failure, and were tested in uniaxial bending. This axial capacity model is implemented in OpenSees as a limit state material model and used as a spring connected to a column end. Removal of the corresponding spring is also implemented in an earlier version of OpenSees. When the drift during a simulation reaches the drift corresponding to axial failure, the corresponding spring and the column, to which the spring is connected, are removed using the element removal algorithm.

6.3.2 Beam-Column Joints

Beam-column joints of old non-ductile RC buildings, e.g., designed in the 1960s, are generally unreinforced without any transverse steel bars. The beams connected to such joints rotate relative to the columns, i.e., right angle between the beam and column is not maintained, due to joint shear failure and corresponding deformation. Joint panel flexibility can be modeled by using a rotational spring located between the beam and column end nodes. It is noted that rigid end offsets are used at the beam and column ends to consider the joint physical dimensions (Fig. 6.4d). The rotational spring is defined by a nonlinear constitutive relationship, which is characterized by a backbone curve and a set of hysteresis rules (Park and Mosalam 2013a). These characteristics are recently developed empirically, based

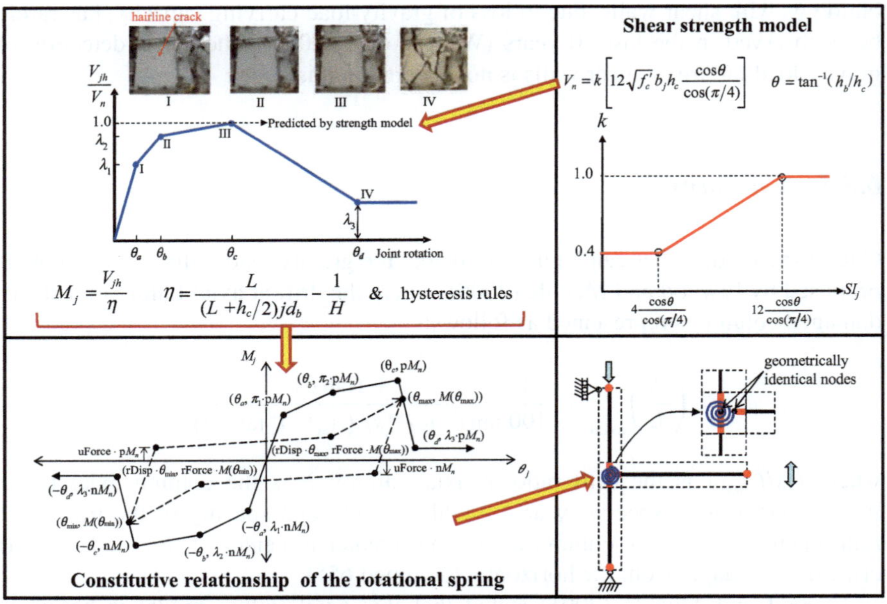

Fig. 6.4 Overview of unreinforced beam-column joint model (Park and Mosalam, 2012c)

on the measured joint responses and visual observations from tests of four-corner beam-column-slab joint specimens (Park and Mosalam 2013b) and verified by comparison with other exterior and interior beam-column joint tests. A strength model is developed to determine the peak force of the backbone curve, which also corresponds to the joint shear strength (Park and Mosalam 2009, 2012a). The practical strength model in Fig. 6.4a, which accounts for the effects of two main parameters, namely, (1) the joint aspect ratio, defined as the ratio of beam to column cross-sectional heights and (2) the beam reinforcement ratio, is verified through its accurate prediction of various beam-column joint test results available in the literature (Park and Mosalam 2012b, c).

The rotational spring, and the constitutive relationship mentioned above, can be used to represent the axial failure and the corresponding removal of a beam-column joint using the proposed extension by Hassan (2011). Hassan proposed an axial capacity model for beam-column joints where the drift, i.e., beam tip displacement normalized by the beam length (refer to the test setup in Park and Mosalam (2012c)), at axial collapse is represented as a function of the axial force and the beam bottom reinforcement strength (Fig. 6.5). Equation (6.2) is suitable to be used to remove a beam-column joint as a part of the progressive collapse algorithm as long as the drift in this equation is replaced by the joint rotation of the considered analytical model (Fig. 6.4c). Note that the difference between the joint rotation and the abovementioned drift in the tests of Hassan (2011) is the sum of the flexural deformation of the beam and the beam displacement due to column rotation

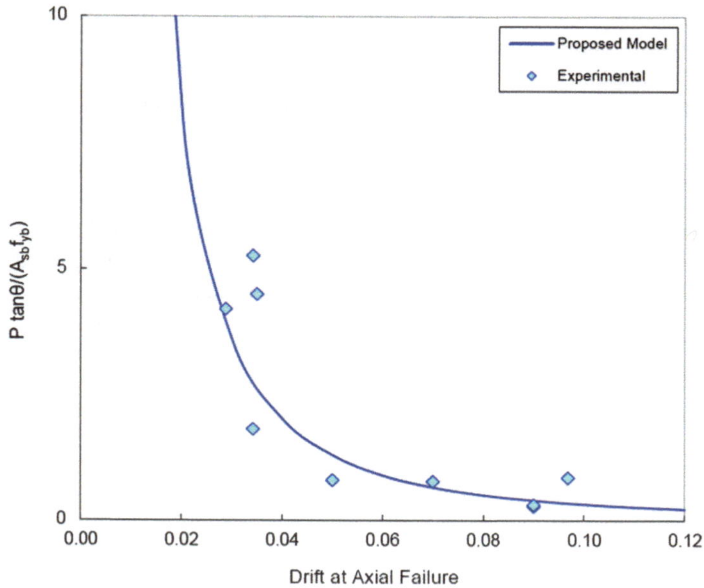

Fig. 6.5 Axial capacity model in Eq. (6.2) for beam-column joints (Hassan 2011)

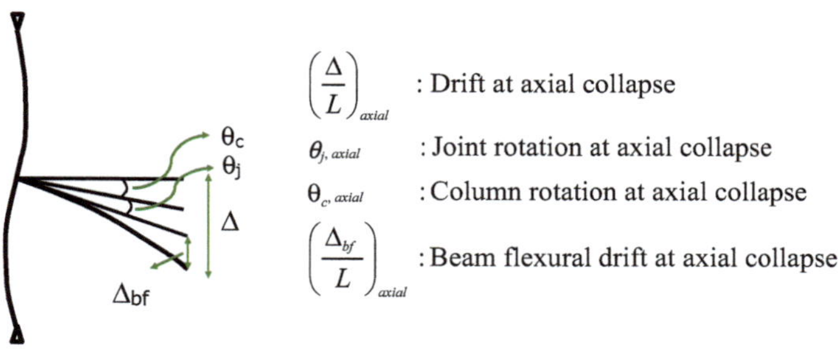

Fig. 6.6 Contribution of different components to beam displacement in a beam-column joint test

(Fig. 6.6). However, it should be noted that this expression is based on a rather small database of joint axial failures. Therefore, more joint axial failure tests are needed to further verify this relationship. Hassan (2011) also mentioned that the case of high axial load on a joint where the beam flexural capacity is much smaller than the direct joint failure capacity is excluded from the application of this model.

$$\theta_{\text{axial}} = 0.057\sqrt{\frac{A_{\text{sb}}f_{\text{yb}}}{P\tan\theta}} \tag{6.2}$$

where θ_{axial} is the joint rotation at the joint axial failure, P is the axial force, A_{sb} and f_{yb} are the respective area and yield strength of the beam bottom reinforcement, and θ is the crack angle.

The removal of a beam-column joint is not implemented in OpenSees yet. However, the idea is similar to the case of column failure, i.e., when the rotation of the spring (representing the joint) during a simulation reaches the rotation corresponding to axial failure defined by Eq. (6.2), the joint (rotational spring and rigid end offsets) is removed using the element removal algorithm. Because the generic removal algorithm is already implemented, including the removal of beam-column joints using the abovementioned criteria is rather straightforward.

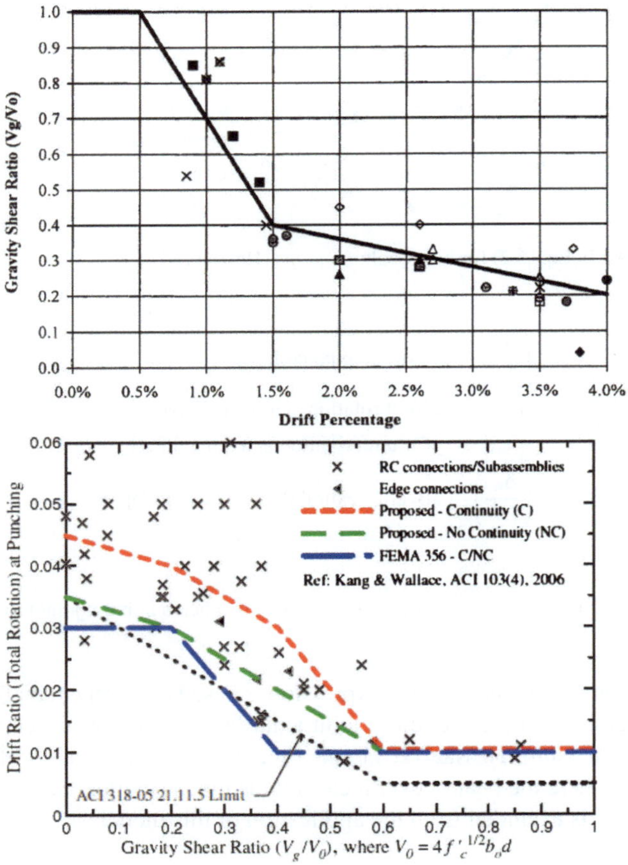

Hueste and Wight (1999)

Elwood et al. (2007)

Fig. 6.7 Detection of gravity load failure for slab-column joints

6.3.3 Slab-Column Joints

Gravity support loss of slab-column joints can be defined with the punching shear failure. In order to detect the punching shear failure, available limit models in literature such as Hueste and Wight (1999) or Elwood et al. (2007) can be used (Fig. 6.7). In these models, drift or plastic rotation values corresponding to the punching shear failure are determined as functions of the gravity shear ratio, defined as the value of the vertical gravity shear divided by the punching shear strength of the joint. Kang et al. (2009) used these limit models for the detection of punching shear failure while conducting analytical simulations to predict the results of shaking table tests. In that regard, they used Option 3, the representation of the post-failure response with degradation, for post-failure modeling.

6.3.4 Unreinforced Masonry Infill Walls

When the seismic vulnerabilities present in the RC system are combined with the complexity of the interaction between the infill walls and the surrounding frame and the brittleness of the unreinforced masonry (URM) materials, the URM infill walls can increase the vulnerability and collapse potential of non-ductile RC buildings. Earthquakes in the last two decades, e.g., 1999 Kocaeli, 2008 Wenchuan, and 2009 L'Aquila earthquakes, led to several observations related to URM infill walls (Mosalam and Günay 2012), which are listed as follows:

1. URM infill walls contribute to the stiffness and strength of the frames as evidenced from the weak/soft story damage of the open-ground story buildings and the torsional response created by the nonuniform distribution of infill walls around the building perimeter.
2. URM infill wall failure is a combination of in-plane (IP) and OOP effects as evidenced from some of the URM infill wall failures taking place at the upper stories instead of the lower stories where the story shear forces are the highest.
3. Failure of infill walls at a story leads to the formation of weak/soft stories during the earthquake, which may result in the failure of a story as evidenced by intermediate story collapses.
4. Infill walls interact with the frame members as evidenced by shear cracks and failure of columns and beam-column joints in infilled bays.

Accordingly, URM infill walls should be modeled to consider these observations. The first two observations can be reflected by employing a practical model that considers IP-OOP interaction of infill walls (Fig. 6.8) (Kadysiewski and Mosalam 2008). The fourth observation can be taken into account by modeling nonlinear shear springs at the column ends to consider the effect of additional horizontal forces transferred from the infill walls to the columns. The third observation can be considered by removal of failed infill walls where detection of failure is based on

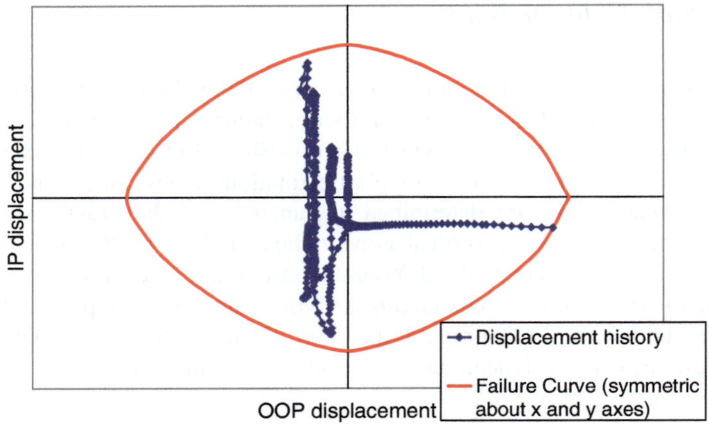

Fig. 6.8 Failure detection for URM infill walls considering IP-OOP interaction

a combination of IP and OOP displacements (OpenSees Wiki 2011; Mosalam and Günay 2015).

Modeling of infill walls considering the element removal due to IP-OOP interaction has been recently used for the investigation of the earthquake response of buildings designed according to modern seismic codes without considering the infill walls in the design process (Mosalam et al. 2013). It is noted that, different from the other discussed elements, failure of infill walls is not directly considered as gravity load failure. It is rather considered to negatively affect the lateral response. However, detection and post-failure modeling of infill walls are still important modeling aspects for the objective of the accurate determination of collapse, because the consequences are likely to affect the gravity load failure of other discussed elements.

6.4 Explicit Modeling of Gravity Systems

Modeling of the gravity system as a single column that accounts for P-Δ effects (Fig. 6.9) is a commonly utilized approach (Liel et al. 2009; Lai and Mahin 2013). Such modeling does not generally account for the strength and stiffness of the gravity system, either because collapse is not of interest or the investigated collapse mechanism is side-sway. For the collapse type investigated herein, namely, the gravity load failure, explicit modeling of the gravity system is essential. As shown in Fig. 6.1, gravity support may not be completely lost after the lateral and axial failures of the primary system. Therefore, explicit modeling of the gravity system generally leads to a more accurate and realistic determination of the global collapse.

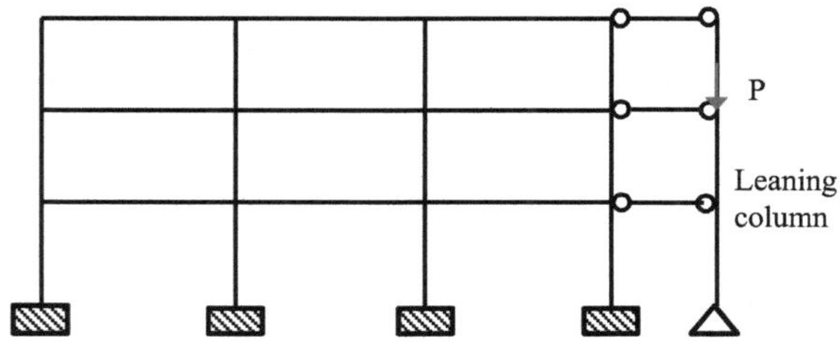

Fig. 6.9 Modeling of the gravity system as a leaning column

6.5 Tornado Diagram Analysis

Non-ductile RC buildings generally involve a significant number of uncertainty sources. Some of these sources are concrete strength, longitudinal and transverse reinforcement yield strength, concrete and masonry modulus of elasticity, masonry compressive and shear strengths, damping ratio, story mass, and ground motion record-to-record variability. Considering all these uncertainty sources in the process of accurate determination of the most collapse-vulnerable buildings is likely to result in an extensive number of collapse simulations. However, uncertainties of some of these parameters may have insignificant effects on the variability of the structural response. Tornado diagram analysis is a practical method used to identify and rank the effect of parameter uncertainties on the response variability (Lee and Mosalam 2006). Considering that time is of essence to rapidly determine the most collapse-vulnerable buildings so that they can be retrofitted or demolished (if needed) before the next big earthquake, tornado diagram analysis comes forward as a suitable method to eliminate the burden of unnecessary simulations by treating the parameters with insignificant effect as deterministic.

The tornado diagram, commonly used in decision analysis, has been used in sensitivity analysis in earthquake engineering by Porter et al. (2002). The diagram consists of a set of horizontal bars, referred to as swings, one for each random variable, i.e., considered parameter. The length of each swing represents the variation in the output, i.e., EDP, due to the variation in the respective random variable. Thus, a variable with larger effect on the EDP has a larger swing than those with lesser effect. In a tornado diagram, swings are displayed in a descending order from top to bottom. This wide-to-narrow arrangement of swings resembles a tornado. In order to determine the swing due to a considered parameter, two extreme values, e.g., tenth and 90th percentiles, corresponding to predefined lower and upper bounds of the assumed probability distribution for the parameter, are selected. The considered EDP is determined as a result of nonlinear response history analysis using the lower and upper bound values of the considered parameter, while the other input random variables are set to their best estimates such as the medians. This process yields two bounding values of the EDP variation for each input parameter. The absolute difference of these two values is the swing of the EDP corresponding to the selected input parameter. This process is repeated for all the input parameters to compute the swings of the EDP (Fig. 6.10). Finally, one builds the tornado diagram

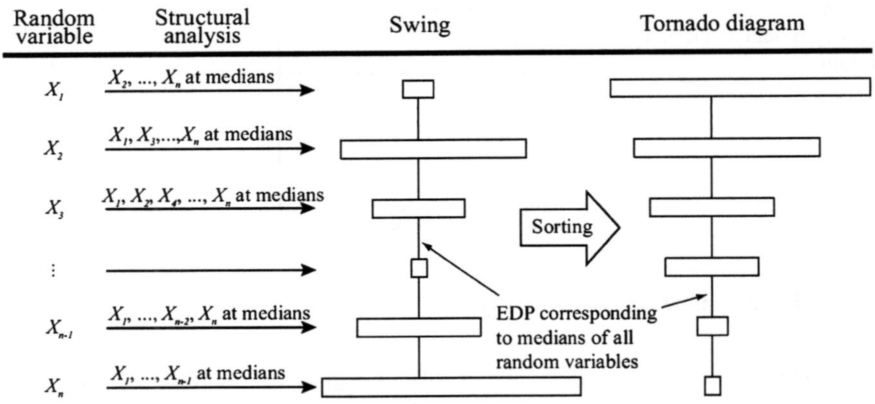

Fig. 6.10 Construction of the tornado diagram (Lee and Mosalam 2006)

by arranging the obtained swings in a descending order as mentioned above. The resulting tornado diagram generally provides an indicative picture for the selection of the necessary random variables to be used in the collapse simulations to develop the fragility curves (Lee and Mosalam 2005).

6.6 Concluding Remarks

Non-ductile RC buildings are one of the main seismic safety concerns worldwide. In order to use the limited monetary resources in an effective manner, these buildings should accurately and reliably be ranked to identify those that are most vulnerable to collapse. This chapter attempted to provide a contribution to the accurate determination of the most collapse-vulnerable non-ductile buildings by discussing the methods from existing literature and exploring the research needs related to gravity load failure modeling and consideration of uncertainty in an efficient manner. These topics are presented to facilitate the accurate and efficient application of refined seismic assessment. Concluding remarks are as follows:

- Non-ductile RC buildings mostly collapse by losing gravity load-carrying capacity much before reaching the lateral displacements that would be experienced in a side-sway collapse mechanism.
- Implicit (non-simulated) modeling of gravity load failure is only adequate in the case where the first element failure determination is sufficient to define global collapse. In all other cases, explicit modeling should be utilized for accurate prediction of global collapse.
- There are various advantages and disadvantages of the methods that can be used for post-failure modeling. Furthermore, the most suitable option may change from one building to another. Advantages and disadvantages presented in this

chapter are expected to provide guidance to the readers for choosing the best starting point and switching between different options.

- There are various methods in literature that can be used for failure detection of the RC elements of a structure. However, there is still a need for enhancement of the failure detection of some of the elements, for example, the non-ductile RC beam-column joints.
- There is an existing element removal algorithm implemented in OpenSees. This algorithm is used in the standard version of OpenSees for the removal of URM infill walls. Removal of columns and beam-column joints are planned to be included in the standard version of OpenSees in the near future.
- Gravity support may not be completely lost after the lateral and axial failures of the primary lateral load-resisting system. Therefore, explicit modeling of the gravity system generally leads to a more accurate and realistic determination of global collapse.
- Non-ductile RC buildings generally involved a significant amount of uncertainties. The presented tornado diagram analysis can be used as a method to handle uncertainty in an efficient and practical manner.

References

Elwood, K. J., Matamoros, A., Wallace, J. W., Lehman, D. E., Heintz, J. A., Mitchell, A., et al. (2007). Update of ASCE/SEI 41 concrete provisions. *Earthquake Spectra, 23*(3), 493–523.

Elwood, K. J., & Moehle, J. P. (2005). Axial capacity model for shear-damaged columns. *ACI Structural Journal, 102*, 578–587.

Grierson, D. E., Xu, L., & Liu, Y. (2005). Progressive-failure analysis of buildings subjected to abnormal loading. *Computer-Aided Civil and Infrastructure Engineering, 20*(3), 155–171.

Hassan, W. M. (2011). *Analytical and experimental assessment of seismic vulnerability of beam-column joints without transverse reinforcement in concrete buildings*. PhD dissertation, University of California, Berkeley, CA.

Holmes, W. T. (2000). *Risk assessment and retrofit of existing buildings*. 12th World Conference on Earthquake Engineering, Auckland, New Zealand. Paper No: 2826.

Hueste, M. B. D., & Wight, J. K. (1999). Nonlinear punching shear failure model for interior slab-column connections. *Journal of Structural Engineering, 125*(9), 997–1007.

Hughes, T. J. R., Pister, K. S., & Taylor, R. L. (1979). Implicit-explicit finite elements in nonlinear transient analysis. *Computer Methods in Applied Mechanics and Engineering, 17*(18), 159–182.

Kadysiewski, S., & Mosalam, K. M. (2008). *Modeling of unreinforced masonry infill walls considering in-plane and out-of-plane interaction* (PEER Report 2008/102, 144 pp.). Berkeley, CA.

Kang, T. H.-K., Wallace, J. W., & Elwood, K. J. (2009). Nonlinear modeling of flat plate systems. *ASCE Journal of Structural Engineering, 135*(2), 147–158.

Lai, J. W., & Mahin, S. A. (2013). *Experimental and analytical studies on the seismic behavior of conventional and hybrid braced frames* (PEER Report 2013/20, 633 pp.). Berkeley, CA.

Lee, T.-H., & Mosalam, K. M. (2005). Seismic demand sensitivity of reinforced concrete shear-wall building using FOSM method. *Earthquake Engineering and Structural Dynamics, 34*(14), 1719–1736.

Lee, T.-H., & Mosalam, K. M. (2006). *Probabilistic seismic evaluation of reinforced concrete structural components and systems* (PEER Report 2006/04, 181 pp.). Berkeley, CA.

Liel, A., Haselton, C., Deierlein, G., & Baker, J. (2009). Incorporating modeling uncertainties in the assessment of seismic collapse risk of buildings. *Structural Safety, 31*(2), 197–211.

Mosalam, K. M., & Günay, M. S. (2012). Behavior and modeling of reinforced concrete frames with unreinforced masonry infill walls. In S. K. Kunnath (Ed.), *Structural engineering and geomechanics*. Oxford, UK: Encyclopedia of Life Support Systems (EOLSS), Developed under the auspices of the UNESCO, EOLSS Publishers. Retrieved February 19, 2013, from http://www.eolss.net

Mosalam, K. M., & Günay, M. S. (2015). Progressive collapse analysis of RC frames with URM infill walls considering in-plane/out-of-plane interaction. *Earthquake Spectra. 31*(2), 921–943.

Mosalam, K. M., Günay, M. S., & Sweat, H. D. (2013). *Simulation of reinforced concrete frames with unreinforced masonry infill walls with emphasis on critical modelling aspects.* 12th Canadian Masonry Symposium, June 2–5, Vancouver.

Mosalam, K. M., Liang, X., Günay, S., & Schellenberg, A. (2013). Alternative integrators and parallel computing for efficient nonlinear response history analyses. In M. Papadrakakis, V. Papadopoulos, & V. Plevris (Eds.), *COMPDYN 2013, 4th ECCOMAS Thematic Conference on Computational Methods in Structural Dynamics and Earthquake Engineering, June 12–14, Kos Island, Greece.*

Mosalam, K. M., Park, S., & Günay, M. S. (2009). Evaluation of an element removal algorithm for shear critical reinforced concrete frames. *COMPDYN, Thematic Conference on Computational Methods in Structural Dynamics and Earthquake Engineering, June 22–24, Greece.*

National Institute of Standards and Technology (NIST). (2010). *Program plan for the development of collapse assessment and mitigation strategies for existing reinforced concrete buildings* (Report No. NIST GCR 10-917-7, NEHRP).

OpenSees Wiki. (2011). http://opensees.berkeley.edu/wiki/index.php/Infill_Wall_Model_and_Element_Removal.

Park, S., & Mosalam, K. M. (2009). *Shear strength models of exterior beam-column joints without transverse reinforcement* (PEER Report 2009/106, 101 pp.). Berkeley, CA.

Park, S., & Mosalam, K. M. (2012a). Parameters for shear strength prediction of exterior beam-column joints without transverse reinforcement. *Engineering Structures, 36*(3), 198–209.

Park, S., & Mosalam, K. M. (2012b). Analytical model for predicting the shear strength of unreinforced exterior beam-column joints. *ACI Structural Journal, 102*(2), 149–160.

Park, S., & Mosalam, K. M. (2012c). *Experimental and analytical studies on reinforced concrete buildings with seismically vulnerable beam-column joints* (PEER Report 2012/03, 224 pp.). Berkeley, CA.

Park, S., & Mosalam, K. M. (2013a). Simulation of reinforced concrete frames with non-ductile beam-column joints. *Earthquake Spectra, 29*(1), 233–257.

Park, S., & Mosalam, K. M. (2013b). Experimental investigation of non-ductile reinforced concrete corner beam-column joints with floor slabs. *ASCE Journal of Structural Engineering, 139*(1), 1–14.

Porter, K. A., Beck, J. L., & Shaikhutdinov, R. V. (2002). Sensitivity of building loss estimates to major uncertain variables. *Earthquake Spectra, 18*(4), 719–743.

Talaat, M., & Mosalam, K. M. (2007). *Computational modeling of progressive collapse in reinforced concrete frame structures* (PEER Report 2007/10, 310 pp.). Berkeley, CA.

Talaat, M., & Mosalam, K. M. (2009). Modeling progressive collapse in reinforced concrete buildings using direct element removal. *Earthquake Engineering and Structural Dynamics, 38*(5), 609–634.

Wallace, J. W., Elwood, K. J., & Massone, L. M. (2008). An axial load capacity model for shear critical RC wall piers. *Journal of Structural Engineering, 134*(9), 1548–1557.

Chapter 7
Probabilistic Seismic Assessment of Existing Buildings: The CNR-DT212 Italian Provisions

Paolo Emilio Pinto and Paolo Franchin

Abstract This chapter briefly illustrates a document recently issued by the Italian National Research Council (CNR), dealing with probabilistic seismic performance assessment of existing buildings. The document, which is aligned with the present international state of the art, is intended to serve the double purpose of providing firmer theoretical bases for the revision of the current European norms on seismic assessment of existing structures and to be practically applicable for cases worth of more rigorous analysis. After a concise overview, this chapter focuses on the specific aspects of limit state quantitative evaluation and of response and capacity modeling.

7.1 Introduction

Seismic assessment of an existing building, designed and built in the absence or with inadequate consideration of the seismic threat, is recognised as a much more challenging problem than designing a new building according to any modern seismic code. The difficulties inherent in the former task are reflected in the current international absence of normative documents possessing the necessary degree of rigour and accuracy.

In the USA, a large multi-annual project funded by FEMA and carried out by the Applied Technology Council has led to the release in 2012 of a comprehensive document: the FEMA-P58 (ATC 2012). Its ambition goes as far as to enabling a probabilistic evaluation of the performances of a structure in terms of the mean annual rate of the occurrence of collapse, of the exceedance of a threshold in terms of direct repair/replacement cost as well as of the indirect cost due to interruption of use. An introductory notice is included in the document, however, stating that "data and [. . .] procedures are not necessarily appropriate for use in actual projects at this time, and should not be used for that purpose. The information contained [. . .] will be subject to further revision and enhancement as the methodology is completed".

P.E. Pinto (✉) • P. Franchin
Department of Structural and Geotechnical Engineering, University of Rome La Sapienza, Via Antonio Gramsci 53, 00197 Rome, Italy
e-mail: paolo.pinto@uniroma1.it; paolo.franchin@uniroma1.it

P. Gardoni, J.M. LaFave (eds.), *Multi-hazard Approaches to Civil Infrastructure Engineering*, DOI 10.1007/978-3-319-29713-2_7

In Europe, the issue of the seismic assessment of buildings is covered in the Eurocode 8 Part 3 (EC8-3): Assessment and Retrofit of Buildings, issued in 2005. It is based on the customary limit state (LS) approach, with the LSs referring to the state of both structural and nonstructural components. Reliability aspects are limited to the choice of the mean return period of the seismic action to be associated with each of LSs and to a global reduction factor to be applied to the material strengths. This factor is a function of the so-called knowledge level acquired on all the geometrical/structural aspects of the building. Only the general methodology of EC8-3 is mandatory, the expressions giving the capacities of the elements to the different LSs having only the status of informative material. This document has been extensively used in European seismic-prone countries since its release, and applications have exposed its insufficient "resolving power", meaning that, depending on the choices that are left to the user, results that are quite distant apart can be obtained. The sources of this observed dispersion of the results have been identified, and efforts are currently under way to arrive at an improved version of the document.

To help in the above direction, the Italian National Research Council (CNR) has taken the initiative of producing a higher-level, fully probabilistic document, DT212 (CNR 2014), for the seismic assessment of buildings, intended to serve the double purpose of providing firmer theoretical bases for the revision of EC8-3, and of being of direct practical applicability for cases worth of a more rigorous analysis. The DT212 has been presented in some detail in Pinto and Franchin (2014). This paper provides a brief overview before focusing on the characterising aspects of the LS quantification and modelling.

7.2 Overview of DT212 Provisions

The DT212 provides the conceptual and operational tools to evaluate the seismic performance of a building in terms of λ_{LS}, the mean annual frequency (MAF) of exceeding an LS of interest (three are defined; see later). The DT212 adopts what is called nowadays the "IM-based approach", which employs the total probability theorem to express λ_{LS} as the integral of the product of the probability of exceedance of the LS conditional to the value $S = s$ of the seismic intensity (denominated as "fragility"), times the probability of the intensity being in the neighbourhood of s. This latter is given by the absolute value of the differential of the hazard function at $S = s$:

$$\lambda_{LS} = \int_0^\infty p_{LS}(s) \left| d\lambda_S(S) \right| \tag{7.1}$$

This approach dates back at least to the early 1980s, e.g. Veneziano et al. (1983), and has been the subject of intensive research since then, the central issue being the optimal related choices of the intensity measure (IM) S and of the ground motions required to compute p_{LS}. Acceleration in the research and dissemination efforts started in the second half of the 1990s when the PEER framework was formulated (Cornell 1996; Cornell and Krawinkler 2000). It can be said that the approach has now attained maturity for practical application, with the main issue identified above being solved through effective ground motion selection procedures (Bradley 2012; Lin et al. 2013).

The evaluation of the MAF in Eq. (7.1) can be carried out in closed form if the hazard curve is fit with a linear or quadratic function in the log-log plane, e.g. Vamvatsikos (2013), and the fragility function is assumed to have a lognormal (LN) shape. Hazard is obtained by standard PSHA: in Italy, one can take advantage of the availability of response spectra for nine different return periods (both median and 16–84 % fractiles), from which the hazard λ_S can be retrieved (Pinto and Franchin 2014). Thus, the task of engineers consists of the selection of ground motions and performance of nonlinear response analyses to evaluate the fragility:

$$p_{LS}(s) = p\left(Y_{LS} \geq 1 \,|S = s\right) = p\left(S_{Y_{LS}=1} \leq S\right) = \Phi\left(\frac{\ln\ S - \mu_{\ln S_{Y=1}}}{\sigma_{\ln S_{Y=1}}}\right) \tag{7.2}$$

Equation (7.2) states that the fragility can be regarded equivalently as the probability that a global LS indicator function Y_{LS} (see later) reaches or exceeds the unit value, given $S = s$, or that the intensity leading to $Y_{LS} = 1$ is lower than the current intensity s. The two parameters of the LN fragility function are evaluated in different ways depending on the strategy employed to perform numerical response analyses. The alternatives are the multiple stripe analysis (MSA) (Jalayer and Cornell 2009) and the incremental dynamic analysis (IDA) (Vamvatsikos and Cornell 2002). In the former case, maximum likelihood estimation can be used to fit the LN fragility to the collected s-p_{LS} pairs. In the latter, which is the main approach put forward in DT212, $\mu_{\ln S_{y=1}}$ and $\sigma_{\ln S_{y=1}}$ are simply obtained from the sample of $S_{Y=1}$ values collected from IDA.

DT212 provides three methods to produce IDA curves, all requiring a 3D model of the structure: (a) full dynamic, (b) static-dynamic hybrid and (c) static. In all three methods, the engineer must select a suite of recorded ground motion records, each with two orthogonal components, with an indicated minimum of 30 time series. Motions can be selected based on PSHA disaggregation results (M, R, ε), but more advanced methods are also considered (Bradley 2012; Lin et al. 2013). The time series are used in inelastic response-history analysis (IRHA) with the full dynamic (complete 3D model) and the hybrid method. In the latter, IRHAs are carried out on equivalent SDOF oscillators obtained from nonlinear static analysis, e.g. with modal patterns, as in Han and Chopra (2006). The two orthogonal components are applied simultaneously in both the full and the hybrid method. In the latter case, the SDOF is subjected to a weighted excitation (Pinto and Franchin 2014). The difference in the static method consists in the fact that the response of the equivalent SDOFs is obtained with the response spectra (median and fractiles) of the selected ground motion records, rather than with a code or other uniform-hazard spectrum.

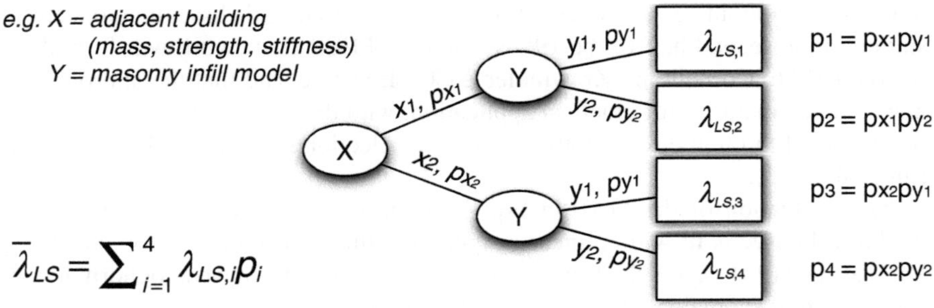

Fig. 7.1 Logic tree on two factors X and Y: x_i and y_i are the considered alternatives and p_{xi} and p_{yi} the associated probabilities

Given that a fully exhaustive (i.e. deterministic) knowledge of an existing building in terms of geometry, detailing and properties of the materials is realistically impossible to achieve, it is required that every type of incomplete information be explicitly recognised and quantified, for introduction in the assessment process in the form of additional random variables or of alternative assumptions. Uncertainties on structure and site are lumped in two classes:

1. Those describing variations of parameters within a single model, described in terms of random variables, with their associated distribution function (mainly material properties, such as concrete and steel strength, or internal friction angle and shear wave velocity, and model error terms). These are usually modelled as LN variables and are approximately accounted for in the evaluation of λ_{LS} by associating one-to-one their samples to the selected motions, with the exception of method c, where response surface is used instead (Pinto and Franchin 2014).
2. Those whose description requires consideration of multiple models.

Type 2 uncertainties include, among others, factors related to the geometry of the structure in areas of difficult inspection, the reinforcement details in important places, the alternative models for the capacity of the elements or the behaviour of the components. These uncertainties are treated with the logic tree technique, where mass probabilities are assigned to the alternative assumptions for each of uncertain factors (Fig. 7.1) and the MAFs obtained with any particular sequence of assumptions (a tree branch) are unconditioned with the branch probabilities, which are simple products of the probabilities in the sequence, due to the assumed independence of the factors (X and Y in the figure).

The search for a balance between the cost for additional information and the potential saving in the intervention is the guiding criterion in the knowledge acquisition process according to DT212. Thus, quantitative minima for the number of elements to be inspected or material samples to be taken are not prescribed. Rather, a sensitivity analysis is required on a preliminary model of the building (a first approximation of the final one). For RC structures, this analysis is of the linear dynamic type (modal with full elastic response spectrum), while for masonry structures, it is nonlinear static with nominal parameter values. In both cases, the results are adequate to expose global modes of response (regular or less regular) and to provide estimates of the member demands, providing guidance on where to concentrate tests and inspections. Further details can be found in Pinto and Franchin (2014).

7.3 Highlights

7.3.1 Limit State Quantification Rules

7.3.1.1 Definition of Limit States

Limit states are defined with reference to the performance of the building in its entirety including, in addition to the structural part, also nonstructural ones like partitions, electrical and hydraulic systems, etc. The following three LSs are considered in DT212:

- Damage limit state (SLD): negligible damages (no repair necessary) to the structural parts and light, economically repairable damages to the nonstructural ones.
- Severe damage limit state (SLS): the loss of use of nonstructural systems and a residual capacity to resist horizontal actions. State of damage is uneconomic to repair.
- Collapse limit state (SLC): the building is still standing but would not survive an aftershock.

Verification of the first and second LS, as well as the corresponding thresholds, is left to the choice of the stakeholder, since they relate to functionality and economic value of damage. On the other hand, verification of the collapse LS, related to the safety of life and content, is mandatory, and the minimum safety level is prescribed by DT212. Adoption of Eq. (7.1) as a measure of seismic performance required the establishment of the safety level in terms of acceptable values for the collapse MAF. Values, depending on building importance class, have been set within DT212 in continuity with the implied safety level in the current Italian code (aligned with the Eurocodes).

As already noted in the Introduction, the above verbal definitions are qualitative and refer to the global performance of the building. The need thus arises for effective quantitative measures of performance that are truly consistent with these LSs' definitions. This has led to the formulation of three scalar LS indicators Y_{LS} described in the following sections, Sects. 7.3.1.2–7.3.1.4.

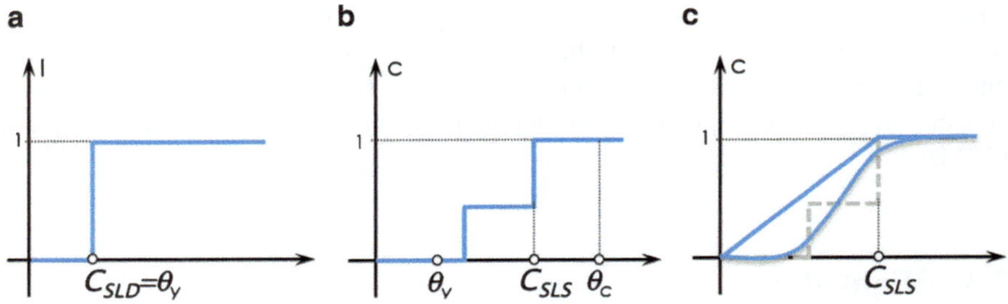

Fig. 7.2 Indicator function (**a**), realistic conventional cost function (**b**) and alternative approximate conventional cost functions (**c**)

7.3.1.2 Light Damage

For the purpose of the identification of the light damage LS, the building is considered as composed by N_{st} structural members and N_{nst} nonstructural components:

$$Y_{SLD} = \frac{1}{\tau_{SLD}} \max\left[\sum_{i=1}^{N_{st}} w_i I\left(\frac{D_i}{C_{i,SLD}}\right);\quad \sum_{j=1}^{N_{nst}} w_j I\left(\frac{D_j}{C_{j,SLD}}\right)\right] \tag{7.3}$$

In the above expression, D and C indicate the appropriate demand and capacity values; demand can be, for example, interstorey drift for partitions and piping, or chord rotation for beams and columns, or floor spectral acceleration for heavy pieces of equipment; the associated capacity must correspond to a light damage threshold, e.g. the yield chord rotation θ_y for structural members, or a drift close to 0.3 % for common hollow-brick partitions (the thresholds associated with nonstructural components must be established based on the specific technology adopted); I is an indicator function taking the value of one when $D \geq C$ and zero otherwise (see Fig. 7.2a), and the w's are weights summing up to one, accounting for the importance of different members/components. Typically, since structural damage must be low, not requiring repair action, while light repair is tolerated for nonstructural components, the weights w_i are simply set to $1/N_{st}$, while the weights w_j may be set proportional to the ratio of the component extension (e.g. for partitions) to the total extension of like components.

The indicator Y attains unity when the max function equals τ_{SLD}, a user-defined tolerable maximum cumulated damage (e.g. something in the range 3–5 %).

7.3.1.3 Severe Damage

For the purpose of the identification of the severe damage LS, the indicator Y is formulated in terms of a conventional total cost of damage to structural and nonstructural elements as

$$Y_{\mathrm{SLD}} = \begin{cases} \frac{1}{\tau_{\mathrm{SLD}}}\left[\alpha_{\mathrm{st}}\sum_{i=1}^{N_{\mathrm{st}}} w_i c\left(\frac{D_i}{C_{i,\mathrm{SLD}}}\right) + (1-\alpha_{\mathrm{st}})\sum_{j=1}^{N_{\mathrm{nst}}} w_j c\left(\frac{D_j}{C_{j,\mathrm{SLD}}}\right)\right] & \text{If } Y_{\mathrm{SLC}} < 1 \\ 1 & \text{If } Y_{\mathrm{SLC}} \geq 1 \end{cases} \tag{7.4}$$

where α_{st} is the economic "weight" of the structural part (i.e. about 20 % in a low- to mid-rise residential building) and $c(D/C)$ is a conventional member/component cost function. The choice of employing a conventional cost rather than attempting the evaluation of an actual economic value of damage, as done, for example, in ATC 2012, is regarded as a convenient compromise in view of the difficulty of establishing reliable cost estimates even at the component level. The reasons for this difficulty are that the repair cost for a component depends on many factors beyond its own damage state, such as the number and location of the damaged components, as well as the modified post-earthquake market conditions. Indeed, demand and supply for repair may be such as to increase the actual cost, in case the number of damaged buildings exceeds the capacity of construction firms.

A conventional cost function can be established in a realistic manner, such as that shown in Fig. 7.2b, which varies stepwise to reflect the fact that the same repair action is needed for an interval of damage, irrespective of the actual damage within the interval. Alternatively, simpler approximate cost functions can be adopted (Fig. 7.2c), as simple as a linear one that starts from zero for $D = 0$ and reaches unity, i.e. the replacement cost for the element, for $D = C_{SLS}$ (with C_{SLS} usually a fraction of the ultimate capacity of the element).

As for the light damage LS, the indicator function attains unity when the quantity within square brackets equals τ_{SLS}, a user-defined fraction of the total building value over which repair is considered economically not competitive with demolition and replacement. Obviously, if collapse occurs, Y_{SLS} is set to 1. Finally, the same considerations on the weights w_i and w_j already made with reference to light damage apply here to severe damage.

7.3.1.4 Collapse

Modelling choices, which determine the numerical response, influence the quantification of the collapse LS (see, e.g. Goulet et al. 2007; Baradaran Shoraka et al. 2013). DT212 prescribes exclusive recourse to nonlinear methods of analysis, accounting for material and geometric nonlinear phenomena. Models of the inelastic response of structural members under cyclic loading of increasing amplitude can be distinguished in two classes, as shown in Fig. 7.3:

- Non-degrading, i.e. stable hysteretic behaviour without degradation of strength but overall degradation of stiffness (Takeda-type models), Fig. 7.3a
- Degrading, where both stiffness and strength degrade with increasing cyclic amplitude down to negligible values, Fig. 7.3b

In the more common case, when non-degrading inelastic response models are adopted, according to DT212, the structural system is described as a serial arrangement of a number of elements in parallel, according to the so-called cut-set formulation, so that the Y variable takes the expression (Jalayer et al. 2007)

$$Y_{\mathrm{SLC}} = \max_{i=1,N_s} \min_{j \in I_i} \frac{D_j}{C_{j,\mathrm{SLC}}} \tag{7.5}$$

where N_S is the number of parallel subsystems (cut-sets) in series and I_i is the sets of indices identifying the members in the ith subsystem. This formulation requires the a priori identification of all the cut-sets, i.e. sets of members whose joint failure induces system failure, in order to find the critical one with certainty. Carrying out this task is in general not immediate and actually quite onerous, even for static problems, the more so in dynamics, since the critical cut-set depends on the dynamic response and changes from record to record, with failure modes that can involve one storey (i.e. weak or soft storey, commonly for existing nonconforming buildings) or multiple adjacent storeys (e.g. Goulet et al. 2007).

This said, the widespread use of the peak interstorey drift ratio $\theta_{\max}$, as an indicator of global structural damage, can be regarded as an approximate application of the formulation in Eq. (7.5), with the storeys being the subsystems in series. This baseline choice is appropriate to detect even multistorey failure modes, since one storey will always necessarily be more strained than the others involved in the failure mode. However, since the non-degrading model does not account for all member failure modes, premature shear failures would not be detected by just monitoring $\theta_{\max}$. Thus, the demand-to-capacity ratios in shear, at least for all columns, must be included in the evaluation of Eq. (7.5) by post-processing the response.

This demand-to-capacity ratio can be formulated either in terms of deformation or force. The latter choice is more common (member acting shear V over ductility-reduced shear strength V_R), since, as discussed later, most available deformation capacity models are established based on data sets that include a smaller proportion of shear failure modes, especially of the brittle type (Fig. 7.4).

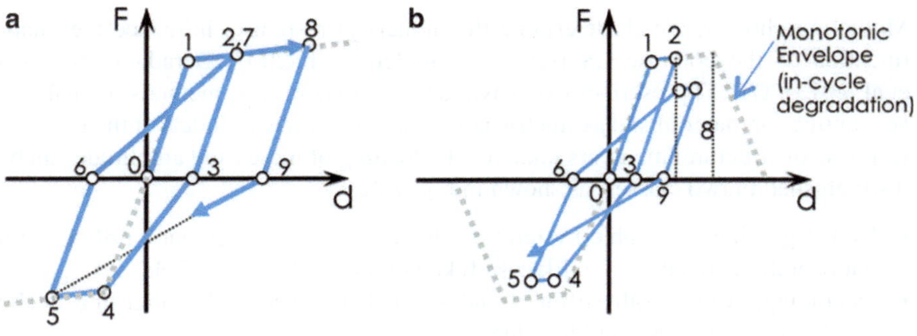

Fig. 7.3 Inelastic response models: non-degrading (**a**) vs degrading (**b**)

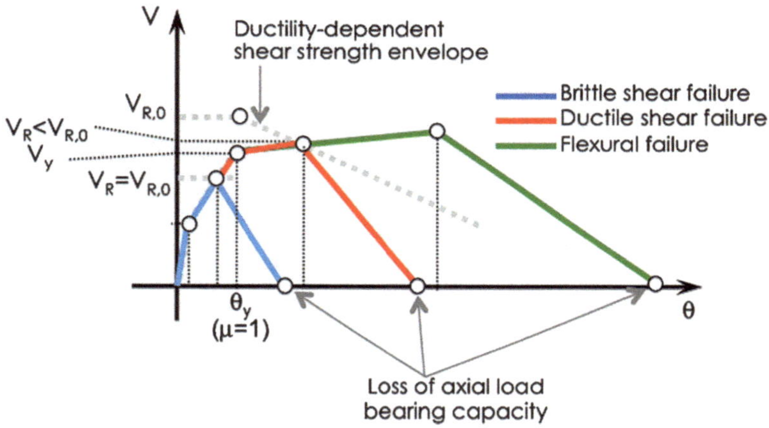

Fig. 7.4 The three failure modes of an RC member

At the other end of the modelling spectrum lies the category of ideal "fully degrading" response models, able to simulate all types of failure; accounting, for instance, for the interaction of bending and shear; etc. With these models, the collapse state $Y = 1$ is identified with the occurrence of the so-called dynamic instability, that is, when the curve intensity-response becomes almost flat. Operatively, in order to identify the point on the curve corresponding to $Y = 1$, one can use the expression

$$Y_{\mathrm{SLC}} = (1 + \Delta) - \frac{S'}{S'_0} \quad \text{with} \quad 0 < S' < S'_0 \tag{7.6}$$

with values for Δ in the interval 0.05–0.10, corresponding to a small residual positive stiffness, in order to avoid numerical problems.

Finally, if the response models are of the degrading type but their formulation cannot account for all possible failure modes, the indicator variable can be expressed as

$$Y_{\mathrm{SLC}} = \max\left[(1 + \Delta) - \frac{S'}{S'_0}; \quad \max_{\mathrm{nsm}}\left(\frac{D}{C}\right)\right] \tag{7.7}$$

which simply indicates that the collapse condition is attained for the most unfavourable between dynamic instability and the series of the "non-simulated (collapse) modes". Typically, this set includes the axial failure of columns. Care should be taken in selecting the columns to be included in the evaluation of Eq. (7.7), limiting it only to those that can really be associated with a partial/global collapse. It can be observed that axial failure of columns and the ensuing loss of vertical load-bearing capacity, associated with concomitant loss of shear capacity,

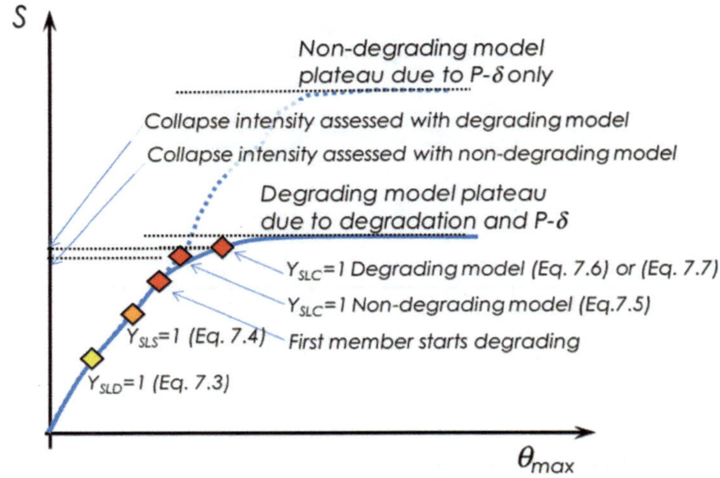

Fig. 7.5 IDA curves as a function of modelling choices

are not included in Eq. (7.5) since this failure occurs at larger deformation than that corresponding to the peak shear strength (see Fig. 7.4).

Figure 7.5 shows an idealised intensity-response relation S vs $\theta_{\max}$, with marks on the points corresponding to the attainment of the LSs according to the above definitions. The solid line corresponds to the response of a degrading model while the dashed one to the response of a non-degrading one. The two coincide until the first member in the structural model starts degrading. From this point on, the non-degrading response cannot be relied upon. Collapse is thus identified by post-processing the response to evaluate Eq. (7.5). The point where this equation yields the value of one is below the corresponding point given by Eq. (7.6) or (7.7), for the reasons discussed above. In other words, the failure criterion adopted with non-degrading response models is characterised by more conservatism.

7.3.2 Modelling Response and Capacity

As mentioned before, DT212 classifies inelastic response models into degrading and non-degrading ones. In general, they all require specification of the monotonic backbone and of a hysteretic rule. Degrading ones, which provide a closer description of reality, intend to follow the cyclic member response up to collapse due to complete exhaustion of its resistance, either in terms of deformation capacity (for monotonic response) or of energy dissipation capacity (for cyclic response). This class of models is schematically represented in Fig. 7.6, which highlights the characterising difference with non-degrading models, i.e. the negative stiffness

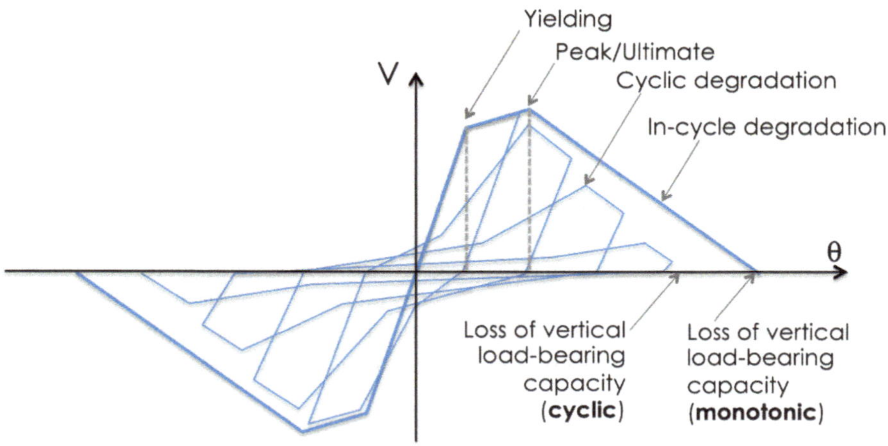

Fig. 7.6 Cyclic and in-cycle components of degradation (response shown is from Ibarra et al. model)

Fig. 7.7 Deformation limits for monotonic and cyclic loading

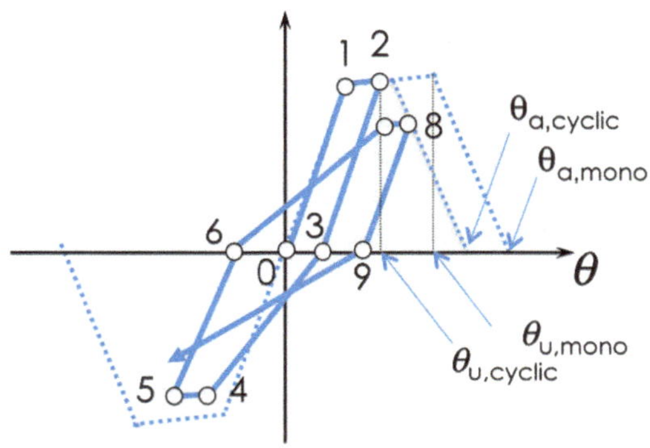

branch in the backbone and cyclic degradation in strength encoded in the hysteretic rule.

Non-degrading response models are a lower-level representation of member behaviour and thus provide a poorer approximation of the actual response (e.g. the difference in the intensities corresponding to global dynamic instability—plateau—in Fig. 7.5). Their use poses the problem of how to reintroduce the neglected degradation in the assessment of the member. "Capacity models", which are empirically derived deformation or force thresholds corresponding to state transitions of the member, are conventionally used for this purpose. These are conceptually the same thresholds that mark transition in a degrading response model, e.g. the yield, peak or axial failure chord rotations in Fig. 7.6, but intend to provide the lower cyclic values for these thresholds (Fig. 7.7). In conclusion, if non-degrading models are chosen, one should use Eq. (7.5) for collapse identification, with peak deformation thresholds $\theta_{u,cyclic}$ that account on the capacity side for the degradation disregarded on the response side. The latter is presently the default route in most cases, since degrading models are still not in the average technical background of engineers and, also, they are still evolving towards a more mature and consolidated state. If degrading models are used, Eq. (7.6) or (7.7) is employed, and the monotonic deformation thresholds, $\theta_{u,mono}$, $\theta_{a,mono}$, etc., are used as input parameters for the response model (together with degradation parameters). In the following, the term capacity model is used to denote in general analytical formulas predicting deformation or force values marking state transitions in the member response, irrespective of whether they are used as thresholds in conjunction with non-degrading response models or as parameters of degrading ones.

7.3.2.1 Response Models

In order to facilitate practical application, DT212 provides a reasoned summary of response models for beam-columns, joints and masonry infills.

In particular, focusing on RC columns, their failure modes have been already schematically shown in Fig. 7.4. The figure illustrates the possible modes of collapse in a monotonic loading condition, in terms of shear force-chord rotation of the member. The plot shows the monotonic response in a pure flexural mode, in green, with the usual I, II and III stages up to ultimate/peak strength, followed by a fourth descending branch to actual collapse, and the shear strength envelope in dashed grey. The latter starts with $V_{R,0}$ and decreases as a function of deformation, measured in terms of ductility μ. Depending on whether the two curves cross before flexural yield, after, or do not cross at all, the member fails in brittle shear, ductile shear or flexure. In all cases, collapse occurs due to loss of vertical load-bearing capacity ($V_R = N_R = 0$) at the end of the degrading branch. In cyclic loading at large amplitude, the response presents a second contribution to degradation, which is cyclic degradation, as shown in Fig. 7.6.

Available models can be classified into mechanical and phenomenological. The state of the art of purely mechanical models is not yet capable of describing the full range of behaviour of RC members illustrated in Figs. 7.4 and 7.6 (especially for brittle and ductile shear collapse). Models of this type are all based on the fibre section discretisation and a beam-column element formulation (stiffness, flexibility or mixed field based). The main advantage of the fibre section model is that biaxial flexure and axial force interaction are correctly described within the range of response where the model is applicable. Problems arise when the member is shear sensitive, since fibre models rely almost exclusively on the plane section assumption, and the behaviour of an RC member failing in brittle or ductile shear is not beam-like. Approximate solutions to overcome this difficulty have been proposed in the last two decades, a relatively recent summary of which can be found in Ceresa et al. (2007). Even if the member is not shear sensitive, the plane section assumption is compromised at the largest response levels, closer to flexural collapse, due to bar buckling and slippage, as well as concrete expulsion from the core. Solutions to model these degradation phenomena within the context of fibre models are available (e.g. among many others Monti and Nuti 1992; Gomes and Appleton 1997; Spacone and Limkatanyu 2000), but their increased complexity is usually paid in terms of computational robustness.

Regarding the use of degrading models, currently, the only viable option is to use phenomenological (e.g. Ibarra et al. 2005) or hybrid models (Elwood 2004; Marini and Spacone 2006). These models, however, also have their limitations and, for instance, rely heavily on the experimental base used to develop them, which is often not large enough (e.g. for the Ibarra et al. model, the proportion of ductile shear and flexural failures dominate the experimental base, resulting in limited confidence on the model capability to describe brittle failures). Further, computational robustness is an issue also with these models.

7.3.2.2 Capacity Models

In parallel with the survey of response models, DT212 provides detailed information on capacity models. Requirements for an ideal set of models are stated explicitly:

- Consistency of derivation of thresholds of increasing amplitude (i.e. yield, peak and axial deformation models derived based on the same experimental tests, accounting also for correlations)
- Support by an experimental base covering the full range of behaviours (different types of collapse, different reinforcement layouts, etc.) in a balanced manner

Such a set of models is currently not available. One set of predictive equations that comes closer to the above requirements and is used for the parameters of the degrading response model by Ibarra et al. (2005) is that by Haselton et al. (2008). As already anticipated, one problem with this model is that brittle shear failures are not represented in the experimental base. Further, the predictive equations provide only mean and standard deviation of the logarithm of each parameter, disregarding pairwise correlation, in spite of the fact that they were established on the same experimental basis (the authors state explicitly that they did not judge the available experiments enough to support the estimation of a full joint distribution). The latter problem requires caution in the use of the equations to establish the member constitutive laws, avoiding non-physical situations. To illustrate this fact, Fig. 7.8 shows the trilinear moment-rotation monotonic envelope according to the Ibarra model, with (marginal) probability density functions for its parameters, as supplied by Haselton et al. (2008). Not all the parameters can be independently predicted at the same time, to maintain physical consistency of the moment-rotation law. For instance, in applications, the rotation increment $\Delta\theta_f$ and $\Delta\theta_a$ can be used (darker PDFs in the figure) in place of θ_f and θ_a, to ensure that situations with $\theta_f > \theta_a$ cannot occur. Care must be taken also in ensuring that K_y is always larger than $K_{40\%}$ (used as an intermediate value between I and II stage stiffness, since the model is trilinear), by assuming, e.g. that they are perfectly correlated.

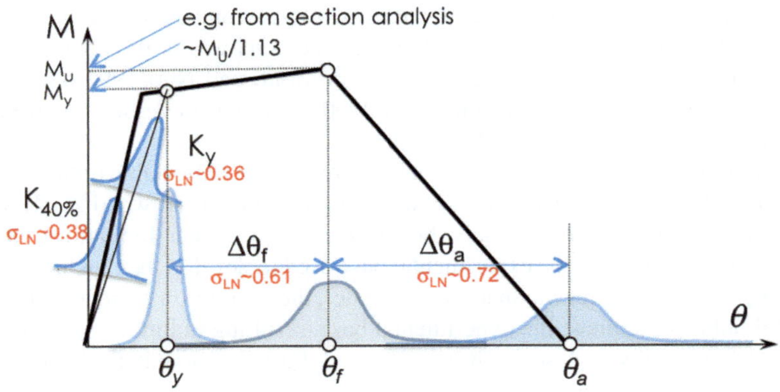

Fig. 7.8 Deformation limits for monotonic loading with schematic indication of the marginal PDF of each parameter

The document provides also equations meant to provide cyclic values of the deformation thresholds, for use in conjunction with non-degrading models, such as those by Biskinis and Fardis (2010a, b), adopted since 2005 in earlier form in Eurocode 8 Part 3 (CNR 2015) and in the latest fib Model Code (fib 2013), and those by Zhu et al. (2007).

7.4 Conclusions

The chapter introduces the latest Italian provisions, issued by the National Research Council as Technical Document 212/2013, for the probabilistic seismic assessment of existing RC and masonry buildings. The characterising traits of the document are:

(a) The systematic treatment of the problem of identification of global LS exceedance, in a manner consistent with the verbal description of the LSs, with the introduction of LS indicator variables differentiated as a function of LS and modelling option.
(b) The explicit probabilistic treatment of all uncertainties, related to ground motion, material properties, modelling, geometry and detailing. In particular, the distinction of uncertainties that can be described within a single structural model via random variables and uncertainties that require the use of multiple models (logic tree) is introduced.
(c) The mandatory use of ground motion time series (preferably recorded) for the description of the seismic motion variability, irrespective of the analysis method (dynamic or static).

It is of course recognised that DT212 does not represent an accomplished final stage of development for such a document, since there are several areas where progress is needed and research is still active. In particular, the most notable research gap is on the inelastic response models and capacity models for collapse. Nonetheless, it is believed that the available analytical tools and models, as well as the theoretical framework, are mature enough for practical application to real buildings, as it is demonstrated by the case studies included in DT212.

Acknowledgement The production of DT212 has been funded by the Department of Civil Protection under the grant "DPC-ReLUIS 2010–2013" to the Italian Network of University Laboratories in Earthquake Engineering (ReLUIS).

References

ATC. (2012). *Seismic performance assessment of buildings* (Report FEMA-P58). Prepared by the Applied Technology Council, Redwood City, CA.

Baradaran Shoraka, M., Yang, T. Y., & Elwood, K. J. (2013). Seismic loss estimation of non-ductile reinforced concrete buildings. *Earthquake Engineering & Structural Dynamics, 42*(2), 297–310.

Biskinis, D., & Fardis, M. N. (2010a). Flexure-controlled ultimate deformations of members with continuous or lap-spliced bars. *Structural Concrete, 11*(2), 93–108.

Biskinis, D., & Fardis, M. N. (2010b). Deformations at flexural yielding of members with continuous or lap-spliced bars. *Structural Concrete, 11*(3), 127–138.

Bradley, B. A. (2012). A ground motion selection algorithm based on the generalized conditional intensity measure approach. *Soil Dynamics and Earthquake Engineering, 40*, 48–61.

CEN. Comité Européen de Normalisation. (2005). Eurocode 8: Design of structures for earthquake resistance - Part 3: Assessment and retrofitting of buildings.

Ceresa, P., Petrini, L., & Pinho, R. (2007). Flexure-shear fiber beam-column elements for modeling frame structures under seismic loading—state of the art. *Journal of Earthquake Engineering, 11*(S1), 46–88.

CNR. (2014). *Istruzioni per la Valutazione Affidabilistica della Sicurezza Sismica di Edifici Esistenti (in Italian)* (Report DT212/2013). Available at: http://www.cnr.it/sitocnr/IlCNR/Attivita/NormazioneeCertificazione/DT212.html.

Cornell, C. A. (1996). Reliability-based earthquake-resistant design: The future. In *Proceedings of the Eleventh World Conference on Earthquake Engineering*. Acapulco, Mexico.

Cornell, C. A., & Krawinkler, H. (2000). Progress and challenges in seismic performance assessment. *PEER Center News, 3*(2), 1–3.

Elwood, K. (2004). Modelling failures in existing reinforced concrete columns. *Canadian Journal of Civil Engineering, 31*, 846–859.

fib, International Federation of Structural Concrete. (2013). *Model Code for Concrete Structures 2010*. Berlin: Ernst & Sohn. ISBN 978-3-433-03061-5.

Gomes, A., & Appleton, J. (1997). Nonlinear cyclic stress-strain relationship of reinforcing bars including buckling. *Engineering Structures, 19*(10), 822–826.

Goulet, C. A., Haselton, C. B., Mitrani-Reiser, J., Beck, J. L., Deierlein, G. G., Porter, K. A., et al. (2007). Evaluation of the seismic performance of a code-conforming reinforced-concrete frame building—from seismic hazard to collapse safety and economic losses. *Earthquake Engineering & Structural Dynamics, 36*(13), 1973–1997.

Han, S., & Chopra, A. (2006). Approximate incremental dynamic analysis using the modal pushover analysis procedure. *Earthquake Engineering & Structural Dynamics, 35*, 1853–1873.

Haselton, C. B., Liel, A. B., Taylor Lange, S., & Deierlein, G. G. (2008). *Beam-column element model calibrated for predicting flexural response leading to global collapse of RC frame buildings* (PEER report 2007/03).

Ibarra, L. F., Medina, R. A., & Krawinkler, H. (2005). Hysteretic models that incorporate strength and stiffness deterioration. *Earthquake Engineering & Structural Dynamics, 34*(12), 1489–1511.

Jalayer, F., & Cornell, C. A. (2009). Alternative non-linear demand estimation methods for probability-based seismic assessments. *Earthquake Engineering & Structural Dynamics, 38*(8), 951–972.

Jalayer, F., Franchin, P., & Pinto, P. E. (2007). A scalar damage measure for seismic reliability analysis of RC frames. *Earthquake Engineering & Structural Dynamics, 36*, 2059–2079.

Lin, T., Haselton, C. B., & Baker, J. W. (2013). Conditional spectrum-based ground motion selection. Part I: Hazard consistency for risk-based assessments. *Earthquake Engineering & Structural Dynamics, 42*(12), 1847–1865.

Marini, A., & Spacone, E. (2006). Analysis of reinforced concrete elements including shear effects. *ACI Structural Journal, 103*(5), 645–655.

Monti, G., & Nuti, C. (1992). Nonlinear cyclic behavior of reinforcing bars including buckling. *Journal of Structural Engineering, 118*(12), 3268–3284.

Pinto, P. E., & Franchin, P. (2014). Existing buildings: The new Italian provisions for probabilistic seismic assessment. In A. Ansal (Ed.), *Perspectives on European earthquake engineering and seismology* (Vol. 34). Geotechnical, Geological and Earthquake Engineering, ISBN 978-3-319-07117-6, doi:10.1007/978-3-319-07118-3

Spacone, E., & Limkatanyu, S. (2000). Responses of reinforced concrete members including bond-slip effects. *ACI Structural Journal, 97*(6).

Vamvatsikos, D., & Cornell, C. A. (2002). Incremental dynamic analysis. *Earthquake Engineering & Structural Dynamics, 31*(3), 491–514.

Vamvatsikos, D. (2013). Derivation of new SAC/FEMA performance evaluation solutions with second-order hazard approximation. *Earthquake Engineering & Structural Dynamics, 42*(8), 1171–1188.

Veneziano, D., Casciati, F., & Faravelli, L. (1983). Method of seismic fragility for complicated systems. In *Proceedings of the 2nd Committee of Safety of Nuclear Installation (CNSI). Specialist Meeting on Probabilistic Methods in Seismic Risk Assessment for Nuclear Power Plants*. Livermore, CA: Lawrence Livermore Laboratory.

Zhu, L., Elwood, K., & Haukaas, T. (2007, September). Classification and seismic safety evaluation of existing reinforced concrete columns. *Journal of Structural Engineering, 133*, 1316–1330.

(CEN) Comité Européen de Normalisation. (2005). Eurocode 8: Design of structures for earthquake resistance – Part 3: Assessment and retrofitting of buildings.

Chapter 8
Multi-Hazard Multi-Objective Optimization of Building Systems with Isolated Floors Under Seismic and Wind Demands

Hussam Mahmoud and Akshat Chulahwat

Abstract Traditionally, structural design standards or retrofit guidelines have been geared toward meeting the demand of individual hazards based on probability of exceedance of a certain event level. Performance requirements of the collective effects of individual hazards, however, acting simultaneously or spatially over time, may significantly increase the potential for substantial damage, collapse, and/or economic and life losses. In addition, performance-based earthquake engineering has recently evolved from its basic concept of defining performance objectives to prevent structural collapse, with acceptable high level of damage, to minimizing structural loss without compromising on performance. One way to achieve this objective is to make traditional seismic force-resisting systems stiffer (which also implies higher strength). However, it is neither effective nor economical to embrace such an approach. Moreover, while stiffening a structure may improve its performance under earthquake loading, the added stiffness may compromise the performance under wind loading. Therefore, it is necessary to create new seismic force-resisting systems that satisfy higher performance goals for multiple hazards and can be easily repaired, with minimal cost, after major events. Motivated by the mentioned objectives, the concept of sliding slab systems is introduced and discussed in this chapter where curved slabs are isolated from their respective bays and utilized to act as tuned mass dampers. The decision on which slab to optimize in order to achieve a superior performance under the multiple hazards of wind and earthquake is arrived at using a nested optimization approach. The optimization strategy and the modifications implemented are discussed in detail. Results of the study highlight the effectiveness of the proposed sliding slab system and the optimization scheme in configuring building systems that can withstand the multiple hazards.

H. Mahmoud (✉) • A. Chulahwat

e-mail: Hussam.Mahmoud@colostate.edu

P. Gardoni, J.M. LaFave (eds.), *Multi-hazard Approaches to Civil Infrastructure Engineering*, DOI 10.1007/978-3-319-29713-2_8

8.1 Introduction

8.1.1 Motivations and Significance

Recent events such as the 2011 Tohoku earthquake in Japan is one such example in which widespread damage to houses properly designed for expected earthquake demand was seen due to additional tsunami loads (Guo et al. 2011). During the same event, the earthquake resulted in a fire at the Fukushima nuclear power plant, which caused meltdown of three of the plant's six nuclear reactors and posed significant national security threat. These are not unprecedented events, several similar scenarios have occurred over the years where structures have been subjected to unexpected multiple hazards. For wind and seismic hazards, which are the focus of this study, recent events have demonstrated the potential for structures to be subjected to moderate or high level of seismic and wind demands during their service life. An example of such includes hurricane Sandy, which hits the east coast of the United States and was classified as the deadliest and most destructive hurricane of the 2012 Atlantic hurricane season. The cost of damage was estimated at $68 billion with at least 286 people killed along the path of the storm in seven countries (Blake et al. 2013). In 2010, prior to the 2012 hurricane, the northeastern region of the United States had experienced a magnitude 5.8 intraplate earthquake with the epicenter located in Louisa County in the state of Virginia. In 2000, the economic losses in the United States averaged approximately $5.4 billion annually from hurricanes and $4.4 billion a year from earthquakes alone. With the ever-increasing population in coastal cities due to migration and rapid economic growth and development, losses from the effect of multiple hazards such as wind and seismic are only expected to grow in the future (Perry and Mackun 2001; Regional Plan Association (RPA) America 2005). Coastal areas in the United States possess a population of approximately 153 million people (over half the country's population), who live in the 673 coastal counties (Crossett et al. 2004).

There is an immediate need for effective and optimized design strategies for structures subjected to multiple possible extreme hazards in their lifetime. In the context of this study, the multiple hazards considered, earthquake and wind, are nonconcurrent and uncorrelated. Duthinh and Simiu (2008, 2014) concluded that the ASCE 7 Standard does not take into account the fact that failure probabilities of structures in regions exposed to both strong wind and strong earthquake hazards may exceed their counterparts in regions exposed to only one of the hazards. Crosti et al. (2011) determined that the areas prone to seismic and wind hazards have a risk of exceedance of design limits up to twice the risk of exceedance in regions with only one dominant hazard. Mardfekri and Gardoni (2015) presented a multi-hazard reliability framework to assess offshore wind turbines under wind and seismic loads. The probabilistic framework incorporated site-specific wind and seismic fragility surfaces which were utilized in 3D finite element analysis to predict the demand under the multiple hazards. Chen (2012) conducted a study on the design of steel frames subjected to hurricane wind and earthquake loading to account for the expected demand of both hazards. The study highlighted the significant difference in design requirements under both loadings where wind-designed buildings are stiffer than those designed for earthquake due to lower lateral load demands. Earthquake-designed structures have much greater strength and ductility requirements due to their capacity for substantial plastic hinge development. Wen and Kang (2001) showed, using life cycle cost estimation, that combined low-consequence high-frequency events could have more significant effect on a life cycle compared to a single high-consequence low-frequency event. The above discussion demonstrates the need for developing an integrated approach for addressing multi-hazard design and assessment of structures. While previous work offers fundamental strategies for multi-hazard risk modeling of structures, achieving resilient systems requires a paradigm shift that would allow for not only multi-hazard optimization of structures but also for easy repairs after extreme events.

8.1.2 Vibration Isolation System: Tuned Mass Dampers

Recent research in seismic design has demonstrated superior performance of systems using replaceable elements such as rocking frames with fuses (Eatherton et al. 2010), frames employing scorpion braces (Gray et al. 2012), base isolation systems (Constantinou et al. 1998), and tuned mass dampers (TMDs) (Moon 2010). TMDs are among the oldest methods of vibration isolation. A typical TMD is a simple passive device that can eliminate undesirable motion of the structure, particularly near resonance (Spencer and Sain 1997). The system generally consists of a mass/inertia, a restoring force mechanism, and an energy dissipation device. The most common type of TMDs is a translational tuned mass damper (TTMD), where the mass is allowed to translate horizontally as the primary structure vibrates. The frequency of translation of the added mass is tuned to the resonance frequency of the structure so when the structure is excited at its fundamental frequency, the added mass resonates out of phase with the structural motion, causing reduction in response of the main structure. The concept of TMD was first developed in 1909 to reduce the rolling motion of ships and mitigate ship hull vibrations (Ormondroyd and Den Hartog 1928). Significant contributions to understanding the theoretical and practical application of TTMD in civil structures have been made since then by various researchers (Warburton and Ayorinde 1980; Randall et al. 1981; Warburton 1982; Tsai and Lin 1993). One major limitation of TMD systems is that they need to be tuned to a specific frequency; thus, they are effective over a small domain of input excitations, as they usually are equipped with a small mass (Chopra 2005; Almazán et al. 2007). The larger the mass used in a TMD, the more effective they are in reducing the vibrations (Li and Zhu 2006; Moon 2010). However, due to design limitations, the size of tuned mass is typically restricted since extra weight requires larger structural elements to be employed, making the system impractical. In addition, space limitations hinder the ability to use larger masses as TMDs, and as a result TMDs typically occupy the roof or top floors of building systems. For example, the TMD employed in Taipei 101 in Taipei, Taiwan, is suspended from the 92nd to the 87th floor. Although their effectiveness has been recognized, the limitation on the mass size or the requirement of large space if large mass is to be used has shown to hinder their applicability in certain cases (Mohebbi et al. 2012).

In this chapter, a multi-hazard optimization framework is discussed which is tested on a previously formulated floor isolation system, referred to as "sliding floor" system. The floor isolation system is optimized under individual hazards initially and then under a multi-objective multi-hazard optimization framework. Finally, the performance of the optimized systems is compared to each other to draw a contrast on the importance of multi-hazard design of structures as opposed to individual hazard design.

8.2 Sliding Floor Isolation System

8.2.1 System Description

In Engle et al. (2015), a new sliding floor isolation system was proposed to overcome the needed limitations of TMD systems, and its performance was evaluated against traditional composite systems under seismic loading. The proposed system comprises of isolated floor slabs that are free to move relative to the frame. The slabs are placed on curved supports, as it allows gravity to reposition the slab back to its original location and thus acts as a self-centering mechanism. To keep the motion of slabs in check, a stiffness element is attached at the ends of the slab, which could be rubber bumpers or any other suitable mechanism. The energy dissipation mechanism is the friction between the slab and the curved support. An elevation view of a single-story system is shown in Fig. 8.1. The logic behind the principle of the sliding system is twofold. First, the sliding system can be considered a combination of two vibration isolation systems—TMDs and base isolation systems. From a global perspective, the slabs tend to act as TMDs for the frame, and from a local perspective, the slabs are isolated from the frame by the concept of

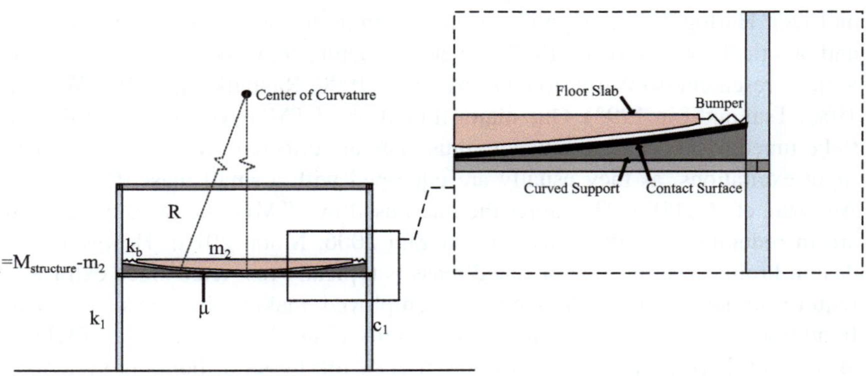

Fig. 8.1 Elevation view of proposed single-story system (Engle et al. 2015)

base isolation. The mass of the entire structure, the frame alone, and the slab are designated as $M_{\text{structure}}$, m_1, and m_2, respectively. The stiffness of the frame and the sliding system are designated as k_1 and k_b, respectively. The internal damping of the frame is designated as c_1, the friction coefficient of the contact surface between the slab and the support is designated as μ, and the radius of curvature of the contact surface is represented as R. In this study, the analysis of the sliding floor system is not restricted to only seismic but extended to wind loading as well to optimize the performance for multi-hazard scenarios.

8.2.2 System of Equations

The sliding system can be employed in a multi-span multi-floor building structure as shown in Fig. 8.2. The idealization of the structure as a spring-mass system is shown in Fig. 8.3. The generalized equations of motion derived for an N-story sliding system structure are shown below in Eq. (8.1). The first N rows of the matrix equation represent the response of the steel frame, and the latter N rows represent the motion of the slabs on each floor. All slabs on each individual floor behave in a similar manner; thus, only one variable is required to describe slabs of each floor. The mass and damping matrix are similar to a typical composite frame structure; however, the stiffness matrix differs in that it includes the stiffness elements of the sliding system and the additional stiffness provided by the curvature of the support on which the slabs are placed. The force vector includes the frictional force generated by the motion of the slab. The system of equations was solved using a combination of modal analysis and Newmark numerical integration (Newmark 1959) to obtain the respective response equations. The derivation of the equations has been described in detail in Engle et al. (2015):

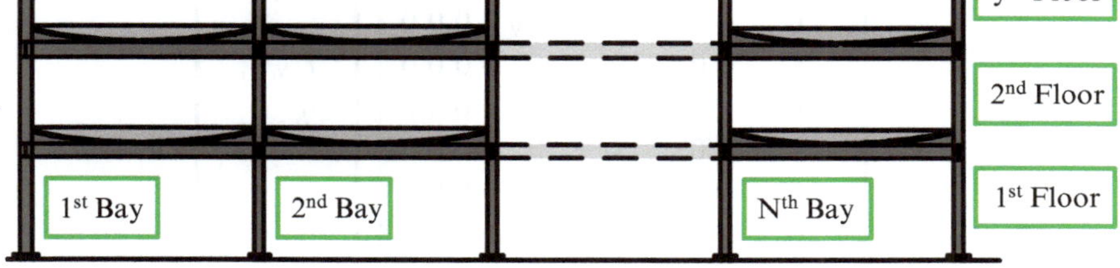

Fig. 8.2 Elevation view of multistory sliding floor system (Engle et al. 2015)

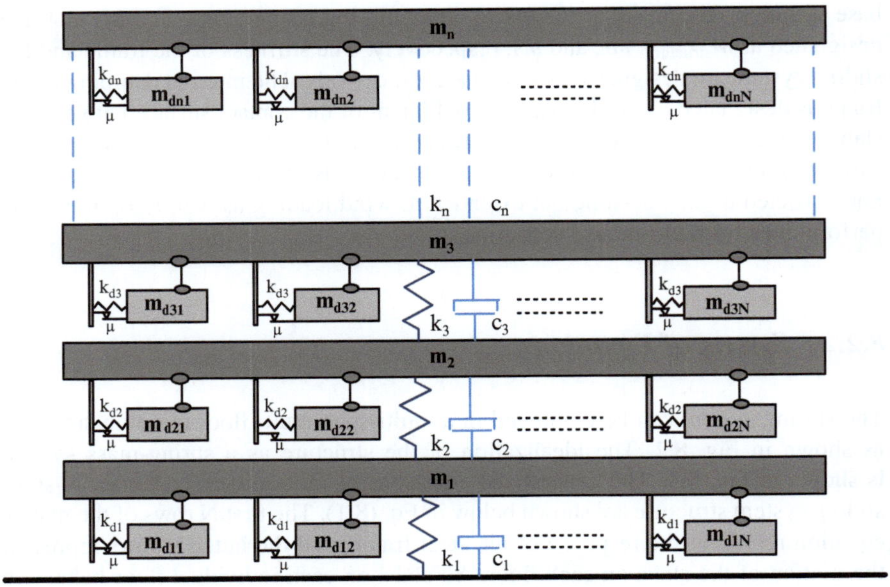

Fig. 8.3 Idealized representation of multistory sliding floor system (Engle et al. 2015)

$$\begin{bmatrix} m_1 & 0 & 0 & 0 & 0 & 0 & 0 & 0 \\ 0 & \cdots\cdots & 0 & 0 & 0 & 0 & 0 & 0 \\ 0 & 0 & \cdots\cdots & 0 & 0 & 0 & 0 & 0 \\ 0 & 0 & 0 & m_N & 0 & 0 & 0 & 0 \\ 0 & 0 & 0 & 0 & m_{N+1} & 0 & 0 & 0 \\ 0 & 0 & 0 & 0 & 0 & \cdots\cdots & 0 & 0 \\ 0 & 0 & 0 & 0 & 0 & 0 & \cdots\cdots & 0 \\ 0 & 0 & 0 & 0 & 0 & 0 & 0 & m_{2N} \end{bmatrix} \begin{pmatrix} \ddot{U}_{1_{k+1}} \\ \vdots \\ \vdots \\ \ddot{U}_{N_{k+1}} \\ \ddot{U}_{N+1_{k+1}} \\ \vdots \\ \vdots \\ \ddot{U}_{2N+1_{k+1}} \end{pmatrix}$$

$$+ \begin{bmatrix} c_1 + c_2 & -c_2 & 0 & 0 & 0\,0\,0\,0 \\ -\,c_2 & \cdots\cdots & \cdots\cdots & 0 & 0\,0\,0\,0 \\ 0 & \cdots\cdots & \cdots\cdots & -c_N & 0\,0\,0\,0 \\ 0 & 0 & -c_N & c_N & 0\,0\,0\,0 \\ 0 & 0 & 0 & 0 & 0\,0\,0\,0 \\ 0 & 0 & 0 & 0 & 0\,0\,0\,0 \\ 0 & 0 & 0 & 0 & 0\,0\,0\,0 \\ 0 & 0 & 0 & 0 & 0\,0\,0\,0 \end{bmatrix} \begin{pmatrix} \dot{U}_{1_{k+1}} \\ \vdots \\ \vdots \\ \dot{U}_{N_{k+1}} \\ \dot{U}_{N+1_{k+1}} \\ \vdots \\ \vdots \\ \dot{U}_{2N_{k+1}} \end{pmatrix} +$$

$$
\begin{bmatrix}
k_1 + k_2 + \left(k_s + \frac{m_5 g}{R}\right) & -k_2 & 0 & 0 & -\left(k_{N+1} + \frac{m_N + 1g}{R}\right) & 0 & 0 & 0 \\
-k_2 & \cdots & \cdots & 0 & 0 & \cdots & 0 & 0 \\
0 & \cdots & \cdots & -k_N & 0 & 0 & \cdots & 0 \\
0 & 0 & -k_N & k_N + \left(k_N + \frac{m_N g}{R}\right) & 0 & 0 & 0 & -\left(k_{2N} + \frac{m_{2N} g}{R}\right) \\
-\left(k_{N+1} + \frac{m_N + 1g}{R}\right) & 0 & 0 & 0 & & & & 0 \\
0 & \cdots & 0 & 0 & 0 & & & 0 \\
0 & 0 & \cdots & 0 & 0 & & & 0 \\
0 & 0 & 0 & -\left(k_2 N + \frac{m_2 g}{R}\right) & 0 & & & \left(k_{2N} + \frac{m_{2N} g}{R}\right)
\end{bmatrix}
\begin{pmatrix}
U_{1_{k+1}} \\ \vdots \\ \vdots \\ U_{N_{k+1}} \\ U_{N+1_{k+1}} \\ \vdots \\ \vdots \\ U_{2N_{k+1}}
\end{pmatrix}
=
\begin{pmatrix}
-m_1 * \ddot{U}_g + \mu * m_5 * g \\ \vdots \\ \vdots \\ -m_N * \ddot{U}_g + \mu * m_{2N} * g \\ -m_{N+1} * \ddot{U}_g - \mu * m_{N+1} * g \\ \vdots \\ \vdots \\ -m_{2N} * \ddot{U}_g - \mu * m_{2N} * g
\end{pmatrix}
\tag{8.1}
$$

8.2.3 *Validation*

The response equations derived for the sliding system were validated on a test structure. An example from Chopra (2005) of a five-story structure was considered, and the key variables of the sliding system were adjusted to emulate a composite frame. The radius of curvature and the stiffness of the sliding slabs were considered to be quite high to negate the motion of slabs, and the coefficient of friction was chosen to be zero. The response of the sliding frame with the modified variables was compared to that of the composite frame. Figure 8.4a shows the comparison for a specific earthquake. It can be seen that the two cases comply quite well with each other. For a much more comprehensive validation, a frequency response analysis was also conducted to validate the response equations as shown in Fig. 8.4b. From both figures it is shown that the response from the equations derived for the sliding system matches well with that of the composite structure.

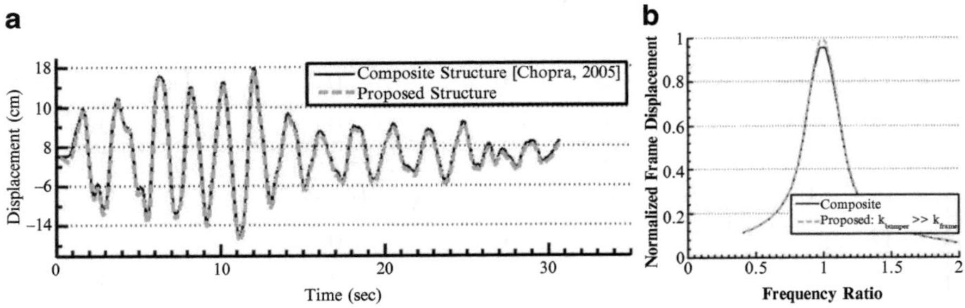

Fig. 8.4 (**a**) Response comparison for validation of calculations and (**b**) verification of isolated structure to composite structure for a range of frequencies (Engle et al. 2015)

8.3 Optimization

8.3.1 Key Parameters

It is commonly known that TMDs need to be tuned for best performance by adjusting their parameters. The proposed sliding system is quite similar to TMD systems as well; hence, in order to maximize the effectiveness of the system, an optimization framework is required. This is particularly important when dealing with different performance objectives as dictated by the seismic and wind demands. The performance of the sliding system is seen to be a function of four key parameters—(1) Stiffness elements at the end of the slabs (k_b), (2) Radius of curvature of the slab support (R), (3) Coefficient of friction of the slab support surface (μ), and (4) Location parameter (S_n). The S_n parameters define the floor level at which the slabs should be suspended. The most optimal performance of the sliding system might require not all slabs to be isolated on all floors (Engle et al. 2015; Mahmoud and Chulahwat 2015); therefore, the location parameter "S_n" is introduced to indicate the presence or absence of isolated slabs on a given floor as it takes only binary values, i.e., 0 (absence) or 1 (presence). The location parameter regulates the parameter values as shown in Eq. (8.2). If S_n is 1, the k_b, R, and μ variables are unaltered, and the slab is therefore sliding. However, if S_n is zero, the variables are altered such that the slabs emulate that of a composite frame. Figure 8.5 shows the physical representation of the location parameter in a matrix form which is simplified to a vector by assuming that if $S_n = 1$, then all slabs on the specific floor will be isolated. A thing to note is that the mass of slabs is not considered as a parameter for the sliding system as it would reduce the practicality of the system:

$$f\ S_n = 1 \rightarrow \begin{matrix} k_b = k_b \\ R = R \\ \mu = \mu \end{matrix} \quad \text{if} \quad S_n = 0 \rightarrow \begin{matrix} k_b = \infty \\ R = \infty \\ \mu = 0 \end{matrix} \tag{8.2}$$

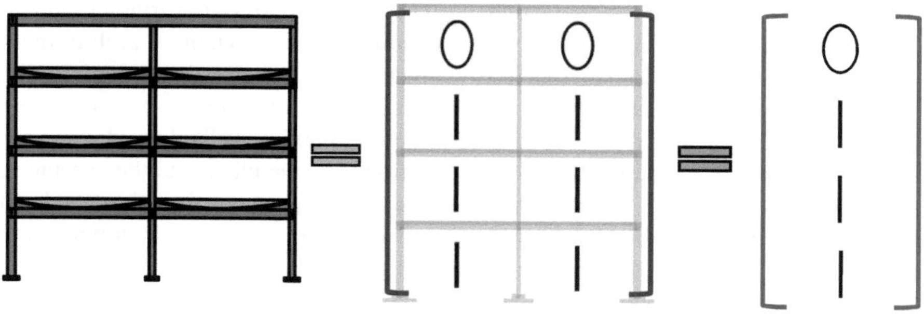

Fig. 8.5 Binary representation of activated stories of structure in matrix and vector form

In order to maximize the effectiveness of the sliding system, the aforementioned parameters have to be optimized using a suitable algorithm. Visual assessment of the equations of motion clearly indicates that it is a complex optimization problem, and the fact that S_n can be considered a vector, the problem is also multimodal in nature, i.e., there are several local minimas present. Therefore, to optimize the performance of the sliding slab floor system, a robust nonlinear-type optimization is required. Evolutionary algorithms, such as genetic algorithm, bee swarm, etc., are considered to be far more effective than gradient-based methods in case of highly nonlinear problems, and as a result they have developed quite a reputation over the years and have been applied to several fields of science and engineering (Fogel et al. 1966; Dellaert and Beer; 1996; Bentley 1999; Bentley and Corne 2001; Davidson 2001). One of the most commonly used evolutionary-type optimizations is the "genetic algorithm," which has also found its application in optimal design of TMD systems (Hadi and Arfiadi 1998; Mohebbi et al. 2012). However, there lies one major drawback in genetic algorithms in that it can be inefficient for highly multimodal problems. For this study a different evolutionary-type optimization was used, which is referred to as "covariance matrix adaptation evolution strategy (CMA-ES)."

8.3.2 Covariance Matrix Adaptation Evolution Strategy (CMA-ES)

The "covariance matrix adaptation evolution strategy" (CMA-ES) is a stochastic method of optimization for continuous, nonlinear, non-convex functions (Hansen 2011). Compared to other conventional optimization techniques, CMA-ES has been found to be more flexible and efficient for nonlinear multimodal problems (Hansen and Kern 2004). CMA-ES differs from traditional evolutionary algorithms as it exploits two key concepts—maximum likelihood principle and evolution paths. Maximum likelihood is a popular estimation method utilized in the field of statistics to make estimations based on given data. The role of maximum likelihood is to maximize the probability of selecting correct values of design variables and to increase the chances of selecting a viable search domain. A viable search domain can be referred to as the domain that always encapsulates the global optima. Initially, a test population is chosen at the beginning of the optimization process by assuming a particular distribution. After the first iteration, the maximum likelihood principle is used to update the parameters of the distribution (i.e., the mean and the variance), such that the likelihood of the correct values of the previous iteration tends to increase. Similarly, to increase the likelihood of the correct search domain, the covariance matrix of the distribution is updated.

The second key concept is the use of evolution or search paths. Evolution paths essentially provide an idea of the direction toward which the most favorable solutions can be found based on the results from the previous step; thus, it gives a correlation between consecutive steps. These evolution paths serve dual purposes. First, they are used in the adaptation process of the covariance matrix instead of focusing on the variance of a single successful step. This enables faster and more efficient improvement in moving toward favorable direction. The second is the use of evolution path to control the step size. The advantage of step-size control is to avoid premature convergence, which is a common problem in most evolutionary algorithms. Premature convergence is basically the convergence on local optimal solutions rather than the global optima. This takes place mainly due to lack of genetic variation among the population generated (i.e., all the members of the population have identical properties). The CMA-ES procedure comprises various features including sampling, selection and recombination, adaptation, evolution path, and step size. The general procedure of CMA-ES, including details of these features, can be found in Hansen 2011. In this study two modifications were applied to CMA-ES to improve its performance, which include elitism and flatness recovery as highlighted in Sect. 8.3.4.

8.3.3 Optimization Framework

From the previous discussion, on the sliding slab system, it is seen that the system is a function of four key parameters—S_n, k_b, R, and μ—which need to be optimized to maximize the potential of the system under any given loading conditions. Normally, an optimization could be directly applied to optimize all variables at the same time; however in this case the relation between the variable S_n and other variables k_b, R, and μ is rather complex. For every possibility of the S_n vector, there exists an optimal set of the other three variables; thus, the two sets of variables can be considered on a separate plane with the S_n vector stacked on top of other variables. Due to this relation, a nested optimization is proposed in this study, which comprises of two layers of CMA-ES algorithm—an inner and outer CMA-ES. In short, the sliding system is a nested optimization problem in which the four variables are not optimized all at the same time. Since k_b, R, and μ are a function of S_n, the outer layer is used for optimization of S_n and the inner layer for the other three variables.

To conduct a multi-hazard optimization, the fitness function can be formulated as a multi-objective function of the earthquake and wind response as shown in Eqs. (8.3), (8.4), (8.5), and (8.6). The fitness function is considered as a weighted sum of the performance functions of the two hazards, where β_1 and β_2 are their respective weights. The two hazard performance functions depend on different performance parameters. The earthquake response is a function of the mean of inter-story drift, and the wind response is a function of the mean of floor acceleration of the complete structure. Both functions are represented as percentage improvement over a reference structure, which in this study is a conventional composite slab-frame system of the same dimensions as that of the sliding slab-frame system. By normalizing or, in other words, by bounding the response functions of both hazards between −1 and 1, it is ensured that the optimization is not biased toward a particular hazard. In addition, the response functions of the respective hazards are supplemented with a penalty function to ensure the relative drift of the slab does not exceed a certain limit:

$$\text{Penalty}(N) = \sum_{n}^{N} \left[\text{if}\left((\text{Slab})_{n,\text{Drift}} > (\text{Limit})_{\text{Drift}}\right)\right] \tag{8.3}$$

$$(\text{Eq})_{\text{Im}} = \frac{(\text{ISD})_{\text{Mean}} - (\text{Comp})_{\text{Eq}}}{(\text{Comp})_{\text{Eq}}} + \alpha\Big(\text{Penalty}(N) \tag{8.4}$$

$$(\text{Wind})_{\text{Im}} = \frac{(\text{FA})_{\text{Mean}} - (\text{Comp})_{\text{Wind}}}{(\text{Comp})_{\text{Wind}}} + \alpha\,(\text{Penalty}(N)) \tag{8.5}$$

$$\text{Fitness function} = \beta_1(\text{Eq})_{\text{Im}} + \beta_2(\text{Wind})_{\text{Im}} \quad \forall\left(\sum \beta_i = 1\right) \tag{8.6}$$

where:

$\alpha = 10^{11}$ (an arbitrarily high value)
$(\text{Eq})_{\text{Im}}$ = percentage improvement in earthquake response of sliding system over a composite frame
$(\text{Wind})_{\text{Im}}$ = percentage improvement in wind response of sliding system over a composite frame
$(\text{ISD})_{\text{Mean}}$ = mean of inter-story drifts of all floors for sliding system
$(\text{FA})_{\text{Mean}}$ = mean of acceleration of all floors for sliding system
$(\text{Comp})_{\text{Eq}}$ = mean of inter-story drifts of all floors for composite frame
$(\text{Comp})_{\text{Wind}}$ = mean of acceleration of all floors for composite frame
β_i = weighting factor for ith hazard
Penalty(N) = penalty function to increase the fitness value substantially if slab drift is in excess
$(\text{Slab})_{n,\text{Drift}}$ = Nth floor slab drift relative to the corresponding floor
$(\text{Limit})_{\text{Drift}}$ = maximum allowable slab drift

The algorithm starts by generating a population sample for the variable S^n, whose value is required to find the necessary structural properties for each case of the population sample. The fitness value of each case is determined by sending the information to the inner layer algorithm, which optimizes the performance of the sliding system based on the information and delivers the optimized values of k_b, R, and μ along with the fitness value for that specific population sample case back to the outer layer CMA-ES. The fitness value is evaluated from Eq. (8.6) in which the performance improvement of the proposed system over that of a composite system, under earthquake and wind demands, is evaluated separately in the algorithm and then combined to obtain a single fitness value. The outer CMA-ES arranges the members of the population sample based on their respective fitness received from the inner layer and assigns ranks to identify the most suitable cases, which are further modified using the elitist strategy. In case the flatness recovery modification is not activated, the algorithm moves on to the updating stage where parameters for sampling of the next generation are evaluated and the existing population is checked for convergence. When flatness recovery is activated, the algorithm checks for convergence before updating the sampling parameters. In case the algorithm shows positive signs of convergence, the step length for the next iteration is updated based on the equation of flatness recovery instead of the general equation of CMA. The process repeats itself until convergence is achieved or, as in the case of flatness recovery, until a certain number of iterations are performed. The complete layout of the framework is shown in Fig. 8.6.

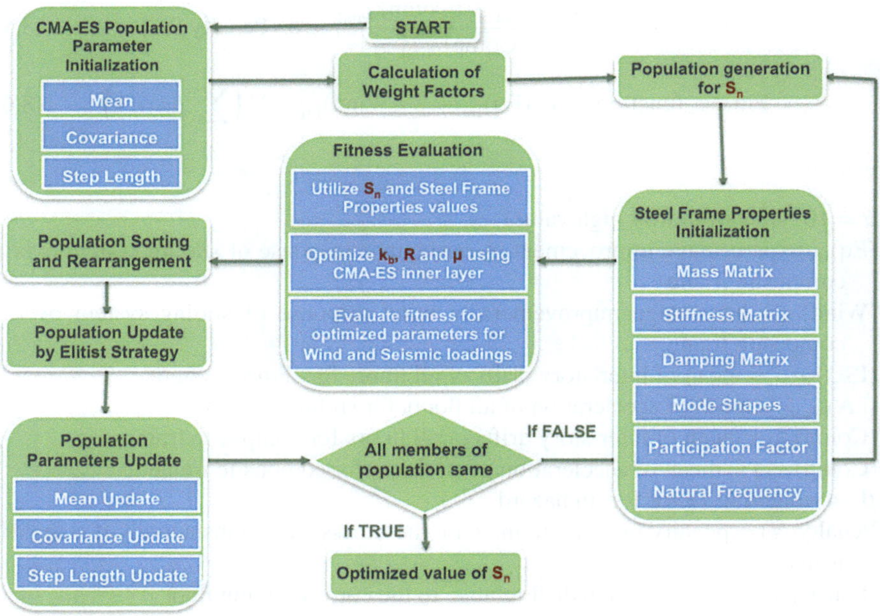

Fig. 8.6 Flowchart of CMA optimization process

8.3.4 Modifications

The nested CMA-ES framework discussed in this study was also used in Engle et al. (2015) but for optimization against earthquake hazards only. In order to utilize the optimization for both earthquake and wind loading, some modifications were made to increase the performance of CMA-ES. Specifically, three key modifications were introduced—type I elitism, type II elitism, and flatness recovery, which act together to improve the search capacity of CMA-ES.

8.3.4.1 Elitist Selection (Elitism)

The elitist selection ensures that the best case is always carried on to the next generation (i.e., the optimum point is not lost). The CMA-ES algorithm generates the search points based on a probabilistic distribution; as a result sometimes the optimum point generated in the previous step may not be generated in the subsequent generations. In the case of the sliding slab system, the sampling function uses a filter to convert the random search points to binary values; as a result the movement of the mean and other parameters tends to occur at a faster rate than usual. Due to this, sometimes it is possible that some solutions may be overlooked or the search may get stuck on a local solution, so the elitist selection ensures all essential information is preserved at every step. To further improve efficiency of the elitist selection, two modifications were made and tested separately. In the first modification (type I), it was assumed that in a particular generation, the best case is identified and selected. The stored value is then compared to the best case from the next generation. In case the stored value turns out to be better, then the worst case for that generation is replaced by the stored value. The second type (type II) was assumed to encapsulate the principle of type I, and in addition, it was assumed that if the best case from the previous generation turns out to be equal or worse than the best case of the next generation, then it replaces the worst case from that generation. Both types are illustrated in Fig. 8.7, where the roman numerals indicate the order of fitness, i.e., "*I*" is the best, and F_I^k is the best fitness value in the *k*th generation of the population sample and *I*th iteration of the algorithm.

8.3.4.2 Flatness Recovery

As CMA-ES converges to a solution, the step length tends to decrease; therefore, if the solution is a local one rather than global, then it results in premature convergence. In order to provide escape from such situations flatness recovery method was proposed, which scales up the step length. By increasing the step length, the chances of finding a better solution tend to increase, and if the algorithm is able to find one, then the entire optimization would shift in the other direction, thereby

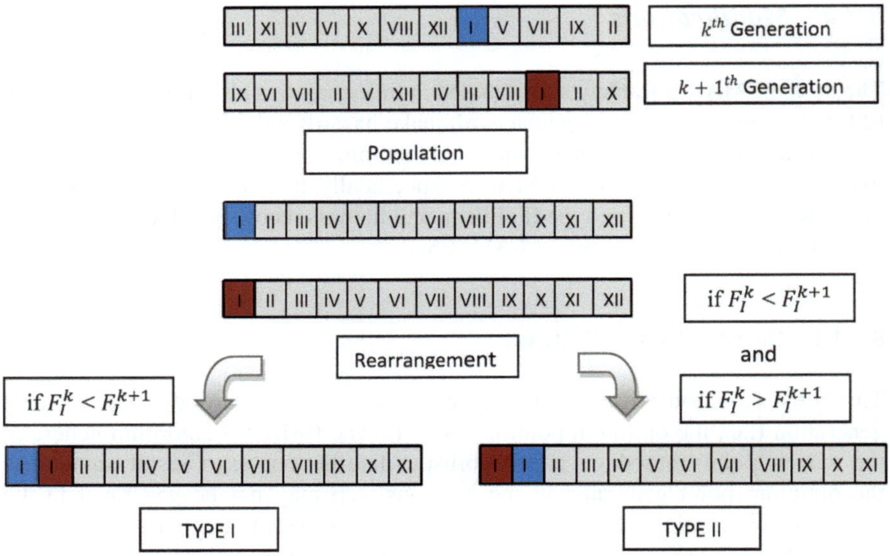

Fig. 8.7 Elitist strategy modifications

preventing premature convergence. However, increasing the step length too much may cause the algorithm to diverge or converge on the wrong solution. As previously mentioned, S_n can take only 0 or 1, so once flatness is achieved, the mean would converge to 0 or 1. In order to shift the optimization to a new solution, the mean would have to be converted from 0 to 1 or vice versa. The step length would have to be chosen such that the probability of generating a number greater than 0.5 is significant. It was seen from some tests that for sigma greater than 2, the increase in probability was not significant; thus, it was set as the maximum limit. The original step length function was substituted with a modified exponential function that is bounded between 0 and 2.

8.3.4.3 Performance Tests

The modifications suggested for improving the performance of CMA-ES were tested on a seven-story three-bay and ten-story five-bay structures under dynamic excitation, which was considered as a sinusoidal excitation of unit amplitude and a frequency equivalent to the natural frequency of the structure being optimized. Since the main purpose of these tests is to evaluate the performance of the algorithm and not the sliding structure, therefore a simple fitness function is used. The optimization was conducted to minimize the top floor drift of the frame. The mass and stiffness for each structure are shown in Table 8.1.

Table 8.1 Structural properties of seven- and ten-story structures

Floor number	Mass (kg)	Stiffness (kN/m)
Structural properties of seven-story three-bay frame		
1–6	156.3×10^3	19.7×10^3
7	138.8×10^3	19.7×10^3
Structural properties of ten-story five-bay frame[a]		
1–9	252.9×10^3	33×10^3
10	227.7×10^3	33×10^3

[a]The ten-story structure is utilized in the multi-hazard performance optimization tests in Sect. 8.4

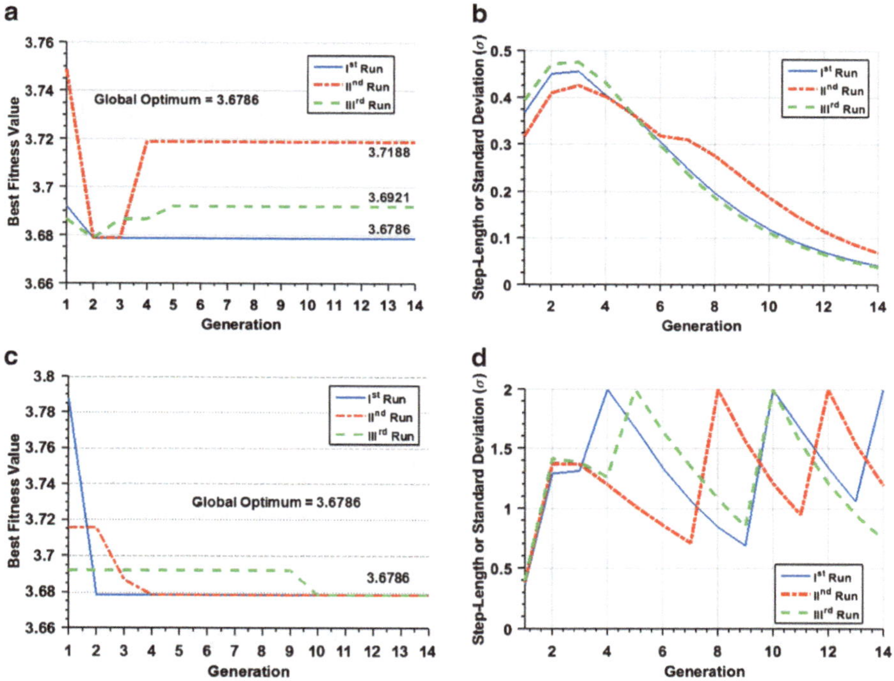

Fig. 8.8 Performance of optimization algorithm for the seven-story test structure (**a**) convergence using CMA-ES only, (**b**) standard deviation using CMA-ES only, (**c**) convergence using the modified CMA-ES, and (**d**) standard deviation using the modified CMA-ES

The performance of the optimization algorithm using the nested CMA-ES with no modifications is evaluated against that which utilizes dual CMA-ES in addition to type I elitism, type II elitism, and flatness recovery. In terms of performance, Fig. 8.8a shows the accuracy and Fig. 8.8b shows variation in step length of the algorithm for the seven-story sliding structure. As seen in Fig. 8.8a, the algorithm converged to local optimas and missed the global optima, which tends to suggest a need for improvement. Figure 8.8b shows a confined search space marked by lower values for the step length or standard deviations. For the case of the modified algorithm, the accuracy in this case, also for the seven-story sliding structure, is seen to be quite better than the predecessors as shown in Fig. 8.8c. From the plot of standard deviation in Fig. 8.8d, it is shown that the algorithm does not converge to a unique solution; however, it still manages to find the global optima. The concept behind the modified algorithm is to increase the standard deviation drastically (Fig. 8.8d) when the algorithm starts to reach flatness, such that the algorithm is forced to generate new search points away from the point of convergence. The step length varies drastically (Fig. 8.8d); as a result the algorithm does not converge to a single point unless it is the global optima or there are no better solutions nearby. This is a significant advantage as it prevents the search from being stuck on a local solution by increasing the probability of finding better solutions. The flatness recovery modification acts as a destabilizing agent, and at the same time, the inherent convergence mechanism of CMA-ES coupled with the properties of elitism modifications tends to act as stabilizing agents and thus momentarily allows for the search space to be expanded to greater distances, thereby increasing the chances of finding a better solution before convergence. As for the computational time of convergence, it is shown to be almost the same as previous cases. The accuracy of the optimization and the variation in step length are very similar in the case of the ten-story structure (results not shown). For the multi-hazard optimization tests discussed in the next section, complete analysis is performed using the modified algorithm on the ten-story structure.

8.4 Multi-Hazard Optimization Tests

The proposed framework is utilized for multi-hazard optimization of the ten-story five-bay sliding slab system. The structural properties are shown in Table 8.1. The beams are W 27×40, while the columns are W 14×109 sections. The internal damping ratio of the frame was assumed to be 5 %, which is a typical value for steel frames; however it suggests significant yielding in the frame. Even though this may interfere in the assessment of sliding system performance, the fact that the fitness function is formed as a percentage improvement over a composite frame counterpart tends to negate the effect of damping. The width of the floor bays was assumed to be 9.14 m, and the height of each story was assumed to be 4.572 m. For analysis, two earthquakes were selected—one near-field earthquake and one far-field earthquake. The earthquake time histories were scaled to match the design code for location of the reference structure according to ASCE 7 (American Society of Civil Engineers 2010) and FEMA (Mahoney and Hanson 2009). The ground motions with their respective properties are summarized below in Table 8.2. The location for analysis was chosen to be Los Angeles, California, and the site was assumed to have soil type D. The wind records were obtained from "Wind Effects on Buildings and Urban Environment" database, created by Tokyo Polytechnic University (2013), to obtain suitable wind pressure coefficients of the test structure followed by design wind pressure based on the location and site conditions (American Society of Civil

Table 8.2 Selected seismic records and properties

Mw	Year	Earthquake name	Station ID	Reference name	Distance (km)	PGA	Scaling factor
6.5	1979	Imperial Valley BC	(HBCR230)	IV-HBCR	2.7	0.775	2.882
6.7	1994	Northridge, Century City	(CCN360)	NR-CCN	25.7	0.222	0.714

Table 8.3 Optimization limits

	R (m)	K_b (kN/m)	μ
Max	68.66	61.25×10^{3}[a]	0.5
Min	11.89	8.75[a]	0.05

[a] An arbitrary value that can be adjusted as needed

Engineers 2010). The coefficients were finally normalized to adjust for the height of the structure.

The variables k_b, R, and μ are unbounded variables; therefore, the upper and lower bounds were set for each parameter. The bounds for k_b and μ were chosen arbitrarily; however for R the upper and lower bounds were calculated by solving the general equation of a circle. The equation was used to calculate the vertical height of the curvature. If the curvature is too small, then the vertical displacement of the slab, produced as the slab travels along the curvature path, would become impractical; so the lower bound was set corresponding to a vertical height of 0.9144 m, and the upper bound was set corresponding to a height of 0.1524 m. The respective upper and lower bounds are shown in Table 8.3. The structure was first optimized for each hazard individually using their corresponding response functions, i.e., Eq. (8.4) for the two earthquakes and Eq. (8.5) for the wind loading. Finally, the two earthquakes were combined one at a time with the wind time history to conduct multi-hazard optimization using Eq. (8.6). The weighting factors for this study were chosen to be 0.50 for both earthquake and wind to assign equal weightage for design. Since this is a preliminary study, the weighting factors were kept constant; however, in general they can be varied to obtain a family of solutions from which the designer can choose an optimal design at their own discretion. Initially it was pointed out that if $S_n = 1$, then all slabs on the specific floor are isolated. However, for the analysis another parameter is introduced "H_n," which determines the number of bays in which slabs are sliding at a given floor if $S_n = 1$. The optimization for each case is conducted for two values of H_n—1 and 5. The optimal values obtained for parameters k_b, R, and μ for each optimization case have been shown in Table 8.4.

Figure 8.9a shows the improvement in performance of the sliding system over a composite frame for different optimization cases under earthquake (HBCR230) and wind loadings. Similarly, Fig. 8.10a shows the performance improvement under earthquake (CCN360) and wind loadings. The ten-story five-bay sliding system was optimized for three cases—earthquake only, earthquake + wind, and wind only (all optimization cases are shown in Table 8.4). The legends in the figures indicate the hazard for which the optimization was performed. The y-axis of the plots shows the type of loading used to evaluate/test the performance improvement. For example,

Table 8.4 Optimized parameters and peak responses

Simulation number	Hazard	H_n	Radius (m)	Bumper stiffness (kN/m)	Coefficient of friction
1	Earthquake (HBCR230)	1	55.88	2.5×10^3	0.32
2	Earthquake (HBCR230)	5	49.784	4×10^3	0.24
3	Earthquake (CCN360)	1	11.89	4.16×10^3	0.05
4	Earthquake (CCN360)	5	11.89	10.73×10^3	0.12
5	Wind	1	11.89	61.25×10^3	0.05
6	Wind	5	11.89	61.25×10^3	0.50
7	HBCR230 + wind	1	11.89	4.33×10^3	0.05
8	HBCR230 + wind	5	40.13	4.68×10^3	0.26
9	CCN360 + wind	1	51.308	4.98×10^3	0.05
10	CCN360 + wind	5	68.66	3.16×10^3	0.05

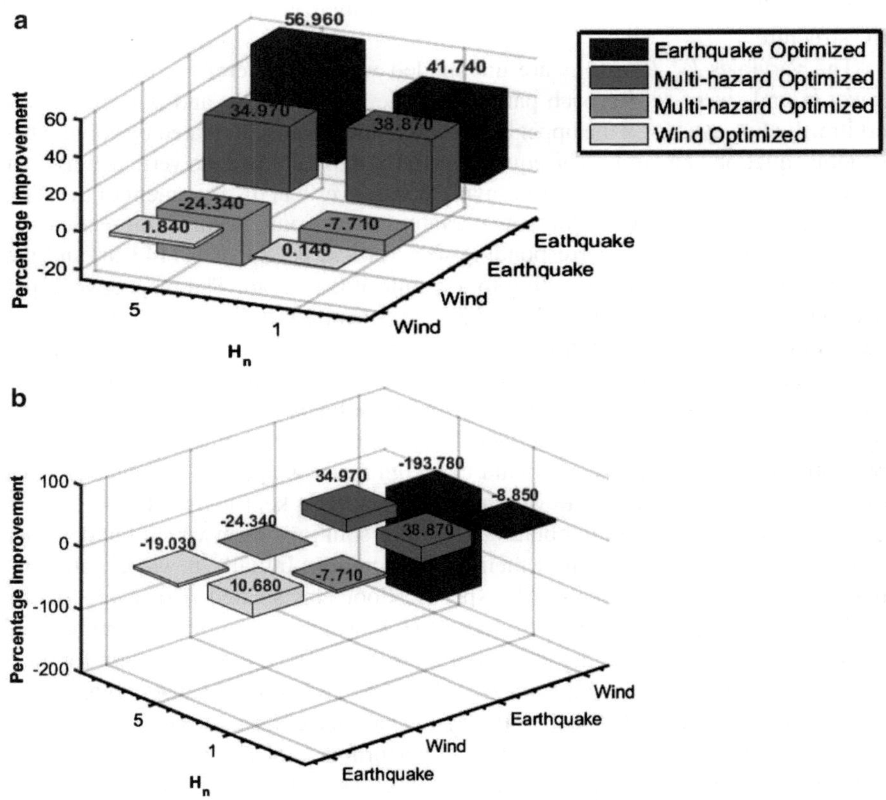

Fig. 8.9 Percentage improvements in performance for different optimization cases (**a**) for earthquake—HBCR230. (**b**) Reverse plot for HBCR230

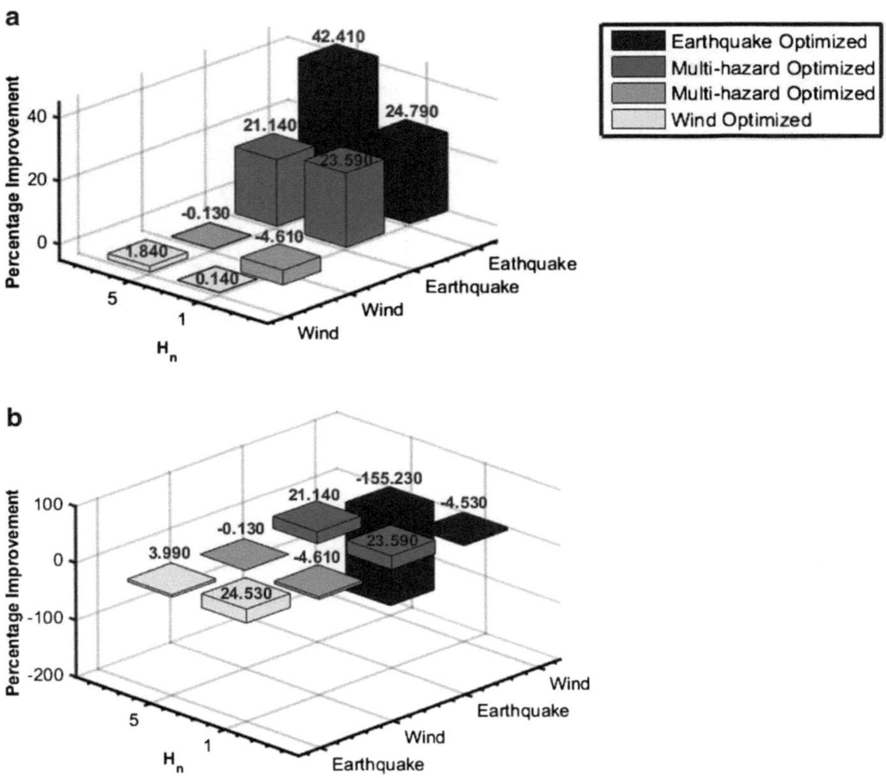

Fig. 8.10 Percentage improvements in performance for different optimization cases (**a**) for earthquake—CCN360. (**b**) Reverse plot for CCN360

in Fig. 8.9a, the "earthquake optimized" in the legend corresponds to performance improvements of 56.96 % for H_n of 5 and 41.74 % for H_n of 1 when tested under earthquake demand. It is clear from the figures that multi-hazard optimization circles around the concept of give-take, i.e., some amount of compromise is required if the structure has to be efficient for both hazards. To obtain a better perspective of the importance of multi-hazard optimization, a "reverse plot" was made for both cases (Figs. 8.9b and 8.10b). A reverse plot is also a performance improvement plot; except the order of hazards is reversed (as can be seen from the *y*-axis), i.e., the earthquake-optimized system is tested under wind and vice versa. The reverse plots when seen along with the performance plots tend to give a much better picture. It can be seen that even though individual optimizations for both earthquake and wind loadings resulted in acceptable performance, when tested against each other, they showed poor performance. Therefore, from a holistic point of view, the multi-hazard designs can be considered the most optimal. Even though in general the multi-hazard designs are better than individual optimization designs, but for the

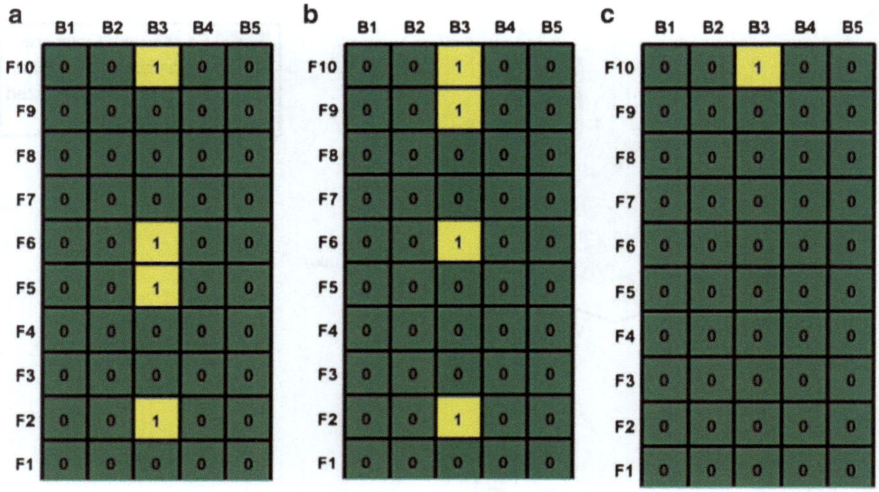

Fig. 8.11 Optimal isolated slab locations for (**a**) CCN360 only, (**b**) CCN360 + wind, and (**c**) wind (when $H_n = 1$)

earthquake CCN360 when $H_n = 1$ case, the individual optimizations showed quite similar response to that of the multi-hazard designs, i.e., in that specific case even the individual optimizations gave an overall stable design for both hazards.

Figures 8.11, 8.12, 8.13, and 8.14 show the optimal isolated slab locations in the ten-story five-bay structure tested. In the plots, the ten-story five-bay structure is represented, where F1–F10 represent floors 1–10 and B1–B5 represent bays 1–5. The plots are a physical representation of the optimized values of the S_n vector with two different values of parameter H_n for each optimization case. The "0" indicates a composite slab and "1" indicates an isolated slab at the specific location. From these plots it is quite evident that all slabs of a structure are not required to be isolated for best performance; thus, it validates the requirement of an optimization algorithm, as also mentioned earlier in the study.

8.5 Conclusions

In this chapter, the concept of using isolated floor slabs in building systems to reduce structural response under extreme loads is discussed. The presence or absence of an isolated slab at a given floor requires the use of a nonlinear optimization algorithm that can handle a large number of variables. To that end, a modified CMA-ES optimization scheme is presented and used in the optimization process of a ten-story five-bay structure under single and combined hazards of winds and seismic loads. The following main conclusions can be drawn from the chapter:

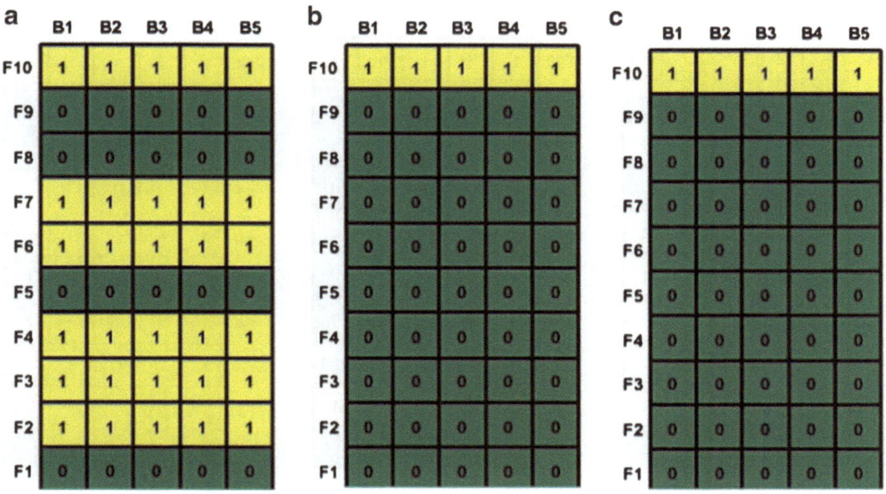

Fig. 8.12 Optimal isolated slab locations for (**a**) CCN360 only, (**b**) CCN360 + wind, and (**c**) wind (when $H_n = 5$)

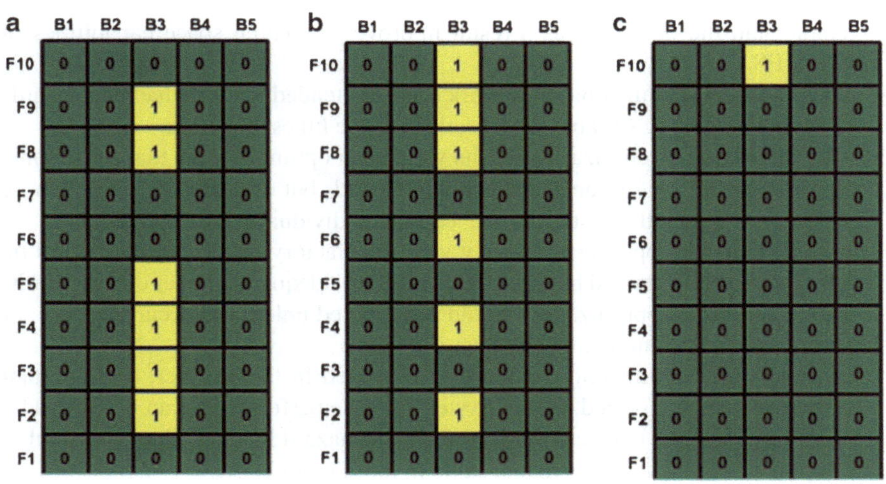

Fig. 8.13 Optimal isolated slab locations for (**a**) HBCR230 only, (**b**) HBCR230 + wind, and (**c**) wind (when $H_n = 1$)

- Even though CMA-ES has been shown to be better than other evolutionary optimizations regarding multimodal problems, the nested CMA-ES without any modifications is still not efficient enough to accurately identify the global optima for the ten-story five-bay structure.
- The three proposed modifications improved the efficiency of CMA-ES significantly without compromising the time cost. Each modification tends to

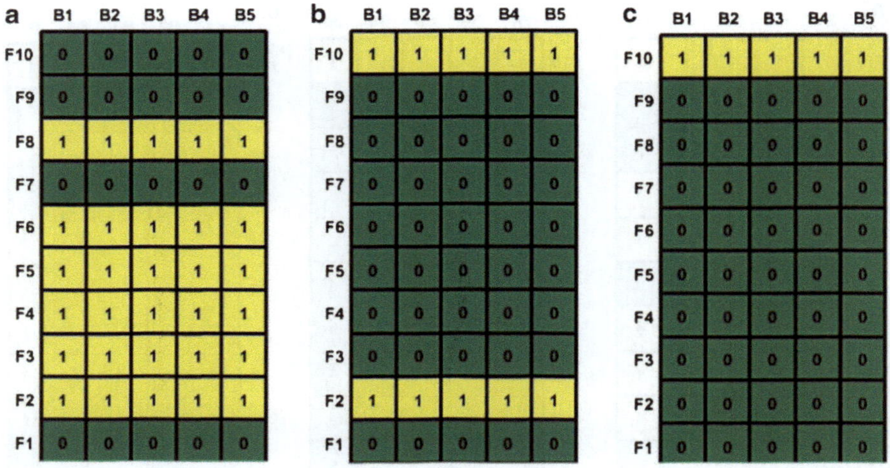

Fig. 8.14 Optimal isolated slab locations for (**a**) HBCR230 only, (**b**) HBCR230 + wind, and (**c**) wind (when $H_n = 5$)

supplement the other and as a result improve the search space capabilities of CMA-ES.

- The proposed optimization framework can be extended to more than two hazards at a time by making suitable modifications to the fitness function.
- The multi-hazard optimization results yielded an optimal design that made some compromise on performance against both hazards but overall gave a much more stable system than the systems obtained from individual optimizations.
- Individual hazard optimizations may yield satisfactory results for the hazard on which they are optimized but may fail to perform adequately under other hazards. Individual hazard optimization should not be used unless the occurrence rate of other hazards is significantly small.
- In light of the optimization framework proposed in this chapter, an important conclusion can be formed, that is, systems whose performance can be altered by just a few variables can be optimized for multi-hazard loading, since the number of optimization variables is manageable in those cases.

References

Almazán, J. L., De la Llera, J. C., Inaudi, J. a., López-García, D., & Izquierdo, L. E. (2007). A bidirectional and homogeneous tuned mass damper: A new device for passive control of vibrations. *Engineering Structures, 29*(7), 1548–1560.

American Society of Civil Engineers (2010). Minimum Design Loads for Buildings and Other Structures. ASCE/SEI 7-10.

Blake, E., Kimberlain, T., Berg, R., Cangialosi, J., & Beven, J., II. (2013). *Tropical Cyclone Report. Hurricane Sandy* (ALI82012). 22–29 October 2012, National Hurricane Center (NHC), NOAA.

Bentley, P. J. (1999). *Evolutionary design by computers*. Morgan Kaufmann Pub.

Bentley, P. J., & Corne, D. W. (2001). *Creative evolutionary systems*. Morgan Kaufmann Pub.

Chen, E. (2012). *Multi-hazard design of mid- to high-rise structures*. Masters thesis, Department of Civil and Environmental Engineering, University of Illinois at Urbana-Champaign.

Chopra, A. K. (2005). *Dynamics of structures*, 4th ed. Upper Saddle River, NJ: Prentice Hall

Constantinou, M. C., Soong, T. T., & Dargush, G. F. (1998). *Passive energy dissipation systems for structural design and retrofit* (1st ed.). Buffalo: Multidisciplinary Center for Earthquake Engineering Research.

Crossett, K., Culliton, T., Wiley, P., and Goodspeed, T. (2004). *Population trends along the coastal United States: 1980-2008.* U.S. Dept. of Commerce. National Oceanic and Atmospheric Administration (NOAA).

Crosti, C., Duthinh, D., & Simiu, E. (2011). Risk consistency and synergy in multi-hazard design. *ASCE Journal of Structural Engineering, 3*(8), 844–849.

Davidson, E. H. (2001). *Genomic regulatory systems: Development and evolution*. Academic Press.

Dellaert, F., & Beer, R. D. (1996). A development model for the evolution of complete autonomous agents. In P. Maes, et al. (Eds.), *From Animals to Animate, Proceedings of the Fourth International Conference on Simulation of Adaptive Behavior* (pp. 393–401). MIT Press.

Duthinh, D., & Simiu, E. (2008). A multihazard approach to the structural safety of buildings exposed to strong winds and earthquakes. In N. Raufaste (Ed.), *Proceedings of the U.S.-Japan Cooperative Program in Natural Resources, Panel on Wind and Earthquake Effects*. Gaithersburg, MD: National Institute of Standards and Technology.

Duthinh, D., & Simiu, E. (2014). Safety of structures in strong winds and earthquakes: Multi-hazard considerations. *Journal of Structural Engineering, 136*(3), 330–333.

Eatherton, M., Hajjar, J., Deierlein, G., Ma, X., & Krawinkler, H. (2010). Hybrid simulation testing of a controlled rocking steel braced frame system. In *Proceedings of the 9th U.S. National and 10th Canadian Conference of Earthquake Engineering*. Toronto.

Engle, T., Mahmoud, H., & Chulahwat, A. (2015). Theoretical development of an innovative hybrid tuned mass damper and isolation floor slab system optimized for vibration control. *Journal of Earthquake Engineering, 19*(8), 1197–1221.

Fogel, L. J., Owens, A. J., & Walsh, M. J. (1966). *Artificial intelligence through simulated evolution*. Wiley.

Gray, B. M., Christopoulos, C., Packer, J., & Carlos, D. O. (2012). A new brace option for ductile braced frames. *Modern Steel Construction*, 40–43.

Guo H., Chen Y., Feng Q., Lin Q., & Wang F. (2011). Assessment of damage to buildings and farms during the 2011 M 9.0 earthquake and tsunami in Japan from remote sensing data. *Chinese Science Bulletin, 56*(20), 2138–2144.

Hadi, M. N. S., & Arfiadi, Y. (1998). Optimum design of absorber for MDOF structures. *Journal of Structural Engineering*, *124*(11), 1272–1280.

Hansen, N. (2011). *The CMA evolution strategy: A tutorial*, Vu le 29. https://www.lri.fr/*hansen/cmatutorial.

Hansen, N., & Kern, S. (2004). Evaluating the CMA evolution strategy on multimodal test functions. X. Yao et al. (Eds.) *Eighth International Conference on Parallel Problem Solving from Nature PPSN VIII, Proceedings* (pp. 282–291), Springer-Verlag Berlin Heidelberg.

Li, C., & Zhu, B. (2006). Estimating double tuned mass dampers for structures under ground acceleration using a novel optimum criterion. *Journal of Sound and Vibration, 298*(1–2), 280–297.

Mahmoud, H., & Chulahwat, A. (2015). Response of building systems with suspended floor slabs under dynamic excitations. *Engineering Structures, 104*, 155–173.

Mahoney, M., & Hanson, R. D. (2009). Quantification of building seismic performance factors. *FEMA P695.*

Mardfekri, M., & Gardoni, P. (2015). Multi-hazard reliability assessment of offshore wind turbines. *Journal of Wind Engineering, 18*, 1433–1450.
Mohebbi, M., Shakeri, K., Ghanbarpour, Y., & Majzoub, H. (2012). Designing optimal multiple tuned mass dampers using genetic algorithms (GAs) for mitigating the seismic response of structures. *Journal of Vibration and Control, 19*(4), 605–625.
Moon, K. S. U. N. (2010). Vertically distributed multiple tuned mass dampers in tall buildings: Performance analysis and preliminary design. *The Structural Design of Tall and Special Buildings, 19*, 347–366.
Newmark, N. M. (1959, July). A method of computation for structural dynamics. *Journal of the Engineering Mechanics Division, 85*, 67–94.
Ormondroyd, J., & Den Hartog, J. P. (1928). The theory of dynamic vibration absorber. *ASME Journal of Applied Mechanics, 50*(7), 9–22.
Perry, M. J., & Mackun, P. J. (2001). *Population change and distribution, 1990–2000, in Census Brief 2000*. US Census Bureau.
Randall, S. E., Halsted, D. M., & Taylor, D. L. (1981). Optimum vibration absorbers for linear damped systems. *ASME Journal of Mechanical Design, 103*, 908–913.
Regional Plan Association (RPA) America. (2005). *A Prospectus: 2005*. New York, NY: Regional Plan Association's National Committee for America 2050.
Spencer, B. F. J., & Sain, M. K. (1997, December). Controlling buildings: A new frontier in feedback. *IEEE Control Systems, 17*, 19–35.
Tsai, H., & Lin, G. (1993). Optimum tuned-mass dampers for minimizing steady-state response of support-excited and damped systems. *Earthquake Engineering and Structural Dynamics, 22*, 957–973.
Warburton, G. B. (1982). Optimum absorber parameters for various combinations of response and excitation parameters. *Earthquake Engineering and Structural Dynamics, 10*(3), 381–401.
Warburton, G. B., & Ayorinde, E. O. (1980). Optimum absorber parameters for simple systems. *Earthquake Engineering and Structural Dynamics, 8*, 197–217.
Wen, Y. K., & Kang, Y. J. (2001). Minimum building life-cycle cost design criteria. I: Methodology. *ASCE Journal of Structural Engineering, 127*(3), 330–337.
Wind Effects on Buildings and Urban Environment. (2013). Tokyo Polytechnic University. Retrieved from http://wind.arch.t-kougei.ac.jp/system/eng/contents/code/tpu

Chapter 9
Energy Efficiency and Seismic Resilience: A Common Approach

Gian Michele Calvi, Luis Sousa, and Cristiana Ruggeri

Abstract To the present date, building retrofit and enhancement interventions tend to focus on either energy efficiency or seismic resilience techniques, highlighting the lack of consistent language and understanding across both fields, as well as the disconnection among stakeholders that arises from the development of seismic risk mitigation independently of sustainable development goals. Although extensive know-how can be identified in both areas, efforts for its joint consideration presented in the literature are based on the evaluation of environmental impacts of expected repairs due to seismic action over a period of time, neglecting the potential of energy efficiency enhancements and, more importantly, the possible benefits of an integrated investment strategy. This chapter presents a proposal for the integrated assessment of energy efficiency and earthquake resilience, according to which environmental and seismic impact metrics are translated into common financial decision-making variables. In this context, similarly to what is a common practice when evaluating the energy and environmental performance of buildings, discrete classes of both earthquake resilience and energy efficiency are proposed, providing a consistent proxy for building classification—*green and resilient indicator* (*GRI*)—as a function of mutual performance parameters. The findings of this chapter highlight the fact that it is possible to directly compare energy efficiency and seismic resilience from a common point of view, as it is plausible to assume the *green* and *resilient* counterparts of the *GRI* classes as a proxy for investment return potential. In addition, it is verified that the *benefit* of a given intervention can only be maximized up to the point in which an additional investment does not result in increased performance. Thus, an integrated approach shall always be

G.M. Calvi (✉)
Istituto Universitario di Studi Superiori (IUSS), Pavia, Italy
e-mail: gm.calvi@iusspavia.it

L. Sousa
RED-Risk Engineering and Design and FEUP - Faculty of Engineering of the University of Porto, Pavia, Italy
e-mail: costa.sousa@fe.up.pt

C. Ruggeri
Studio Calvi, Pavia, Italy
e-mail: Cristiana.ruggeri@studiocalvi.eu

P. Gardoni, J.M. LaFave (eds.), *Multi-hazard Approaches to Civil Infrastructure Engineering*, DOI 10.1007/978-3-319-29713-2_9

advantageous with respect to the investment in only *earthquake resiliency* or *energy efficiency*, devising an investment strategy in a way that simultaneously maximizes "individual" benefits and its integrated result.

9.1 Introduction

Resources are usually the absent guest in all tables where the issue is to define the performance level to be met when designing new structures and much more so when the issue is the upgrading of existing constructions (Calvi 2013). However, at present, most interventions for building enhancement focus either on energy efficiency or seismic strengthening, considering the maximization of benefits independently of the integrated correlation between both variables.

While the application of performance-based concepts has been applied to the specific case of seismic design using a decision support system (e.g., Caterino et al. 2008; Zareian and Krawinkler 2012), little attention has been devoted to existing buildings to what concerns seismic resilience (Calvi 2013) and energy efficiency counterparts.

The aforementioned circumstances highlight the necessity of a sensitivity study that enables the addressing of potential benefits arising from an integrated approach including disaster risk management and climate adaptation. With a starting point in state-of-the-art knowledge and best practices in both aspects of building assessment and enhancement, the present document intents to provide a proposal for an innovative model that combines the mentioned interventions, in order to maximize the financial and technical efficiency of the two practices under a comprehensive methodology. In this context, any prioritization scheme obviously implies the possibility of correcting performance level and time constraints to available resources, thus giving an answer to the problem of comparing resources and needs (Calvi 2013). Nonetheless, sounder approaches allow the definition of prioritization schemes that somehow associate the target performance to a level of "risk rating" (e.g., Grant et al. 2007). Since the obstinate pursuing of a predetermined level of performance may result in irrational cost-benefit ratios, the herein proposed model shall provide guidance on why and in which situations should energy efficiency and disaster-resilient practices be integrated and, more importantly, how should interventions be prioritized to achieve the highest return to investment.

Three real case study buildings representing distinct typologies—residential, educational, and healthcare—are modeled and analyzed in order to demonstrate the correlation between cost and benefit associated with the joint investment in energy sustainability and seismic resilience. To this end, climate and seismic hazard are established for both the original design locations and additional sites, considered in order to replicate conditions different from that of the original design. In this context, it is decided to verify the validity of such methodology through the analysis of the mentioned representative buildings, allowing the definition of the appropriate guidelines for a definitive model.

9.2 State of the Art

9.2.1 Earthquake Resilience

A commonly accepted framework in the assessment of building-specific resilience to seismic action is established in a formulation derived from the SAC project (Cornell et al. 2002) and included in different forms in FEMA and ATC documents (Calvi 2013), according to which seismic *risk* is evaluated in a probabilistic manner. This formulation addresses several types of uncertainties conjugated in a triple integral, as follows:

$$\lambda\left[\mathrm{DV}\middle|D\right] = \iiint p\left[\mathrm{DV}\middle|\mathrm{DM}, D\right]\cdot p\left[\mathrm{DM}\middle|\mathrm{EDP}, D\right]\cdot p\left[\mathrm{EDP}\middle|\mathrm{IM}, D\right] \cdot \lambda\left[\mathrm{IM}\middle|D\right] \mathrm{dIM}\cdot \mathrm{dEDP}\cdot \mathrm{dDM}$$

It is understood that such formulation, which will not be herein discussed in detail—addressing the reader to the works of Porter et al. (2002) and Haselton et al. (2007)—may appear quite difficult and inapplicable. However, as depicted in Fig. 9.1, it is rather a conventional formal expression that combines four steps of analysis—seismic hazard, structural response, damage assessment, and loss estimation—each one characterized by epistemic and aleatoric uncertainties associated with the ultimate estimation of $\lambda\,[\mathrm{DV}|D]$, the rate of exceedance of a

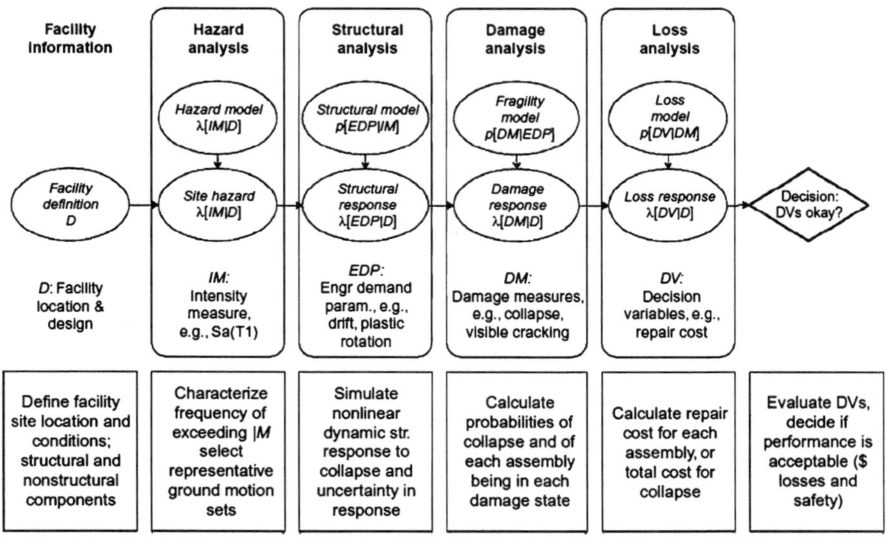

Fig. 9.1 PEER methodology to estimate losses in a probabilistic framework (Porter et al. 2002; Haselton et al. 2007)

certain decision variable (DV) that allows understanding risk through a probabilistic standpoint, for a given reference time span.

9.2.1.1 Multi-criteria Decision and Seismic Risk Mitigation

The inclusion of probabilistic *seismic risk* assessments into the multi-criteria decision for building intervention is not extensively available in the literature. In most cases in which such criteria are considered, the associated parameters are related to the deterministic seismic intensity at which a certain damage is attained, rather than with the probabilistic building expected *loss* due to seismic-induced damage over a given time period. In the work of Caterino et al. (2008, 2009), whose philosophy is also applied in the research campaign of Tesmafatian et al. (2010), the *seismic risk* decision variables are incorporated into "significant damage risk" and "damage limitation risk" evaluation criteria for different retrofit techniques. These are designated by the author as the "risk of significant damage attainment" and "risk of nonstructural damage to repair," respectively, reflecting, in fact, the influence of an intervention scheme in the building resistance, expressed as the deterministic level of seismic intensity at which the two established damage states are achieved.

Notwithstanding the above, seismic risk analysis has been widely employed in single criteria selection of retrofit techniques by means of an accepted risk tolerance and respective decision-making tool, generally reported as *cost-benefit* analysis, *life-cycle* costing, deterministic and stochastic *net present values*, or *multi-attribute* utility-based procedures (Tesmafatian et al. 2010). However, in addition to the fact that only economic factors are considered, all the previously mentioned sensitivity analyses depend on the definition of "performance levels" and "retrofitting objectives," which reflect the extent to which the building intervention will upgrade its response to a given level of seismic intensity.

Most approaches to the definition of performance levels are derived from the scheme originally proposed for new constructions in the Vision 2000 document (SEAOC 1995). This document introduced the concept of varying the expected performance for different return period ground motions as a function of the relevance of the facility. The resulting matrix is well known and not essential in this context, while the numerical definition of four intensity levels in terms of their return period or probability of exceedance—*frequent* (50 % in 30 years), *occasional* (50 % in 50 years), *rare* (10 % in 50 years), and *very rare* (5 % in 50 years)—is still used as a possible reference. Even more relevant is the attempt of numerically defining the expected performance as a function of a single parameter, the inter-story drift, as shown in Table 9.1.

The concepts presented in the Vision 2000 document were elaborated and extended in FEMA 356 (2000), but the next conceptual step was essentially due to the SAC project (Cornell et al. 2002), where the modeling of uncertainties was proposed, which inspired the development of a series of ATC documents culminated with the ATC (2011), still in a draft form.

Table 9.1 Vision 2000 (SEAOC 1995) definition of performance levels

Performance level	Performance description	Story drift (%)
Fully operational	Continuous service, negligible damage	0.2
Operational	Safe for occupancy, light damage, repairs for nonessential operations	0.5
Life safety	Moderate damage, life safety protected, repair possible but may be impractical	1.5
Near collapse	Severe damage, collapse prevented, falling of nonstructural elements	2.5

		Target Building Performance Levels			
		Operational Performance Level (1-A)	Immediate Occupancy Performance Level (1-B)	Life Safety Performance Level (3-C)	Collapse Prevention Performance Level (5-E)
Earthquake Hazard Level	50%/50 year	a	b	c	d
	20%/50 year	e	f	g	h
	BSE-1 ≈10%/50 year	i	j	(k)	l
	BSE-2 ≈2%/50 year	m	n	o	(p)

Fig. 9.2 Relationship between earthquake hazard and building performance, from FEMA 356

Currently, the specifications for the required structural response improvement can be found in several commonly employed code standards for building assessment. These include the European EC8, part 3 (CEN 2005); FEMA 356 (2000) and ASCE 41-06 (2006), used in the United States; and the New Zealand Society of Earthquake Engineering—NZSEE (2006)—recommendations on "earthquake risk buildings."

According to the aforementioned code standards, the building assessment and retrofit are required to comply with specific damage states considered acceptable for established seismic action levels. In this context, following the previously mentioned prescriptions primarily established by the Vision 2000 document, it is a common practice to expect the building to perform with moderate damage for a "current" seismic event, generally characterized by a 10 % probability of being exceeded in 50 years and sustain collapse under a "rare" event with a usually accepted 2 % probability of being exceeded in 50 years, as stated in FEMA 356 and ASCE 41-06. Accordingly, as presented in Fig. 9.2 (reproduced from FEMA 356), this philosophy defines a *Basic Safety Objective* (BSO) that approximates the earthquake risk to life safety levels traditionally considered acceptable.

Seismic codes for the earthquake-resistant design and retrofit of structures are fundamental for the mitigation of seismic risk. As duly mentioned, codes provide guidance for engineers on the analysis of the effects of earthquake ground motions on structures, as well as on the required configuration and detailing for improved seismic performance. However, the experience of the 1994 Northridge earthquake and the 1995 Great Hanshin (Kobe) earthquake demonstrated that both economic *losses* and human casualties could be considerable, even if the non-collapse objective had been met for many structures (Bommer and Pinho 2006).

It is thus clear that an approach based on code-defined performance objectives for specific levels of ground motion intensity is unable to fully take into account the seismic hazard characteristics of the site in question, as well as the expected *loss* for a period of time (i.e., the expected level of *loss* in a period of time arising from a probabilistic seismic risk analysis).

From a design point of view, a risk-based methodology has been an object of research, such as the risk-targeted approach presented by Luco et al. (2007), which considers only the no-collapse limit state in a probabilistic formulation for definition of design seismic demand. Furthermore, the degree of seismic detailing required to meet performance objectives other than collapse at lower intensities is investigated by Bommer et al. (2005), by means of an iterative procedure to determine the benefit of different strengthening scenarios in resulting earthquake *losses*.

In addition, Crowley et al. (2012) developed a proposal for calibration of codes for performance-oriented design, according to the quantitative comparison of incremental costs of improving the seismic resistance of a building with the associated avoided *losses*, expressed as the expected annual loss for a given region (northwest area of Turkey). The proposed procedure presents a similar philosophy to a methodology developed at the PEER (Porter 2003) for probabilistic-based seismic design, taking into account the *loss* in terms of costs, casualties, and downtime for increasing levels of hazard from a cost-benefit viewpoint. However, the PEER approach is applicable only for building-specific design and thus does not consider the convolution of hazard and vulnerability on an urban scale, such as the work of Crowley et al. (2012).

To what concerns the seismic assessment and retrofit of existing buildings in the context of a risk-based methodology, the most recent effort is related to the development of *next-generation performance-based seismic design procedures and guidelines for application to new and existing buildings*, under the scope of ATC-58 project. As stated in the related publication FEMA P-58-1 (FEMA 2012), the next-generation design (and assessment) procedures address performance objectives as statements of the acceptable risk of incurring casualties, direct economic *loss* (repair costs), and occupancy interruption time (downtime) associated with repair or replacement of damaged structural and nonstructural building elements. Accordingly, the building performance is determined through a complete seismic risk analysis, combining all its features for an alternative and more meaningful way of communicating risk with stakeholders and other decision-making entities with regard to improvement of seismic resilience of buildings.

As additionally stated in FEMA P-58-1 (FEMA 2012), from a decision-making point of view, time-based assessments can be used to provide the average annual value of *loss*, which is particularly useful in the context of *benefit-cost* studies for different retrofit strategies or building design. Hence, as more thoroughly presented in the remainder of this document, the average annual value of a performance measure—*loss* ratio—is considered as input to determine a reasonable investment for improved seismic resistance in a building, comparing the *net present value* (NPV) of average annual costs that are avoided throughout the building life cycle, versus the costs associated with providing an enhanced seismic resistance.

9.2.2 Energy Efficiency

Starting from the well-known reality in which buildings nowadays generate approximately 40 % of the world's carbon emissions, it is clear how the construction of more environmentally friendly buildings and the renovation of existing buildings will play a critical role in reducing these effects. Furthermore, it is generally recognized that a sustainable policy in the construction field is strongly advised.

Sustainability can be defined as the ability to meet the needs of the present without compromising the ability of future generations to meet their own requirements, creating and maintaining the conditions under which humans and nature can exist in productive harmony, making sure that we have and will continue to have the water, materials, and resources to protect human health and our environment. In this context, green building philosophy is the practice of creating structures and using processes that are environmentally responsible and resource efficient throughout a building's life cycle.

The market related to green construction has grown dramatically since the year 2000, and it is forecasted to continue to grow, despite of the expected decline in the overall construction market. It has been estimated that the total value of green construction was $10 billion in 2005, increasing to $49 billion by 2008. In addition, by the end of this current year, it estimates that the market could grow to as much as $96–140 billion (USGBC 2011).

As building owners select more environmentally friendly designs for their buildings, the demand for "green" services will continue to rise. Similarly, owners are aggressively retrofitting buildings in their existing portfolio to take advantage of reduced operating costs and to maintain or increase the value of their property. To this end, efficient energy use, sometimes simply called *energy efficiency* (EE), is the goal to reduce the amount of energy required to provide products and services. Taking into account that since 1994 energy prices have raised drastically (more than 30 %), it is clear that reduction of consumption for a building must be one of the principal aims in design and retrofitting. It is strategic though to have in mind what is going to be the return period or the break-even point related to the initial investment done to enhance the building efficiency.

For building owners, energy efficiency offers the opportunity to lower operating costs, enhance building quality, and increase financial returns. Standards such as LEED—a green building tool that addresses the entire building life cycle recognizing best-in-class building strategies—and Energy Star, an Environmental Protection Agency voluntary program to identify and promote energy-efficient products and buildings in order to reduce energy consumption, reflect and foster increasing interest in making buildings more energy efficient. However, the vast majority of EE opportunities remain unfinanced due to split incentives, insufficient credit, and limited data, among other reasons (USGBC 2011).

As with virtually all economic problems, the economics of energy efficiency is at its heart a question of balancing of costs and benefits. For the individual energy user, this involves weighting the higher initial cost of purchasing energy-efficient solutions with the expected benefits of future cost savings. Since energy efficiency is strictly related to environmental pollution associated with energy production—particularly carbon dioxide emissions from fossil fuels—it is important to realize in advance not only the cost-benefit in terms of economic savings but also in terms of the positive environmental impact, which is obviously an important benefit for the all community.

9.2.3 Integrated Approach

Given the fact that natural hazard risk mitigation has evolved independently of consideration for sustainable development goals, there tends to be a disconnection among stakeholders, as well as lack of consistent language and understanding across the field of natural hazard mitigation and those embracing sustainable development concepts (Tapia and Padgett 2012). Recent disasters have highlighted the need for a paradigm shift that allows the understanding of buildings' life-cycle costs from the "earthquake-induced environmental impact" point of view (Comber et al. 2012), so as to develop an overarching framework that enables the consideration of both perspectives.

It is important, in the present context, to adequately relate seismic performance and the economic/environmental benefits that can be realized by performance-based design and assessment procedures such as the ones presented in Sect. 9.2.1.1 of the present document. Hence, a simple way of approaching the problem is to consider the influence of the design/retrofit on seismic risk in such a way that enables the building's life span that is relied upon from an environmental point of view. Although valid from a descriptive perspective, the latter is clearly incomplete, since rather than the simple influence of building collapse into the expected environmental performance over a given expected period, the goal of an integrated approach is to adequately incorporate seismic risk analysis and its environmental impacts into a life-cycle assessment, considering the probabilistic distribution of possible building damage over a period of time.

A common practice in conducting an environmental life-cycle assessment (LCA) on a building includes a consideration of the impacts stemming from the first construction. Considerations are also often made for additional impacts stemming from building maintenance and energy usage throughout its life span, but very rarely the potential additional environmental impacts that could arise from repairing or demolishing after a natural disaster are included in an LCA, if ever at all (Comber et al. 2012). To this end, discussions have begun to surface in the literature that highlight the conceptual environmental impacts of individual buildings (Kneer and Maclise 2008a, b) and regional impacts (Burton et al. 2011) resulting from seismic damage. In fact, in areas of high seismicity, an opportunity exists for the structural engineering community to carefully consider the level of risk and employ seismic damage mitigation strategies that effectively minimize the probabilistic integrated life-cycle impacts.

Frameworks incorporating natural hazard risks and their mitigation into the LCA to assess sustainability of infrastructures have recently begun to emerge. Itoh et al. (2005) proposed a model to determine life-cycle environmental impacts and costs from construction and maintenance phases as well as from losses and recovery after an earthquake event, using CO_2 emissions as the environmental impact indicator. In addition, natural hazard considerations were integrated into the LCA to quantify environmental among other metrics of sustainability (social and economic) by Padgett et al. (2009) and expanded by Dennemann (2009). The sustainability metrics considered in the latter for the specific case of non-seismically designed bridges are life-cycle cost (due to repair/replacement from seismic exposure), functionality/downtime and expected number of fatalities due to bridge failure over structure life, and energy usage associated with repair and reconstruction of damaged structures.

Two recent studies by Ghosh et al. (2011) and Tapia et al. (2011) address a number of shortcomings of the aforementioned works. These studies propose frameworks for quantifying life-cycle cost, embodied energy, and CO_2 emissions for structures subjected to aging, deterioration, and seismic action. LCA is thus simplified by only including the repair and replacement phases of these structures, although in a probabilistic fashion. The inherent results underscore the extent to which natural hazard mitigation can reduce environmental- and economic-related sustainability metrics as well as the importance of their evaluation (Tapia and Padgett 2012).

The aforementioned research campaigns constitute valid, yet simplified, attempts to join seismic risk and environmental impacts in a comprehensive integrated approach. However, they are based on the specific case of bridges, which greatly differ from the object of this study in the structural parameters involved.

A recent study developed by Comber et al. (2012) has led to the development of a method for quantifying the impacts of seismic damage to the full building system including nonstructural components and contents that maintains a level of approximation considered appropriate given the nature of the two procedures that are being paired together: seismic *loss* estimation and LCA. The seismic *loss* assessment methodology selected for incorporation into the authors' approach is the Advanced Engineering Building Module (AEBM) of the Hazards US (HAZUS) methodology, which was developed by FEMA (FEMA 1999). The latter is explored in assessing the expected seismic *losses* in building components and contents, paired with an extensive component database that enables the link between component damage and the correspondent environmental impact. A probabilistic seismic hazard assessment was conducted in order to annualize the environmental impacts due to seismic damage for a case study building (with different design schemes), expressed in terms of equivalent embodied annual carbon footprint. It seems clear that the presented methodology does not take into account energy efficiency in the consideration of the annualized environmental impact. Differently, it incorporates such impacts as resulting from the expected seismic damage (and inherent repairs), excluding the carbon footprint associated with consumed energy over the same time period. However, comparing impacts from annualized seismic damage with its energy use counterpart is a possible and valuable exercise, as exemplified in the work of Comber et al. (2012). In fact, considering the energy efficiency improvement currently being targeted by the US federal government, buildings such as the ones assessed in the referred study are expected to benefit from a 20 % decrease in energy usage through efficiency enhancements (DOE 2006), which is in the same order of magnitude as the probabilistically realized reduction resulting from a seismic retrofit scheme, as verified by Comber et al. (2012), for a specific case study. According to the latter, this case study shows that in some buildings, be them new construction or seismic retrofit projects, the increase in initial environmental impact that can result from constructing the building to an enhanced seismic performance objective may result in a net overall reduction throughout the building's life span, due to the achieved reduction in the expected annual level of CO_2 emissions resulting from seismic damage repair operations. Therefore, in order to achieve a more accurate sense of what type of seismic system should be used to minimize a building's full life-cycle impact, a consideration of the expected seismic damage over time should be incorporated into the LCA (Comber et al. 2012).

From a building performance analysis and *loss* assessment point of view, the previously presented HAZUS methodology imparts a number of simplifications that limit the applicability scope of the proposed methodology. In this context, the recently proposed ACT-58 method (FEMA 2012) for performance-based seismic design and retrofit of buildings constitutes a comprehensive and generally applicable methodology for the determination of building-specific *loss* due to seismic action. Thus, the purpose of the related ATC-86 project (Court et al. 2012) is to develop a performance-based environmental impact assessment methodology to integrate into the P-58 procedures and its companion Performance Assessment Calculation Tool (PACT). Moreover, the ATC-86 approach should account specifically for the environmental repercussions of the probable earthquake damages and repairs predicted by the P-58 method (Court et al. 2012).

Life-cycle assessment provides a natural framework for this purpose, traditionally accounting for environmental impacts over an entire building life cycle from cradle to grave. Although the latter does not typically account for earthquake effects, it incorporates it by taking advantage of the P-58 powerful tool for predicting earthquake damage and its consequences. Hence, it offers a unique opportunity to quantify the probable earthquake impacts and add them to the full building life-cycle assessment. It consists of a procedure for measuring the environmental performance of products or processes over their full life cycles from cradle to grave. In this context, environmental metrics typically include global warming potential, embodied and operational energy, natural resource consumption, waste generation, and a broad range of environmental pollutant impacts. To this end, the International Standards Organization in its ISO 14040 series (ISO 2006a, b) provides appropriate life-cycle assessment guidelines (Court et al. 2012).

As in the case of the work of Comber et al. (2012), the environmental implications of retrofit interventions under the scope of a life-cycle assessment are determined according to life-cycle inventories (LCIs) that list all the energy flows associated with material components during their life cycle, quantifying average material emissions for different material processes. In the context of the P-58 methodology, the environmental impact measures are combined with a bill of materials and processes associated with earthquake damage cleanup and repair, so as to generate measures for the building as a whole. Such matters will be addressed in the P-58 and ATC-86 methodologies (Court et al. 2012), in order to establish how and to which extent design or retrofit of buildings for better seismic performance could reduce the damage-related impacts, depending on the rate of occurrence of damaging earthquakes during a building's service life.

It should be mentioned, however, that the aforementioned project (ATC-86) is currently in a development stage, which makes the access to its definitive conclusions and proposed methodology unavailable.

9.3 Seismic Resilience and Energy Efficiency: A Model for a Common Approach

9.3.1 Scope Definition and Initial Assumptions

Following the literature review examples analyzed in Sect. 9.2.3, a proposal for the integrated assessment of energy efficiency and earthquake resilience is presented. As aforementioned, the more recent efforts to account for both types of demands tackle the problem of integrating the building environmental performance as a secondary variable, depending on the level of earthquake *loss*. Under this assumption, the inherent performance measures reflect the annualized carbon footprint due to the expected level of repair or retrofit operations resulting from earthquake damage.

Although the above can provide valid information in the context of a decision-making process, it does not directly reflect the main subject under scrutiny, i.e., the potential economic benefits of an integrated investment approach on seismic and energy efficiency building improvement. To this end, the consideration of a multidisciplinary procedure in which environmental and economic variables are incorporated in the investment decision parameters is tackled in this document, as further presented.

In the context of seismic *risk* assessment, one of the most important outputs is the frequency that the possible levels of *loss* will be equaled or exceeded on an annual basis, as illustrated in Fig. 9.3, in which the so-called *loss exceedance curve* is graphically represented.

However, even though particularly useful information can be derived from the latter, the communication of seismic *risk* with stakeholders, insurance companies, governmental agencies, and other decision-making entities requires the definition of a performance metric that can adequately translate the mentioned information in a concise and readily applicable fashion. Accordingly, the concept of *expected*

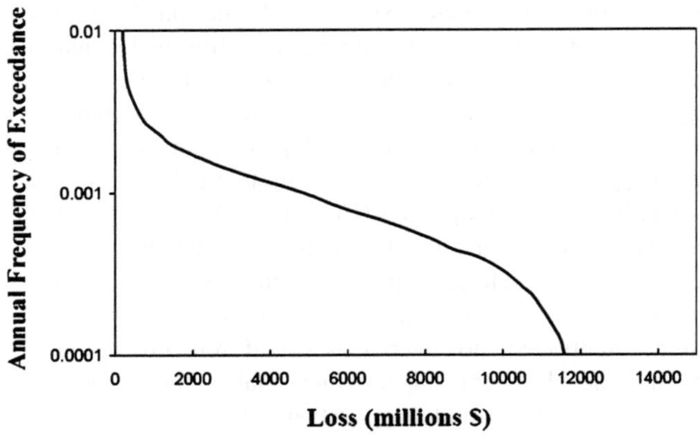

Fig. 9.3 Illustrative example of a loss exceedance curve

annual loss is usually considered as the final output of the assessment, calculated as the expected value of the referred *loss exceedance curve*, i.e., the average value of *loss* that a building (or group of buildings) will sustain annually over its life span due to seismic action.

According to the aforementioned, by approaching the present problem simply from a seismic resilience point of view, the selection of *expected annual loss* as a comparison parameter for a cost-benefit analysis seems particularly adequate. As subsequently enunciated, an analog reasoning can be employed when incorporating the energy efficiency analysis in the equation, by defining an *energy expected annual loss* (EAL_E) that can directly be compared with its seismic counterpart (EAL_S). Thus, if one chooses to consider the building value (or cost to replace) as the common ratio denominator, the *energy loss* can be determined as the ratio between cost of consumed energy and building value, as follows:

$$\text{seismic loss} = \frac{\text{cost to repair}}{\text{total building value}}; \quad \text{energy loss} = \frac{\text{energy cost}}{\text{total building value}}$$

As further presented in detail (see Sect. 9.3.2.1), it is common practice to classify building energy efficiency according to its expected annual energy consumption, which is of particular interest in the present context. Accordingly, the mentioned value of *energy expected annual loss* (EAL_E) can be determined as the ratio between the average annual cost of consumed energy and total building value, rendering a compatible base of analysis and comparison:

$$EAL_E = \frac{\text{mean annual energy cost}}{\text{total building value}}; \quad EAL_S = \frac{\text{expected seismic loss}}{\text{total building value}}$$

9.3.2 Framework Proposal

The herein presented framework consists of an initial model for a cost-benefit analysis methodology to test different scenarios for tackling energy efficiency and earthquake resilience. Starting from a classification system according to both "green" and "resilient" indicators, initial condition and "upgraded" configurations are considered and building performance evaluated.

The investment required for a given building performance improvement is related to its benefit over time through appropriate indicators and return ratios, allowing for the evaluation of different possibilities for strengthening and refurbishment, based on current best practices. The appraised results shall therefore lead to the identification of potential financial benefits from implementing an integrated approach, as well as pointing out possible challenges to be faced in its practical application.

9.3.2.1 Building Performance Classification

In order to establish the base for building assessment and improvement, a consistent proxy for building classification is required. To do so, building performance—from earthquake resilience and energy efficiency points of view—shall be divided into two directly comparable and cumulative components of a comprehensive *green and resilient* indicator (*GRI*).

The main advantage of such classification is the possibility of separately analyzing the building in terms of earthquake resilience and energy efficiency, determining the advantages and associated costs of improving its capacity to a preestablished level of performance, determined by the *GRI*. As proposed in the work of Calvi (2013), similarly to what is done when evaluating the energy and environmental performance of buildings, discrete classes of earthquake resilience might be defined, in accordance with the predicted value of *expected annual loss*. Therefore, since both components of *EAL* are expressed in terms of analog quantities (see Sect. 9.3.1), the cost and benefit of the separated approaches can be conjugated.

For the purpose of the present exercise, as a starting point, the preliminary classification system proposal is based on a comprehensive approach in which the energy efficiency classification in Italy (D.P.R. 2009), which is based on European legislation on the matter, is further "translated" into categories based on the previously enunciated *energy expected annual loss* (EAL_E). The Italian building classification and its regional specifications (DGR Emilia-Romagna n.1366/2011) provide indications for different types of buildings, according to the total (annual) energy consumption due to heating and water supply. As presented in Fig. 9.4, eight classes of performance are established for residential buildings. The indicator used (EP—primary energy) is the amount of energy needed to heat the building, according to the calculation criteria of the UNITS 11300 (2008a, b, 2012), expressed in kWh/m^2.

Fig. 9.4 Italian building energy classification system, Emilia-Romagna region (DGR Emilia-Romagna n.1366/2011) for residential buildings

Energy Class	***Residential*** **- kWh/m²**
A+	$EP_{tot} < 25$
A	$25 < EP_{tot} < 40$
B	$40 < EP_{tot} < 60$
C	$60 < EP_{tot} < 90$
D	$90 < EP_{tot} < 130$
E	$130 < EP_{tot} < 170$
F	$170 < EP_{tot} < 210$
G	$EP_{tot} > 210$

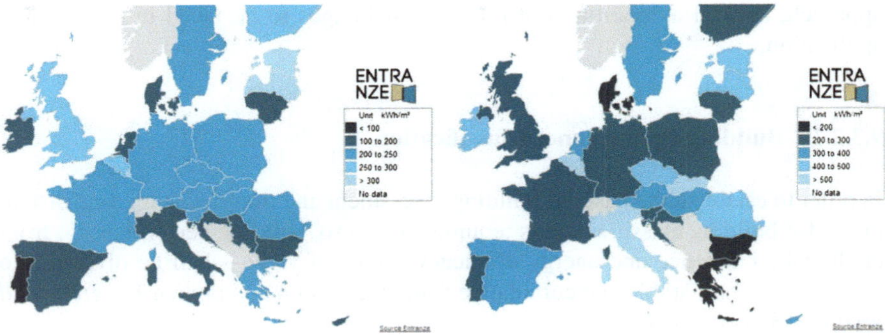

Fig. 9.5 European countries' average annual energy consumption for residential (*left*) and non-residential (*right*) buildings

As presented in Fig. 9.5, the aforementioned classification satisfactory agrees with the annual energy consumption data from the European residential building portfolio, provided by the ENTRANZE project, financed by the European Union (ENTRANZE project: http://www.entranze.eu/), in which the 27 EU members are scrutinized. The referred data corresponds to the year 2008, so as to consider data not affected by the 2009 international financial and economic crisis, as stated in the project data interface (ENTRANZE project database interface: http://www.entranze.enerdata.eu/).

However, to what concerns the nonresidential building data, it is clear that the classes presented in Fig. 9.5 are not able to adequately characterize the expected levels of building energy consumption. Thus, in order to outline a more representative set of classification indicators, taking into account both the residential classification system (Fig. 9.4) and the information presented in Fig. 9.5, the following energy classes are considered:

Taking into account the latest indication of average construction cost values in Europe, provided by the European statistical bureau—EUROSTAT—in its report *A decade and more of monthly construction statistics* (Fassbender 2007), and current values of energy cost (EUR/kWh), the energy classes presented in Fig. 9.6 can be expressed in terms of EAL_E ratios, as follows:

$$EAL_E = \frac{\text{mean annual energy cost [EUR]}}{\text{total building value [EUR]}} = \frac{\text{mean net annual energy cost } [\text{EUR/m}^2]}{\text{construction cost } [\text{EUR/m}^2]}$$

where the *mean net annual energy cost* stands for the average cost of consumed energy divided by the building area and construction cost is represented by the ratio between building value (or "cost to replace") and its construction area. Therefore,

$$EAL_E = \frac{\text{EP}_{\text{TOT}}\,[\text{kWh/m}^2] \bullet \text{unit energy cost [EUR/kWh]}}{\text{construction cost } [\text{EUR/m}^2]}$$

According to the above, the *GRI* classification, particularly in what concerns its *green* indicator, can be considered as presented in Fig. 9.7.

Given the fact that the values obtained in the above procedure for the *green* indicator can furthermore be considered reasonable assumptions from an earthquake resilience classification point of view, to what concerns the typical values of EAL_S

Fig. 9.6 Adopted energy classification system, for residential and nonresidential buildings

Energy Class	***Residential* -kWh/m²**
A+	$EP_{tot} < 25$
A	$25 < EP_{tot} < 40$
B	$40 < EP_{tot} < 80$
C	$80 < EP_{tot} < 130$
D	$130 < EP_{tot} < 170$
E	$170 < EP_{tot} < 250$
F	$250 < EP_{tot} < 400$
G	$EP_{tot} > 400$

Fig. 9.7 *GRI* classification proposal for energy efficiency

***Green* Indicator**	**EAL_E (%)**
A+	$EAL < 0.50$
A	$0.50 < EAL < 0.75$
B	$0.75 < EAL < 1.50$
C	$1.50 < EAL < 2.50$
D	$2.50 < EAL < 3.50$
E	$3.50 < EAL < 4.50$
F	$4.50 < EAL < 7.50$
G	$EAL > 7.50$

Fig. 9.8 *GRI* classification proposal for energy efficiency and seismic resilience

GRI Classification (*Green* and *Resilient* Indicators)	EAL_E or EAL_S (%)
A+	$EAL < 0.50$
A	$0.50 < EAL < 0.75$
B	$0.75 < EAL < 1.50$
C	$1.50 < EAL < 2.50$
D	$2.50 < EAL < 3.50$
E	$3.50 < EAL < 4.50$
F	$4.50 < EAL < 7.50$

registered for reinforced concrete buildings, it is decided to assume a unique *GRI* for both seismic performance and energy efficiency assessments, for simplification purposes, as a function of EAL_S and EAL_E, respectively. Thus, the classes A+ to G presented in Fig. 9.7 are at this point referred to as parameters of the *GRI* (*green and resilient* indicator), as definitively presented in Fig. 9.8.

As previously mentioned, the presented building classification intents to serve the purposes of the present exercise and shall, therefore, be considered as a preliminary indication that requires further refinement. Nonetheless, the presented proposal consists of a valid preliminary proxy for building energy efficiency and seismic resilience, allowing for a cost-benefit analysis of the advantages/shortcomings of an integrated *GRI* improvement approach, as further presented.

9.4 Seismic Resilience and Energy Efficiency: Model Application

9.4.1 Case Studies

Three buildings are analyzed according to different scenarios of initial design, seismic and energy efficiency demands, and strengthening/refurbishing possibilities, so as to assess the influence of the interested variables in building performance over its expected life cycle. Moreover, a general overview of the advantages/shortcomings of an integrated approach, as well as the circumstances in which its application is deemed to be beneficial, is presented, reflecting the influence of different seismic and energy efficiency demands in the buildings' *expected annual loss* (EAL_E and EAL_S), according to the selected "building-specific environment":

- Location of the building (climate zone in which the building is constructed and seismic hazard at the site)
- The function of the building itself (residential, healthcare, and public administration, in the present case)

Fig. 9.9 3D representation of case study A

It shall furthermore be mentioned that, for the purpose of assessing the building performance in terms of both earthquake resilience and energy efficiency in conditions different from those foreseen in design stages, additional locations have been considered. Therefore, in addition to the design locations presented in Sects. 9.4.1.1, 9.4.1.2, and 9.4.1.3, supplementary hypothetical building locations have been established for each of the further presented case studies, as defined by the climate and seismic hazard characteristics of Istanbul (Turkey), Yerevan (Armenia), and Messina (Italy).

9.4.1.1 Case Study A

The first case study refers to a residential complex constructed in Pavia, Italy, constituted by five seven-story buildings with a 30 m wide square plan configuration, as illustrated in Fig. 9.9. Each building features an inner courtyard of 18 m on each side bounded by a gallery that provides access to individual apartments by vertical connections.

The reinforced concrete structural system complies a solution in which precast columns are conjugated with cast-in-place concrete slabs and central core. Furthermore, all floors are made of reinforced concrete solid slabs and the cover is made of metal carpentry.

The "skin" of the buildings has been provided with several devices in order to minimize the heat loss. The final exterior surface, constructed of pre-colored silicate paste, guarantees the minimization of thermal bridges around the casing, ensuring the total continuity of technological heat-insulating layers in the classic points of disconnection. On the upper floors, a ventilated wall with aluminum substructure and mantle brick was implemented, in order to replace the thermal coat. In addition,

Fig. 9.10 3D representation of case study B

the windows made of PVC maximize energy performance and ensure at the same time lightness and maneuverability due to its reduced weight.

The heat and air conditioning system is based on the use of geothermal energy, resorting to water-to-water heat pumps that provide hot and cold water for heating, hot water for sanitary use, and refrigerated water. In addition, an integrated photovoltaic system has been implemented on the roof of the buildings, allowing the production of a peak power of approximately 195.5 kWp. The electricity produced is used for the maintenance of heating, cooling, and domestic hot water production.

Each apartment is provided with a satellite counter to estimate the real consumption of heat and cold water, together with electricity, and a building management system is installed in order to overview and control the functionality of the all generation systems.

9.4.1.2 Case Study B

Case study B represents a school constructed in Emilia-Romagna, Italy (Fig. 9.10) in which the structural solution establishes an approach toward the optimization of the building process. It involves the construction of a continuous reinforced concrete foundation slab, above which the building structural system will be implemented. The lateral load-resisting system is materialized by composite cast-in-place columns and precast reinforced concrete beams in longitudinal and transversal directions. The columns are circular with outer jacket in stainless steel having an outer diameter of 600 mm and thickness equal to 6.3 mm, which fulfills the function of both formwork and confinement.

Particular attention was given to the building envelope concerning the choice of thermal performance of materials. The window frames proposed are highly efficient, and matt surfaces are materialized by pre-insulated sandwich panels with the addition of mineral wool and counter wall. In addition, highly efficient heat pumps are used for winter and summer acclimatization, in order to achieve an

Fig. 9.11 3D representation of case study C

energy class of excellence. The technological solutions are all characterized by a high standard of quality, through the use of leading brands leader in the field of air conditioning and heating.

9.4.1.3 Case Study C

Case study C refers to a healthcare building constructed in Milan, Italy, illustrated in Fig. 9.11 by a 3D building representation. The building is constituted by a main body of four floors inscribed in a plan configuration of approximately $104 \times 70\ m^2$ and total height of 20.1 m.

The vertical elements and lateral load-resisting system are materialized by reinforced concrete columns and walls of variable dimensions.

The building envelope is designed in order to obtain an A building energy class, using high-performance technology solutions in both parts of opaque shell, as well as low-emission windows with argon gas for the transparent casing. The heat and air conditioning system is based on the use of geothermal energy, resorting to water-to-water heat pumps that provide hot and cold water for heating, hot water for sanitary use, and refrigerated water. In addition, highly performing heat exchangers ensure the air treatment and provide new air across the building.

9.4.2 Seismic Hazard

The aim of hazard analysis for a site is to estimate the rate of exceedance of a given ground motion intensity measure, i.e., that with a given annual probability of exceedance or average return period. The intensity measure traditionally used to represent the seismic hazard has been the peak ground acceleration (PGA), possibly associated with a spectral shape to immediately estimate a structure acceleration.

Once made clear that displacements are more relevant than accelerations, it may appear rational to shift to a displacement response spectrum as key intensity measure. Someone may state that there is no difference, since displacement spectrum ordinates (S_Δ) can be derived from pseudo-acceleration spectrum ordinates (S_A) as a function of the period of vibration (T): $S_\Delta(T) = (T/2\pi)^2 S_A(T)$. This is a misleading observation, since peak displacement and corner period are essentially affected by the long-period part of the acceleration spectrum, where accelerations are low and consequently little attention is paid (Calvi 2013).

The discussion becomes more complex when the problem to be addressed is how to define the seismic intensity parameter based on seismic sources and their assessed potential to induce given magnitude earthquakes with specific recurrence intervals. The standard approach is a site-specific *probabilistic seismic hazard analysis* (PSHA), based on the separate consideration of all possible earthquakes that could occur within a given region, their frequency of occurrence, and the levels of ground motion intensity they could produce at a given site.

For the purpose of this exercise, the outcome of the FP7 European project SHARE (Giardini et al. 2013) has been utilized for the implementation of the corresponding PSHA model, which has been used for the computation of uniform hazard spectra for the locations referred in Sect. 9.4.1. To this end, it should be mentioned that due to the lack of publicly available information regarding seismic hazard results for Armenian territory (to the authors' knowledge), Yerevan and Istanbul are hereby considered identical for seismic analysis purposes.

9.4.3 Energy Demand

The most commonly used methodology for the definition of energy demand is stationary and based on heating degree days. Heating degree day (HDD) is a measurement designed to reflect the demand for energy needed to heat a building and is defined as a measure of the coldness of the weather experienced. A day's average temperature gives some idea of how much energy will be required on that day for heating purposes. The calculation procedure assumes some energy will be required on any day that has an average outdoor temperature of less than 18 °C (65 °F), computing the heating needs for each day by subtracting the day's average temperature from 18 °C. The result is the number of heating degrees for that day or HDDs.

The energy demand for cooling, which varies significantly as a function of the climatic region and building use, has also been taken into account in all the case studies presented in Sect. 9.4.1. In extreme cases such as healthcare facilities, the cooling demand can indeed be very high in the summer period and possibly necessary even in the winter season. On the other hand, school buildings are usually associated with low or negligible cooling demands, given the fact that building use is interrupted in the summer period.

Italy is divided in six climatic zones from A to F, as shown in Fig. 9.12, based on the values of degree days. The Italian code, such as the majority of the European

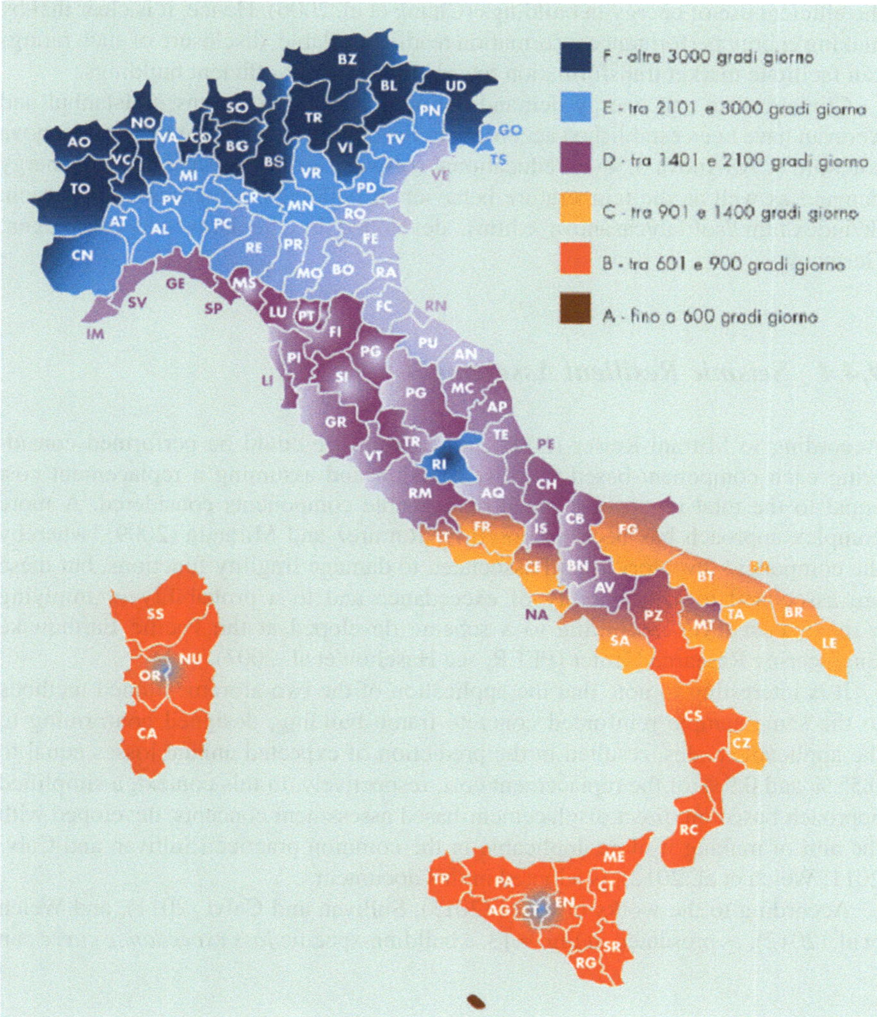

Fig. 9.12 Italian climatic zones, referred to degree day units

codes, defines both the climatic zone and the time interval in which heating system can be used, in order to reduce energy consumption.

According to the above, it is clear how benchmarking energy use can provide a mechanism for measuring how efficiently a building uses energy in relation to other similar buildings or modeled simulations of a building built to code or desired standard. Benchmarking models developed from energy efficiency indicators are valuable tools for both the government and the private sector in managing energy consumption. Some governments have used these tools to formulate policies for the efficient use of energy in buildings (Chung et al. 2006). Hence, it is clear that by making energy performance information readily available, disclosure of such ratings can facilitate market transformation toward more energy-efficient buildings.

Climatic data and energy demand definition for the locations of Istanbul and Yerevan have been established according to the definition provided by the Casanova software (CASAnova 3.3—an educational software for heating and cooling energy demand as well as the temperature behavior of buildings: http://nesa1.uni-siegen.de/index.htm?/softlab/casanova_e.htm), developed by the University of Siegen, Germany.

9.4.4 Seismic Resilient Assessment

According to Mitrani-Reiser (2007), a loss estimate could be performed considering each component-based fragility function and assuming a replacement cost equal to the total repair cost of all damageable components considered. A more complex approach has been proposed by Ramirez and Miranda (2009), whereby the component inventory is still associated to damage fragility functions, but these are associated to a probability of exceedance and to a probability of implying a repair cost level, according to a scheme developed at the Pacific Earthquake Engineering Research Center (PEER, see Haselton et al. 2007).

It is interesting to note that the application of the two aforementioned methods to the same sample reinforced concrete frame building, designed conforming to the applicable codes, resulted in the prediction of expected annual losses equal to 0.55 % and 0.9 % of the replacement cost, respectively. In this context, a simplified approach based on direct displacement-based assessment concepts, developed with the aim of making it more applicable in the common practice (Sullivan and Calvi 2011; Welch et al. 2012), is adopted in this document.

According to the works of Calvi (2013), Sullivan and Calvi (2011), and Welch et al. (2012), as presented in Fig. 9.13, a building-specific *loss exceedance curve* can

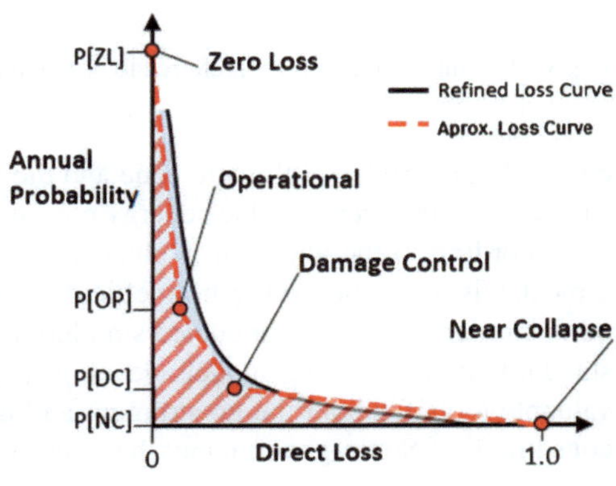

Fig. 9.13 Schematic representation of simplified building loss curve (from Welch et al. 2012)

be approximated by a simplified *loss exceedance curve* defined by four points that represent the annual probability of exceedance of particular damage states.

These damage states can adequately represent the building behavior, which is intuitively apprehended by the fact that lower *loss* ratios, usually related to operational damage states (building is considered fully or nearly fully operational), are related to lower seismic intensities, which have a higher probability of being exceeded, and a more severe damage state is expected to be related to both higher *loss* ratio and lower annual probability. Hence, by defining the annual probability of exceedance and *loss* ratios associated with the referred damage states (zero loss, operational, damage control, and collapse, as presented in Fig. 9.13), it is possible to completely compute the loss exceedance curve of a particular building, subjected to a given site seismic hazard.

The definition of the aforementioned damage states requires a relatively extensive discussion and will not be presently addressed for the sake of synthesis, referring the reader to the aforementioned publications. However, different retrofit techniques will influence the building performance at different damage states. Accordingly, given the fact that the *expected annual loss* is given by the area of the *loss curve* (Fig. 9.13), improving the building performance to less severe performance objectives (such as "operational" and "damage control"), therefore reducing its annual probability of being exceeded, will have a significantly greater impact in the final value of *expected annual loss* than the prevention of collapse or near collapse.

In the aforementioned framework, the objective of loss analysis is to calculate the probable repair cost for each level of damage state defined above. In principle, a loss estimate should include death toll, repair cost, and downtime consequences (the well-known *3D approach*; see Fajfar and Krawinkler 2004). However, one chooses to neglect the value of human life in the appraisal of seismic-induced *losses*. To this end, the seismic *loss* can strictly be evaluated in terms of the ratio of *cost to repai*r to *cost to rebuild*, to which the associated indirect (downtime) *losses* are appended. For the sake of simplification, indirect (or downtime) *losses* are not considered in the present exercise.

Similarly to what has been performed in the work of Calvi (2013), for each assessed building and specified location, site seismicity has been assessed, and a complete adaptive pushover analysis has been run. For each damage state and corresponding return period, drift and floor acceleration have been computed (as average values), considering each floor, and the corresponding damage has been calculated. In its extreme simplification, one can assume that the cost of repair will be proportional to damage, for example, associating 75 % of the value of the building to nonstructural content and 25 % to structures, as proposed by Calvi (2013). The nonstructural part has been considered 80 % drift sensitive and 20 % acceleration sensitive, and each story has been assumed of equal value. Furthermore, assumptions on floor acceleration-associated and drift-associated nonstructural damage fragility, as well as drift-associated structural damage fragility, shall be referred to the work of Calvi (2013), as illustrated in Fig. 9.14.

Again, this can be modified and elaborated, and probabilistic aspects can be included without losing generality. In this framework, however, the attention will

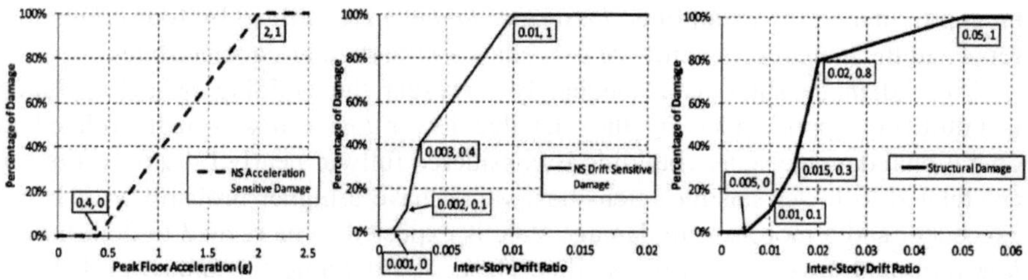

Fig. 9.14 Assumptions on floor acceleration-associated nonstructural damage (*left*, weight 20 %), drift-associated nonstructural damage (*center*, weight 80 %), and drift-associated structural damage (*right*), from Calvi (2013)

be focused on cost of different strengthening measures and on their comparison with a reduction in the expected repair costs, to obtain some cost-benefit evaluation to drive the strengthening choices.

9.4.4.1 Implemented Improvements

The subject of selecting the appropriate technical solution lies on the evaluation of the available techniques and its repercussion in a particular feature of building performance, as well as social, economic, and historical environment. As stated by Thermou and Elnashai (2006), the selection of the rehabilitation scheme and the level of intervention are a rather complex procedure, due to the fact that many aspects of different nature come into play when deciding which strategy to employ.

As supported by Calvi (2013), the possible alternative choices to be considered are schematically subdivided into three conceptual categories: the modification of damage and collapse modes (Elnashai and Pinho 1998; Thermou et al. 2007), the introduction of isolation systems (Filiatrault and Christopoulos 2006), and the reduction of the displacement demand by added damping or tuned masses (Sadek et al. 1997). In this context, this section provides a brief overview of the strengthening measures considered in this exercise (see Sect. 9.5), highlighting its relative effects, with no attempt to depict a complete and thorough presentation, impossible within the space constrains of an article:

- *Strengthening elements* ("retrofit 1"): A basic strategy to improve the seismic response of a building counts on the application of capacity design principles to eliminate all possible sources of brittle failures. In this context it is thus possibly required to increase some element strength in a selected way, to favor ductile damage modes. For example, it is typical to increase the shear strength of columns and beams to obtain flexural failure modes, to increase the strength of external joints, and to increase the flexural strength of columns to shift the formation of plastic hinges to the beams. This last example does not aim to avoid a brittle collapse, but to prevent the formation of a soft story. In some case (possibly academic), weakening ductile modes have been considered instead of strengthening brittle ones. These kinds of intervention tend to modify in a significant way the last part of the pushover curve (and possibly the associated vertical deflected shape), increasing the displacement capacity of the structure, but may have a negligible effect on the first part of the curve (the modification of stiffness up to yielding is not significantly affected) and on the yield strength (since, in general, shear and flexural failure modes have similar strengths). Typical interventions are based on an external jacketing or wrapping of an element or part of it, using carbon or glass fibers, steel plates, or thin layers of reinforced concrete. It is obvious that the cost of strengthening of an element will vary significantly, as a function of its geometry and of the applied technique; therefore, estimates provided by Calvi (2013) are presently considered.

- *Locally increasing the deformation capacity* ("retrofit 2"): If it is assumed that all the possible brittle failure modes have been eliminated by a proper application of capacity design principles, i.e., by an appropriately selected local element strengthening, the displacement capacity of the structure can be limited by insufficient curvature (and consequently rotation) capacity in critical section of columns and beams. An insufficient rotation capacity of columns can only be detected in the case of a soft story formation or exclusively at the column base. Note that a soft story mechanism is not always unacceptable; it depends on the associated story rotation capacity (including second-order effects) and the associated global displacement capacity (Calvi 2013). These kinds of intervention are normally based on confining measures, to avoid bar bucking and increase the compression deformation capacity of concrete. Fiber wrapping and steel encasing are thus again the typical choices, which can, however, be limited to the critical zones of the elements. The effects on strength and stiffness will be even more negligible than in the previous case, and the effects will be still limited to the last part of the pushover curve. The cost per structural member is therefore lower, but of the same order of magnitude, of that discussed at the previous point (Calvi 2013). In the discussion of relative merits of different strengthening choices, no distinction will be made about these two kinds of strengthening for what concerns costs.
- *Inserting additional elements* ("retrofit 3"): A conceptually different approach to seismic strengthening relies on the insertion of additional elements reacting to horizontal actions. This is normally based on steel-braced frames or concrete walls (possibly obtained by strengthening masonry panels) that could be inserted in the interior of a building or outside it. If the primary reaction system of the original building was already made by walls, the purpose could be to increase strength and stiffness and to regularize the torsional response, thus reducing the expected damage even at relatively low displacement demand. If the original structural system was based on frames, the introduction of much stiffer elements may completely change the response, arriving at the limit that the original frame can give a negligible contribution to the response, and the only fundamental requirement will be that its displacement capacity will be larger than the displacement demand associated with the response of the new shear wall

system (Calvi 2013). In all cases two issues will have to be carefully considered: the capacity of the foundations corresponding to the new shear walls and the capacity of the horizontal diaphragm to transmit the action, globally and locally. To estimate the potential average cost of this kind of intervention, the approach proposed by Calvi (2013) will be assumed.

- *Capacity protection of existing structure* ("retrofit 4"): The insertion of an isolation system, at the base or at some height of the building, can often be a last-recourse intervention to improve the seismic performance of the building. The essence is that in this way the maximum shear that will pass through the system is governed by the system capacity (see, as a general reference, Filiatrault and Christopoulos 2006). As an example, imagine using friction pendulum devices: the maximum shear is essentially governed by the stick-slip value, which is typically in the range of two times the dynamic friction (Calvi et al. 2010). Using devices with a dynamic friction in the range of 3–5 %, it is likely that the maximum base shear force on the building will be in the range of 6–10 % of gravity. For earthquake events, or portions of events, that will not induce this level of acceleration, the structure is responding like a fixed base structure, while for any value of acceleration exceeding this value, the difference in base shear, and consequently in structural drift demand, will be marginal (Calvi 2013).

In general, the global cost is proportional to the plan area of the building, rather than to its volume, and may be as high as about 1200 EUR/m^2 for the most complex cases, including uplifting (Calvi 2013):

- *Introducing tuned masses* ("retrofit 5"): An example of intervention oriented to the reduction of demand is the introduction of a tuned mass. The general concept is simple: if the building can be regarded essentially as a single degree of freedom system, with most of its mass associated to the first mode of vibration, adding a tuned mass that vibrates with a similar period of vibration, but in the opposite phase, will induce a favorable reduction of shear force at all instants (Calvi 2013). This applies when the system responds essentially elastically. A complete description of the approach for seismic application is obviously more complex, and it can be shown that the maximum efficacy can be obtained for a specific ratio ($f = T_S/T_{TM}$) between the first period of vibration of the building (T_S) and the period of vibration of the tuned mass (T_{TM}) (see Sadek et al. 1997).

Again, it is difficult to give reliable figures of cost; however, as proposed by Calvi (2013), a rough estimate of a mean cost may be around 15 EUR per cubic meter of building.

For each one of the five strengthening strategies, a pushover curve has been recalculated, obtaining the following effects (Calvi 2013):

- For the case of member strengthening, the curve does not vary in the first part, but larger displacement capacity is achieved.
- Adding a shear wall system increases the shear capacity significantly, reducing the displacement capacity.

- Inserting a base isolation system does not induce changes in the first part of the curve, but the limitation of base shear avoids brittle failure modes.
- The additional damping case is not featured in this step, since its effect is considered on the demand side, reducing it, rather than modifying its capacity.

9.4.5 Energy Efficiency Assessment

The directive on the energy performance of buildings (EPBD) of the European Parliament and Council on energy efficiency of buildings came into force on 4 January 2003 and was deemed to be implemented by the EU member states by late 2006. Inspired by the Kyoto protocol, this directive forced every member state to meet minimum requirements in terms of energy efficiency both in the new constructions and in the renewal of any building. In reality, different approaches are currently used in European state members to encourage energy efficiency in building codes for new buildings and major renovation or refurbishment. Most of them are based on energy efficiency indicators, which are basically used to establish the energy consumption performance level of energy-consuming systems.

The importance of the external climate is tremendous when it comes to energy efficiency decisions; in fact the different climatic zones in which a building will be considered to be located strongly influence not only the choices related to the energy system but also the strategic selection of the most appropriate solution to enhance the building's overall energy performance. The Italian classifications actually used to label the existing and new buildings do not yet take into account the energy consumption due to water supply and to cooling systems, since cooling is not compulsory except in healthcare facilities. However, the Italian code requires that 55 % of the total energy for water heating shall be supplied by means of renewable energies (such as solar panels or heat pump) (D.P.R. 2009).

Notwithstanding the above, the UNITS (2008b, 2012) based on the law requirements of the European Union in terms of energy efficiency establishes a simplified method to calculate a ratio for water and cooling consumption, which has been used in this work in order to estimate the overall amount of consumed energy, allowing a more realistic result in terms of return period of the initial investment. The energy needed for cooling is strictly related to the use of the building, its location, and the decision-makers' motivation to invest. To this end, it is considered reasonable that in temperate climates, primary and secondary schools are not equipped with air conditioning systems, since the warmest period corresponds with the inactivity period of the school itself.

Similarly to that presented in the previous section for the earthquake resilience assessment, since the expected annual *loss* in energy efficiency terms is determined according to an analog ratio, cost-benefit analysis can likewise be performed, being expressed in terms of the correspondent break-even point for different solutions.

9.4.5.1 Implemented Improvements

According to the directive on the energy performance of buildings, enhancing the energy efficiency of building means to provide substantial improvement measures both on the envelope of the building and HVAC system, which are responsible for the indoor climatic conditions, as dictated by the so-called building-plant system. In this context, the referred envelope can be acknowledged as the "skin" of the building, since it works as the interface between the interior and the outdoor environment (including walls, roof, and foundation). By acting as a thermal barrier, the building envelope plays an important role in regulating interior temperatures and strongly contributes in determining the amount of energy required to maintain thermal comfort. Minimizing heat transfer through the building envelope is undoubtedly crucial for reducing the need for heating and cooling (C2EF 2012).

Different intervention can be done while retrofitting the buildings, by using architectural and heating/cooling solutions to reduce energy losses. Usually, an energy analysis of the existing building must be done in order to point out the weaknesses of the overall energy performance. The results of a careful energy diagnosis are of great help in determining the most effective solution to be used. Local climate is, of course, an important determinant for identifying the design features that will result in the greatest reductions of energy needs. These may include such measures as south-facing windows in cool climates and shading to avoid summer sun in hot climate. The use of very high-performance doors and windows is one of the principal goals to achieve, in order to increase the general performance of the buildings, in general both in new and retrofit construction.

It is clear that the *fenestration* (which includes windows, exterior doors, and skylights) influences both the lighting and the HVAC requirements of a building. In addition to design considerations (the placement of windows and skylights affects the amount of available natural light), materials and installation can affect the amount of energy transmitted through the windows, doors, or skylight, as well as the amount of air leakage around the window components. Thus, new materials, coatings, and designs contribute to improved energy efficiency of high-performing windows, doors, and buildings (DOE 2006).

Some of the energy performance advances based in windows include multiple glazing, the use of two or more layers of glass or other films for insulation, which can be further improved by filling the space between the panes with a low-conductivity gas, such as argon, and low-emissivity (low-e) coatings, which reduce the flow of infrared energy from the building to the environment. Shading, shutters, and reflection can also greatly reduce sun penetration of windows and other glass areas, as schematically presented in Fig. 9.15.

In residential buildings, using optimum window design and glazing specifications is estimated to reduce energy consumption from 10 to 50 % below accepted practice in most climates. Moreover, in commercial buildings, an estimated 10–40 % reduction in lighting and HVAC costs is attainable through improved fenestration (Arvizu 2011).

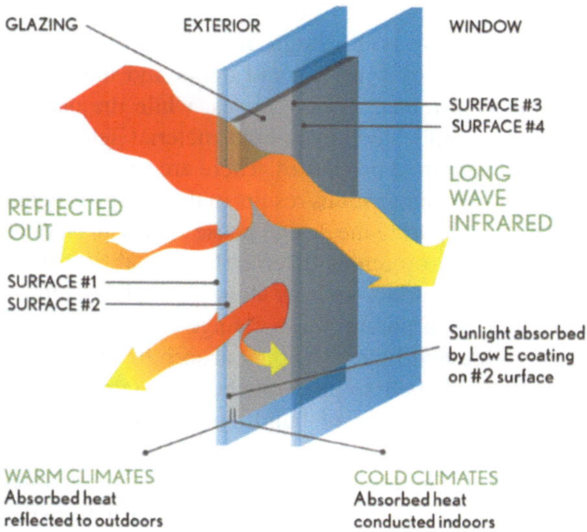

Fig. 9.15 Schematic representation of insulated glass solution

Another fundamental issue in design and retrofitting stages is the enhancement of buildings' insulation, due to the provided resistance to heat flow, reduction in the amount of energy needed to sustain the interior desired temperature in the winter and summer. A variety of insulation options exist, including blanket, concrete block, insulating concrete forms, spray foam, rigid foam, and natural fiber insulation. Usually insulation is discussed in terms of its ability to resist heat flow, which depends on the R-value which is a measure of thermal resistance of the material and is the mathematical inverse of the U value (overall heat transfer coefficient that describes how well a building element conducts heat). The higher is the R-value of an element, the better is the insulation of the element itself. Adding insulation strategically will improve the efficiency of the building; however, it is only effective if the building is properly sealed. Sealing cracks and leaks prevents air flow and is crucial for effective building envelope insulation. Leaks can generally be sealed with caulk, spray foam, or weather stripping (C2EF 2012).

Another important way to provide high level of efficiency is to focus on the roof design and materials. This can drastically reduce the amount of air conditioning required in hot climates by increasing the amount of solar heat that is reflected, rather than absorbed, by the roof. High-performing roofs are estimated to reduce the demand for peak cooling by 10–15 % (Arvizu 2011). In addition to the reduction of consumption, the roof can offer several opportunities for installing on-site generation systems that use renewable energy (solar, photovoltaic, etc.). In addition, it may include supplementary layers such as a root barrier and drainage or irrigation systems, which enhance the rainwater absorbing properties (http://en.wikipedia.org/wiki/Rainwater); provides better insulation (meaning an effective reduction for energy costs); minimizes the acoustic impact; helps in maintaining the waterproof of the ceiling; and gives a contribute to lower urban air temperatures (C2EF 2012).

In strict analogy with the roof, the amount of energy lost or retained through walls is influenced by both design and materials. Design considerations affect clearly the placement of windows and doors and the size and location of which can be optimized to reduce energy losses, while the use of the appropriate material can be more complicated, since both material selection and wall insulation affect the building's thermal properties. There are several different solutions that can be used in order to obtain virtuous results in the energy efficiency of the envelope, if they are to be applied during the design phase. This can be obtained based also in the thermal mass of the construction material (Levy et al. 2003). Thermal mass materials include traditional materials such as stone and adobe and cutting-edge products, such as those that incorporate phase change materials (PCMs). PCMs are solid at room temperature and liquefy as they absorb heat; the absorption and release of energy through PCMs help to moderate building temperature throughout the day.

Energy retrofitting of buildings is more difficult, given the massive measures required to obtain a well-performing building envelope. Therefore, in order to avoid severe workloads, one of the main used strategies is to cover up the entire building with insulating panels (coated structures) and substitute the fenestration with new improved frames in order to reduce the fuel consumption. Together with the envelope, a great importance must be given to the air conditioning and heating systems. It is shown that heating, ventilating, and air conditioning (HVAC—heat ventilation and air conditioning systems) account for 39 % of the energy used in commercial buildings. It is clear that by improving the control of HVAC operations and the efficiency of machines, significant savings can be achieved. The use of high-performance HVAC equipment can result in considerable energy, emissions, and cost savings (10–40 %) (Arvizu 2011).

Buildings constructed for healthcare purposes are well known for being particularly "energy consuming," particularly due to the high standard of IAQ (indoor air quality) established in local codes. The recent Commercial Buildings Energy Consumption Survey (CBECS) conducted by the Energy Information Administration (EIA) shows that the average hospital in North America consumes nearly 250 % more energy than the average commercial building (ASHRAE 2011a). In an average hospital, for example, lighting consumes a large portion of the overall energy budget. Therefore, the design should include an energy-efficient lighting design and efficient lighting fixtures as well as evaluating opportunities for dimming controls and multilevel switching systems. In the many areas where the design team brings quality daylight into the space, lighting controls can be used to regulate the output of electric lights to optimize the quality of the visual environment, while saving significant amounts of energy. Many different measures and design solutions can be employed in order to achieve goals of consumption reduction in this kind of facilities, as, for example, *photovoltaic technology*, *cogeneration* (cogeneration heat and power—CHP systems) to simultaneously generate electricity and useful heat, geothermal heat pump (GHP) from earth's natural heat to provide heating and cooling and water heating, and solar panels for heating water and other purposes, such as "renewable energy."

The use of a suitable and well-designed air conditioning system plays also a very important role. Solutions such as VAV (volume air variable), low-temperature radiant systems, and free cooling and free heating solutions based on the indoor quality of the air can lead to significant savings with no relevant up-front investments (ASHRAE 2011b). Furthermore, using new energy-efficient lighting, such as LED and occupancy sensors to avoid the waste of lighting when there is an absence of people together with lighting control systems with photo sensors which determine the lighting levels and dimmer lights to meet the minimum required level, will also provide a great improvement to the final bill in all kinds of buildings, with exception of residential dwellings, in which energy consumption due to lighting is relevant (ASHRAE 2011c).

9.5 Seismic Resilience and Energy Efficiency: Results and Conclusions

9.5.1 Result Integration and Comparison

In order to evaluate the influence of different measures for building retrofit and/or energy efficiency improvement, as well as the more general impact of improving the *green* and *resilient* components of *GRI*, a cost-benefit analysis is proposed.

The *benefit-cost* ratio at a given time over the building life cycle is expressed by the ratio between (a) the economic benefit determined by the difference between *net present value* (NPV) of accumulated *loss—expected annual loss* (*EAL*)— for the initial and improved states and (b) the cost associated with the referred improvement:

$$\frac{\text{Benefit}}{\text{Cost}} = \frac{\text{NPV}_{\text{Existing}} - \text{NPV}_{\text{Retroft}}}{\text{Cost}_{\text{Retroft}}}$$

The *net present value* (NPV) of a given value of *loss* for a specific time is intended to account for the interest rate—*r*—one would expect to pay if capital was borrowed from others or the rate of return on investment that would be expected (Calvi 2013), as follows:

$$\text{NPV} = \frac{\text{Value}}{(1 + r)^t}$$

At this stage, for simplicity purposes, it is decided not to explicitly consider the influence of the interest rate in the NPV calculations (i.e., *r* equal to zero), when incorporating the results of both approaches. Although the latter can be considered as a drawback of the present exercise, it is considered as a suitable assumption, since the main objective is the validation of the main guidelines for a more extensive exercise.

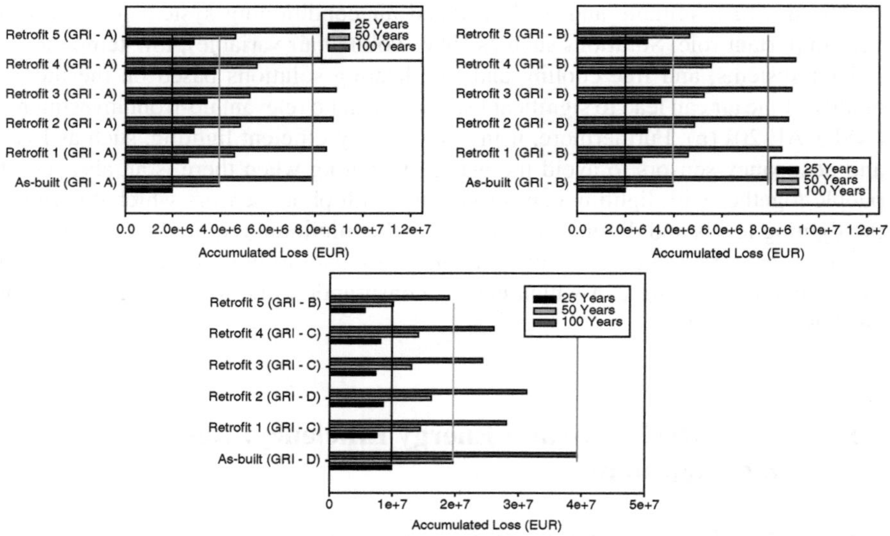

Fig. 9.16 Earthquake resilience assessment results. Building A: design location—Pavia (*upper left*), Messina (*upper right*), and Istanbul/Yerevan (*bottom*)

According to the above, the referred financial benefit indicators can be combined in order to determine the time at which the return of the initial investment will be achieved—the *break-even point*—for a particular building and intervention, in which detailed information regarding the analytical procedure is provided:

$$t_{\mathrm{break}\times\mathrm{even}} = \frac{\mathrm{Cost}_{\mathrm{ratio}}}{\mathrm{EAL}_{\mathrm{INITIAL}}\,\mathrm{EAL}_{\mathrm{IMPROVED}}}$$

$\mathrm{Cost}_{\mathrm{ratio}}$ stands for the cost to improve the building behavior for seismic resilience or energy efficiency, divided by the total building value.

9.5.2 Seismic Resilience and Energy Efficiency Assessment

For what concerns the evaluation of energy efficiency and seismic resilience assessments, it shall be mentioned that the results are gathered in a graphical fashion, for comparison purposes, according to the results of accumulated *loss* at periods of 25, 50, and 100 years, as depicted in Figs. 9.16, 9.17, 9.18, 9.19, and 9.20.

Figures 9.16, 9.17, 9.18, and 9.19—referring to case studies A, B, and C, respectively—illustrate the accumulated seismic *loss* resulting from the integration of EAL_S over periods of 25, 50, and 100 years, in which the results from an "as-built" situation are directly compared with the outcome of retrofit strategies 1–5 presented in Sect. 9.4.4.1.

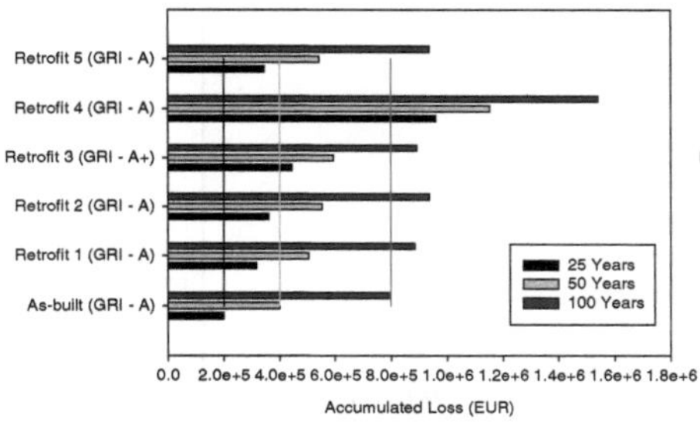

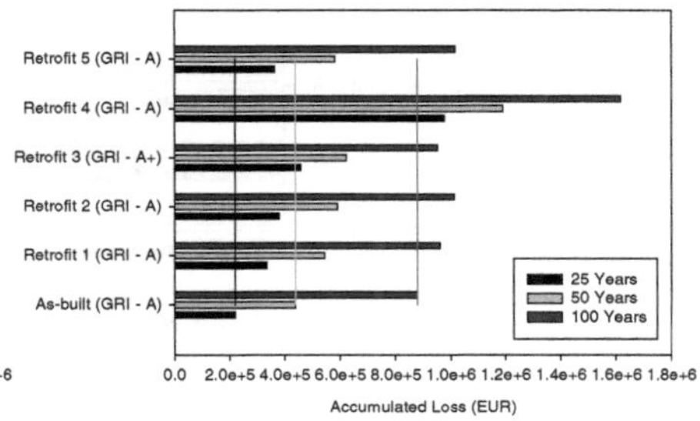

Fig. 9.17 Earthquake resilience assessment results. Building B: design location—Emilia-Romagna (*left*) and Messina (*right*)

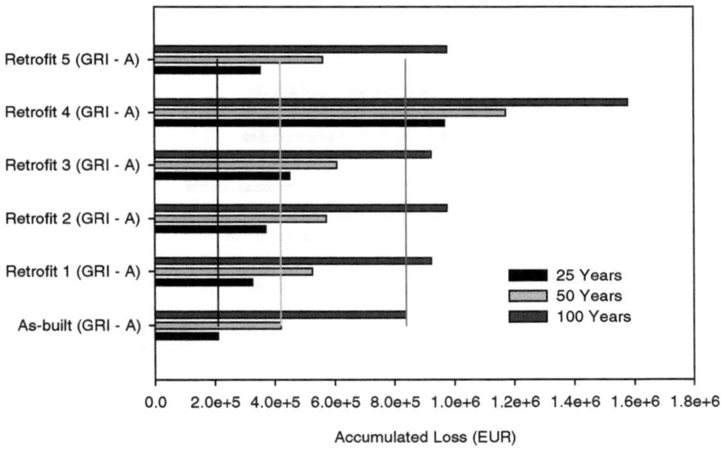

Fig. 9.18 Earthquake resilience assessment results. Building B: Istanbul/Yerevan

Similarly, energy *loss* accumulated over the aforementioned periods is depicted in Fig. 9.20, where the results referring to an "as-built" configuration are illustrated against the outcome of the enhanced situation for all the foreseen design locations (see Sect. 9.4.1).

9.5.3 Integrated Approach

For the purposes of integrated approach assessment, the most effective seismic retrofit scenario (see Figs. 9.16, 9.17, 9.18, and 9.19) in terms of *benefit-cost ratio* is selected for each of the buildings and locations. Accordingly, as presented in Figs. 9.21, 9.22, and 9.23, *benefit-cost* ratios for periods of 25, 50, and 100 years are calculated for earthquake resilience and energy efficiency improvement situations, as well as for the combined intervention.

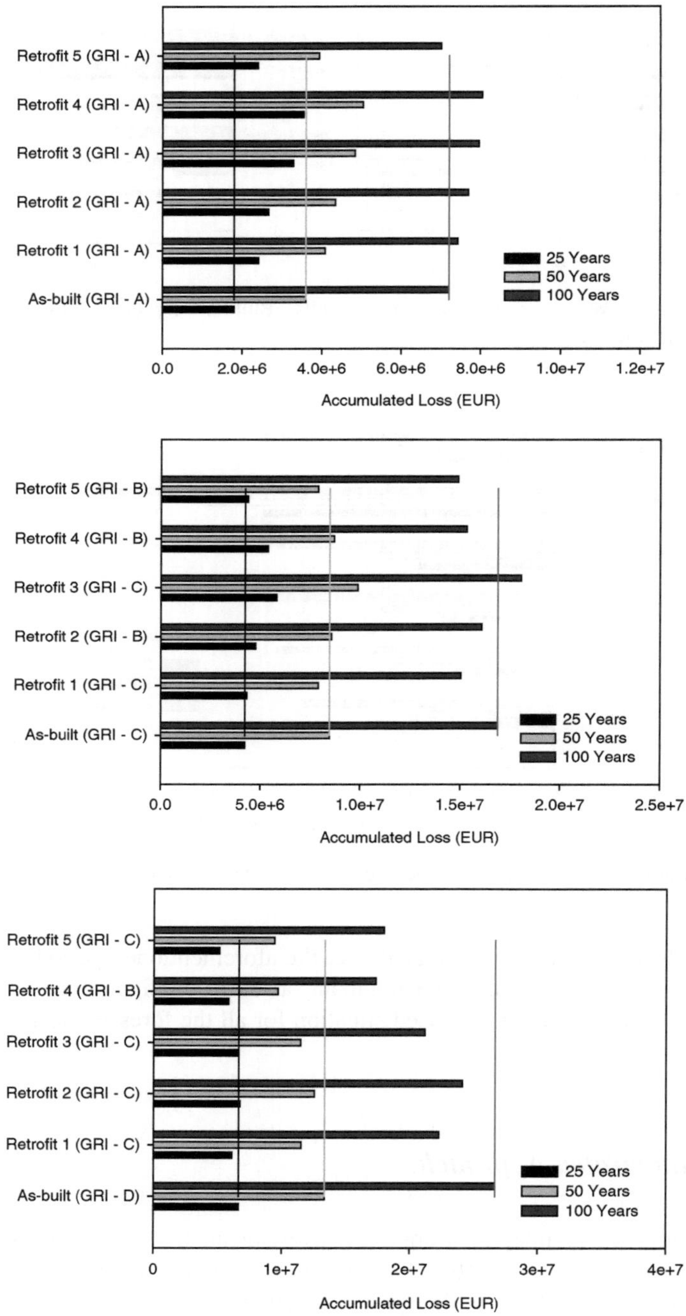

Fig. 9.19 Earthquake resilience assessment results. Building C: design location—Milan (*upper*), Messina (*middle*), and Istanbul/Yerevan (*bottom*)

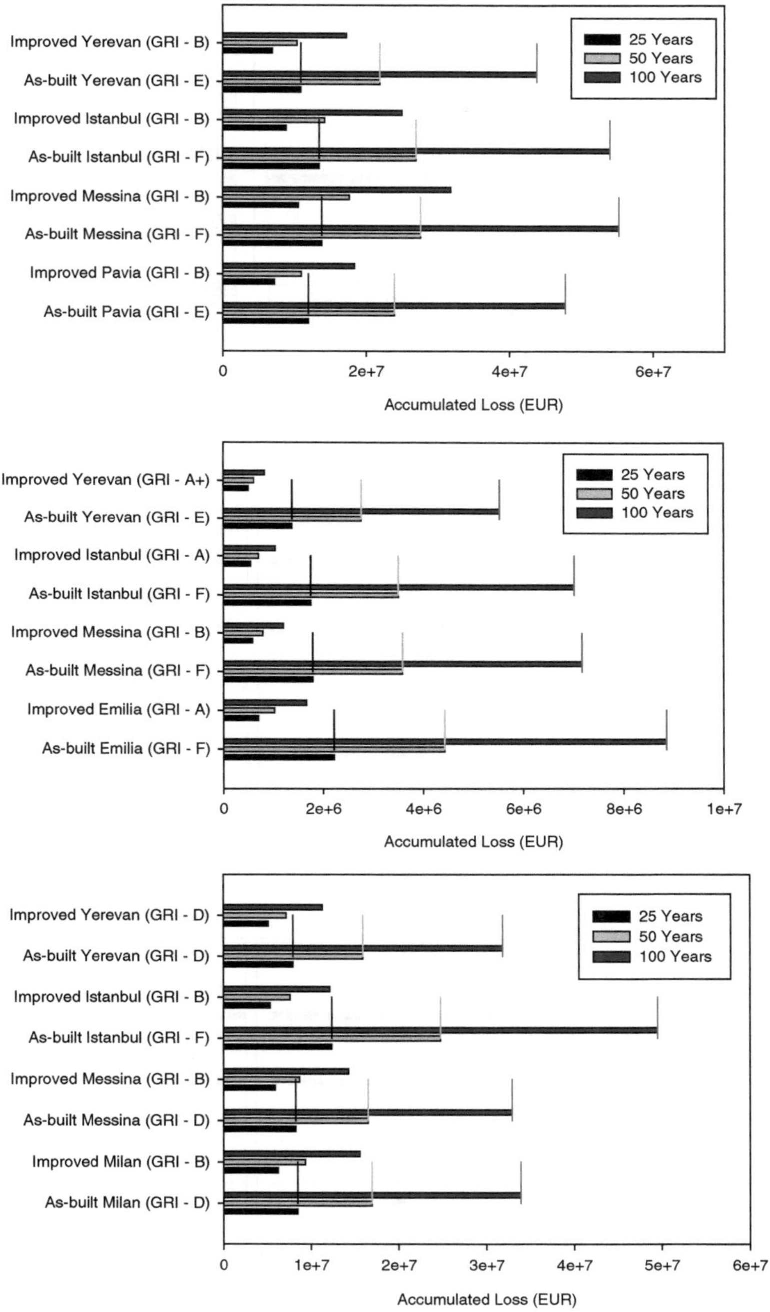

Fig. 9.20 Energy efficiency assessment results. Building A (*upper*), building B (*middle*), and building C (*bottom*)

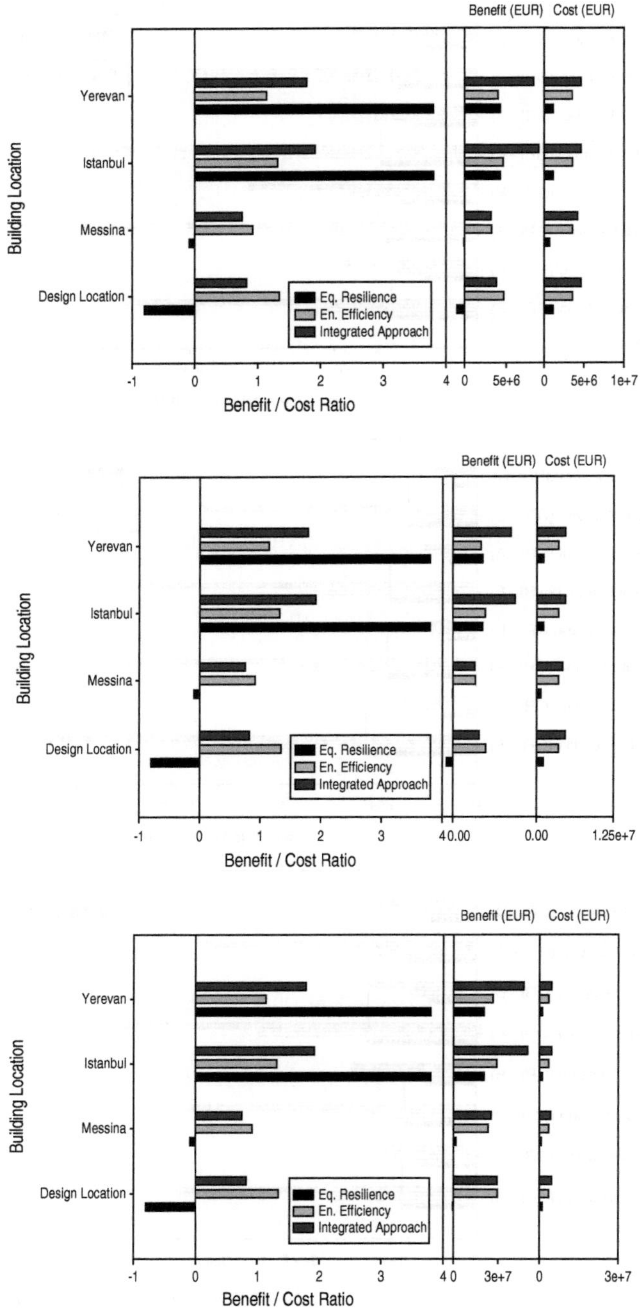

Fig. 9.21 Case study A—cost-benefit ratios for 25 (*upper*), 50 (*middle*), and 100 (*bottom*) years

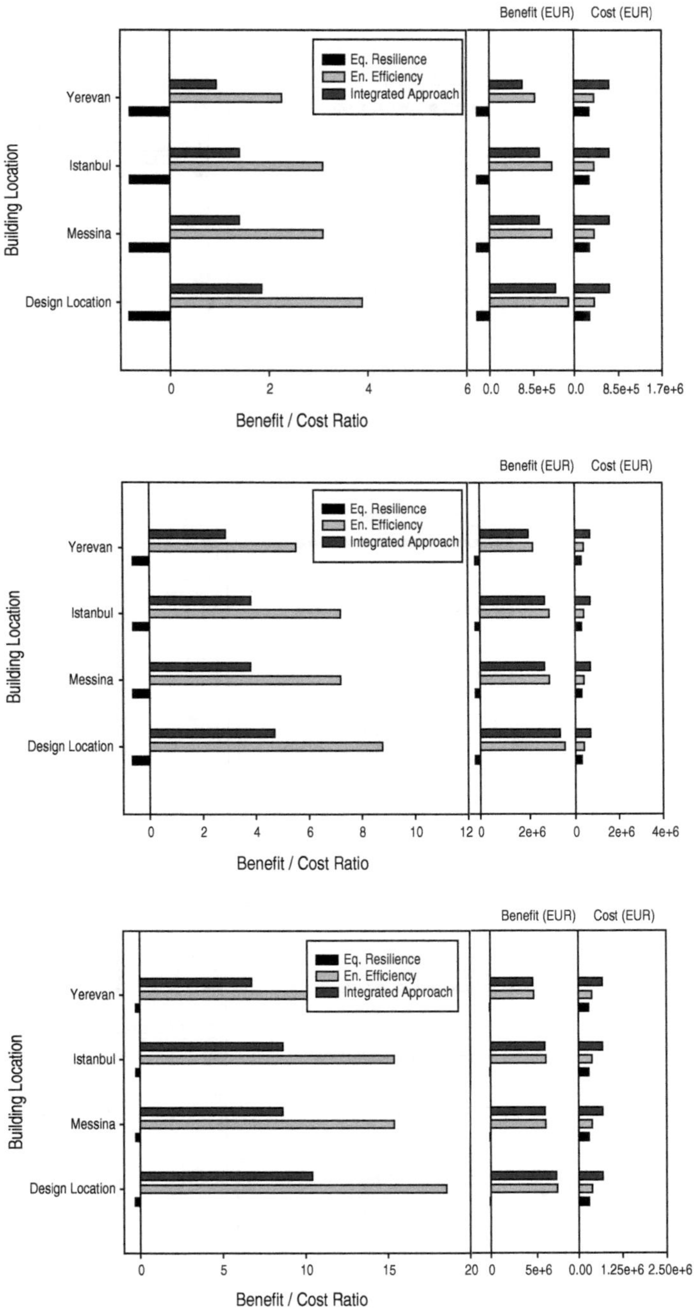

Fig. 9.22 Case study B—cost-benefit ratios for 25 (*upper*), 50 years (*middle*), and 100 (*bottom*) years

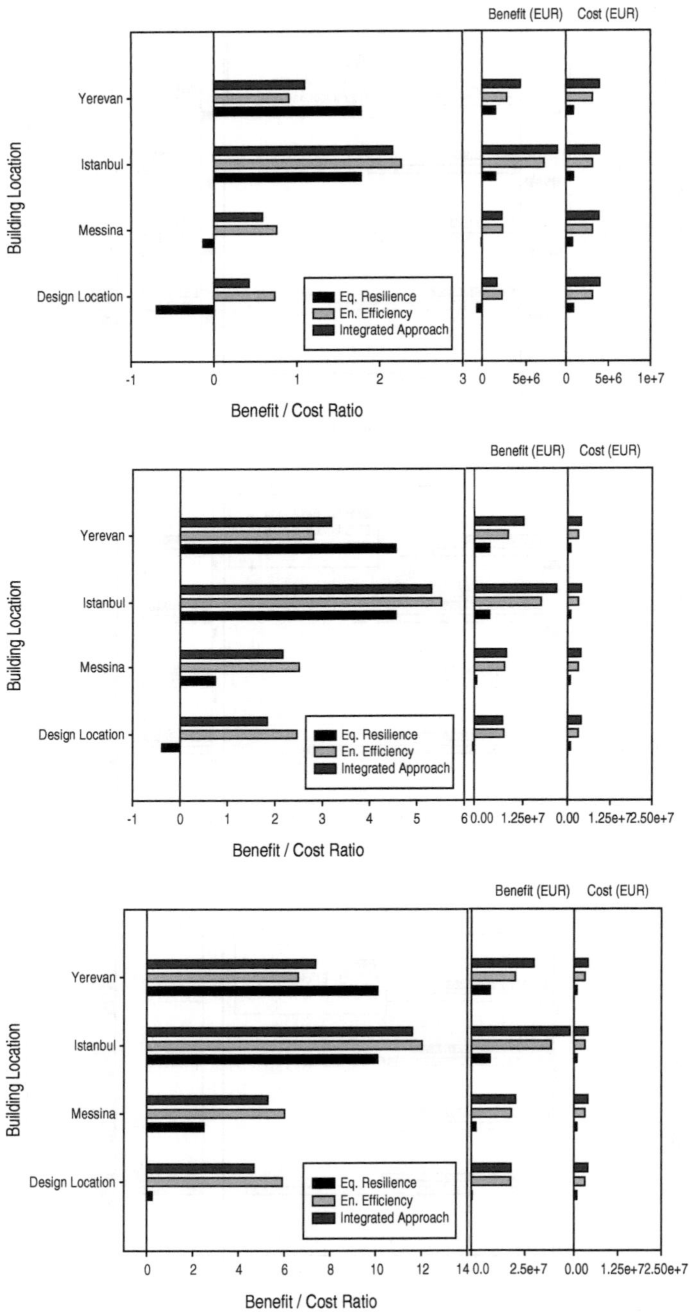

Fig. 9.23 Case study C—cost-benefit ratios for 25 (*upper*), 50 (*middle*), and 100 (*bottom*) years

As previously mentioned in this manuscript, the evaluation of the interest rate influence on the appraised results is not subject of detailed scrutiny. Thus, the subsequently presented results, which reflect a value of r equal to zero, are referred as the basis for the present comparison exercise. Moreover, as intuitively can be inferred, a negative *benefit-cost* ratio means in reality that at the time T (years) assessed, the reduction of *EAL* due to building behavior enhancement is not sufficient to compensate the initial investment required for the mentioned improvement (i.e., T < *break-even* point).

9.5.3.1 Conclusions

As presented in Figs. 9.20, 9.21, 9.22, and 9.23, addressing the matter of assessing the integrated advantages is not straightforward. As can be depicted in the mentioned figures, an analysis based on *benefit-cost* ratios cannot be dissociated from the correspondent absolute values of economic *benefit* and investment *cost.* To this regard, the following considerations can be gathered:

- As clearly illustrated in Fig. 9.20 (case study A), for the cases of Istanbul and Yerevan, significantly higher benefit-cost ratios can be attained for a particular intervention, even if similar values of economic *benefit* are presented for both approaches. In the case of building A, this is explained by the higher initial *cost* to improve the building from an energy efficiency point of view.
- Improving the behavior of a building with initial high *GRI* classification will likely lead to a very high investment return period, as expressed by the negative values of *benefit-cost* ratio determined for 25, 50, and 100 years for building B, as particularly evident in Figs. 9.21 and 9.22.
- The location of the building greatly influences the final result of the exercise in terms of both *resilience* and *energy* points of view. As illustrated in Figs. 9.20, 9.21, 9.22, and 9.23, for all the buildings analyzed, it is possible to depict a considerable trend of variation in *benefit* depending on the building considered construction site. This is related to the relationship between the demands at design and idealized locations, which essentially reflects the suitability of the design and retrofitted conditions to the expected demands. Thus, a building with poor *GRI* classification (not adequately designed), on either of the deemed approaches, will *a priori* attain greater benefit from a given improvement intervention than a building with high *GRI*.
- Distinct *energy* and *resilience benefit-cost* ratios for a given building/site configuration might be registered even for similar values of benefit, in cases in which the levels of initial investment differ significantly (see Fig. 9.20) or in situations in which considerable differences in benefit correspond to similar values of initial investment (see Fig. 9.21).

According to the aforementioned, it seems reasonable to assume a prioritization scheme in which the investment is directed to the maximum *benefit-cost* ratio (i.e., higher possible reduction of *EAL* for the minimum amount of initial cost).

To this end, the appraised results highlight the advantages of the considered *GRI* classification scheme, due to the fact that the higher *benefit-cost* ratio between seismic resilience and energy efficiency for a given integrated intervention tends to correspond to the approach with "poorer" *GRI* index (i.e., the one which can benefit more from a behavior improvement intervention).

The role of the idealized *GRI* classification should be appropriately emphasized under this context. Since the *GRI* classes are common for *green* and *resilient* indicators, it is not only possible to directly compare energy efficiency and seismic resilience from an *EAL* point of view, as it is plausible to assume that the GRIs of *green* and *resilient* approaches represent a proxy for investment return potential; i.e., "poorer" GRI classes are associated with higher potential for investment return and benefit maximization, as enunciated in the previous paragraph.

However, considering only one of the approaches (*resilience* or *energy*) might not be the most adequate solution, due to the fact that the *benefit* of a given intervention (*green* or *resilient*) can only be maximized up to the utmost possible level of building behavior improvement or, in other words, up to the point in which an additional investment does not result in *EAL* reduction, and therefore, the *benefit-cost* ratio is maximum. Thus, at least from a theoretical point of view, an integrated approach should always be advantageous with respect to the investment in only *earthquake resiliency* or *energy efficiency* improvement, dividing the investment in such a way that simultaneously maximizes the *benefit-cost* ratio of the "individual" improvements in *green* and *resilient* perspectives and its integrated result.

9.5.4 Future Developments

Under the previously mentioned context, a fully representative analysis campaign is submitted for further developments. According to the latter, all the applicable combinations of building configuration, retrofit and improvement techniques, seismic hazard and energy demands, as well as financial and economic parameters (such as interest rates and downtime *losses*) shall be convoluted in order to create a set of results that can adequately constitute the base for the development of integrated approach guidelines.

As presented in Sect. 9.2.3, recent efforts have been developed toward the explicit consideration of environmental impacts of earthquake action in a comprehensive approach. Thus, although only financial aspects are considered in the present exercise, an interesting further development would be the consideration of additional decision-making variables, directly related to the environmental counterpart of energy efficiency and earthquake *loss*, as well as the correlation between both.

In addition, it is required to identify possible challenges to be faced in the implementation of an integrated approach. They are expected to include mainly technical (related to the regional characteristic "know-how" and material/technical resources), bureaucratic (due to the possible necessity of institutional coordination between different decision-making agencies), and legislation issues (relative to the applicability of different measures, under the context of regional code standards). However, the comprehensive compilation of such a set of constraints would require an extensive research procedure that cannot be performed in the present exercise and will, therefore, be recommended as the subject of further developments.

Acknowledgments Part of this work has been developed under the scope of the World Bank's Regional Urban, Water and Disaster Risk Management program, and referred Innovation Grant KP-P143217-KMPD. In this context, the assistance and support of the project manager, Mr. Sergio Dell'Anna, are acknowledged.

References

Arvizu, D. (2011). Direct solar energy. In *IPCC special report on renewable energy sources and climate change mitigation*. Cambridge: Cambridge University Press.

ASCE/SEI 41. (2006). *Seismic rehabilitation of existing buildings*. Reston, VA: American Society of Civil Engineers.

ASHRAE Design Guide. (2011a). *Advanced energy design guide for small hospitals and healthcare facilities*. Atlanta, GA: ASHRAE.

ASHRAE Design Guide. (2011b). *Advanced energy design guide for small office buildings*. Atlanta, GA: ASHRAE.

ASHRAE Design Guide. (2011c). *Advanced energy design guide for K-12 school buildings*. Atlanta, GA: ASHRAE.

ATC. (2011). *Guidelines for seismic performance assessment of buildings* (uncompleted draft). Redwood City, CA: Applied Technology Council.

Bommer, J. J., Pinho, R., & Crowley, H. (2005). Using displacement-based earthquake loss assessment in the selection of seismic code design levels. In *Proceedings of the 9th International Conference on Structural Safety and Reliability, Rome, Italy*.

Bommer, J. J., & Pinho, R. (2006). Adapting earthquake actions in Eurocode 8 for performance-based seismic design. *Earthquake Engineering & Structural Dynamics, 35*(1), 39–55.

Burton, H., Deierlein, G., & Lepech, M. (2011). Assessing the scale of environment impacts from a major California earthquake earthquake. *Proceedings of the 80th Structural Engineers Association of California (SEAOC) Annual Convention*. Las Vegas, Nevada

Calvi, G. M. (2013). Choices and criteria for seismic strengthening. *Journal of Earthquake Engineering, 17*, 769–802. doi:10.1080/13632469.2013.781556.

Calvi, G. M., Pietra, D., & Moratti, M. (2010). Criteri per la progettazione di dispositivi di isolamento a pendolo scorrevole. *Progettazione Sismica, 2*(3), 7–30.

Caterino, N., Iervolino, I., Manfredi, G., & Cosenza, E. (2008). Multi-criteria decision making for seismic retrofitting of RC structures. *Journal of Earthquake Engineering, 12*(4), 555–583.

Caterino, N., Iervolino, I., Manfredi, G., & Cosenza, E. (2009). Comparative analysis of multi-criteria decision-making methods for seismic structural retrofitting. *Computer-Aided Civil and Infrastructure Engineering, 24*, 432–445.

C2EF—Center for Climate and Energy Solution. (2012). *Climate techbook: Solar system*. Arlington, VA: C2EF—Center for Climate and Energy Solution.

CEN. (2005). *Eurocode 8: Design of structures for earthquake resistance. Part 3: Assessing and retrofitting of buildings*. EN 1998-3. Brussels, Belgium: Comité Européen de Normalisation.

Chung, W., Hui, Y. V., & Miu Lam, Y. (2006). Benchmarking the energy efficiency of commercial buildings. *Applied Energy, 83*(1), 1–14.

Comber, M., Poland, C., & Sinclair, M. (2012). Environmental Impact Seismic Assessment: Application of Performance-Based Earthquake Engineering Methodologies to Optimize Environmental Performance. *Structures Congress, 2012*, 910–921. doi: 10.1061/9780784412367.081.

Cornell, C., Jalayer, F., Hamburger, R., & Foutch, D. (2002). Probabilistic basis for 2000 SAC Federal Emergency Management Agency steel moment frame guidelines. *Journal of Structural Engineering, 128*(Special Issue: Steel Moment Frames After Northridge—Part II), 526–533.

Court, A., Simonen, K., Webster, M., Trusty, W., & Morris, P. (2012). Linking Next-Generation Performance-Based Seismic Design Criteria to Environmental Performance (ATC-86 and ATC-58). *Structures Congress, 2012*, 922–928. doi: 10.1061/9780784412367.082.

Crowley, H., Silva, V., Bal, I., & Pinho, R. (2012). Calibration of seismic design codes using loss estimation. *Proceedings of the 15th WCEE, Lisbon, Portugal.*

Dennemann, K. L. (2009). *Life-cycle cost-benefit (LCC-B) analysis for bridge seismic retrofits civil and environmental engineering*. Houston, TX: Rice University.

D.P.R. (2009) n. 59 recante; Attuazione dell'articolo 4, comma 1, lettere a) e b), del decreto legislativo 19 agosto 2005, n. 192, e successive modificazioni, concernente attuazione della direttiva 2002/91/ce sul rendimento energetico in edilizia.

Elnashai, A. S., & Pinho, R. (1998). Repair and retrofitting of RC walls using selective techniques. *Journal of Earthquake Engineering, 2*(4), 525–568.

Fajfar, P., & Krawinkler, H. (Eds.). (2004). *Performance-based seismic design concepts and implementation. Report 2004/05.* Richmond, CA: Pacific Earthquake Engineering Research Center (PEER).

Fassbender, J. (2007). A decade and more of monthly construction statistics. ISSN. Catalogue number: KS-SF-07-129-EN-N.

FEMA. (1999). *Earthquake loss estimation methodology—HAZUS 99.* Washington, DC: Federal Emergency Management Agency and National Institute of Buildings Sciences.

FEMA. (2000). *Prestandard and commentary for the seismic rehabilitation of buildings* (FEMA Publication No. 356). Prepared by the American Society of Civil Engineers for the Federal Emergency Management Agency, Washington, DC.

FEMA. (2012). *Seismic performance of buildings. Volume 1—Methodology* (FEMA Publication No. P-58-1). Prepared by the Applied Technology Council, Redwood City, CA.

Filiatrault, A., & Christopoulos, C. (2006). *Principles of passive supplemental damping and seismic isolation.* Pavia, Italy: IUSS Press.

Ghosh, J., Tapia, C., & Padgett, J. (2011). Life-cycle analysis of embodied energy for aging bridges subject to seismic hazards. *Proceedings of the 11th International Conference on Applications of Statistics and Probability in Civil Engineering, ETH Zurich.* Boca Raton, FL: CRC Press.

Giardini, D., Woessner, J., Danciu, L., Cotton, F., Crowley, H., Grünthal, G., et al. (2013). *Seismic hazard harmonization in Europe (SHARE).* Online Data Resource, doi:10.12686/SED-00000001-SHARE.

Grant, D. N., Bommer, J. J., Pinho, R., Calvi, G. M., Goretti, A., & Meroni, M. (2007). A prioritization scheme for seismic intervention in school buildings in Italy. *Earthquake Spectra, 23*(2), 291–314.

Haselton, C. B., Goulet, C. A., Beck, J. L., Deierlein, G. G., Porter, K. A., Stewart, J. P., et al. (2007). An assessment to benchmark the seismic performance of a code-conforming reinforced concrete moment-frame building. *Report 2007/12.* Richmond, CA: Pacific Earthquake Engineering Research Center (PEER).

ISO. (2006a). *14040:2006 life cycle assessment—Principles and framework.* Geneva: ISO.

ISO. (2006b). *14044:2006 environmental management—Life cycle assessment—Requirements and guidelines*. Geneva: ISO.

Itoh, Y., Wada, M., & Liu, C. (2005). Lifecycle environmental impact and cost analyses of steel bridge piers with seismic risk. *9th International Conference on Structural Safety and Reliability, Rome, Italy*. Rotterdam: Millpress.

Kneer, E., & Maclise, L. (2008a). Consideration of building performance in sustainable design: A structural engineer's role. *Proceedings of the 77th Structural Engineers Association of California (SEAOC) Annual Convention*. Hawaii, U.S.A.

Kneer, E., & Maclise, L. (2008b, December). Disaster resilience as sustainable design. *Structural Engineer*, ZweigWhite Information Services LLC, 9(11).

Levy, J. I., Nishioka, Y., & Spengler, J. D. (2003). The public health benefits of insulation retrofits in existing housing in the United States. *Environmental Health: A Global Access Science Source*, *2*, 4. doi:10.1186/1476-069X-2-4.

Luco, N., Ellingwood, B. R., Hamburger, R. O., Hooper, J. D., Kimball, J. K., & Kircher, C. A. (2007). Risk targeted versus current seismic design maps for the conterminous United States. In *Proceedings of the 76th Structural Engineers Association of California (SEAOC) Annual Convention*. Squaw Creek, U.S.A.

Mitrani-Reiser, J. (2007). *An ounce of prevention: Probabilistic loss estimation for performance based earthquake engineering*. PhD dissertation, California Institute of Technology, Pasadena, CA.

NZSEE. (2006). Assessment and improvement of the structural performance of buildings in earthquakes. *Recommendations of a NZSEE Study Group on earthquake risk buildings*. New Zealand: NZSEE.

Padgett, J. E., Ghosh, J., & Tapia, C. (2009). Sustainable infrastructure systems subjected to multiple threats. *TCLEE Conference, Oakland, CA*.

Porter, K. (2003). An overview of PEER's performance-based earthquake engineering methodology. *Proceedings of the Ninth Conference on Application of Statistics and Probability in Civil Engineering, San Francisco, 2003*.

Porter, K. A., Beck, J. L., & Shaikhutdinov, R. V. (2002). *Investigation of sensitivity of building loss estimates to major uncertain variables for the Van Nuys testbed* (Report 2002/03). Richmond, CA: Pacific Earthquake Engineering Research Center (PEER).

Ramirez, C. M., & Miranda, E. (2009). *Building specific loss estimation methods and tools for simplified performance-based earthquake engineering* (Report 171). Stanford, CA: John A. Blume Center, Stanford University.

Sadek, F., Mohraz, B., Taylor, A. W., & Chung, R. M. (1997). Method of estimating the parameters of tuned mass dampers for seismic applications. *Earthquake Engineering and Structural Dynamics*, *26*, 617–635.

SEAOC. (1995). *Vision 2000—A framework for performance-based engineering*. Sacramento, CA: Structural Engineers Association of California.

Sullivan, T. J., & Calvi, G. M. (2011). Considerations for the seismic assessment of buildings using the direct displacement-based assessment approach. *2011 ANIDIS Conference, Bari, Italy*.

Tapia, C. & Padgett, J. (2012). Examining the Integration of Sustainability and Natural Hazard Risk Mitigation into Life Cycle Analyses of Structures. *Structures Congress, 2012*, 1929–1940. doi:10.1061/9780784412367.169.

Tapia, C., Ghosh, J., & Padgett, J. (2011). Life cycle performance metrics for aging and seismically vulnerable bridges. *Proceedings of the 2011 Structures Congress, Las Vegas, Nevada*. Reston, VA: ASCE.

Tesmafatian, S., Sadiq, R., & Najjaran, H. (2010). Decision making under uncertainty—An example for seismic risk management. *Risk Analysis, 30*(1), 78–94.

Thermou, G., & Elnashai, A. (2006). Seismic retrofit schemes for RC structures and local–global consequences. *Earthquake Engineering and Structural Dynamics, 8*, 1–15.

Thermou, G. E., Pantazopoulou, S. J., & Elnashai, A. S. (2007). Design methodology for seismic upgrading of substandard reinforced concrete structures. *Journal of Earthquake Engineering, 11*(4), 582–606.

DOE—United States Department of Energy. Federal Energy Management Program. (2006). *Guiding principles of federal leadership in high performance and sustainable buildings*. Retrieved May 19, 2011, from http://www1.eere.energy.gov/femp/regulations/eo13514.html

UNI/TS 11300. (2008a). Prestazioni energetiche degli edifici—Parte 1: Determinazione del fabbisogno di energia termica dell'edificio per la climatizzazione estiva ed invernale.

UNI/TS 11300. (2008b). Prestazioni energetiche degli edifici—Parte 2: Determinazione del fabbisogno di energia primaria e dei rendimenti per la climatizzazione invernale e per la produzione di acqua calda sanitaria.

UNI/TS 11300. (2012). Prestazioni energetiche degli edifici—Parte 4: Utilizzo di energie rinnovabili e di altri metodi di generazione per la climatizzazione invernale e per la produzione di acqua calda sanitaria.

USGBC (2011) *Green Job Study*. Prepared by Booz Allen Hamilton for the U.S. Green Building Council. Washington DC.

Welch, D. P., Sullivan, T. J., & Calvi, M. (2012). Towards a direct displacement-based loss assessment methodology for RC frame buildings. *Proceedings of the 15th World Conference on Earthquake Engineering*. Lisbon, Portugal.

Zareian, F., & Krawinkler, H. (2012). Conceptual performance-based seismic design using building level and story-level decision support system. *Earthquake Engineering and Structural Dynamics, 41*(11), 1439–1453.

Part IV
Fire, Blast, Shock and Impact

Chapter 10
Probabilistic Evaluation Framework for Fire and Fire Following Earthquake

Negar Elhami Khorasani, Maria Garlock, and Paolo Gardoni

Abstract This work provides a probabilistic framework to evaluate the performance of a structure under fire and fire following earthquake, by studying response of the structure for several limit states and incorporating uncertainties in demand and capacity parameters. The multi-hazard framework is then applied to a steel moment resisting frame (MRF) to evaluate the structural performance of the MRF under post-earthquake fires. The study develops probabilistic models for key quantities with uncertainty including fire load, as well as yield strength and modulus of elasticity of steel at elevated temperatures. The MRF is analyzed under several fire scenarios and fire locations. Results show that the location of fire in the frame (e.g., lower vs. upper floors and interior vs. exterior bays) affects the element response. Compartments in the interior bays reach limit states faster than those on the perimeter of the frame, and upper floors reach limit states sooner than lower floors. The post-earthquake damage does not affect the structural response under fire for the considered limit states, but post-earthquake fire increases the drift demand on columns located at the perimeter of the structure.

10.1 Introduction

The available codes and guidelines on design of structures for fire are based on performance evaluation at the component level and using deterministic approaches, while fire events and performance of structures at elevated temperatures involve

N.E. Khorasani (✉)
Department of Civil, Structural and Environmental Engineering, University at Buffalo, Buffalo, NY, USA
e-mail: negarkho@buffalo.edu

M. Garlock
Department of Civil and Environmental Engineering, Princeton University, Princeton, NJ, USA
e-mail: mgarlock@princeton.edu

P. Gardoni
Department of Civil and Environmental Engineering, University of Illinois at Urbana-Champaign, Urbana, IL, USA
e-mail: gardoni@illinois.edu

P. Gardoni, J.M. LaFave (eds.), *Multi-hazard Approaches to Civil Infrastructure Engineering*, DOI 10.1007/978-3-319-29713-2_10

considerable uncertainties. Measured data specifically indicates large uncertainties in the intensity and characteristics of fire load and of material properties at elevated temperatures. A risk-informed performance-based design that models the uncertainties can optimize safety, efficiency, and the overall cost of the design. Researchers have recently been focusing on applying risk-based decision-making optimization techniques (De Sanctis et al. 2011), using reliability analysis (Gue et al. 2013), and developing probabilistic frameworks (He 2013) to design structures for fire.

Aside from fire itself, fire following earthquake (FFE) is also an extreme hazard, which is a low probability, high consequence event. In a recent study on historical FFE events, 20 cases from seven different countries were collected, 15 of which occurred between 1971 and 2014 (Elhami Khorasani and Garlock 2015). Fire that followed the earthquake in a majority of these cases caused considerable damage. Based on our current design guidelines, buildings are generally designed for individual extreme events (i.e., earthquakes or fires only), but the structural response of buildings under cascading multi-hazard fire and earthquake events is not evaluated during the design process, and this is something to consider for resiliency planning in densely populated regions (Elhami Khorasani 2015; Elhami Khorasani et al. 2015c). Previous studies on the response of structures in FFE scenarios analyze the problem using deterministic approaches and model thermal and seismic analyses in different programming environments (Della Corte et al. 2003, 2005; Yassin et al. 2008) due to the limited available tools which can analyze both loading events. Most commercially available finite element programs require extensive computational resources to model and run seismic and thermal analysis sequentially.

This work is a step toward developing guidelines to include uncertainties in the new generation of performance-based design, for fire and cascading multi-hazard events such as post-earthquake fires. Any guideline for fire design includes three steps: (1) determining the design fire (demand model, design fire), (2) performing thermal analysis (demand model, thermal analysis), and (3) performing a structural analysis that considers the thermal load (capacity model, structural analysis). Given the three steps, this work (1) provides probabilistic models for fire load and mechanical properties of steel at elevated temperatures using literature data, (2) develops a framework to evaluate performance of structures under fire and FFE incorporating the developed probabilistic models, and (3) applies the framework to a 9-story moment resisting frame (MRF) and studies response of the MRF under FFE by modeling the seismic and thermal analyses in one programming environment.

10.2 Probabilistic Models

This section provides the developed probabilistic models for fire load density (demand), yield strength and modulus of elasticity of steel (capacity), and a quick overview of the mathematical procedure needed to develop the models.

10.2.1 Background

A brief overview of the mathematical procedure used to develop probabilistic models in this work is discussed in this section. Gardoni et al. (2002) proposed a formulation that developed probabilistic models by correcting deterministic models with correction terms calibrated using observed experimental data. The deterministic formulations may come from codes or standards, and correction terms are added to remove the inherent bias in the deterministic model. A model error is included and calculated to account for the remaining inaccuracy of the model. The general form of the model is shown in Eq. (10.1):

$$C(\mathbf{x}, \boldsymbol{\Theta}) = \widehat{c}(\mathbf{x}) + \gamma(\mathbf{x}, \boldsymbol{\theta}) + \sigma\varepsilon \tag{10.1}$$

where $C(\mathbf{x}, \boldsymbol{\Theta})$ is the quantity of interest (or a transformation into a new space using transformations like the natural logarithm, square root function, or the logistic function); $\boldsymbol{\Theta} = (\boldsymbol{\theta}, \sigma)$, in which $\boldsymbol{\theta} = (\theta_1, \theta_2, \ldots)$, denote the set of unknown model parameters; $\hat{c}(\mathbf{x})$ is a selected deterministic model that is expressed as a function of the variables $\mathbf{x}$ ($\hat{c}(\mathbf{x})$ is transformed accordingly if $C(\mathbf{x}, \boldsymbol{\Theta})$ is transformed); $\gamma(\mathbf{x},\boldsymbol{\theta})$ is a correction term for the bias inherent in the deterministic model; and $\sigma\varepsilon$ is the model error that captures the remaining scatter in the residuals, where ε is a random variable with zero mean and unit variance and σ represents the standard deviation of the model error. The formulation in Eq. (10.1) is general and can also be used when there are no deterministic models available; in such case, $\hat{c}(\mathbf{x}) = 0$.

Equation (10.1) is based on three assumptions: (1) ε follows a standard normal distribution (normality assumption), (2) σ does not depend on $\mathbf{x}$ (homoskedasticity assumption), and (3) $\gamma(\mathbf{x},\boldsymbol{\theta})$ can be added to $\hat{c}(\mathbf{x})$ instead of being, for example, a multiplicative term (additivity assumption). These assumptions can typically be satisfied by considering an appropriate variance-stabilizing transformation of the original quantity of interest (Box and Cox 1964) and verified by using diagnostic plots (Rao and Toutenburg 1997). The selection of the variance-stabilizing transformation is often guided by the physical range of the quantity of interest, and the transformation is selected so that the transformed quantity ranges from $-\infty$ to $+\infty$. For example, if the physical quantity is nonnegative, then the natural logarithm or the square root function can be used as a transformation. If the quantity of interest is between 0 and 1, then the logistic function can be used (Stone 1996).

Different procedures, such as linear regression, nonlinear regression, or the maximum likelihood method, can be used to calculate the unknown model parameters $\boldsymbol{\Theta}$. However, a Bayesian approach may also be used when prior information is present (Box and Tiao 1992). In this work, a Bayesian updating framework is used to estimate the unknown parameters $\boldsymbol{\Theta} = (\boldsymbol{\theta}, \sigma)$. The details of the mathematical formulation are provided in Gardoni et al. (2002). When using that approach, as opposed to a purely empirical model, the probabilistic model includes the physical understanding that is often behind the deterministic models. Also, the newly developed probabilistic model is based on an already existing model; therefore, it is easier for the engineering practice to accept and apply the new model.

10.2.2 Fire Load

The temperature-time evolution of fire in this work is calculated using the Eurocode1 (EC1) formulation (CEN 2002), which depends on geometric characteristics of the compartment, thermal properties of the boundary of enclosure, and fire load density. Among the three factors, determination of fire load density most strongly influences the fire temperature. A sensitivity study by Guo et al. (2013) confirms that the value of fire load density has a significant influence on the structural response to fire.

The authors collected and studied available surveys in the literature on fire load density in office buildings (Elhami Khorasani et al. 2014). The results of four surveys on office buildings from different countries, and using different surveying methods, were compared to older data from eight other surveys, as well as the design values suggested by codes. The available surveys showed a wide range in the recorded mean fire load density values (348–852 MJ/m^2), which implies that a considerable amount of uncertainty exists in predicting fire load density. Another important observation was that survey results showed a correlation between fire load density and the room size and use. The majority of data showed that meeting rooms or relatively large offices have a smaller fire load density than storage rooms, file areas, or smaller offices in an office building.

Data from a US survey with the most comprehensive collection for fire load density in office buildings (Culver 1976) were used to generate a new probabilistic model for fire load density q in office buildings. The model for q includes the effect of room size A_f on the fire load density and consists of two equations: Eq. (10.2a) for lightweight categories (general offices, clerical, etc.) and Eq. (10.2b) for heavy weight categories (library, storage, and file rooms):

$$q = \exp\left[6.951 - 0.0047\left(A_f \times 10.76\right) + 0.5712\varepsilon\right] \tag{10.2a}$$

$$q = \exp\left[8.252 - 0.0081\left(A_f \times 10.76\right) + 0.5508\varepsilon\right] \tag{10.2b}$$

where q is in units of MJ/m^2, A_f is the room size in m^2, and ε is a random variable that follows the standard normal distribution. Eqs. (10.2a) and (10.2b) represent the characteristic fire load density value ($q_{f,k}$) in the EC1 formulation, which is discussed in Annex E of EC1 (CEN 2002). The equations can therefore be used with EC1 to develop fire characteristic time-temperature curves.

10.2.3 Mechanical Properties at Elevated Temperatures

This section provides probabilistic models for mechanical properties of steel at elevated temperatures based on available data in the literature. The two major mechanical properties of steel when performing structural analysis at elevated temperatures are yield strength at a strain equal to 2 % and modulus of elasticity. The two properties are part of the Eurocode3 (EC3) formulation (CEN 2005), which will be used in this work to evaluate performance of a structure at elevated temperatures. It should be noted that EC3 uses yield strength at a strain equal to 2 %, as opposed to yield strength at 0.2 % offset, to define the constitutive material model of steel at elevated temperatures.

The reported analytical equations for the parameters at elevated temperatures are in terms of normalized values, meaning that the parameter at every temperature is a factor of the parameter at the ambient temperature. The normalized parameters for yield strength and modulus of elasticity are assumed to vary between one and zero (one at ambient and zero at temperatures above 1000 °C). The reported measured data are also normalized values between zero and one, with occasional values larger than one at lower temperatures. Values larger than one imply that the strength of the specimen is larger than the assumed strength at the room temperature.

A study by the National Institute of Standards and Technology (NIST) (Luecke et al. 2011) provides measurements of yield strength at a strain equal to 2 % at elevated temperatures. Those data and the EC3 model (Eq. 10.3) are used to develop a probabilistic model for the normalized 2 % yield strength ($k_{y,2\%,T}$). The data in Fig. 10.1 is normalized based on the 0.2 % offset yield strength of steel at the ambient temperature, and the EC3 model consists of eight linear functions for different temperature ranges. The developed probabilistic model is provided in Eq. (10.4) and shown in Fig. 10.1. In Eq. (10.4), $\widehat{k}^{*}_{y,2\%,T} = \left(\widehat{k}_{y,2\%,T} + 10^{-6}\right)/1.7$ where $\widehat{k}_{y,2\%,T}$ is the normalized 2 % yield strength based on EC3 (Eq. 10.3), $\text{logit}\left(\widehat{k}^{*}_{y,2\%,T}\right) = \ln\left[\frac{\widehat{k}^{*}_{y,2\%,T}}{1-\widehat{k}^{*}_{y,2\%,T}}\right]$, T is temperature in Celsius, and ε is the standard normal distribution. In this model, $\gamma(\mathbf{x},\boldsymbol{\theta})$ (in Eq. 10.1) consists of three terms, and $\sigma = 0.43$:

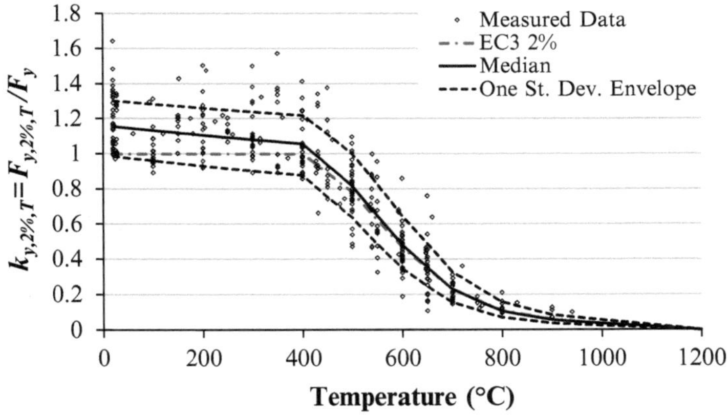

Fig. 10.1 Proposed probabilistic model for yield strength at a strain of 2 % using EC3 (Eq. 10.3)

$$\widehat{k}_{y,2\%,T} = \frac{F_{y,2\%,T}}{F_y} = \left\{ \begin{array}{ll} 1.0 & T < 400\,°\mathrm{C} \\ -2.2 \times 10^{-3}T + 1.88 & 400°\mathrm{C} \leq T < 500°\mathrm{C} \\ -3.1 \times 10^{-3}T + 2.33 & 500°\mathrm{C} \leq T < 600°\mathrm{C} \\ -2.4 \times 10^{-3}T + 1.91 & 600°\mathrm{C} \leq T < 700°\mathrm{C} \\ -1.2 \times 10^{-3}T + 1.07 & 700°\mathrm{C} \leq T < 800°\mathrm{C} \\ -5 \times 10^{-4}T + 0.51 & 800°\mathrm{C} \leq T < 900°\mathrm{C} \\ -2 \times 10^{-4}T + 0.24 & 900°\mathrm{C} \leq T < 1200°\mathrm{C} \\ 0.0 & 1200°\mathrm{C} \leq T \end{array} \right\} \tag{10.3}$$

$$k_{y,2\%,T} = 1.7 \times \frac{e^{\left[logit\left(\hat{k}^{*}_{y,2\%,T}\right)+0.412-0.81\times10^{-3}T+0.58\times10^{-6}T^{1.9}+0.43\times\varepsilon\right]}}{e^{\left[logit\left(\hat{k}^{*}_{y,2\%,T}\right)+0.412-0.81\times10^{-3}T+0.58\times10^{-6}T^{1.9}+0.43\times\varepsilon\right]}+1} \tag{10.4}$$

Figure 10.1 shows the measured data and median of the proposed probabilistic model ($\varepsilon = 0$ in Eq. 10.4) based on the logistic transformation, along with the one standard deviation confidence interval of the models in the transformed space ($\varepsilon = \pm 1$ in Eq. 10.4). The model asymptotically approaches zero at higher temperatures, while the confidence interval is also closing faster at higher temperatures, reflecting smaller dispersion of data.

Similar to yield strength, data from the study by NIST (Luecke et al. 2011) are used to develop the probabilistic model for normalized modulus of elasticity at elevated temperatures ($k_{E,T}$). In developing the probabilistic model, the deterministic base, $\hat{c}(\mathbf{x})$, in Eq. (10.1) is set to zero, and using the logistic function, one arrives at Eq. (10.5) with three terms for $\gamma(\mathbf{x},\boldsymbol{\theta})$ and σ of 0.36. Figure 10.2 shows the proposed models and one standard deviation confidence interval (in the transformed space) in relation to the measured data. The figure shows that the standard deviation envelope for the proposed model follows the scatter of data and is the smallest at low and high temperatures:

$$k_{E,T} = 1.1 \times \frac{e^{\left(2.54-2.69\times10^{-3}\times T-2.83\times10^{-6}T^{2}+0.36\times\varepsilon\right)}}{e^{\left(2.54-2.69\times10^{-3}\times T-2.83\times10^{-6}T^{2}+0.36\times\varepsilon\right)}+1} \tag{10.5}$$

Both of the proposed models have important advantages over the existing models: they are single continuous curves, as opposed to the available deterministic models such as EC3, and they are unbiased and account for uncertainty. More detail about derivation and application of the models using the EC3 formulation can be found in Elhami Khorasani et al. (2015a).

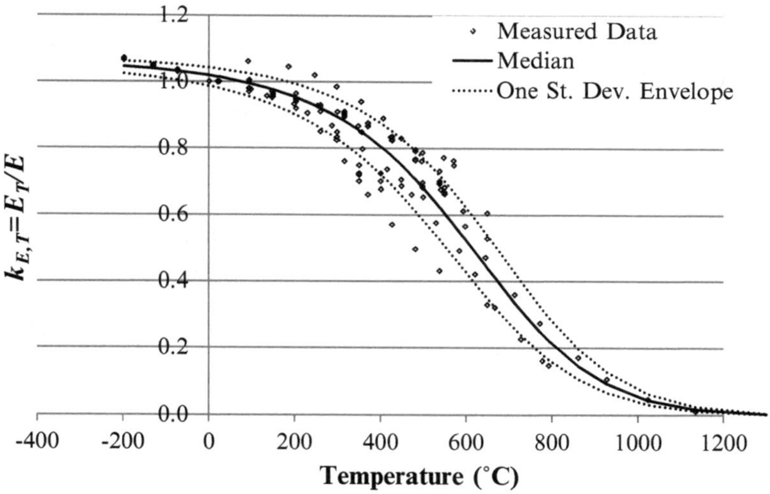

Fig. 10.2 Proposed probabilistic models for modulus of elasticity with no deterministic base

10.3 Methodology

One of the objectives of this work is to develop a framework to evaluate the performance of a MRF under both fire-only and FFE scenarios. In the fire-only scenario, the MRF is intact with no prior damage, while the MRF has gone through nonlinear seismic analysis in an FFE scenario and may have permanent residual deformations before the fire starts. The comparison between the two events shows the influence of the earthquake on performance of the frame during fire.

The steps to perform post-earthquake fire analysis of the MRF are as follows: (1) select an earthquake scenario, (2) select a fire scenario, (3) perform seismic structural analysis, (4) change certain model constraints to allow for thermal expansion, and (5) perform structural-fire analysis. Uncertainties in the above process are grouped in steps 1 and 2 as part of demand and step 5 as part of capacity. A routine Monte Carlo simulation can be used for reliability analysis where the process is repeated multiple times to incorporate the uncertainties. The frame structure in this work is analyzed for one ground motion in step 1 (i.e., deterministic assumption), while multiple fire scenarios considering uncertainties in fire load and fire location (step 2) are modeled, and variability in material properties at elevated temperatures (step 5) are included. Fire load and material properties are randomly generated using the developed probabilistic models provided in Eqs. (10.2a, 10.2b), (10.4), and (10.5).

Four different engineering design parameters (EDP) related to the beam are defined to evaluate performance of the MRF. The performance of the columns was also considered, and no limit state was detected in any of the cases; this is due to large and heavy column sections of the MRF. Therefore, the results do not include

Table 10.1 Defined EDPs and the corresponding limit states

EDP	Limit state
Plastic hinges	3 plastic hinges
Pseudo-velocity	0.254 mm/s (0.01 in/s)
Tension force	20 % P_u of the column
Deflection	*L*/20

column performance. Similar results were observed in a study by Keller and Pessiki (2015). A corresponding limit state is defined for each considered EDP as shown in Table 10.1 and discussed as follows:

1. *Plastic hinges*: In a MRF with moment connections, three plastic hinges in a beam form a mechanism, in which case the beam eventually becomes unstable and loses its capacity to provide lateral restraint to the column.
2. *Pseudo-velocity*: The pseudo-velocity is calculated as the rate of displacement of the beam and is used as a measure of instability at elevated temperatures. The EDP is defined as the pseudo-velocity of the beam at the beam mid-span. Based on previous studies, and values provided in the study by Usmani et al. (2003), a limiting value of 0.254 mm/s (0.01 in/s) is defined for pseudo-velocity of the beam.
3. *Tension force:* Large tension forces can develop in the beam during the cooling phase of the fire and consequently cause connection failure. The finite element model does not capture connection failure. Therefore, based on a sample calculation of connection capacity, a limit state of 20 %P_u of the column is defined for the maximum tension force in the beam before a connection fails (connections should generally be able to carry at least 2 %P_u to allow for tensile forces due to column out of plumbness). The limiting value of 20 %P_u is based on conservative calculations for a prototype 9-story building that will be discussed in Sect. 10.4 and can be refined based on particular designs under study.
4. *Deflections*: Excessive deflections can cause instability, and the limit state is defined as *L*/20 (*L* is the span length) (BRE 2005).

During the structural analysis of the frame at elevated temperatures, the concrete slab is not modeled, and non-composite behavior is assumed since the concrete slab goes into tension and cracks (Quiel and Garlock 2010a). However, it is assumed that there remains sufficient mechanical locking between the cracked concrete and shear studs to prevent lateral torsional buckling of the beams.

The MRF is modeled in the programming environment OpenSees, which is a finite element program and object-oriented software for nonlinear analysis of structures under seismic loadings, primarily developed at the University of California, Berkeley (McKenna and Fenves 2006). The seismic and thermal analyses of the MRF are both performed in OpenSees, using the recently added thermal module (Jiang et al. 2015). The new thermal module was modified by Elhami Khorasani et al. (2015b) to enhance the thermal analysis by allowing strain reversals, a seamless transition from seismic to thermal analysis, and including reliability analysis. The current constitutive material model for steel at elevated temperature was modified to incorporate the effect of plastic strain during both heating and cooling phases of fire. This modification facilitates sequential seismic and thermal analyses. Also, the current reliability module was adjusted to incorporate uncertainties in the thermal analysis.

10.4 Case Study: 9-Story MRF

This work applies the framework discussed above to study the performance of an example MRF under fire and FFE. The geometry and building description of the prototype MRF is based on the SAC steel project (SAC 2000). The SAC project considered different building heights, location, and both stiff and soft soil. The MRF in the present study is a 9-story frame that is assumed to be located on stiff soil in downtown Los Angeles and has plan and elevations that are based on SAC buildings. As the seismic design provisions have been updated in the code since the SAC steel project, the MRF is redesigned based on ASCE7-10 specifications (ASCE 2010).

The building geometry consists of a square plan with 5 bays, each at 9.14 m (30 ft), in either direction. Columns and girders are spaced at 9.14 m (30 ft) and beams are spaced at 3.05 m (10 ft) intervals. The floor beams provide lateral support for the MRF girders. The 9-story building has a typical floor height of 3.96 m (13 ft) with a basement height of 3.66 m (12 ft) and ground floor height of 5.50 m (18 ft). The building consists of 4 MRFs, one on each side, and placed such that biaxial bending is avoided at corners. The MRFs in the two orthogonal directions are identical. The columns are pinned at the foundation and laterally braced at the ground level. The frame has a first mode period of 1.75 s. Figure 10.3 shows plan and elevation of the 9-story structure.

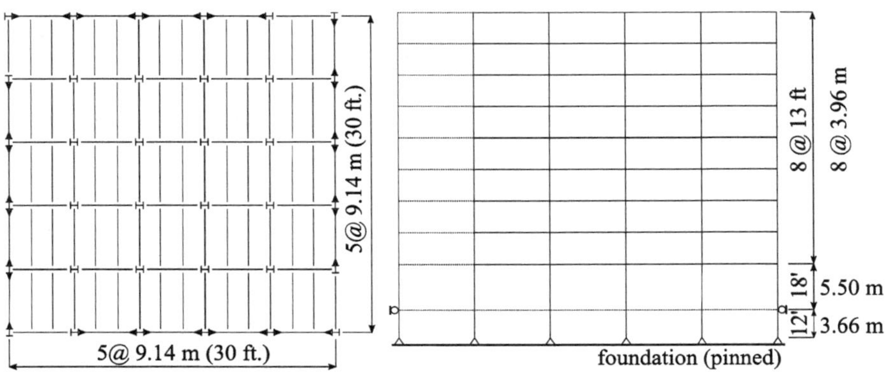

Fig. 10.3 Plan and elevation of the 9-story MRF

Given the MRF example, the following provides details of each step, discussed in Sect. 10.3, to perform post-earthquake fire analysis applied to the case study:

Step 1: Select an Earthquake Scenario In this study, the earthquake scenario is kept as deterministic and the case study is completed for one ground motion. In general, the variability in ground motions should be considered, and the analysis should be repeated for various earthquakes in order to capture uncertainty in the earthquake scenario. The ground motion selected for this study is the 1989 Loma Prieta, CA earthquake ground motion that was recorded at station 47381 Gilroy (Array #3), and will be referred to as Gilroy earthquake. Only one component of the ground motion (the G03090 component) is applied since two-dimensional models are being used. The location was on stiff soil, and the earthquake had a magnitude of 6.9 with the closest distance from a fault rupture zone of 14.4 km. The hazard level chosen for this study is the maximum considered earthquake (MCE) (the 2 % in 50-year earthquake). The scale factor for the Gilroy ground motion is calculated to be 2.78 for the MCE level. The scaling procedure is based on the work of Somerville et al. (1997).

Step 2: Select a Fire Scenario As discussed in Sect. 10.3, the frame structure is analyzed for multiple fire scenarios considering uncertainties in fire load and fire location. The probabilistic fire load density developed in Sect. 10.2.1 is used to generate random fire load density q. This study assumes a single compartment 6.1 m deep by 9.1 m wide (20 ft deep by 30 ft wide) in every floor that is subject to fire. Given the floor area of the compartment, Eq. (10.2a), developed for lightweight compartments (general office space), is a better choice. Therefore, the full fire temperature-time history is constructed using q values from Eq. (10.2a), the EC1 formulation (CEN 2002), and based on the work of Quiel and Garlock (2010a) to resemble an actual fire event. Given that the fire occurs after an earthquake, it is assumed that the compartment has no functional active firefighting measures and that the passive fire protection has been damaged enough to render it ineffective. Previous research by Braxtan and Pessiki (2011a, b) showed that spray fire resistive material, as passive fire protection, can delaminate or dislodge during inelastic seismic response and extensive damage can be expected especially in the beam plastic hinge regions. In addition, steel has a high thermal conductivity, and a local damage to passive fire protection can cause the structural element to heat up during fire.

Figure 10.4a shows the probability density function (PDF) for the maximum fire temperature reached in the compartment based on 50,000 random realizations of q generated using Eq. (10.2a) based on random generations of ε. The generated fire scenarios show that the 90 percentile maximum fire temperature value ($T_{\max}$) is approximately 1000 °C. Figure 10.4b shows the cumulative distribution function (CDF) for $T_{\max}$ larger than 1000 °C. The MRF in this study is analyzed for 50 randomly generated fire temperature-time curves with $T_{\max}$ larger than 1000 °C.

Four different fire locations are assumed in the frame by varying floors and bays of the fire compartment, shown in Fig. 10.5 where "B" stands for bay and "F" stands

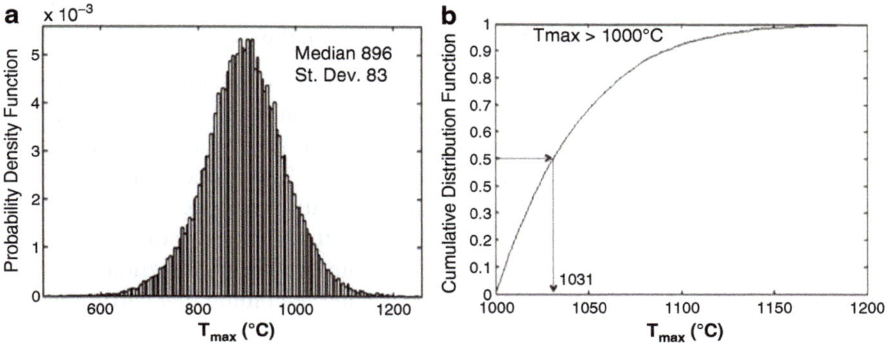

Fig. 10.4 (**a**) PDF for $T_{\max}$, (**b**) CDF for $T_{\max}$ above 1000 °C

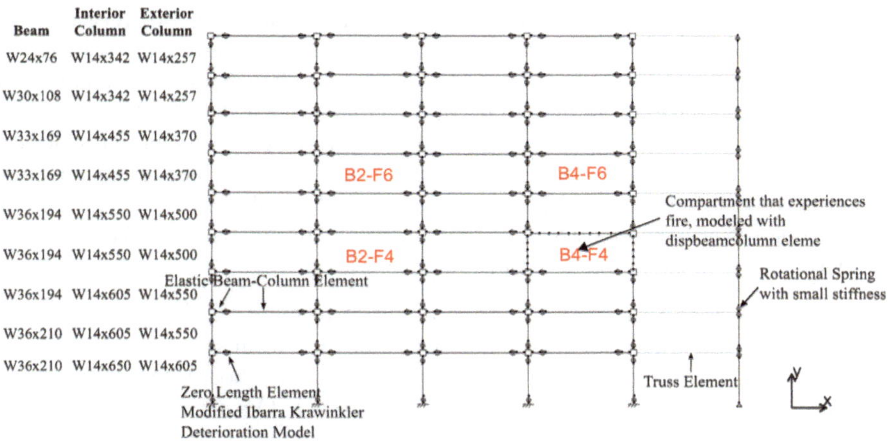

Fig. 10.5 Analytical model of the MRF frame in OpenSees

for floor. The locations of compartment allow evaluating the effect of story level (floor 4 vs. floor 6) and the effect of bay location (interior bay 2 vs. exterior bay 4). In a complete probabilistic analysis, uncertainty in the location of fire in the building should be considered. The aim of this research is to demonstrate the methodology while investigating post-earthquake fire performance of a tall building. Therefore, four compartments that are representative of interior/exterior bays and lower/higher stories for the 9-story building are selected as example studies.

The heat transfer analysis is performed for all the considered cases to obtain steel temperatures for the beams, perimeter columns, and interior columns of the compartment under study. Heat transfer for the two columns and a beam in the fire compartment is mainly per the closed-form solution developed by Quiel and Garlock (2010b) based on a lumped-mass method. The procedure is slightly modified for the beam to include the effect of the concrete slab (which acts as a heat sink) on the top flange temperature. An empirical equation developed by Ghojel and Wong (2005) is used to calculate the heat flux between the top flange and the slab.

Step 3: Perform Seismic Structural Analysis Figure 10.5 shows the analytical model for the 9-story frame in OpenSees and the design sections of the MRF. The concentrated plasticity concept with rotational springs is used to model the nonlinear behavior of the 9-story MRF under dynamic loading. The frame is modeled with elastic beam-column elements that are connected with zero-length elements. The zero-length elements serve as the rotational springs that follow a bilinear hysteretic response. Panel zones are also modeled to capture the shear distortion in beam-column joints (Gupta and Krawinkler 1999). A leaning-column that carries gravity load is linked to the frame to simulate P-Delta effects.

Step 4: Change Model Constraints to Allow for Thermal Expansion The model constraints need to be adjusted when transferring from seismic to thermal analysis. During the seismic analysis, a constraint is placed on the nodes of every floor to ensure that they move together horizontally, representing the effect of concrete slab diaphragm in the composite structure. However, after the seismic analysis is completed, the constraint on the nodes of the compartment that would experience fire is removed. This is explained in the previous research by Quiel and Garlock (2010a), which showed the steel in the composite girder during the thermal analysis experiences a faster increase in temperature than the slab. The steel expands at a faster rate than concrete, which eventually results in cracking of concrete, thus rendering the slab negligible for structural response.

Step 5: Perform the Structural-Fire Analysis In performing an efficient and proper FFE analysis, the procedure must seamlessly transition from seismic to thermal in the OpenSees environment, and the modeling technique should be applicable to both dynamic and thermal analysis. Thermal modeling in OpenSees is only possible with the *dispBeamColumnThermal*-type element (Jiang and Usmani 2013), which is defined using fibers and considers plasticity along its length. Meanwhile, an efficient seismic model discussed above uses various other element types, including zero-length deterioration spring elements to capture nonlinear behavior. The 9-story frame in this work is modeled with springs and elements applicable for seismic analysis, except for the beams and columns of the fire compartment that are assumed to be heated (which will be modeled with *dispBeamColumnThermal* elements), as shown in Fig. 10.5. The authors tested the thermal elements for dynamic loading, and the results showed that the elements could reasonably capture the nonlinear dynamic behavior during an earthquake.

The two major parameters with uncertainties in the structural-fire analysis of the frame are yield strength and modulus of elasticity of steel at elevated temperatures. Equations (10.4) and (10.5) are used to randomly generate $k_{y,2\%,T}$ and $k_{E,T}$, for the beam and each column in the fire compartment. The material properties are randomly generated for every considered fire scenario.

10.5 Results

This section provides results of the probabilistic study for the 9-story frame under fire-only and FFE scenarios. As explained in Table 10.1 (Sect. 10.3), four limit states are considered for performance evaluation of the structure: (1) formation of three plastic hinges (PH), (2) pseudo-velocity (PSV), (3) tension force, and (4) deflection. The selected earthquake is the Gilroy ground motion scaled to the maximum considered earthquake (Gilroy-MCE). The thermal loading scenarios are obtained based on 50 randomly generated fire curves, applied to four fire compartments. The randomly generated material properties for the beam and columns, $k_{y,2\%,T}$ and $k_{E,T}$, are kept the same in the four compartments for comparison purposes (e.g., so that the effects of location can clearly be observed).

Figure 10.6 identifies the limit states reached under 50 fire scenarios for fire-only and FFE-Gilroy, respectively. The plots show results based on the bay under study (bay 2 and bay 4) and the two considered floors (floor 4 and floor 6). Also, the plots differentiate between the cases that the program stops converging during the heating

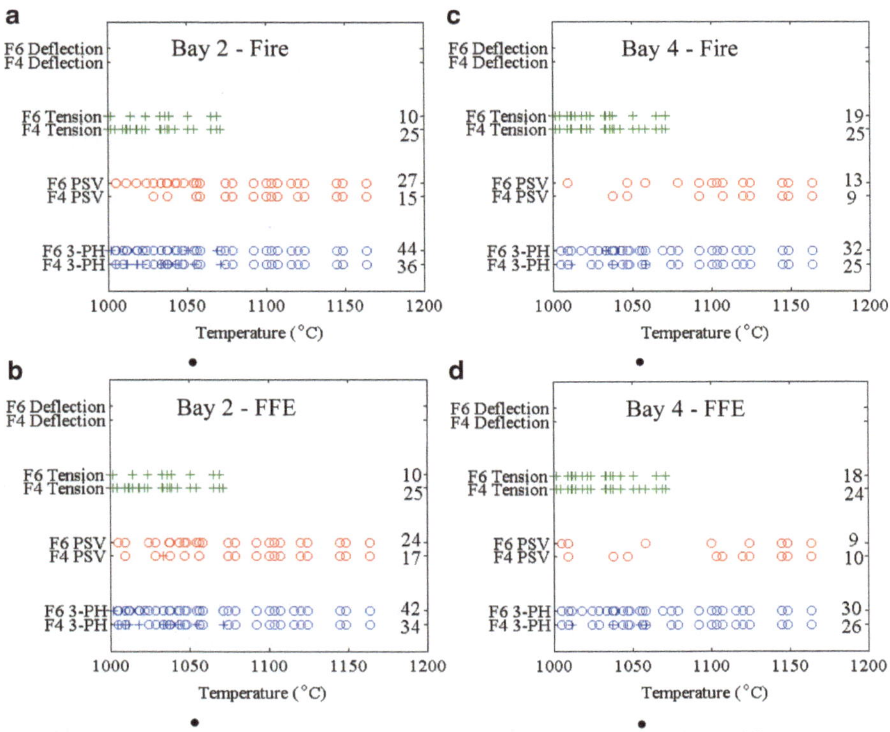

- "o" : cases that the program stops converging during the heating phase of fire
- "+" : cases that the program continues through cooling phase of of fire

Fig. 10.6 Plots of limit states reached for fire-only and FFE-Gilroy scenarios phase of fire (circle markers) versus cases that continue through the cooling phase of the temperature-time curve of fire (plus-sign markers). Non-convergence does not imply collapse of the structure, but means that the beam element is not locally stable any longer. On the right margin of each plot, the total number of cases reaching the limit state is indicated.

The results show that, considering all the scenarios, at least 50 % of the cases reach the three plastic hinge limit state during the heating phase (25 out of 50 cases in B4-F4), while the number could reach up to 88 % (44 out of 50 cases in B2-F6). All cases that reach the PSV limit state form three plastic hinges, whereas formation of three plastic hinges does not guarantee reaching the PSV limit state. In addition, the program stops converging in all cases that reach the PSV limit state (implying that the beam is locally unstable and OpenSees can no longer advance the analysis). The results show that large tension forces can develop during the cooling phase, which may lead to potential connection failures. Overall, the upper floors reach limit states for plastic hinges and PSV more frequently than the lower floors due to the smaller beam sizes. The interior bays reach limit states more frequently than the exterior bays due to adjacent restraints. Finally, response of the MRF under fire in a seismically induced damaged state is similar to having a fire within an undamaged MRF.

Another parameter that is studied, and is affected by the earthquake, is the interstory drift of compartments during fire. The analyses showed that the exterior bays experience drifts that are larger than the interior bays and fire causes more drift on floor 6 than floor 4. The maximum recorded drift reached after the earthquake and during the fire was approximately 3.5 % in the B4-F6 compartment. Overall, the earthquake does not increase the probability of reaching a limit state, but affects the drift demands during the fire event.

10.6 Summary and Conclusions

This work has provided a procedure to evaluate performance of a 9-story steel moment resisting frame (MRF) subject to fire and fire following earthquake (FFE) within a probabilistic framework. As part of the framework, parameters with uncertainties were identified, and probabilistic models were developed for fire load density, as well as for yield strength and modulus of elasticity of steel at elevated temperatures. The developed models were based on literature data, in conjunction with available deterministic equations in codes and standards. The proposed models are unbiased and account for uncertainties. The procedure used to develop the models can be applied to derive probabilistic models for other material properties and parameters with uncertainty.

The framework, together with the developed models, was applied to evaluate performance of a 9-story MRF under fire and FFE. The frame was modeled in OpenSees, and both seismic and thermal analyses were performed in one seamless programming environment. Four fire locations and 50 fire scenarios were considered to investigate the effects of fire intensity and location on the performance of the MRF. The results were compiled based on four engineering design parameters related to the beam, including formation of three plastic hinges, pseudo-velocity, tension force, and deflection. Performance of the columns was not included in the results, as analysis showed that the columns did not reach any limit state.

Results show that interior bays reached limit states more often than exterior bays due to the added restraint of the adjacent structure and upper floors were more vulnerable than lower floors due to the smaller section sizes. Overall, post-earthquake damage does not affect the fire performance of the MRF for the considered design parameters. During fire, the only parameter affected by the post-earthquake initial condition was the inter-story drift. The residual drift after the earthquake increased the total drift during fire, but the total drift did not exceed 4 %. This work focused on the performance of the MRF under FFE, assuming that fire occurs on the building perimeter where the MRF is located. However, it is equally likely to have a fire ignition inside the building where gravity frames are located. Gravity frames have considerably smaller sections compared to the MRF, and it is the subject of future work of this research. Finally, the proposed framework can be extended to evaluate structures under other multi-hazard scenarios. The case study in this work was performed for one earthquake scenario; however, the procedure can be expanded to include uncertainty in ground motion. Also, the framework can be adopted for fire following blast scenarios, a similar cascading multi-hazard loading event.

References

ASCE. (2010). *Minimum design loads for buildings and other structures* (ASCE/SEI 7-10). Reston, VA: Author.

Box, G. E. P., & Cox, D. R. (1964). An analysis of transformations (with discussion). *Journal of Royal Statistical Society, Series B, 26*, 211–252.

Box, G. E. P., & Tiao, G. C. (1992). *Bayesian inference in statistical analysis*. Reading, MA: Addison-Wesley.

Braxtan, N. L., & Pessiki, S. (2011a). Bond performance of SFRM on steel plates subjected to tensile yielding. *Journal of Fire Protection Engineering, 21*(1), 37–55.

Braxtan, N. L., & Pessiki, S. (2011b). Post-earthquake fire performance of sprayed fire resistive material on steel moment frames. *Journal of Structural Engineering, 137*(9), 946–953.

BRE: Building Research Establishment Ltd. (2005). *The integrity of compartmentation in buildings during fire* (Project report No. 213140(1)).

CEN. (2002). *Eurocode1: Actions on structures, Part 1-2: General actions – actions on structures exposed to fire*. Brussels, Belgium: European Committee for Standardization (CEN).

CEN. (2005). *Eurocode3: Design of steel structures, Part 1-2: General rules – structural fire design* (ENV 1993-1-2:2001). Brussels: European Committee for Standardization (CEN).

Culver, C. (1976). *Survey results for fire loads and live loads in office buildings*. Washington, DC: National Bureau of Standards.

De Sanctis, G., Fischer, K., Kohler, J., Fontana, M., & Faber, M. H. (2011). A probabilistic framework for generic fire risk assessment and risk-based decision making in buildings. In *Proceedings of the 11th International Conference on Application of Statistics and Probability in Civil Engineering*, ICASP11. August 1–4, 2011, Zurich, Switzerland.

Della Corte, G., Landolfo, R., & Mazzolani, F. M. (2003). Post-earthquake fire resistance of moment resisting steel frames. *Fire Safety Journal, 38*, 593–612.

Della Corte, G., Faggiano, B., & Mazzolani, F. M. (2005). On the structural effects of fire following earthquake. In *Improvement of buildings' structural quality by new technologies: Proceedings of the Final Conference of COST Action C12* (pp. 20–22). Innsbruck, Austria.

Elhami Khorasani, N. (2015). *A probabilistic framework for multi-hazard evaluations of buildings and communities subject to fire and earthquake scenarios*. PhD dissertation, Princeton University.

Elhami Khorasani, N., Garlock, M. E. M., & Gardoni, P. (2014). Fire load: Survey data, recent standards, and probabilistic models for office buildings. *Journal of Engineering Structures, 58*, 152–165.

Elhami Khorasani, N., Gardoni, P., & Garlock, M. E. M. (2015a). Probabilistic evaluation of steel structural members under thermal loading. *Journal of Structural Engineering*, doi:10.1061/(ASCE)ST.1943-541X.0001285.

Elhami Khorasani, N., Gernay, T., & Garlock, M. E. M. (2015c). Modeling post-earthquake fire ignitions in a community. *Submitted to Fire Safety Journal, Elsevier.*

Elhami Khorasani, N., & Garlock, M. E. M. (2015). Overview of fire following earthquake: Historical events and community responses. *Submitted to International Journal of Disaster Resilience in the Built Environment.*

Elhami Khorasani, N., Garlock, M. E. M., & Quiel, S. E. (2015b). Modeling steel structures in OpenSees: Enhancements for fire and multi-hazard probabilistic analysis. *Journal of Computers and Structures, 157*, 218–231.

Gardoni, P., Der Kiureghian, A., & Mosalam, K. M. (2002). Probabilistic capacity models and fragility estimates for reinforced concrete columns based on experimental observations. *Journal of Engineering Mechanics, 128*(10), 1024–1038.

Ghojel, J. I., & Wong, M. B. (2005). Three-sided heating of I-beams in composite construction exposed to fire. *Journal of Constructional Steel Research, 61*, 834–844.

Guo, Q., Kaihang, S., Zili, J., & Jeffers, A. (2013). Probabilistic evaluation of structural fire resistance. *Fire Technology, 49*(3), 793–811.

Gupta, A., & Krawinkler, H. (1999). *Seismic demands for performance evaluation of steel moment resisting frame structures* (Technical Report 132). Stanford, CA: The John A. Blume Earthquake Engineering Research Center, Department of Civil Engineering, Stanford University.

He, Y. (2013). Probabilistic fire-risk-assessment function and its application in fire resistance design. *Procedia Engineering, 62*, 130–139.

Jiang, J., & Usmani, A. (2013). Modeling of steel frame structures in fire using OpenSees. *Journal of Computers and Structures, 118*, 90–99.

Jiang, J., Jiang, L., Kotsovinos, P., Zhang, J., Usmani, A., McKenna, F., et al. (2015). OpenSees software architecture for the analysis of structures in fire. *Journal of Computing in Civil Engineering, 29*(1).

Keller, W. J., & Pessiki, S. (2015). Effect of earthquake-induced damage on the sidesway response of steel moment-frame buildings during fire exposure. *Earthquake Spectra, 31*(1), 273–292.

Luecke, W. E., Banovic, S., & McColskey, J. D. (2011, November). *High-temperature tensile constitutive data and models for structural steels in fire* (NIST Technical Note 1714). U.S. Department of Commerce, National Institute of Standards and Technology.

McKenna, F., & Fenves, G. L. (2006), *Opensees 2.4.0, Computer Software*. UC Berkeley, Berkeley, CA [http://opensees.berkeley.edu].

Quiel, S. E., & Garlock, M. E. M. (2010a). Parameters for modeling a high-rise steel building frame subject to fire. *Journal of Structural Fire Engineering, 1*(2), 115–134.

Quiel, S. E., & Garlock, M. E. M. (2010b). Closed-form prediction of the thermal and structural response of a perimeter column in a fire. *The Open Construction and Building Technology Journal, 4*, 64–78.

Rao, C. R., & Toutenburg, H. (1997). *Linear models, least squares and alternatives*. New York, NY: Springer.

SAC Joint Venture. (2000, September). *FEMA 355C: State of the art report on system performance of steel moment frames subject to earthquake ground shaking*. Federal Emergency Management Agency.

Somerville, P., Smith, N., Punyamurthula, S., Sun, J., & Woodward-Clyde Federal Services. (1997). *Development of ground motion time histories for phase 2 of the FEMA/SAC steel project* (BD-97/04). SAC Joint Venture.

Stone, J. C. (1996). *A course in probability and statistics*. Belmont, CA: Duxbury.

Usmani, A. S., Chung, Y. C., & Torero, J. L. (2003). How did the WTC towers collapse: A new theory. *Fire Safety Journal, 38*, 501–533.

Yassin, H., Iqbal, F., Bagchi, A., & Kodur, V. K. R. (2008). Assessment of post-earthquake fire performance of steel frame building. In *Proceedings of the 14th World Conference on Earthquake Engineering*. Beijing, China.

Chapter 11
Progressive Collapse Resistance for Steel Building Frames: A Cascading Multi-Hazard Approach with Subsequent Fire

Spencer E. Quiel

Abstract The design of building frames for progressive collapse resistance typically includes a two-phase sequence: (1) the initial hazard that causes localized damage and (2) the subsequent response of the structure to redistribute loads and bridge over the damaged areas. However, recent events have shown that local damage to building frames is commonly followed by a fire which subsequently ignites near the location of the damage. In current practice, little consideration if any is given to fire as a cascading hazard for progressive collapse-resistant design. Fire exposure for a damaged structure could be detrimental to the short-term stability of that structure and may pose a significant threat to the safe evacuation of building occupants. This chapter explores the effects of fire following an extreme event that causes failure of one column on the perimeter of a steel building frame—this damage scenario represents the typical design case for assessing progressive collapse resistance. Two prototype office buildings are considered: a low-rise (5-story) new construction and a high-rise (38-story) preexisting structure. The approach focuses on implementation of the US Government guidelines for progressive collapse resistance and assumes that the extreme event not only damages one column but also damages active fire protection (i.e., sprinklers) in the vicinity of the structural damage. Results of these studies include estimates of the time to collapse initiation and a correlation between the level of remaining passive fire protection (i.e., fire-resistive materials applied to the structural elements) and the collapse time. The goal of this chapter is to raise awareness of potential fire hazards that may follow extreme events and provide guidance for the assessment of these hazards.

S.E. Quiel (✉)
Department of Civil and Environmental Engineering, Lehigh University,
117 ATLSS Drive, Bethlehem, PA 18015, USA
e-mail: squiel@lehigh.edu

P. Gardoni, J.M. LaFave (eds.), *Multi-hazard Approaches to Civil Infrastructure Engineering*, DOI 10.1007/978-3-319-29713-2_11

11.1 Introduction

The ability of a structure to resist a disproportionately large collapse due to localized damage (i.e., progressive collapse) has become a topic of increasing concern within the building community in the wake of structural collapses worldwide over the last 50 years. Design concepts for progressive collapse resistance have been developed for building codes and criteria in response to these events, which include the 1968 collapse at Ronan Point in the UK (Pearson and Delatte 2005) and the progressive collapse of the Murrah Federal Building resulting from the 1995 Oklahoma City bombing (Corley et al. 1998). These methodologies have improved the design of structures to resist a collapse that is disproportionate to the level of damage sustained during an extreme event; however, additional recent collapses have shown that a fire caused by the extreme event at the location of damage may critically affect the progressive collapse resistance of the structure. Fire as a cascading hazard is typically not considered in current progressive collapse design guidance but can be critical to the structure's ability to withstand all hazards due to an extreme event. The investigation of the collapse of the World Trade Center (WTC) Twin Towers by the National Institute of Standards and Technology (NIST 2005) concluded that the fire following the aircraft impact and explosion caused the progressive collapse of the damaged structure. As was the case for the Twin Towers, an extreme event can also cause damage to the active and passive fire protection systems, allowing the fire to have even more adverse impact on the performance of the damaged structure. Steel-framed structures are a concern in this case because they rely heavily on these protection measures, typically more so than concrete structures, to limit the temperature increase of the structural elements and prevent further structural deterioration.

This chapter discusses a cascading multi-hazard approach to evaluate the performance of a structure that is damaged by an extreme event and is subsequently subjected to a resulting fire. Two prototype steel building frames are considered for this study. The first is based on an actual five-story office building that was recently designed for progressive collapse resistance according to criteria established by the US Department of Defense (Quiel and Marjanishvili 2012). The building has a conventional steel-framed design and therefore represents a common example of low-rise office construction. The second prototype is the One Meridian Plaza (1MP), a 38-story office building in Philadelphia, PA, that was constructed in 1968. In 1991, the 1MP suffered a severe multistory fire (Chubb et al. 1991); though it did not collapse, the building suffered extensive damage and was subsequently demolished in 1999 following several years of postfire investigation and litigation (Dexter and Lu 2001). The 1MP's structural system consists of steel moment-resisting frames and is an example of existing high-rise construction.

The objective of this chapter is to evaluate steel structures that meet the current design provisions for progressive collapse resistance (which assumes the structure to be at ambient temperature) for exposure to fire following the emergence of local damage. Several design scenarios are considered in which varying levels of passive fire protection are applied to the structural elements. The studies that provide the basis for this chapter with regard to the low-rise and high-rise prototypes are available in Quiel and Marjanishvili (2012) and Neal et al. (2012), respectively.

11.2 Multi-Hazard Approach

The concept of analyzing structures that have been damaged by an extreme event and experienced a subsequent fire has been previously considered by several other researchers. Among those studies are Della Corte et al. (2003), Zaharia and Pintea (2009), and Khorasani et al. (2015), each of which examined the performance of a steel moment-resisting frame (MRF) under fire after it had sustained permanent deformations due to seismic loading. Other examples include studies by Chen and Liew (2005) and Quiel and Marjanishvili (2013), which evaluated the fire resistance of steel columns that had been damaged by blast. The studies presented in this chapter take a similar approach by evaluating the response of a steel building frame to fire once it has been damaged by blast or impact. As will be discussed later, the prototypes discussed in this chapter both meet the current progressive collapse resistance design criteria published by the US Government (DoD 2013). To account for the effects of multiple cascading hazards, the proposed approach analyzes the response of a building frame to the following, in order: (1) removal of a key structural element via blast or impact, occurring on a time scale of milliseconds; (2) the redistribution of load and the onset of stable plastic deformations as the structure's progressive collapse resistance is engaged, on a time scale of seconds; and (3) the development of a severe fire near the location of member removal, on a time scale of minutes. This sequence of events is illustrated in Fig. 11.1. The approach focuses on the behavior of the damaged structure once the members targeted for removal have already been removed, and the analyses therefore begin in phase (2) described above. It is assumed that the building sustains no additional structural damage due to the initial extreme event beyond the targeted removed members (which is consistent with the current progressive collapse resistance criteria (DoD 2013)) and that the fire starts just after the onset of stable plastic deformations in the structural frame near the removal location.

11.2.1 Evaluating Resistance to Progressive Collapse

Progressive collapse occurs when relatively localized damage causes a chain reaction of failures that eventually lead to the collapse of a disproportionately large part of the structure. The most recent DoD criteria (2013) specify levels of progressive collapse resistance as a function of the building's occupancy and height. According to these classifications, the prototype structures considered for

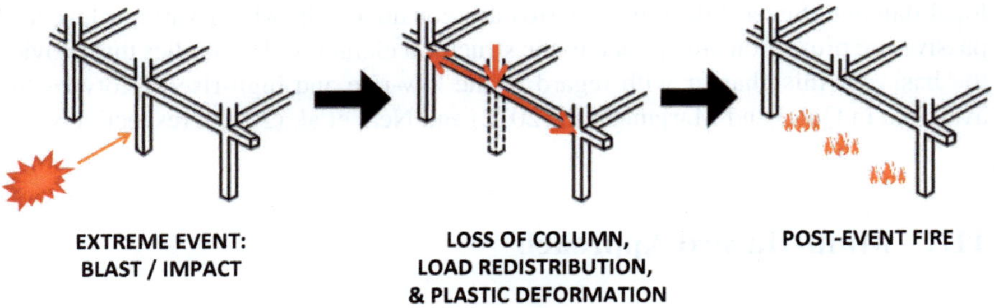

Fig. 11.1 Three-phase sequence of the cascading multi-hazard approach

this study were evaluated for progressive collapse resistance using the alternate path method (APM), in which the structure must bridge across elements that are "removed" to represent local damage. Based on concepts first proposed by Ellingwood and Leyendecker (1978), the current guidelines for progressive collapse mitigation published by the US Government (GSA 2013; DoD 2013) rely primarily on the use of APM and other direct design methods by which collapse resistance to specific damage scenarios is evaluated. Both documents use a multi-deterministic approach to implement selective strengthening and improved structural continuity via seismic-based detailing to prevent progressive collapse. In the current threat-independent context, progressive collapse resistance is implicitly quantified by the ability of the damaged structure to avoid collapsing when subjected to a factored combination of the in situ gravity loads.

For structural steel frames, the alternate path method according to the DoD criteria (DoD 2013) specifies that the structure must bridge over columns that are damaged and ineffective due an extreme event. To simulate the damage, column removal is mandated as one at a time in one-story lengths at several plan and elevation locations. When a column is removed, beam-to-beam continuity over that column is assumed to be preserved. The required plan locations of column removal include perimeter columns at the middle of a long or short side of the building, a corner, and at any location where a building's perimeter geometry changes significantly. In elevation, columns must be removed at the first story above grade, the story above a column splice, the story at mid-height, and the story directly below the roof. Perimeter columns are particularly susceptible to blast or impact from threats originating outside the building, and they have fewer pathways for load redistribution than interior columns. Removal of interior columns is typically not considered unless they are located in an area of the building, such as a public lobby, which has been identified by the building owner/operator as susceptible to potential hand-carried or package explosive threats.

A study by Marjanishvili and Agnew (2006) compared the currently available methodologies, which vary in their complexity, for conducting progressive collapse analysis: linear static (LS), nonlinear static (NS), linear dynamic (LD), and non-linear dynamic (ND). While each approach has trade-offs in terms of efficiency and accuracy, the study highlighted the effectiveness of NS and ND procedures because they realistically account for the nonlinear, plastic response of the structure to the damage scenario (Marjanishvili and Agnew 2006). The use of ND analysis for evaluating progressive collapse resistance is increasingly common in practice because it explicitly captures the dynamic effects of loads on the damaged structure, which typically results in greater design efficiency compared to NS analysis (Kim et al. 2009). The progressive collapse evaluations discussed in this chapter were therefore performed using ND analysis.

The DoD criteria require the use of a 3D model to capture all potential pathways for load redistribution following the column removal. In practice, beam elements with discrete hinges are commonly used to construct these models. The floor slabs are typically modeled as rigid lateral diaphragms but are not considered to compositely contribute to the collapse resistance of the steel frame. In the current state of practice, this simplifying assumption is regarded as conservative, and new research has indicated that composite action between the floor slabs and the steel framing can provide significant increases to collapse resistance (Hadjioannou et al. 2013). The studies described in this chapter will implement the current approach via beam element models in SAP2000 (CSI 2009) and will consider the weight but not the flexural stiffness contribution of the floor slabs.

Gravity and lateral loads are applied according to the following load combination (DoD 2013):

$$1.2D + 0.5L + L_{\mathrm{LAT}} \tag{11.1}$$

where lateral load L_{LAT} is applied to the perimeter of each floor in one direction at a time, perpendicular to the building face. L_{LAT} is calculated as 0.2 % of the total load acting on that floor and represents a nominal pushover load for notional engagement of the lateral system. The gravity load portions of Eq. (11.1) are consistent with those recommended by ASCE 7-10 for extreme events (ASCE 2010). At the start of each analysis, the undamaged structure (i.e., with no columns yet removed) is initialized for gravity and lateral loads. Once equilibrium is reached, the targeted column is then rapidly removed, and the frame is analyzed until a subsequent state of equilibrium is achieved. In accordance with the DoD criteria (DoD 2013), it was assumed that all elements are at ambient temperature throughout the onset of progressive collapse resistance.

The current APM approaches are performed on a “pass/fail” design basis—a structure is deemed adequate for collapse resistance if it is able to bridge over the removed column by meeting the specified performance criteria and inadequate if it cannot. The DoD criteria stipulate that the performance criteria for the analysis of elements in the damaged frame are divided into two categories according to ASCE/SEI 41-06 (ASCE 2007): force controlled and deformation controlled. Components under high axial load ($P/P_{\mathrm{CL}} > 0.5$, where P_{CL} is the lower-bound axial load capacity) are classified as force controlled. These components (i.e., the majority of columns) must have a demand-to-capacity ratio (DCR) less than unity for both (a) combined axial load and biaxial bending and (b) shear. The capacity for force-

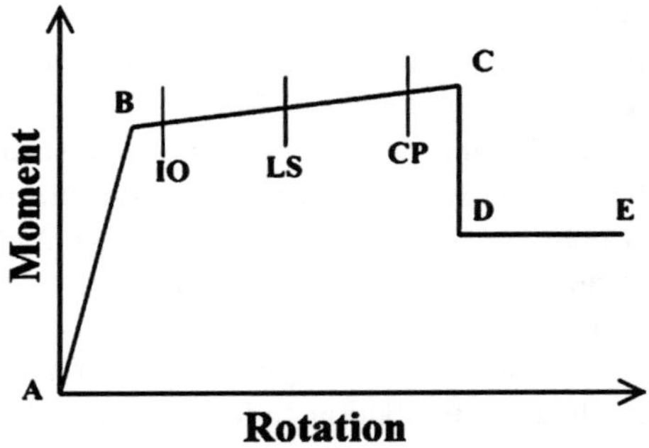

Fig. 11.2 Representative backbone curve for the plastic hinge model (ASCE 2007; CSI 2009)

controlled elements accounts for lower-bound strength of the material (in this case, as defined by Chapter 5 in ASCE 41-06 (ASCE 2007)) as well as all appropriate strength-reduction factors according to the material-specific design code (in this case, the AISC Steel Construction Manual (AISC 2010)).

Components with low axial load ($P/P_{CL} \leq 0.5$) are classified as deformation controlled. These components (i.e., beams, girders, and lightly loaded columns) are capable of developing significant plastic deformations without collapsing. Plastic deformations are assumed to be primarily in flexure, and hinge models are derived as recommended in ASCE/SEI 41-06 (ASCE 2007) and shown in Fig. 11.2. Limits of plastic rotation for deformation-controlled steel members are provided in the DoD criteria and Chapter 5 of ASCE/SEI 41-06. Three thresholds of plastic rotation are defined in the following order of severity: immediate occupancy (IO), life safety (LS), and collapse prevention (CP). Beams and girders subjected to flexure or a combination of flexure plus axial tension will experience a relatively ductile mode of eventual collapse, and the collapse prevention (CP) rotation limits are therefore used for these elements. Lightly loaded columns and other elements with a combination of flexure and compression will experience a more brittle collapse mode, and life safety (LS) rotation limits are therefore used. Plastic moment capacity and rotational limits of deformation-controlled elements account for the expected strength of the material.

11.2.2 Evaluating Resistance to Subsequent Fire

Extreme events (such as blast or impact) that may result in a column removal as described in the previous section are commonly followed by fire near the location of the structural damage. Even if a structure has been designed to resist progressive collapse at ambient temperature, its progressive collapse resistance will depend on its ability to resist a subsequent fire in its damaged state. The time lapse to the onset of fire-induced collapse of the damaged structure may prove critical to achieving building egress following an extreme event.

It is assumed that the fire starts immediately following the onset of stable permanent deformation due to the column removal. Della Corte et al. (2003) and Khorasani et al. (2015) implemented a similar two-phase approach to determine the fire resistance of a steel MRF structure that was damaged by seismic loading. Once a structural configuration that meets the progressive collapse resistance criteria is obtained, structural fire analysis via time-series integration is performed using SAFIR, a software developed at the University of Liege specifically for the analysis of structures under fire (Franssen 2005). SAFIR uses uncoupled analyses of, first, a thermal model of each member exposed to fire and, second, a structural model of the frame composed of those members. Thermal and structural material properties for steel according to Eurocode (CEN 2001, 2002) were used for the computational analyses discussed in this chapter. These properties are applicable for most conventional grades of hot-rolled or mild structural steel, which comprises the framing of both prototypes considered in this chapter.

The perimeter MRF provides most of the progressive collapse resistance for the removal of perimeter columns due to load bridging—the interior framing can only provide some cantilevered support to the damaged frame on the perimeter. Therefore, the structural fire model of the damaged frame includes only the 2D portion of the perimeter MRF. This model conservatively assumes that all load redistribution will take place in the portions of the MRF above and immediately adjacent to the removed column. Analysis of the 2D model requires less computational effort than a full 3D analysis of the frame and provides adequate accuracy when the composite contribution of the floor slab is neglected. Future research will explore the impact of composite action on not only the progressive collapse resistance of steel building frames but also the resistance to subsequent fire.

Three-noded 2D non-torsional fiber-beam elements are used to represent the structural elements in SAFIR. Each column length is modeled with 15 fiber-beam elements, and each girder is modeled with 30 elements (i.e., roughly a 1-ft discretization). The same loading used for progressive collapse analysis is also used for structural fire analysis, and the fire starts once the loads are initialized and the column is removed. Before structural analysis of the frame via time-series integration can be performed, a thermal model of each element must be analyzed to obtain its temperature-time history. Two-dimensional heat transfer is calculated over the cross sections of each element, which are discretized into solid thermal elements or "fibers." Each flange is modeled with 96 fibers, and each web is modeled with 48 fibers. Levels of discretization used for the structural and thermal models are consistent with those used by Quiel and Garlock (2010) and Garlock and Quiel (2008), respectively. The fiber-beam elements are capable of capturing plastic behavior and out-of-plane buckling but not local plate buckling. Both prototype frames considered in this chapter are composed of heavy wide-flanged sections with compact web and flange plates, and the fire-induced failure modes of these structures will be primarily controlled by plastic deformations rather than local or global instability. The role of post-plastic local buckling in fire-exposed steel elements is a potential topic for future investigation.

It is likely that the automatic sprinkler system near the location of column removal will also be damaged by the extreme event to the point of inoperability. It is therefore assumed that the fire will initiate at the location of column removal and burn in both of the adjacent bays with no extinguishment from the damaged sprinkler system. It is also assumed in this chapter that the fire will not spread horizontally or vertically beyond this location, although the study of such scenarios may be warranted in future research. For example, the ability of undamaged interior framing to resist collapse due to a fire started by the extreme event at the building perimeter could be explored. Also, the fire could spread to floors above the damaged column via partial damage openings in the exterior façade or floor diaphragms. However, the performance of the interior members under fire is considered to be outside the scope of this study. Also, the automatic sprinkler systems in the compartments surrounding those on fire may still be operational and potentially suppress or slow the spread of the fire.

The ASTM E119 standard fire curve (ASTM 2012) is used to represent the fire temperature-time history for this study. This curve assumes an initial rapid growth of temperature followed by a gradual, indefinite temperature increase. The curve does not include a decay phase (i.e., when the fire burns out) or consider compartment properties (fuel, ventilation, etc.)—rather, it is used to ensure eventual failure by exposing the damaged structure to a prolonged period of high temperature. Use of the ASTM E119 curve to represent a compartment fire is generally considered to be a conservative approximation compared to a realistic fire. Structural fire analysis of the damaged frame for exposure to a realistic compartment-based fire model, which includes a decay phase and indicates whether the frame would survive through burnout, has been explored previously by Neal (2011). The duration and maximum temperature for this type of fire curve will depend on the compartment geometry (specifically volume and ventilation) as well as fuel load. A detailed survey and discussion of fire load density in office buildings is provided by Khorasani et al. (2014).

11.3 Prototype Descriptions

The low-rise prototype is a five-story office building with approximately rectangular dimensions of 310 ft by 110.5 ft by 75 ft (length by width by height) and typical bay dimensions of 30 ft by 40.25 ft. The typical framing plan of the building at each floor is shown in Fig. 11.3a. The structural steel frame is composed of wide-flanged steel sections (A992 Grade 50) as shown in Fig. 11.4a. At each occupied floor, the steel framing supports a 3-in. reinforced lightweight concrete floor slab poured over a 3.5-in.-deep metal deck. At the roof, the interior bays supporting mechanical equipment also have the same slab over the metal deck; the remainder of the roof is covered by the metal deck only. Gravity loads are consistent with typical floor loading for an office building: 70 psf dead load (which includes the weight of the floor slab) and 100 psf live load. Lateral resistance due to seismic and wind load requirements

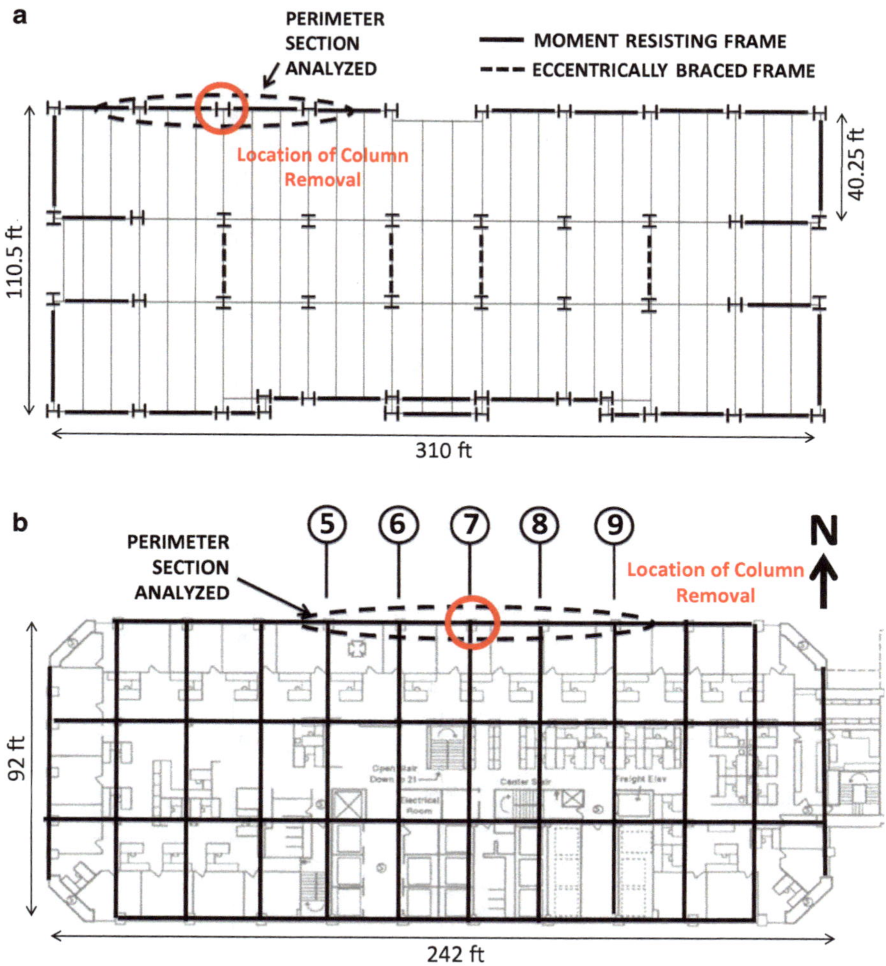

Fig. 11.3 Plan of the two prototype building frames, showing the location of column removal: (**a**) low rise (5-story) and (**b**) high rise (38-story 1MP)

is provided by several moment-resisting frames (MRFs) at the building perimeter as well as four eccentrically braced frames in its interior, as shown in Fig. 11.3a. The façade was composed primarily of 7¾-in.-thick precast concrete panels with cut window openings, and the remainder was comprised of a glazed curtain wall system. The gravity load due to the façade is applied as a line load to the perimeter of the floor framing.

The low-rise prototype building is designed with an automatic sprinkler system as its primary fire protection. Due to its occupancy and height (resulting in a Type IIB construction classification), the building code only requires passive applied fire protection for structural elements that support elevators or stairwells when a

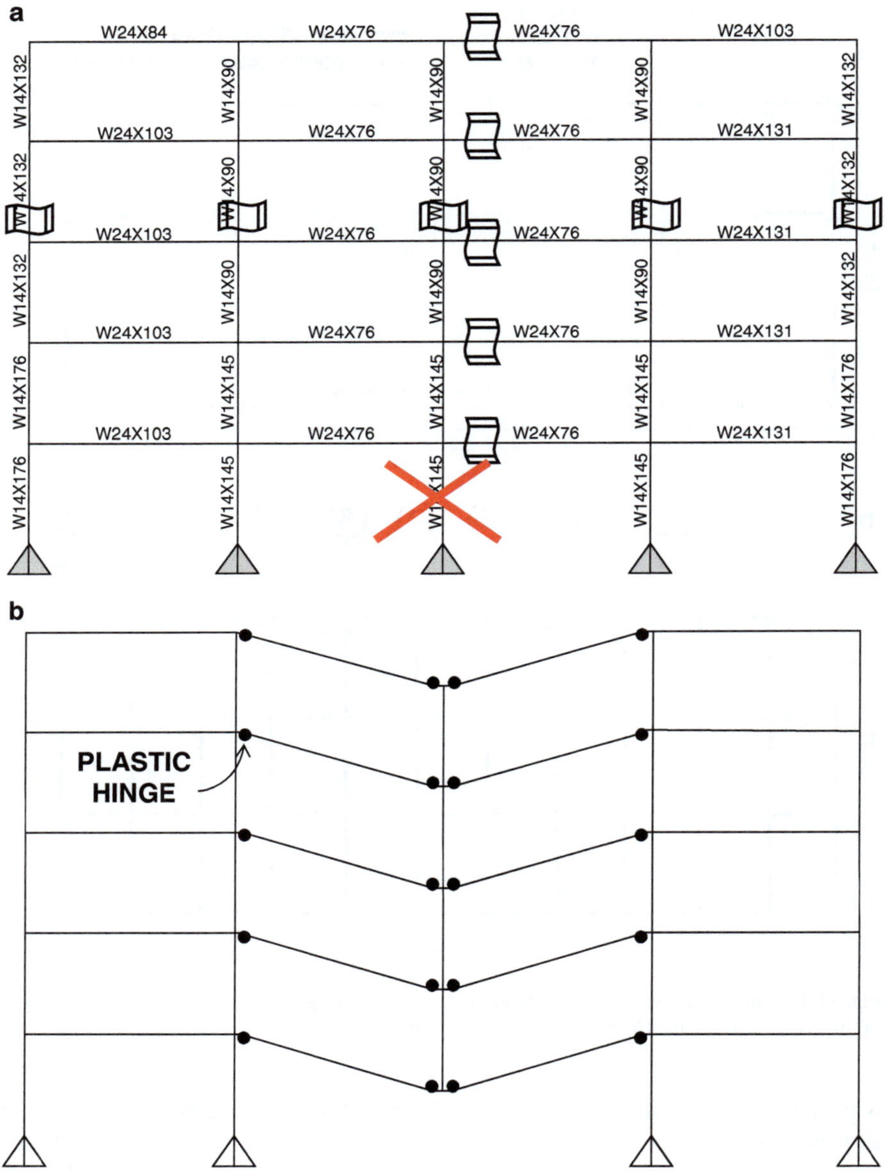

Fig. 11.4 Elevation of the low-rise perimeter MRF model for progressive collapse: (**a**) before and (**b**) after column removal (deformations are shown 5× magnified)

sprinkler system (i.e., active fire protection) is installed (ICC 2009). Passive fire protection for all other structural members is therefore not required. This is not uncommon for this type of office building, which represents a large portion of the steel-framed low-rise office inventory in the USA.

The One Meridian Plaza (1MP) in Philadelphia, PA, is the high-rise prototype building used in this study. On February 23, 1991, at approximately 8 pm, a fire started on the 22nd floor of 1MP. Firefighting activities were hampered by a loss of electrical and emergency power, inadequate water pressure, and other issues (Chubb et al. 1991). Consequently, the fire spread to the 30th floor where it was stopped by the automatic sprinkler system, which was not yet installed in the other fire-burning floors. The fire burned for more than 18 h and completely burned out 8 floors near the top of this 38-floor structure (Chubb et al. 1991). No structural collapse ensued and the building was dismantled in 1999.

The 1MP was rectangular in plan and approximately 242 ft by 92 ft as shown in Fig. 11.3b. The building construction was structural steel (A36) with a slab over a metal deck. The structural system consisted of several moment-resisting frames (MRFs) running in the north–south and east–west directions on every column line as shown in the floor plan in Fig. 11.3b. Similar gravity loads as those used for the low-rise prototype were considered when analyzing models of the 1MP (Garlock and Quiel 2007). The construction required 2-h fire-rated girders and 3-h fire-rated columns, specified by the Philadelphia Department of Licenses and Inspections as equivalent to BOCA Type 1B construction (Chubb et al. 1991). All structural steel beams and metal decks were protected with spray-on fire-resistive material (SFRM). The columns were protected with the same SFRM as well as gypsum plaster boards. Enclosed private offices were typically located along the building perimeter except along the south wall where the core was located. Most of the other space was open.

11.3.1 Progressive Collapse Analysis Cases

This chapter focuses on the column removal locations shown in plan on Fig. 11.3 (i.e., at the middle of a perimeter MRF span). This scenario represents one of the most typical and critical cases for evaluating progressive collapse resistance. The removal of columns at or adjacent to a corner would also be considered as critical cases during a complete progressive collapse evaluation; however, only a single plan location is considered for this demonstration. As shown in Figs. 11.5 and 11.6, both prototypes are subjected to a column removal at the first floor, which is particularly vulnerable to vehicle-borne or pedestrian-carried threats. As shown in Fig. 11.7, the high-rise prototype is also subjected to a column removal at the 21st floor. This second removal case is included for two reasons. First, a tall building may be susceptible to damage via column removal due to aircraft impact or the detonation of small explosive threats in the building interior. Second, the framing sizes are smaller at higher floors (see Figs. 11.6 and 11.7), which may therefore be more susceptible to collapse due to fire.

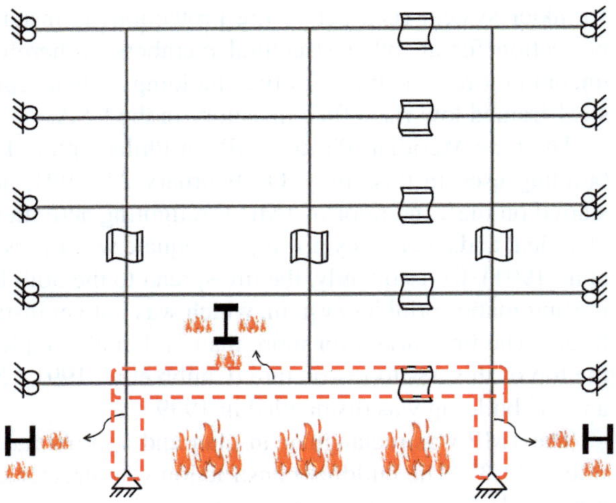

Fig. 11.5 Elevation of the low-rise perimeter MRF model after column removal at the ground floor and subsequent fire ignition

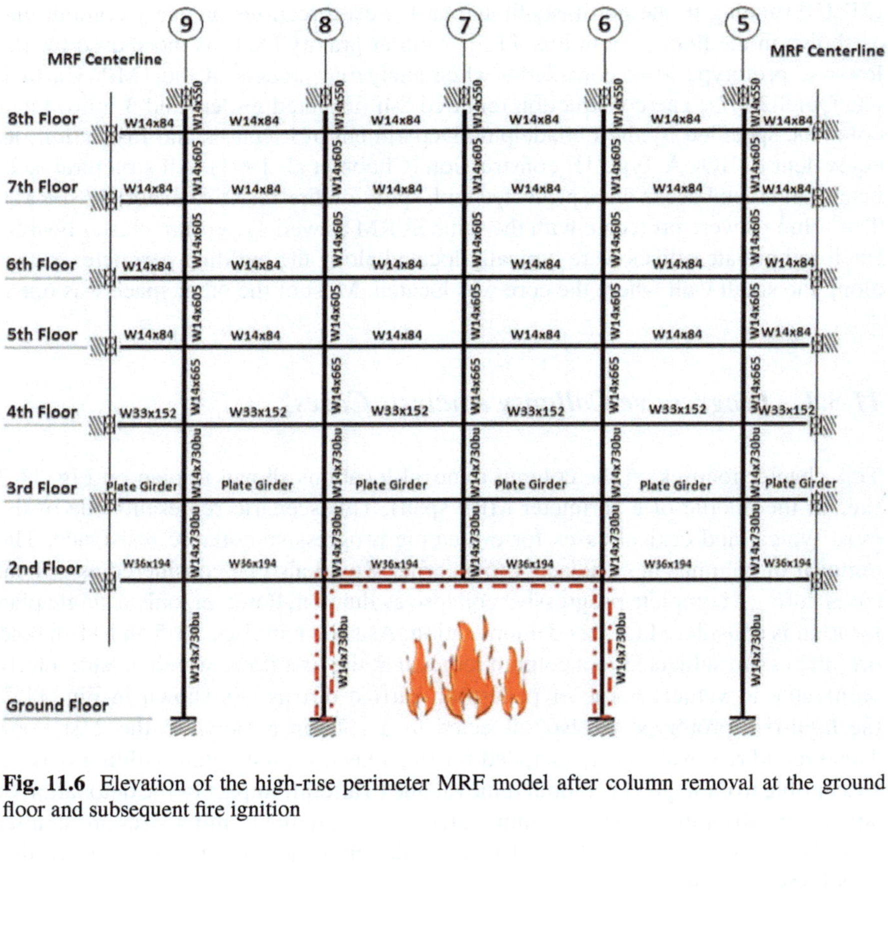

Fig. 11.6 Elevation of the high-rise perimeter MRF model after column removal at the ground floor and subsequent fire ignition

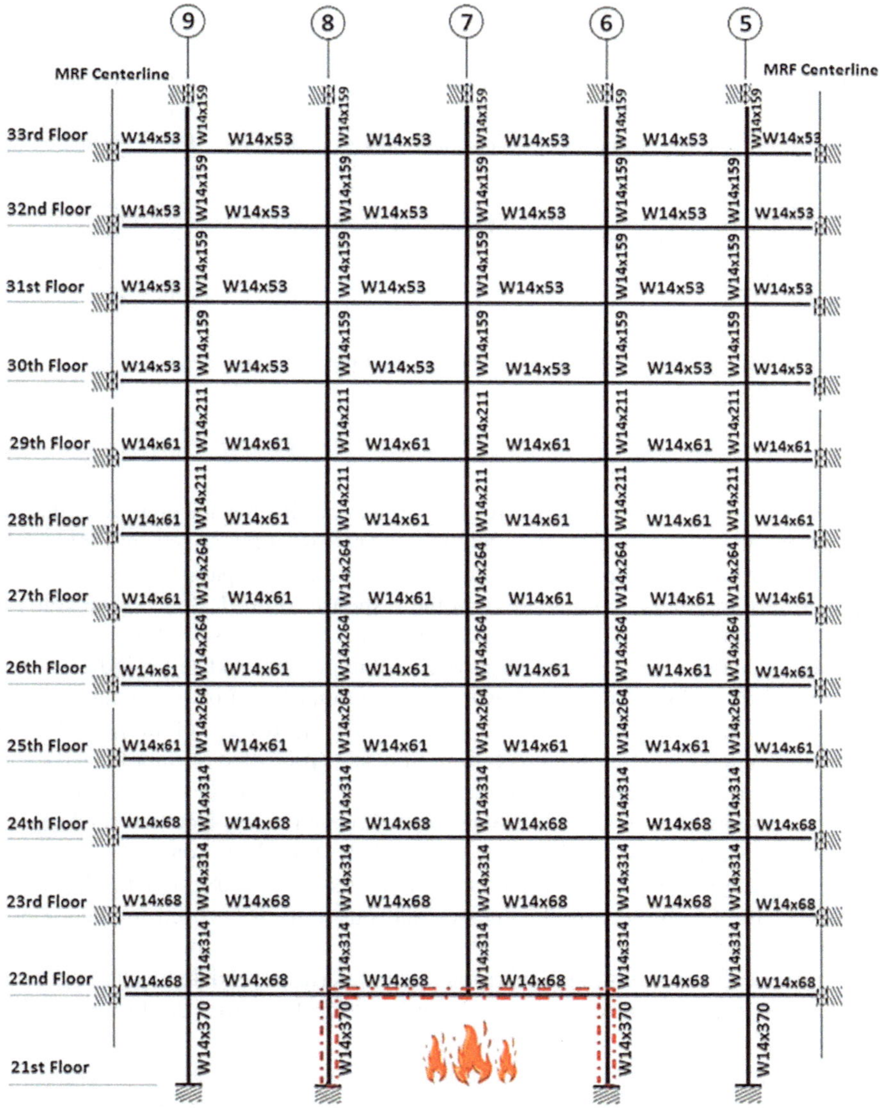

Fig. 11.7 Elevation of the high-rise perimeter MRF model after column removal at the 21st floor and subsequent fire ignition

11.3.2 Structural Fire Analysis Cases

Due to their construction types, the two prototypes have different base configurations of passive fire protection. When the structure is damaged and a column

Table 11.1 Summary of analysis cases

Case name	Prototype	Location	Fire protection rating	
			Girder	Column
LR-GuCu	Low rise (5 stories)	Ground—floor 5 (all floors)	None	None
LR-Gp1Cu			1 h	None
LR-Gp1Cp1			1 h	1 h
HR-L-GuCu	High rise (38 stories)	Ground—floor 8 (lowest floors)	None	None
HR-L-Gp2Cu			2 h	None
HR-L-GuCp3			None	3 h
HR-L-Gp2Cp3			2 h	3 h
HR-U-GuCu	High rise (38 stories)	Floors 21–33 (upper floors)	None	None
HR-U-Gp2Cu			2 h	None
HR-U-GuCp3			None	3 h
HR-U-Gp2Cp3			2 h	3 h

is removed, it is possible that passive fire protection, if present, may be damaged and potentially ineffective. Several analysis scenarios, summarized in Table 11.1, are considered for each prototype to evaluate the effect of passive fire protection (applied to the girders and columns) on the response to fire following the column removal. The first two letters denote whether the analysis case is for the low rise (LR) or high rise (HR). For the high-rise prototype, the next letter indicates the location of the column removal: **l**ower (at the ground floor) or **u**pper (at the 21st floor). Next, the letters in caps denote member type (**g**irder or **c**olumn), the lower-case letter denotes whether fire protection is present (**p**rotected or **u**nprotected), and the number denotes hourly rating (i.e., **1** h) for protected sections. For the low-rise prototype, hourly fire protection ratings were achieved using specified thicknesses of the same spray-on fire-resisting material (SFRM) used to protect the members supporting the prototype's elevators and stairwells (ICC 2008). For the high-rise prototype, the fire protection thicknesses and materials were used in accordance with those reported by Chubb et al. (1991). When included in the thermal models of the steel cross sections, each fire protection material is modeled as having three fibers through its thickness so that realistic heat transfer across the fire protection can be calculated.

Where present, the passive fire protection is assumed to be intact and fully effective on members adjacent to the column removal. Although SFRM and gypsum boards are used here to obtain an hourly rating, typical commercial SFRMs are relatively brittle and may perform poorly when subjected to impact or vibration. Passive fire protection materials that are more resistant to impact or vibration, such as intumescent paint, should be considered in practice to provide fire protection to frames that are designed to resist progressive collapse. Even so, a system with realistic, localized structural damage may still experience nonuniform damage to passive fire protection on intact structural elements. The consequence of said damage to the passive fire protection warrants further research. However, the simplifying assumption of undamaged passive fire protection is used in this study to assess the relative impact when the fire-induced temperature increases of the structural members are uniformly mitigated.

Figures 11.5, 11.6, and 11.7 show that only the columns and the girder directly adjacent to the bays on fire (i.e., those encircled in dashed lines) are modeled as exposed to fire. For each element, only the faces that are directly oriented toward the compartments on fire are assumed to be directly heated by the fire's temperature-time history. Specifically, the heated columns are exposed to fire on two sides, and the heated girders are exposed on three sides (with their top face shielded by the slab). The two unexposed sides of the column are modeled as exposed to ambient temperature. The slab is included in the thermal model of the girder to account for heat transfer between them but not in the structural model of the girder because it will have little effect on the structural solution for the 2D frame (Quiel and Garlock 2010).

11.4 Analysis Results

11.4.1 Progressive Collapse Results

Figure 11.4 shows the elevation of the low-rise perimeter MRF both before and after the column removal as calculated by the 3D SAP2000 model. P/P_{CL} for the perimeter girders is low, and therefore, the girders bridging over the removed column develop plastic hinges in response to the column removal as shown in Fig. 11.4b. The rotation of each hinge complies with plastic rotation limits specified in ASCE/SEI 41-06 (ASCE 2007). P/P_{CL} for the perimeter columns adjacent to the removed column is high, and all of these members have a DCR less than unity for combined axial load and biaxial bending as well as shear. Therefore, the prototype MRF design meets the criteria for progressive collapse resistance according to the UFC (DoD 2013).

Based on design practices at the time of its construction, the high-rise prototype has moment frames on every bay at every column line as shown in Fig. 11.3b. Due to the redundant nature of this building system, a 2D progressive collapse analysis was performed to improve computational efficiency under the assumption that it would be conservative compared to a 3D analysis. The 2D analysis indicates that the high-rise perimeter frame satisfies the DoD progressive collapse criteria for column removals at both the ground floor and 21st floor (Neal 2011). Plastic hinge patterns for the high-rise prototype are similar to those shown in Fig. 11.4b for the low-rise prototype.

11.4.2 Low-Rise Structural Fire Results

Figure 11.8 shows the average temperature in the steel cross section of the fire-exposed (a) girder and (b) column in the low-rise prototype both with and without a 1-h rating of applied fire protection. Figure 11.9 shows the (a) deflection of the girder above the removed column and (b) axial displacement of the undamaged adjacent column at the second floor. Note that the girder deflection shown in Fig. 11.9a is measured relative to its connection to the column at the second floor. As expected, Fig. 11.8 shows that the rate of increase of steel temperature is significantly slowed by the presence of fire protection. Similarly, Fig. 11.9 shows that the rate of girder deflection or column axial displacement decreases with increasing fire protection.

Figure 11.9a shows that analysis cases LR-GuCu and -Gp1Cp1 experience a gradual increase of deflection as the heated girder weakens until reaching ~16 in. At this point in the analyses, deflections rapidly increase to about 19 in. According to the Eurocode material model for mild steel (CEN 2001), plasticized steel fibers will begin to unload at 15 % strain until they reach their ultimate limit at 20 % strain, after which they can no longer carry load. Figure 11.10 shows the maximum strains for the heated girder at the second floor and for the cool girders in all other floors above for cases LR-GuCu and -Gp1Cp1. The heated girder's maximum strain reaches 15 % just before the sudden increase in deflection shown in Fig. 11.9a and then rapidly increases to ultimate 20 % strain as it fails and redistributes its loads to the cool girders. The maximum strain in the cool girders shows a corresponding increase when the heated girder unloads. The temperature in the heated girder is 275 °C for both Gu and Gp1 at this point during fire exposure (see the open circle points in Fig. 11.8a), which corresponds to an 18 % decrease in stiffness and no decrease in strength according to Eurocode reduction factors for steel (CEN 2001).

Once the heated girder sheds its load, deflections continue to increase until the cool girders on the floors above become overloaded, resulting in a second sudden increase in deflection (see Fig. 11.9a) and overall failure of the frame. "Failure" in this study is defined when the analysis fails to converge. During this period, Fig. 11.10 shows that the cool girders' maximum strain increases gradually until it reaches 15 % and then rapidly progresses to ultimate strain, at which point the frame fails. Figures 11.8a and 11.9a show that the heated girder's average temperature (~350 °C) and maximum deflection (~23 in.) for cases LR-GuCu and LR-Gp1Cp1 were very similar when their analyses terminate (i.e., the time at which the frame fails). The column displacements for these cases show no sudden increase in Fig. 11.8b at any point in the analysis, indicating that failure initiated in the girder.

For analysis case LR-Gp1Cu, the sudden decrease of column displacement in Fig. 11.9b indicates that failure initiated in the unprotected column. The girder deflection in Fig. 11.9a is similar to that for case LR-Gp1Cp1 until the unprotected column fails and shows no sudden increase. As shown in Fig. 11.8b, the average column temperature for case LR-Gp1Cu at column failure is just greater than 500 °C, at which point steel has experienced a 25 % decrease in strength (CEN

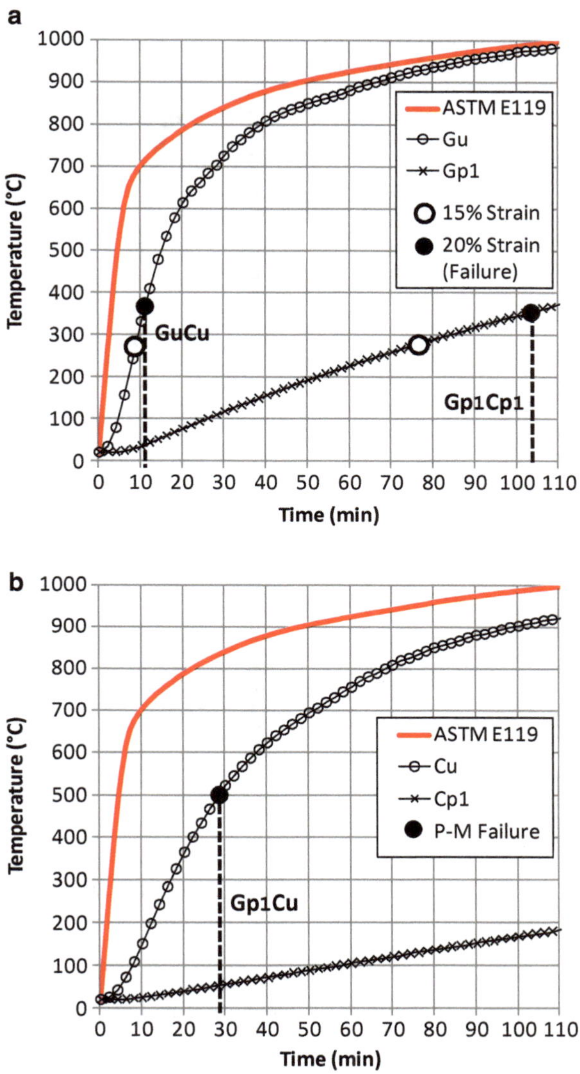

Fig. 11.8 Low-rise results: average temperature in the steel cross section of the fire-exposed (**a**) girder and (**b**) column for varying levels of fire protection

2001). Figure 11.11 shows a plot of the axial load (P) and bending moment (M) in the unprotected column for every time step in the LR-Gp1Cu analysis. The P-M points have been normalized by the yield load (P_y) and plastic moment (M_p), respectively, that correspond to the Eurocode strength reduction for the Cu temperature case (see Fig. 11.8b) at that time step. Also shown is the normalized P-M capacity at the time of column failure ($t = 29$ min). The plot shows that the column's axial load ratio increases due to increasing temperature (and decreasing strength) until the P-M curve reaches the P-M capacity envelope and the column

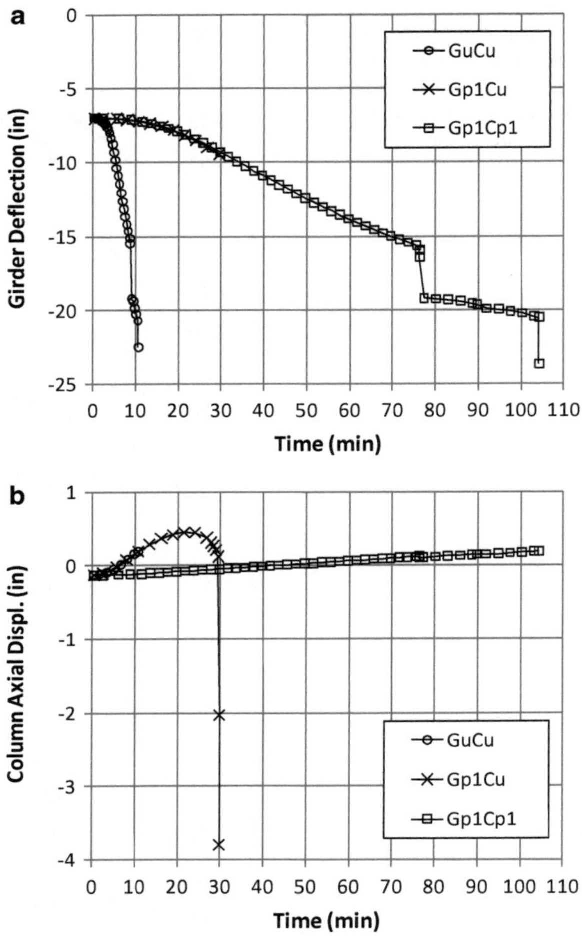

Fig. 11.9 Low-rise results: (**a**) girder deflection directly above the removed column and (**b**) axial displacement of the undamaged columns at the second floor

fails. Note that the *P-M* capacity envelope has shifted because the unprotected column has developed a thermal gradient through its cross section (Garlock and Quiel 2008), which is caused by uneven heating of the column as shown in Fig. 11.5.

As shown in Table 11.2, the damaged frame in analysis case LR-GuCu fails after just 11 min of fire exposure due to rapidly increasing downward deflection of the girders above the removed column. This result provides little time for building egress following the extreme event before the damaged frame collapses. A 1-h fire protection rating for the girder bridging over the removed column, even with columns unprotected (LR-Gp1Cu), provides nearly an additional 20 min for evacuation. However, the LR-Gp1Cu case experiences a sudden failure of its heated columns, which should be avoided if possible. A 1-h fire protection rating for both the girder and column in case LR-Gp1Cp1 provides over an hour and a half of time

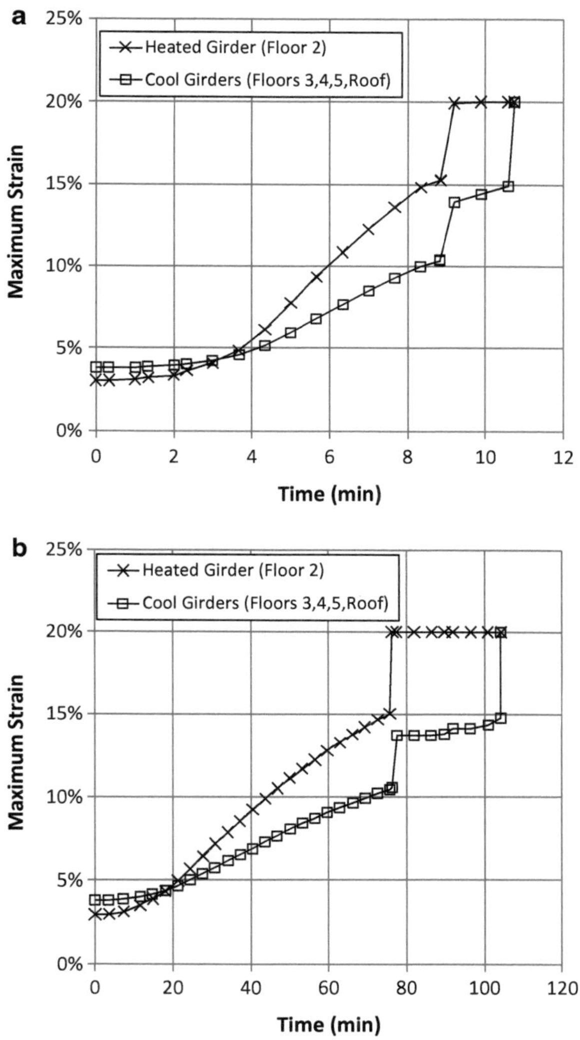

Fig. 11.10 Low-rise results: maximum girder strains for analysis cases (**a**) GuCu and (**b**) Gp1Cp1

to failure and switches the mode of failure initiation back to the girders, whose sagging failure is more ductile than the column failure. This fire protection case provides more time not only for egress but also for firefighting efforts to prevent the frame from reaching an advanced state of fire-induced damage. For exposure to a realistic fire, the damaged frame may also avoid collapse if burnout (i.e., when the fire temperatures decay, having consumed most of the available fuel) is achieved within the first 1.5 h of fire exposure.

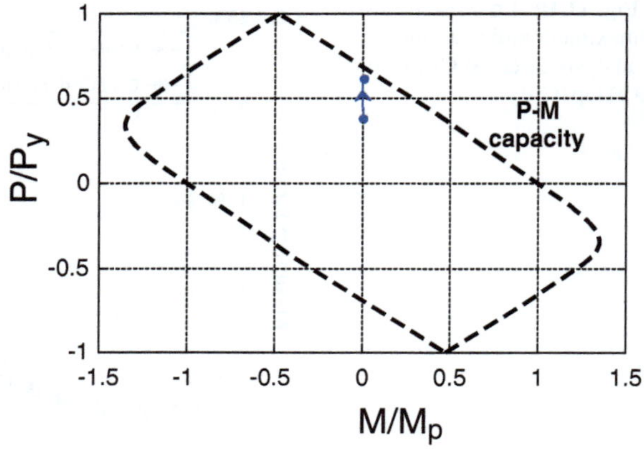

Fig. 11.11 Low-rise results: combined *P-M* performance for the unprotected column in analysis case Gp1Cu at failure ($t = 29$ min)

Table 11.2 Summary of structural fire analysis results

Case name	Time to failure	Failed element	Avg. temperature at failure		Girder deflection at failure
			Girder	Column	
LR-GuCu	11 min	Girder	345 °C	160 °C	22.5 in.
LR-Gp1Cu	29 min	Column	114 °C	513 °C	9.5 in.
LR-Gp1Cp1	104 min	Girder	356 °C	174 °C	23.7 in.
HR-L-GuCu	17 min	Girder	437 °C	67.6 °C	9.0 in.
HR-L-Gp2Cu	173 min	Girder	411 °C	574 °C	8.6 in.
HR-L-GuCp3	17 min	Girder	437 °C	20 °C	8.8 in.
HR-L-Gp2Cp3	170 min	Girder	404 °C	53 °C	8.5 in.
HR-U-GuCu	71 min	Girder	890 °C	631 °C	4.3 in.
HR-U-Gp2Cu	97 min	Column	270 °C	745 °C	2.8 in.
HR-U-GuCp3	92 min	Girder	949 °C	51.4 °C	4.4 in.
HR-U-Gp2Cp3	>4 h	*No failure*	*No failure*	*No failure*	*No failure*

11.4.3 High-Rise Structural Fire Results

Figure 11.12a compares the thermal response of the girder in the lower floor to that of the upper floor, and Fig. 11.12b does the same for the columns. These plots show that there is essentially no difference in the girder temperatures given the location, but the column temperatures differ if they are unprotected. The larger built-up columns at the ground floor take more time to heat up compared to the upper floor columns. For this reason, the structural analysis results show that some cases in the upper floors terminate due to column failure, whereas failure in the lower floors is characterized only by the girder failure.

The results from each analysis case are summarized in Table 11.2. Fire protection significantly affects the temperatures reached in the members and thus has an important effect on the structural response. For Gu cases, the girder always fails before the column because it experiences a faster temperature increase (due to

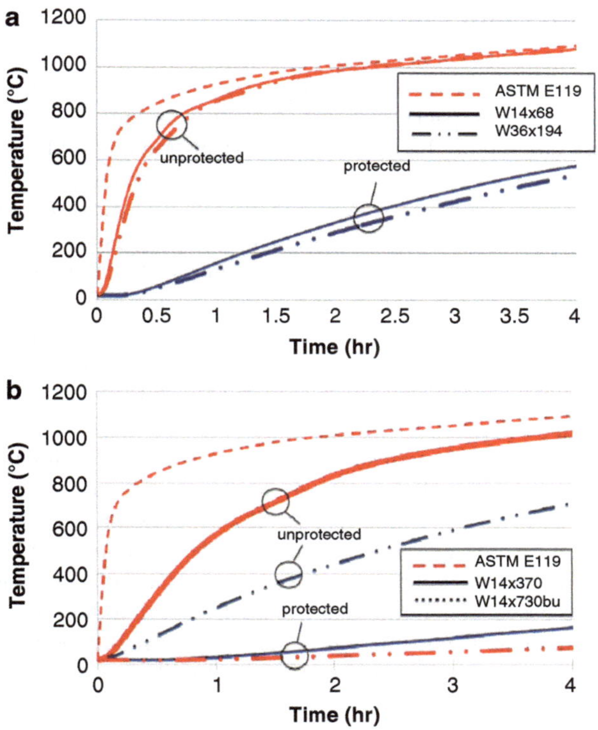

Fig. 11.12 High-rise results: thermal response of (**a**) floor 2 girder and (**b**) ground floor column compared to their counterparts at the 21st floor

having an additional face exposed to fire and having less mass) than the column (even if the column is unprotected). For Gp cases, the mode and time of failure are affected significantly by both the location and the fire type.

Table 11.2 shows that when the girder is unprotected (Gu cases), the fire in the upper floor location results in a longer survival time compared to the lower floors. In all cases, the girder is the source of failure. Since the upper floors are redistributing loads from fewer floors above them, the girders at the upper floors can withstand a longer fire exposure when bridging over the removed column. For GpCu cases, the fire in the upper floor results in a *shorter* survival time compared to the lower floors. In these cases, the failing element changes from column failure for the upper cases to girder failure for the lower cases. The larger lower column experiences a slower temperature increase than the lighter upper column, which leads to a change in failure mode. Collectively, these results highlight the potential sensitivities of steel high-rise frames to fire following local damage based on the location of the damage.

11.5 Summary and Future Work

This chapter evaluates fire as a cascading hazard that follows an initial extreme event (such as blast or impact). The initial event is assumed to cause failure of a single one-story column on the perimeter of a steel-framed building in accordance with typical progressive collapse resistance analysis strategies, and the fire then ignites at the location of damage. Two prototype buildings were considered, both of which complied with the design requirements for progressive collapse resistance that are published by the US Department of Defense (DoD 2013). The structural fire analyses conducted in this study indicate that if the passive fire protection is damaged and the active fire protection is inoperable, then the damaged frame may have less than 20 min of resistance to a subsequent fire. When passive fire protection is intact, the structures remain stable for much longer durations depending on the level of protection.

It is clear, based on the studies in this chapter, that fire following an extreme event represents a significant threat to a structure's collapse resistance. Further studies are warranted to determine the correlation between fire intensity, passive fire protection rating, the degree of damage to the structure and fire protection, performance criteria, and collapse time. These correlations have potential to lead to a performance-based design tool that can be used by building owners and safety officials to determine how much time is realistically available for evacuation before progressive collapse occurs when the structure is locally damaged. Findings in this paper highlight the significance and utility of a cascading, multi-hazard approach for assessing the risk of progressive collapse and ultimately improving structural life safety.

References

AISC. (2010). *Steel construction manual* (14th ed.). Chicago, IL: Author.

ASCE. (2007). *ASCE/SEI 41-06: Seismic rehabilitation of existing buildings*. Reston, VA: Author.

ASCE. (2010). *ASCE/SEI 7-10: Minimum design loads for buildings and other structures*. Reston, VA.

ASTM. (2012). *Designation: E119-12a – Standard test methods for fire tests of building construction and materials*. Conshohocken, PA: American Society for Testing and Materials (ASTM) International.

CEN. (2001). *Eurocode 3: Design of steel structures, Part 1.2: General rules – Structural fire design (ENV 1993-1-2:2001)*. Brussels.

CEN. (2002). *Eurocode 1: Actions on structures Part 1-2: General actions on structures exposed to fire (EN 1991-1-2:2002)*. Brussels.

Chen, H., & Liew, J. Y. R. (2005). Explosion and fire analysis of steel frames using mixed element approach. *Journal of Engineering Mechanics, 131*(6), 606–616.

Chubb, M., Jennings, C., & Routley, J. (1991). *Highrise office building fire – One Meridian Plaza* (U.S. Fire Administration/Technical Report Series). Washington, DC: Federal Emergency Management Agency (FEMA).

Corley, W. G., Mlakar, P. F., Sozen, M. A., & Thornton, C. H. (1998). The Oklahoma City bombing: Summary and recommendations for multihazard mitigation. *Journal of Performance of Constructed Facilities, 12*(3), 100–112.

CSI. (2009). *SAP2000, version 14, integrated software for structural analysis and design, analysis reference manual*. Berkeley, CA.

Della Corte, G., Landolfo, R., & Mazzolani, F. M. (2003). Post-earthquake fire resistance of moment resisting steel frames. *Fire Safety Journal, 38*, 593–612.

Dexter, R. J., & Lu, L.-W. (2001). The effects of a severe fire on the steel frame of an office building. *Engineering Journal, 38*(4), 167–175.

DoD. (2013). *Design of buildings to resist progressive collapse* (Unified Facilities Criteria (UFC) 4-023-03, 14 July 2009, including Change 2 – 1 June 2013). Washington, DC: US Department of Defense (DoD).

Ellingwood, B., & Leyendecker, E. V. (1978). Approaches for design against progressive collapse. *Journal of the Structural Division, 104*(3), 413–423.

Franssen, J.-M. (2005). SAFIR: A thermal/structural program for modeling structures under fire. *Engineering Journal, 42*(3), 143–158.

Garlock, M. E. M., & Quiel, S. E. (2007). The behavior of steel perimeter columns in a high-rise building under fire. *Engineering Journal, 44*(4), 359–372.

Garlock, M. E. M., & Quiel, S. E. (2008). Plastic axial load and moment interaction curves for fire-exposed steel sections with thermal gradients. *Journal of the Structural Engineering (ASCE), 134*(6), 874–880.

GSA. (2013). *Alternate path analysis & design guidelines for progressive collapse resistance*. Washington, DC.

Hadjioannou, M., Donahue, S., Williamson, E. B., Engelhardt, M. D., Izzuddin, B., Nethercot, D., et al. (2013). Experimental evaluation of floor slab contribution in mitigating progressive collapse of steel structures. In *Proceedings of the 5th International Conference on Safety and Security Engineering (SAFE 2013)*. Rome, Italy, Sept. 17–19.

ICC. (2009). *2009 International Building Code*. Country Club Hills, IL.

ICC. (2008). *ES Report (ESR) 1649*. Country Club Hills, IL.

Khorasani, N. E., Garlock, M. E. M., & Gardoni, P. (2014). Fire load: Survey data, recent standards, and probabilistic models for office buildings. *Journal of Engineering Structures, 58*, 152–165.

Khorasani, N. E., Garlock, M. E. M., & Quiel, S. E. (2015). Modeling steel structures in OpenSees: Enhancements for fire and multi-hazard probabilistic analyses. *Computers and Structures, 157*, 218–231.

Kim, T., Kim, J., & Park, J. (2009). Investigation of progressive collapse-resisting capability of steel moment frames using push-down analysis. *Journal of Performance of Constructed Facilities, 23*(5), 327–335.

Marjanishvili, S., & Agnew, E. (2006). Comparison of various procedures for progressive collapse analysis. *Journal of Performance of Constructed Facilities, 20*(4), 365–374.

Neal, M. C. (2011). *Evaluating fire following blast in high-rise buildings designed to resist progressive collapse*. MS thesis, Princeton University, Princeton, NJ.

Neal, M. C., Garlock, M. E. M., Quiel, S. E., & Marjanishvili, S. M. (2012). Effects of fire on a tall steel building designed to resist progressive collapse. In *Proceedings of the 2012 ASCE Structures Congress*. Chicago, IL, March 29–31.

NIST. (2005). *NIST NCSTAR 1: Final report of the National Construction Safety Team on the collapse of the World Trade Center Twin Towers*. Gaithersburg, MD.

Pearson, C., & Delatte, N. (2005). Ronan point apartment tower collapse and its effect on building codes. *Journal of Performance of Constructed Facilities, 19*(2), 172–177.

Quiel, S. E., & Garlock, M. E. M. (2010). Parameters for modeling a high-rise steel building frame subject to fire. *Journal of Structural Fire Engineering, 1*(2), 115–134.

Quiel, S.E., Marjanishvili, S.M. (2012). Fire resistance of a damaged steel building frame designed to resist progressive collapse. *Journal of Performance of Constructed Facilities (ASCE), 26*(4), 402–409.

Quiel, S. E., & Marjanishvili, S. M. (2013). Response of steel columns to fire following blast. In *Proceedings of the 15th International Symposium on the Interaction of the Effects of Munitions with Structures (ISIEMS15)*. Potsdam, Germany, Sept. 17–20.

Zaharia, R., & Pintea, D. (2009). Fire after earthquake analysis of steel moment resisting frames. *International Journal Steel Structures, 9*, 275–284.

Chapter 12
Strategies for Enhancing Fire Performance of NSM FRP-Strengthened Reinforced Concrete Beams

Venkatesh K.R. Kodur

Abstract Fiber-reinforced polymers (FRPs) are widely used in strengthening and retrofitting of reinforced concrete structural members to enhance flexural and shear capacity of such members. Two types of strengthening techniques, namely, externally bonded (EB) FRP and near-surface mounted (NSM) FRP, are used in field applications. FRP, as a material, exhibits poor fire resistance properties, and this can be a limiting factor in the application of FRP strengthening in buildings. There is limited guidance in codes and standards for fire design of NSM FRP-strengthened concrete members. This paper discusses strategies for achieving required fire resistance in NSM FRP-strengthened reinforced concrete beams. Results from numerical and experimental studies are utilized to quantify the influence of various parameters on the fire resistance of NSM FRP-strengthened concrete beams. Guidelines for achieving optimum fire resistance in NSM FRP-strengthened concrete beams are presented.

12.1 Introduction

In recent years, fiber-reinforced polymers (FRPs) are widely used in strengthening and rehabilitation of reinforced concrete (RC) structural members in built infrastructure. This is mainly due to numerous advantages that FRP possesses over other traditional materials, which include high strength to weight ratio, good corrosion resistance, and ease of application. In the case of RC columns, FRP wrapping on the exterior of concrete columns can significantly increase the strength and ductility of these columns. FRP strengthening in beams (and slabs), for enhancing flexural and shear capacity, is typically carried out by applying FRP laminates to the surface of a concrete beam (and slab), and this is designated as externally bonded (EB) reinforcing technique. Filling the FRP with adhesives or grout ensures that FRP

V.K.R. Kodur (✉)
Department of Civil and Environmental Engineering, Michigan State University, East Lansing, MI, USA
e-mail: kodur@egr.msu.edu

P. Gardoni, J.M. LaFave (eds.), *Multi-hazard Approaches to Civil Infrastructure Engineering*, DOI 10.1007/978-3-319-29713-2_12

rebar, strip, or tape is well anchored to concrete and acts as an effective tensile or shear reinforcement in resisting loading on the concrete member (De Lorenzis and Teng 2007).

Alternatively, FRP strips or rods can be inserted into a pre-cut groove(s) on the concrete cover of a RC beam (or slab), which are then filled with an epoxy adhesive or cementitious grout, and this type of strengthening is referred to as near-surface mounted (NSM) technique.

When used in building applications, FRP-strengthened reinforced concrete (RC) beams have to meet fire resistance requirements specified in building codes. Reinforced concrete beams, without strengthening, often satisfy fire resistance ratings without the need for external fire protection. However, when these concrete beams are strengthened with an external FRP system, the fire response of strengthened beams can be much different than that of unstrengthened RC beams. With the increasing use of FRP in strengthening of structural members in buildings, concern has developed regarding the behavior of FRP-strengthened concrete members in the event of fire exposure, and this is mainly attributed to the fact that FRP is combustible and also susceptible to rapid deterioration of mechanical and bond properties at elevated temperature (Bisby et al. 2005; Ahmed and Kodur 2011; Gamage et al. 2006).

In the past decade, a number of experimental and analytical studies have been carried out to develop an understanding on the behavior of EB and NSM FRP-strengthened RC members at room temperature. Based on these studies, guidelines have been developed for structural design of FRP-strengthened members (Bakis et al. 2002; ACI 2007; CSA 2007; FIB 2001). However, there have been only limited experimental and numerical studies on the fire resistance of EB FRP-strengthened concrete members (Williams et al. 2008; Ahmed and Kodur 2011). Thus, there is limited guidance available in codes and standards for the fire design of EB FRP-strengthened RC members (Kodur et al. 2007).

In the case of NSM FRP-strengthened concrete members, there is very scarce information related to fire performance (Yu and Kodur 2014a). Thus, there are no guidelines for evaluating fire resistance of NSM FRP-strengthened RC members. This paper provides guidelines to enhance fire resistance of NSM FRP-strengthened RC beams.

12.2 Behavior of Fire-Exposed NSM FRP-Strengthened Beams

External bonding of FRP is the method of choice in retrofitting and strengthening of concrete beams over other strengthening techniques, such as through steel plates or through external post-tensioning. Flexural strengthening of RC beams is usually achieved by applying FRP sheets/laminates on the beam soffit (tension face) as shown in Fig. 12.1, while shear strengthening is achieved through the provision of EB FRP sheets on the two side faces of a beam. Such application of FRP helps

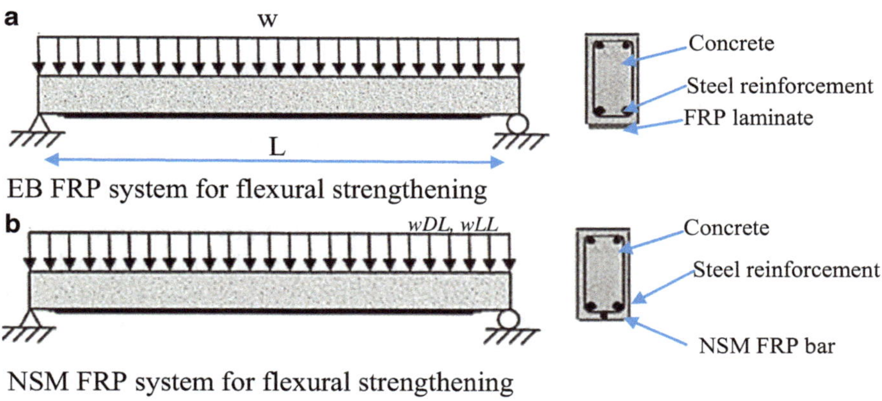

Fig. 12.1 Comparative illustration of FRP strengthening systems in RC beams. (**a**) EB FRP system for flexural strengthening. (**b**) NSM FRP system for flexural strengthening

Fig. 12.2 Variation of strength with temperature in different materials

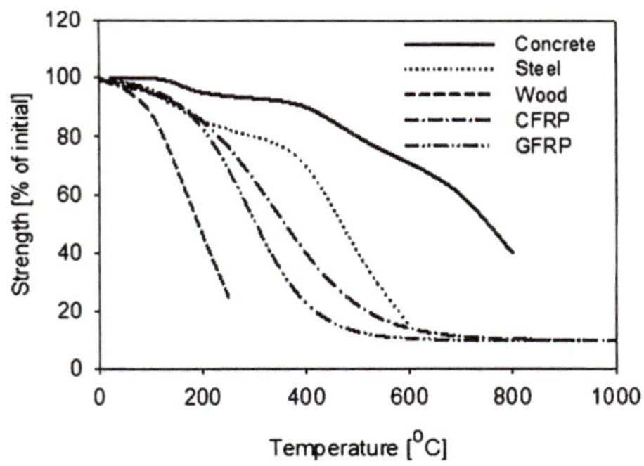

considerably in enhancing flexural and shear capacity of RC beams at ambient conditions.

In the case of NSM FRP technique, slots are cut into the concrete cover of an RC member, and then an FRP rebar, strip, or tape is inserted into the slot, which is filled with an epoxy adhesive or cementitious grout. Compared to externally bonded (EB) FRP strengthening, NSM strengthening can utilize more of the strength of FRP because of better bond adherence (Barros et al. 2007; Oehlers et al. 2008; Rashid et al. 2008). Also, since NSM FRP reinforcement is embedded within the concrete, NSM-strengthened concrete structures are less prone to adverse damage to FRP resulting from fire, acts of vandalism, mechanical damage, and aging effects.

However, under fire conditions, the strength and stiffness properties of FRP (as well as steel rebars in the beam) degrade with rise in temperature (Wang and Kodur 2005; Wang et al., 2007; Yu and Kodur 2014b). Figure 12.2 shows the variation of strength properties with temperature in concrete, steel, wood, and FRP. It can be seen that the strength of FRP (glass (GFRP) or carbon (CFRP)) degrades at a higher rate as compared to concrete and steel. The stiffness properties of FRP also deteriorate at a higher rate with temperature as compared to that of steel and concrete. This degradation in strength and stiffness properties leads to loss of flexural (or shear) capacity in an EB or NSM FRP-strengthened RC beam.

In addition to strength and stiffness properties of FRP, bond is a critical factor in EB/NSM FRP-strengthened RC beams since adhesive is used to strengthen the beam. At elevated temperatures, bond between FRP reinforcement and concrete degrades with temperature and significantly influences the behavior of FRP-strengthened RC beams (Ahmed and Kodur 2011). This bond degradation results in ineffective transfer of forces from concrete to FRP. When the temperatures of FRP/adhesive reaches glass transition temperature (T_g), bond properties of adhesive deteriorates considerably due to softening of adhesive (Yu and Kodur 2014c). The glass transition temperature of FRP varies with type of epoxy and is in the range of 60–85 °C for commercially available FRP strips and bars. At this stage, the bond will no longer be effective to transfer stresses from concrete to FRP fibers, leading to debonding of FRP. Therefore, thermomechanical properties of FRP play a critical role on the behavior of EB/NSM FRP-strengthened RC beam.

The degradation of strength and stiffness properties, combined with loss of bond properties, leads to degradation in moment capacity of the NSM FRP-strengthened beam with fire exposure time. In addition, the degradation of moment capacity is influenced by a number of factors including type of fire exposure, type of strengthening (EB or NSM), loading, support conditions, insulation thickness and properties, and high-temperature thermal and mechanical properties of concrete, steel, and FRP (Ahmed and Kodur 2010; Yu and Kodur 2014b).

Typically strength failure of an NSM FRP-strengthened RC beam occurs when moment due to applied loading exceeds decreasing flexural capacity of the beam. The time to reach failure is referred to as fire resistance of the beam, and this fire resistance can be enhanced by slowing the temperature rise in NSM FRP and steel rebars through provision of fire insulation.

12.3 Factors Governing Fire Resistance

In the last decade, both experimental and numerical studies have been carried out by various researchers to evaluate the response of FRP-strengthened RC members under fire conditions. Notable studies on fire resistance of EB FRP-strengthened RC beams were carried out by researchers Williams et al. (2004), Kodur et al. (2007), Hawileh et al. (2009), and Ahmed and Kodur (2011). Data generated from fire tests was utilized to develop finite element models for tracing the fire response of EB FRP-strengthened RC beams (Kodur and Ahmed 2010). The validated models were applied to conduct a set of parametric studies to quantify the influence of critical factors on fire response of EB FRP-strengthened RC beams.

Recently a few studies have been carried out on fire performance of NSM FRP-strengthened RC beams (Yu 2013; Yu and Kodur 2014a). Data from the studies clearly indicate that an NSM FRP-strengthened RC beam exhibit higher fire resistance than an EB FRP-strengthened RC beam. Also, while EB FRP-strengthened beams need to be provided with external fire insulation to achieve practical fire resistance ratings required in buildings, NSM FRP-strengthened RC beams in many cases can provide required fire resistance without any external fire insulation.

The response of two typical FRP-strengthened RC beams (EB and NSM) is compared with that of an unstrengthened RC beam in Fig. 12.3. For this comparison, three similar RC beams are selected, namely, an unstrengthened RC beam (Beam I-1), an EB FRP-strengthened RC beam (Beam I-2), and an NSM FRP-strengthened RC beam (Beam I-3) as shown in Fig. 12.3b. When these beams are exposed to fire (ASTM E119 2012), cross-sectional temperatures increase with fire exposure time. The rate of rise in steel rebar and FRP temperatures dictate the moment capacity of these beams at any given fire exposure time. The moment capacity at critical sections of the beam was computed using the same equations as that at room temperature, but accounting for temperature-induced degradation in strength of concrete, steel rebars, and FRP reinforcement. In the case of Beam I-2 (Beam with EB FRP), FRP temperature increases very quickly as compared to Beam I-3 with NSM FRP reinforcement (see Fig. 12.3c). Due to different levels of temperature in rebar and FRP, the decrease in moment capacity in these three beams also varies (see Fig. 12.3d).

In the case of unstrengthened beam (Beam I-1), moment capacity decreases gradually with time and is mainly dependent on loss of strength in reinforcing steel rebars with increase in temperature. In the case of EB FRP-strengthened RC beam (Beam I-2), moment capacity degradation is initially rapid due to high temperatures in FRP laminate, and once FRP burns out, the degradation of moment capacity follows that of unstrengthened RC beam. In the case of NSM FRP-strengthened RC beam (Beam I-3), the moment capacity decreases at a slower rate throughout the fire exposure time. This can be attributed to slower strength loss in FRP due to slower temperature rise in NSM FRP, facilitated by the presence of concrete cover over the FRP reinforcement.

When the degrading moment capacity falls below moment due to applied loading, failure of the beam occurs, and the time to reach this point is the fire resistance of the beam. The load level on a beam during fire exposure is about 50 % of the capacity of the beam, and thus the load to be carried by FRP-strengthened RC beam is higher than that on an unstrengthened RC beam. This fire resistance of EB FRP-strengthened RC beam (Beam I-2) can be much lower than unstrengthened RC Beam I-1, since the strengthened beam has to carry increased loading under fire conditions. However, fire resistance of NSM FRP-strengthened RC beam (Beam I-3) is higher than that of beams I-1 and I-2, due to slower degradation of moment capacity with fire exposure time.

Results from previous studies infer that fire performance of EB FRP- and NSM FRP-strengthened RC beams is influenced by a number of factors, and many of these factors are interdependent (Ahmed and Kodur 2010; Yu and Kodur 2014a). Systematic parametric studies were carried out using a macroscopic finite element-

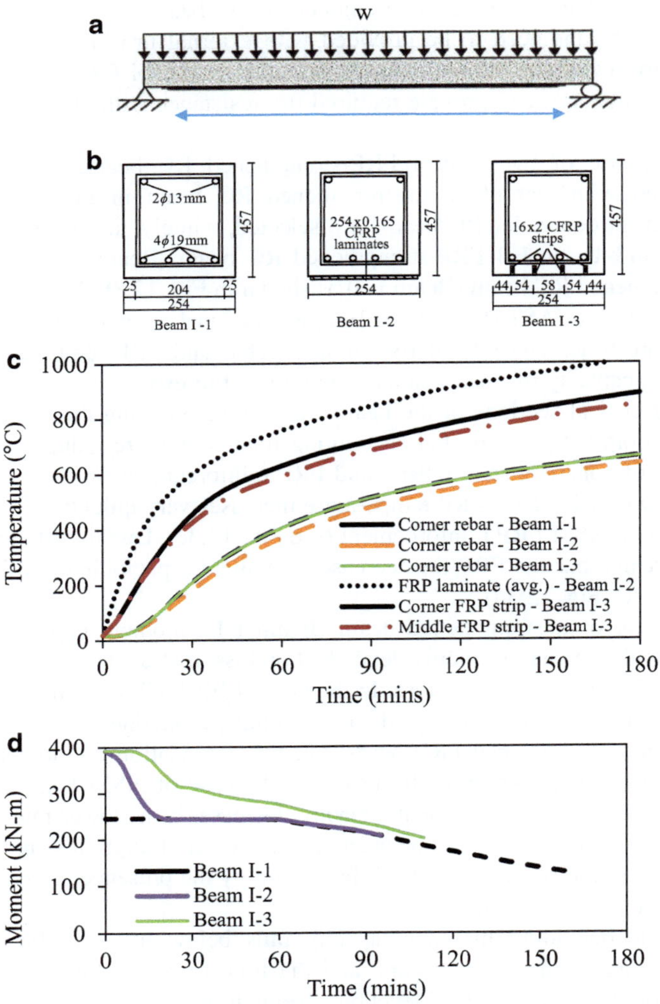

Fig. 12.3 Typical response of an EB and NSM FRP-strengthened concrete beam under fire exposure. (**a**) Layout of beam. (**b**) Cross section of three beams. (**c**) Temperature rise in three beams. (**d**) Variation of moment capacity

based model, and the fire resistance of FRP-strengthened RC beams was evaluated by varying critical factors over a wide range (Yu 2013). The main factors influencing the fire resistance of FRP-strengthened RC beams are type of strengthening, extent of strengthening, load level, fire scenario, axial restraint stiffness and its location, concrete strength, aggregate type, bond strength, thermal properties of fire insulation, and insulation thickness and its geometric configuration. Based on the results from parametric studies, the effect of each of these factors on fire resistance of EB and NSM FRP-strengthened beams is quantified (Ahmed and Kodur 2010; Kodur and Yu 2013). The critical factors that influence fire resistance of NSM FRP-strengthened RC beams are as follows:

- Type of strengthening: The type of strengthening (EB or NSM) influences the fire response of FRP-strengthened RC beams. A concrete beam strengthened with NSM FRP reinforcement yields fire resistance that is lower than conventional concrete beam, but higher than that of similar concrete beam strengthened with EB FRP laminate. This is due to the fact that an FRP-strengthened RC beam is expected to carry a higher service load, as compared to an unstrengthened RC beam.
- Location of FRP reinforcement: Unlike in EB FRP-strengthened RC beams, the location of NSM FRP rebars on the soffit of the beam has an influence on the fire resistance of concrete beams strengthened with NSM FRP. Provision of NSM FRP rebars close to the middle of the beam soffit yields slightly higher fire resistance than when NSM FRP rebars are located closer to the bottom corners of the beam.
- Extent of strengthening: The extent of strengthening or the level of contribution of NSM FRP reinforcement to the overall flexural capacity of the strengthened beam has an influence on the fire resistance of concrete beams strengthened with NSM FRP. Beams with relatively higher strengthening will have lower fire resistance than the case of beams with lower strengthening, since the NSM FRP reinforcement experiences higher strength degradation with rise in temperature.
- Load level: An NSM FRP-strengthened beam experiences higher load level after NSM FRP ruptures under increased temperatures in a fire situation. This leads to early strength failure of the beam under fire conditions. Therefore, higher level of applied loading on an NSM FRP-strengthened beam results in lower fire resistance (Yu 2013).
- Axial restraint: The development of fire-induced axial restraint in NSM FRP-strengthened RC beams enhances fire resistance. This is attributed to arch action generated in the beam as a result of fire induced restraint force.
- Fire scenario: Under most practical design fire scenarios, NSM FRP-RC beams achieve higher fire resistance than under standard fire exposure (ASTM E119 2012). This is mainly due to beneficial effect of cooling phase of fire exposure (decay phase).
- Bond degradation: The bond between concrete and NSM FRP rebars deteriorates considerably at temperatures close to glass transition temperature (T_g). This leads to debonding of NSM FRP, thus, decreasing moment capacity of the beam and lowering the fire resistance of the beam.
- Fire insulation: Application of external fire insulation is an effective mechanism to delay temperature rise in FRP beyond certain limiting temperature and thus enhance fire resistance of an EB or NSM FRP-strengthened RC beam. However, fire insulation configuration and its thickness are the key parameters that govern the fire resistance of EB or NSM FRP-strengthened RC beams. NSM FRP-strengthened RC beam with a U-shape fire protection (insulation) exhibits a much higher fire resistance than that of a beam with insulation only at the beam soffit (Yu 2013).

12.4 Strategies for Enhancing Fire Performance

Data generated in recent studies can be utilized to develop guidelines for achieving required fire resistance in NSM FRP-strengthened RC beams (Kodur et al. 2007; Kodur and Ahmed 2013; Yu 2013; Yu and Kodur 2014a). The following strategies can be effective in enhancing the fire performance of NSM FRP-strengthened RC members:

- In NSM FRP-strengthened RC beams, unlike in EB FRP-strengthened RC beams, it is possible to get 1–2 h of fire resistance without any external fire insulation. However, to develop such fire resistance information, rational calculations have to be carried out (Yu and Kodur 2014a). If more than 2 h of fire resistance is needed, fire insulation is to be applied on NSM FRP-strengthened RC beams.
- Unlike NSM FRP-strengthened RC beams, EB-FRP-strengthened concrete beams, in most cases, require suitable fire insulation to achieve required fire performance under increased service loads. The performance of protected (fire insulated) FRP-strengthened concrete beams at high temperatures can be similar to, or better than, that of conventional RC members. As noted earlier, an FRP-strengthened RC beam will have higher service loading during fire exposure as compared to an unstrengthened RC beam.
- For achieving optimum fire resistance in NSM FRP-strengthened RC beams, NSM FRP rebars should be placed (inserted) closer to the middle of the beam soffit, rather than at the corner of the bottom surface.
- NSM FRP-strengthened concrete beams provide satisfactory fire resistance, even though the glass transition temperature of the NSM FRP polymer matrix is exceeded early into fire exposure. Thus, reaching of matrix glass transition temperature (T_g) in NSM FRP reinforcement during fire does not lead to failure of an NSM FRP-strengthened concrete beam.
- Fire insulation is an important consideration for achieving fire resistance of 2 h or more in NSM FRP-strengthened RC beams. Proper detailing of insulation is needed to keep the temperatures low not only in NSM FRP but also in tension steel reinforcement to achieve optimum fire resistance. To develop such insulation detailing schemes, rational calculations have to be carried out (Yu and Kodur 2014a).
- The five possible insulation configurations that can be adopted in NSM FRP-strengthened RC beams with rectangular and T cross sections are shown in Figs. 12.4 and 12.5 (Yu 2013). The first option, of having NSM FRP-RC beam without any fire protection, may be feasible for a few practical situations in buildings where 1–2 h of fire resistance is required. The second insulation configuration, shown in Figs. 12.4b and 12.5b, is to protect the beam at bottom soffit only and may not be effective. In this configuration, corner rebar temperature increases after certain time due to heat transfer from two sides of the beam and thus increase in fire resistance is only marginal as compared to bare NSM FRP-strengthened RC beams. In the case of configuration 3 (refer to Figs. 12.4c and 12.5c), the fire insulation is provided at the beam soffit and

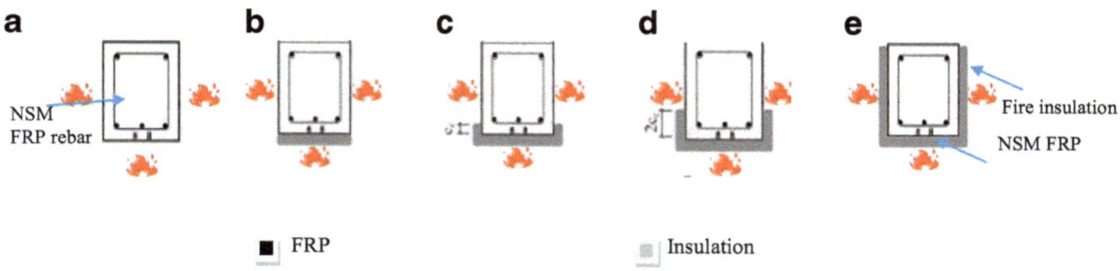

Fig. 12.4 Illustration of geometric configuration schemes for fire insulation in NSM FRP-strengthened rectangular RC beams. (**a**) Bare NSM FRP-RC beam. (**b**) Insulation at bottom only. (**c**) Insulation extending to Cc on sides. (**d**) Insulation extending to 2 × Cc on sides. (**e**) Insulation on bottom and sides

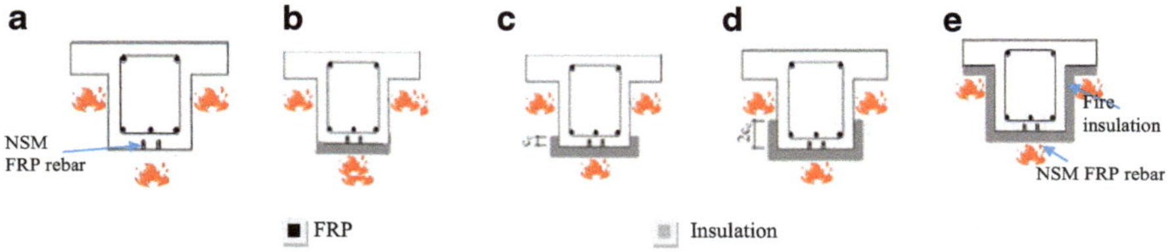

Fig. 12.5 Illustration of geometric configuration schemes for fire insulation in NSM FRP-strengthened RC T-beams. (**a**) Bare NSM FRP RC beam. (**b**) Insulation at bottom only. (**c**) Insulation extending to Cc on sides. (**d**) Insulation extending to 2 × Cc on sides. (**e**) Insulation on bottom and sides

extending to a depth equal to concrete cover thickness (Cc) on both sides of NSM FRP-strengthened RC beam. In this case, resulting temperatures in steel rebars for a given insulation thickness are lower than that in configuration 2. However, there is only slight improvement in fire resistance of NSM FRP-strengthened beam. In the case of configuration 4 (refer to Figs. 12.4d and 12.5d), the strengthened RC beam is protected on three sides (bottom and two sides) where depth of insulation on two sides of the beam is held equal to twice the concrete cover thickness (2 × Cc) to rebars. This configuration is most optimal as it provides effective insulation to overall beam cross section by minimizing temperature rise in NSM FRP interface, as well as in flexural steel reinforcement for considerable time, thus, enhancing fire resistance of the beam. Finally, in configuration 5 (refer to Figs. 12.4e and 12.5e), insulation is provided along the entire two side faces and beam soffit. This insulation scheme is not practically viable and may not be cost effective as well (Kodur and Yu 2013).

- For typical commercially available fire insulation (such as VG-EI insulation), an optimum insulation thickness of 20 mm is required to achieve fire endurance of 3 h, while 40 mm thickness is needed to achieve fire resistance of 4 h in NSM FRP-strengthened RC beams (Yu 2013). This VG-EI fire insulation product is a proprietary insulation material and comprises of vermiculite-gypsum spray on insulation material with a top coating of EI intumescent paint (Williams et al. 2006).

- To protect against sudden and complete loss of effectiveness of NSM reinforcement during fire, the strengthened (with NSM FRP strips or rebars) service load on an RC beam should not exceed the design strength of an unstrengthened (pre-existing) RC beam. This requirement is similar to a strengthening limit requirement currently suggested by ACI Committee 440 and also provides a measure of protection against poor installation practices or fire induced spalling of concrete (Kodur et al. 2007; Bisby et al. 2005).
- Computer models or rational calculation methods available in literature can be applied to develop effective insulation schemes for achieving optimum fire resistance in NSM FRP-strengthened RC beams. Also, these rational calculation methods can be applied to demonstrate that NSM-strengthened RC beams can achieve 2 h fire resistance without any external fire insulation.

12.5 Conclusions

The results presented in this chapter can be adopted as guidance to meet optimum fire resistance of NSM FRP-strengthened RC beams. Based on this information, the following conclusions can be drawn:

- NSM FRP-strengthened RC beams can provide up to 2 h of fire resistance without any supplementary fire insulation. However, these beams need to be provided with external fire insulation in order to achieve fire resistance ratings of 2 h or more.
- The insulation detailing schemes presented in this paper can be adopted as guidance for achieving fire resistance of more than 2 h in NSM FRP-strengthened RC beams.
- The most effective fire insulation scheme for achieving optimum fire resistance in an NSM FRP-RC beam is the one that extends from the beam soffit to both sides of the beam cross section up to a depth equal to twice the clear concrete cover thickness to steel reinforcement.
- For typical commercially available fire insulation (such as VG-EI insulation), an optimum insulation thickness of 20 mm is required to achieve fire endurance of 3 h, while 40 mm thickness is needed to achieve fire resistance of 4 h.

References

ACI 440R-07. (2007). *Report on fiber-reinforced polymer (FRP) reinforcement for concrete structures*. Detroit, MI: ACI Committee 440, American Concrete Institute.

Ahmed, A., & Kodur, V. (2011). The experimental behavior of FRP-strengthened RC beams subjected to design fire exposure. *Engineering Structures, 33*, 2201–2211.

Ahmed, A., & Kodur, V. K. R. (2010). *Factors governing fire resistance of FRP-strengthened reinforced concrete beams*. Las Vegas, NV: Composite & Polycon, American Composites Manufacturers Association (ACMA).

American Society for Testing and Materials. (2012). *Standard test methods for fire tests of building construction and materials*. West Conshohocken, PA: ASTM E119.

Bakis, C. E., Cosenza, E., Lesko, J. J., & Machida, A. (2002). Fiber-reinforced polymer composites for construction—state-of-the-art review. *Journal of Composites for Construction, 6*, 73.

Barros, J. A. O., Diasa, S. J. E., & Limaa, J. L. T. (2007). Efficacy of CFRP-based techniques for the flexural and shear strengthening of concrete beams. *Cement and Concrete Composites, 29*(3), 203–217.

Bisby, L., Green, M., & Kodur, V. K. R. (2005). Response to fire of concrete structures that incorporate FRP. *Progress in Structural Engineering and Materials, 7*, 136–149.

Canadian Standards Association (CSA) S806-07. (2007). *Design and construction of building components with fibre-reinforced polymers*. Rexdale, Canada: CAN/CSA S806-02.

De Lorenzis, L., & Teng, J. G. (2007). Near-surface mounted FRP reinforcement—an emerging technique for strengthening structures. *Composites Part B: Engineering, 38*, 119–143.

Fédération Internationale du Béton (FIB). (2001). *Design and use of externally bonded FRP reinforcement for RC structures*. Bulletin 14, Lausanne, Switzerland.

Gamage, J., Al-Mahaidi, R., & Wong, M. B. (2006). Bond characteristics of CFRP plated concrete members under elevated temperatures. *Composite Structures, 75*, 199–205.

Hawileh, R. A., Naser, M., Zaidan, W., & Rasheed, H. A. (2009). Modeling of insulated CFRP-strengthened reinforced concrete T-beam exposed to fire. *Engineering Structures, 31*, 3072–3079.

Kodur, V. K. R., & Ahmed, A. (2010). A numerical model for tracing the response of FRP strengthened reinforced concrete beams exposed to fire. *Journal of Composites for Construction, 14*, 85.

Kodur, V., & Ahmed, A. (2013). Guidelines for achieving optimum fire resistance in FRP-strengthened reinforced concrete beams. In *Proceedings of the 2013 Structures Congress*(pp. 1120–1130). Reston, VA: ASCE.

Kodur, V. K. R., & Yu, B. (2013). Evaluating the fire response of concrete beams strengthened with near-surface-mounted FRP reinforcement. *Journal of Composites for Construction, 17*, 517–529.

Kodur, V. R., Bisby, L. A., & Green, M. F. (2007). Preliminary guidance for the design of FRP-strengthened concrete members exposed to fire. *Journal of Fire Protection Engineering, 17*, 5–26.

Oehlers, D. J., Haskett, M., Wu, C., & Seracino, R. (2008). Embedding NSM FRP plates for improved IC debonding resistance. *Journal of Composites for Construction, 12*, 635–642.

Rashid, R., Oehlers, D. J., & Seracino, R. (2008). IC debonding of FRP NSM and EB retrofitted concrete: plate and cover interaction tests. *Journal of Composites for Construction, 12*, 160–167.

Wang, Y. C., & Kodur, V. (2005). Variation of strength and stiffness of fibre reinforced polymer reinforcing bars with temperature. *Cement and Concrete Composites, 27*(9), 864–874.

Williams, B., Bisby, L., Kodur, V. K. R., Green, M., & Chowdhury, E. (2006). Fire insulation schemes for FRP-strengthened concrete slabs. *Composites Part A: Applied Science and Manufacturing, 37*, 1151–1160.

Williams, B., Kodur, V., Green, M. F., & Bisby, L. (2008). Fire endurance of fiber-reinforced polymer strengthened concrete T-beams, *ACI structural Journal, 105*(1), 60.

Williams, B., Kodur, V., Bisby, L., & Green, M. F. (2004). The performance of FRP-strengthened concrete slabs in fire.

Yu, B. (2013). *Fire response of reinforced concrete beams strengthened with NSM FRP reinforcement*. PhD dissertation, Michigan State University, East Lansing, MI, USA.

Yu, B., & Kodur, V. R. (2014a). Fire behavior of concrete T-beams strengthened with near-surface mounted FRP reinforcement. *Engineering Structures, 80*, 350–361.

Yu, B., & Kodur, V. K. R. (2014b). Effect of temperature on strength and stiffness properties of near-surface mounted FRP reinforcement. *Journal of Composites Part B: Engineering, 58*, 510–517.

Yu, B., & Kodur, V. K. R. (2014c). Effect of high temperature on bond strength of near-surface mounted FRP reinforcement. *Journal of Composite Structures, 110*, 88–97.
Wang, Y.C., & Kodur, V. (2005). Variation of strength and stiffness of fibre reinforced polymer reinforcing bars with temperature. *Cement and Concrete Composites, 27*(9), 864–874.
Wang, Y. C., Wong, P. M. H., & Kodur, V. K. R. (2007). An experimental study of mechanical properties of FRP and steel reinforcing bars at elevated temperatures. *Composite Structures, 80*(1), 131–140.

Chapter 13
Experimental and Analysis Methods for Blast Mitigating Designs in Civil Infrastructure

Lauren K. Stewart and Bradley J. Durant

Abstract The incorporation of blast and shock loading into a multi-hazard framework for civil infrastructure requires consideration of the mechanical and structural behavior of the component or system outside of the traditional quasi-static and dynamic regimes and into the impulsive loading regime. In dealing with high-rate loading of this nature, material and structural response must often be evaluated experimentally in order to produce basic mechanical properties, initial design validation, and final acceptance for implementation into existing and new infrastructure. The following provides a brief summary of the fundamentals of blast loading and the extension of these types of loads into various analysis methods. Further, the chapter includes a discussion of experimental techniques used to obtain information on behaviors important to analysis. Among these is a relatively new method for experimentation using hydraulic actuators known as blast generators. Using a case study of a curtain wall system for blast response, various analysis and experimental procedures are used to highlight the process of designing and validating systems for blast mitigation.

13.1 Explosive Events

The generation of a shock load can stem from a variety of potential explosion phenomena which include unconfined high explosives (HE), vapor cloud explosions (VCEs), confined explosions, condensed-phase explosions, exothermic chemical reactions, boiling liquid expanding vapor explosions (BLEVEs), and pressure-volume (PV) ruptures (Center for Chemical Process Safety 2012). Blast loading on civil structural components most commonly results from two main explosive

---

L.K. Stewart (✉)
Georgia Institute of Technology, 790 Atlantic Drive, Atlanta, GA 30332, USA
e-mail: lauren.stewart@ce.gatech.edu

B.J. Durant
Protective Technologies, 2510 South Church St, Paris, TX 75460, USA
e-mail: bdurant@protectivetechnologiesllc.com

P. Gardoni, J.M. LaFave (eds.), *Multi-hazard Approaches to Civil Infrastructure Engineering*, DOI 10.1007/978-3-319-29713-2_13

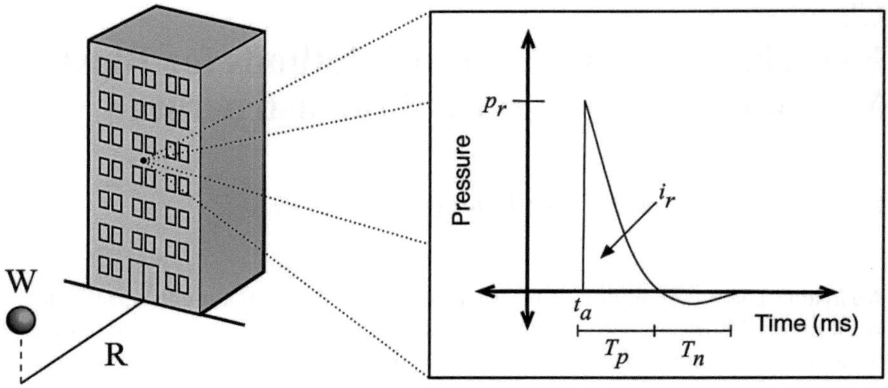

Fig. 13.1 Typical unconfined air blast environment with spherical charge of weight, W, at a standoff distance from the structure, R, and corresponding reflected pressure-time history at a point on the structure

scenarios: those involving detonation of a HE from military or terrorist events and those involving the ignition of a flammable vapor in VCE scenarios, which are more common in petrochemical and manufacturing applications.

13.1.1 High Explosives

High explosives are explosive materials that detonate (as opposed to deflagrate) and are typically products of organic substances such as glycerin, toluene, or phenol (Cook 1958). Trinitrotoluene, or TNT as it is typically referred, is an example of a high explosive along with other common explosives such as PETN, RDX, and C-4. Blast loads from these explosives are most simply quantified considering the typical air blast environment shown in Fig. 13.1. The unconfined environment includes a spherical charge of TNT with weight, W, at a distance away known as the standoff, R. Other types of explosive (i.e., C-4, ANFO, etc.) can be converted to an equivalent weight of TNT, W_{TNT}, using a ratio of specific energies as shown in Eq. 13.1, where Q_{EXP} is the specific energy of the explosive of interest, Q_{TNT} is the specific energy of TNT, and W_{EXP} is the weight of the explosive being converted (Mays and Smith 1995; Smith and Herrington 1994).

$$W_{TNT} = \frac{Q_{EXP}}{Q_{TNT}} W_{EXP} \tag{13.1}$$

Once the explosive is detonated, a shock wave is formed and it quickly moves outward from the charge source toward the structure (Cooper 1996). At the time the wave hits the structure, it undergoes reflection when the forward moving air molecules in the blast wave are brought to rest and then reflect off the structure. These molecules are then further compressed by the moving molecules behind, inducing a reflected overpressure on the structure. An example of a typical reflected pressure-time history, $p(t)$, at a point on the structure is also shown in Fig. 13.1, where p_r is the peak reflected overpressure. Values for peak reflected overpressure can be found in many ways including using equations derived by Rankine and Hugoniot (Hugoniot 1889; Rankine 1870) and using graphical, empirical based methods in Unified Facilities Criteria (UFC) 3-340-02 (Command 2008) and CONWEP (Hyde 1991).

The shape of the pressure-time history curve associated with a HE scenario is characterized by a steep increase in pressure followed by an exponential decay, the shape of which is often represented by the Friedlander equation (Friedlander 1946). The time of arrival, t_a, is the time at which the shock front arrives at the structure. The duration of the positive phase, T_p, is the time in which the pressure is above ambient. Similarly, the duration of the negative phase, T_n, is the time in which the pressure is below ambient. These durations are on the order of milliseconds and are typically much less than the period of the structure of interest, thus are impulsive in nature.

In addition to peak reflected pressure, specific reflected impulse, i_r, is also an important blast wave parameter. Specific impulse, which from herein will be referred to as impulse, is the area under the pressure-time curve from arrival time to the end of the positive phase as given by the basic integral equation in Eq. 13.2. For ductile structural components such as steel or reinforced concrete columns, impulse is the single most important parameter for describing the blast loading. For brittle components, such as window glazing, peak pressure is often the describing parameter.

$$i_r = \int_{t_a}^{T_p - t_a} p(t)dt \tag{13.2}$$

Scaled distance, Z, which is defined in Eq. 13.3, is used to present blast wave data and comparisons for a wide range of explosive environments (Cranz et al. 1944; Hopkinson 1915). It should be noted that cube root scaling is used for spherical charges and if a charge is cylindrical, then square root scaling is often more appropriate (Blair and Duvall 1954; Devine 1962). For the purposes of this chapter, all charges will be assumed to be spherical, unless noted otherwise. Scaled distance often also helps provide a means to classify various blast scenarios. Informally, these classifications are given in Table 13.1.

$$Z \equiv \frac{R}{\sqrt[3]{W}} \tag{13.3}$$

Table 13.1 Informal blast classifications in terms of scaled distance

Classification	Scaled distance range
Far-field explosion	$Z > 10$
Near-field explosion	$3 < Z < 10$
Near-contact explosion	$0 < Z < 3$
Contact explosion	$Z = 0$

13.1.2 Vapor Cloud Explosions

A vapor cloud explosion (VCE) is one type of a fuel-air explosion. It results from the ignition of a flammable mixture of vapor, gas, aerosol, or mist, in which flame speeds accelerate to sufficiently high velocities to produce significant overpressure (Center for Chemical Process Safety 2012). In general, VCEs are often associated with the storage of a large quantity of gas or liquid such as those in found refineries, manufacturing plants, and storage tanks for disposal purposes.

Vapor cloud explosions can only exist under very specific circumstances. In general, five conditions must be met before a VCE with overpressures capable of damaging infrastructure can exist (Gugan 1979; Lee and Moen 1980; Strehlow 1973):

1. The concentration of the resulting vapor cloud must be flammable.
2. An ignition source must be present.
3. A cloud of sufficient size must be present before ignition.
4. Turbulence caused from the interaction of the front and obstacles is required to achieve the speeds required for explosion as opposed to flash fire.
5. Confinement or partial confinement (i.e., that caused by equipment or structural components) must exist to generate sufficient flame speeds.

One relatively common method, due to its simplicity, for determining the characteristics of the loads applied to a structure from a VCE load involves a correlation between the effects from the VCE with those from HE (Bounds 2010; Forbes 1999). More specifically, the method involves a correlation between the weight of the fuel in the VCE, W_F, and those from an equivalent weight of TNT, W_{TNT}. This relation is shown in Eq. 13.4 and scales the weight of fuel by a ratio of the heat of combustion of the fuel, H_F, and the detonation energy of TNT, H_{TNT}, and parameter, α_e, that accounts for the equivalency based on energy. The approaches to the quantifications of these parameters often vary between agencies. The differences in approaches are mainly in the determination of the portion of fuel included (W_F) and the TNT equivalency energy (α_e) (Center for Chemical Process Safety 2012).

$$W_{TNT} = \alpha_e \frac{H_F}{H_{TNT}} W_F \tag{13.4}$$

While relatively simple to compute, the use of this method has some shortcomings mainly due to the differences between VCE and HE as far as blast attenuations, pulse shapes, and other parameters. More complex methods have been developed in order to address the shortcomings of the equivalent TNT approach. These methods, such as the TNO Multienergy Method (Van den Berg 1985) and the Baker-Strehlow-Tang (BST) (Baker et al. 1996; Tang and Baker 1999), are more advanced and able to account for phenomena such as cloud dispersion, obstacles, and partial confinements with the immediate vapor cloud region.

13.2 Blast Analysis Methods

Once the loads from the explosion have been quantified, there are three main methods (or slight variations thereof) for analyzing structures subjected to blast loads: single-degree-of-freedom (SDOF) analysis, Lagrangian finite element analysis (FEA), and complex methods which couple computational fluid dynamics (CFD) of the system with the computational structural dynamics (CSD). As with most applications, these methods often trade accuracy for computational efficiency and vice versa.

13.2.1 Single Degree of Freedom Analysis

The least complex and most common analysis method for predicting structural response to blast loads consists of reducing the structural component to an SDOF system as represented in Fig. 13.2. The equation of motion of the SDOF system can be written in terms of the resistance of the structure and the blast load as a function of pressure as given in Eq. 13.5. Because the loads are impulsive, large structural components do not react to small variations in time, and therefore, the pressure-time history is often approximated as linear. For a simply supported member with no axial load, the component is loaded with the blast pressure until a hinge forms at the midspan (location of maximum flexural stress). The resistance of the component to the blast load can be, in the simplest cases, modeled as elastic, perfectly plastic where the transition from elastic to plastic behavior is the resistance to the formation of the plastic hinge. The load mass factors, K_{LM}, in the equation are functions of the boundary conditions and the loadings applied. The factors for some of the most common boundary and load combinations can be found in (Biggs and Biggs 1964). Solving the equation of motion requires the use of a numerical scheme such as those outlined in (Bounds 2010; Chopra 1995). Software such as SBEDS (Protective Design Center 2006) and ARA SDOF (Applied Research Associates 2010) is also available to generate resistance functions for a variety of structural components.

$$K_{LM}m\ddot{x} + R(x) = p(t) \tag{13.5}$$

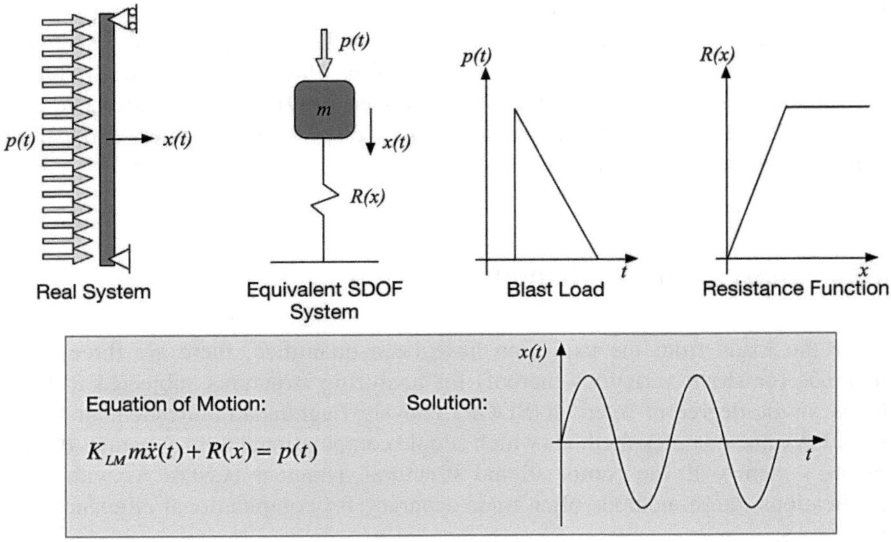

Fig. 13.2 Single Degree of Freedom (SDOF) analysis method for blast loading whereby the component is represented as an equivalent mass and spring system. The spring has resistance described, in this sense, by an elastic-plastic relation where the transition is found using the maximum flexural capacity of the component. A numerical scheme is then used to determine the response (i.e. displacement, rotation) of the component

SDOF methods can, when used properly, provide an efficient and accurate means of estimating the response of structural components to blast loads. However, as with all tools, these methods along with their pressure-impulse (PI) curve counterparts (Li and Meng 2002) can produce misleading results when used outside of the proper context (El-Dakhakhni et al. 2009; Oswald 2008). In general, the response of the component is most accurately predicted in far-field explosive scenarios when the load on the structure is relatively low and uniform. In situations where the charge is close and/or the pressures vary greatly, SDOF methods are usually not appropriate. Furthermore, the formulation of the SDOF resistance function is based upon a flexural mode. Should the component exhibit a different response mode (i.e., shear or combined modes), the resistance against those modes must be verified independently.

13.2.2 Finite Element Analysis

In the instances where simplified methods are not appropriate, continuum-based finite element analyses are often conducted. In this case, the loads generated from the methods described in the Sect. 13.1 are most often represented as pressures (or forces) normal to each element. Many commercial software programs (ABAQUS User's Manual 2000; Birnbaum et al. 1987) exist that are capable of conducting FEA for blast including LS-DYNA (LSTC 2007), which is used in the subsequent sections of the chapter. LS-DYNA is a three-dimensional, explicit, Lagrangian finite element code that uses a central difference time integration method. LS-DYNA is a general purpose, transient, dynamic finite element program that is often used in problems with blast and impact loading. It is able to utilize state-of-the-art constitutive models to represent material behaviors of interest, specifically steel and concrete with strain rate effects. Additionally, LS-DYNA is able to directly impart blast loads through the *BLAST command with sole inputs of charge weight and standoff, computing the loads on each element inside the software framework.

Because blast loads can induce highly nonlinear responses in the continuum, it is important that the finite element analyses have an appropriate scheme to handle large deformations. Many methods exist to create computational approaches to simulate these types of behavior with a stable response. Common approaches include the addition of element erosion based on failure strain and particle methods (i.e., SPH or element to particle conversions). The use of these schemes should be verified to sufficiently represent the physics of the specific problem.

13.2.3 Coupled Fluid and Structural Dynamic Analysis

For some specific cases involving blast loads, the interactions between the air "flow" and the structure are particularly important to include. Such problems include those with intricate geometries and those in which the structure may breach. In these instances, it is important to allow the structural behavior to adapt to the flow and vice versa. For these cases, applying a prescribed pressure load onto the elements is not adequate. Coupled fluid and structure (CFD and CSD) calculations are necessary to correctly predict these responses. Software including LS-DYNA ALE (Arbitrary Lagrangian-Euleran) (Schwer 2010), ALE3D (Nichols 2007), CTH (McGlaun et al. 1990), and FEFLO (Löhner et al. 2001) are all commonly used for coupled computations of this nature. These methods require an input for the equations of state for the explosive, some of which are included in technical reports from Lawrence Livermore National Laboratories (Fried 1994).

13.3 Experimental Methods

The development of simplified and complex computational models often necessitate experiments to provide data to correctly describe material behavior and system-level response for the rates of loadings exhibited by blast. At the material level, Hopkinson/Kolsky bars (Hopkinson 1914; Kolsky 1949) and gas guns (Fowlers et al. 1970) are often employed to produce material parameters for tension, compression, and torsion. For larger system response quantification, impact pendulums/drop towers (Lekan et al. 1996), shock tubes (Resler et al. 1952), explosives, and hydraulic systems are common. The latter two are described in more detail in the following sections.

13.3.1 Live Explosive Testing

Live explosive testing or "field" testing involves, in most cases, the detonation of HE in a remote location such that the component or structure is then loaded by the blast waves. Arguably, field testing provides the closest means of generating loadings and structural responses most closely related to an actual terrorist or military event, specifically because the loads imparted are those generated by live explosives. Often times, for component level testing, these experiments require the construction of a robust reaction structure and can be particularly expensive. In some rare cases, such as experiments conducted by the Defense Threat Reduction Agency (DTRA) (Luccioni et al. 2004; Rinehart et al. 2010), tests involving full building structures are achieved. Instrumentation for field testing requires careful mounting, shielding, and cabling to ensure data is recorded and the instrumentation survives the harsh environment of the blast. Pressure gages are most commonly used to record the pressure-time history in the free field and the reflected pressures on the structure. Time of arrival pins and witness panels are used to measure and record fragmentation velocities. High-speed cameras can also provide details through the duration of the test if the view is not obscured by the fireball or dust clouds.

13.3.2 Hydraulic Blast Generator

A relatively new method to produce blast-like shock loading on full-scale specimens is with ultrafast hydraulic actuators, known as blast generators (Gram et al. 2006; Hegemier et al. 2006; Stewart et al. 2014). Currently, a small number of these systems exist in locations such as the Georgia Institute of Technology (Georgia Tech) and University of California, San Diego (UCSD). The blast generator, shown in the testing frame and schematic form in Fig. 13.3, was designed to produce impulsive loads on large-scale structures by impacting the specimen with a mass in a controlled manner. The loading is accomplished using ultrafast, computer-controlled hydraulic actuators with a combined hydraulic/high-pressure nitrogen energy source. The blast generators consist of a hydraulic actuator, control valves, accumulators, and transducers. Initially, nitrogen is compressed in the pressure accumulator along with high-pressured oil. A servo-controlled high-flow valve controls the oil flow into the actuator. Once the valve has been opened at the desired rate and amount, the oil forces the piston rod/impacting mass assembly to drive outward toward the specimen. A smaller servo-controlled valve controls the outflow

Fig. 13.3 Hydraulic blast generator shown in schematic form on the left and in the testing frame located at the Georgia Institute of Technology on the right

Fig. 13.4 Urethane programmer used as a loading medium to transfer and tailor the shock pulse applied on the specimen. The front face has a pyramidal geometry which was specifically designed to achieve a shock pulse similar to that of a blast load

of the oil and thus is able to retract the actuator after impact, thus controlling the impact duration. The force required to retract the actuator is supplied by pressurized nitrogen gas in a deceleration chamber, which is specifically calibrated before each test.

The actuators are used in conjunction with an appropriate impacting module, which is attached to the piston and assist in the appropriate loading conditions for various shock loads. The impacting module consists of a steel or aluminum mass, a thin aluminum backing plate, and, in most cases, a nonlinear, urethane material called a programmer (Fig. 13.4). The programmer is used to transfer the energy and momentum of the module to the specimen (Freidenberg et al. 2013). The programmer's pyramid geometric and material properties help tailor the duration and magnitude of the pressure, and thus the impulse, to be representative of a blast-like pulse.

The computer-based control system controls the blast generator's position during test setup. During the test, it provides the valve commands to generate the desired impact velocity and timing of the BGs. After the test, it retracts the BGs and returns them to a safe state. A computer model is used to determine the setup values for oil and nitrogen pressures, initial position, and valve command to achieve the desired impact velocity.

In general, there are two options for imparting the shock loadings using the blast generator, which are differentiated by the mode in which the impacting mass is attached to the piston rod (Stewart et al. 2014). In the first method, the mass remains attached to the rod and the hydraulics are used to both accelerate and decelerate the mass. This method allows for the precise control of duration and load on the specimen. In the second method, the mass is initially pushed by the piston and allowed to fly freely once the piston retracts. This method results in a pure collision, totally driven by the fundamental transfer of momentum between the mass and the specimen.

13.4 Design and Analysis of System for Curtain Wall Applications

Successful design, validation, and implementation of blast-resistant design into infrastructure often requires a combination of analysis and experimental techniques discussed in the previous sections. In the following, many of methods are applied to the context of the design of a blast-resistant curtain wall comprised of various cementitious sheathing material and metallic studs, shown in Fig. 13.5. In this case, the walls were designed to meet the General Services Administration (GSA) and other blast standards and build upon decades of research conducted on the subject (DiPaolo and Woodson 2006; Salim and Townsend 2004).

13.4.1 Characterization of Sheathing Material

The curtain wall designs were approached by first considering the behavior of the sheathing to impulsive loads. Because the sheathing spans in between the steel studs, it is important to be able to quantify the resistance of the sheathing and, equally important, the forces transferred from the sheathing into the stud system. An experiment was developed to systematically determine the resistance of the sheathing material in both flexure and shear to a uniform blast load using a hydraulic blast generator, discussed in Sect. 13.3.2.

The setup for the flexure experiment is shown in Fig. 13.6. It consisted of mounting two load cells to a large concrete reaction block and then attaching a 0.91 m×0.76 m×5 cm (36×30×2 in.) steel plate. Rounded steel supports were fabricated and bolted to the steel plate. The test specimen panels were then placed

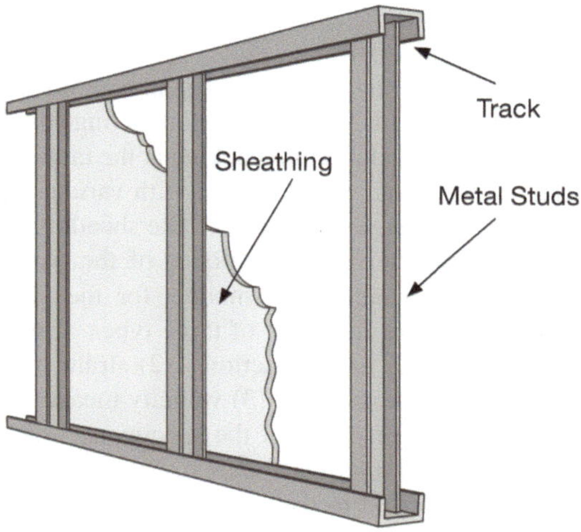

Fig. 13.5 Typical curtain wall design consisting of steel stud system and sheathing material. Typically, the curtain wall systems are attached to the structure, but do not carry floor or roof loads. In a blast event the loading is transferred through the sheathing material, into the studs, and through the connections onto the supporting structure

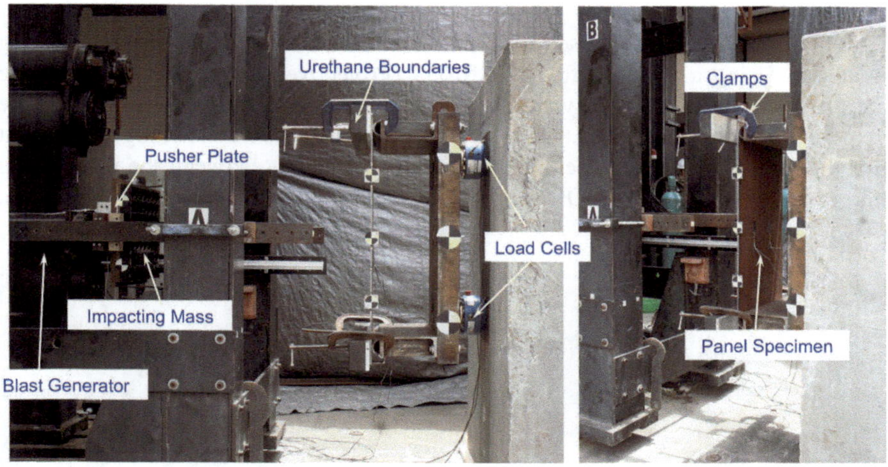

Fig. 13.6 Sheathing experiments using the hydraulic blast generator for development of a resistance function for the flexural response of the panel. The setup consisted of rounded support fixtures to simulate a simply supported condition

on the rounded steel supports to simulate simply supported (roller) boundary conditions. Urethane pads mounted on a 1.3 cm (0.5 in.) aluminum backing were gently clamped to the specimen to just hold the specimen flush against the roller supports while providing minimal rotational restraint. This connection also prevented rebounding of the panel off the supports once the unloading behavior initiated. The setup for the shear experiment consisted of the same mounting plate as the flexural experiments, but the rounded supports were replaced with a system that clamped the sheathing panel onto the steel supports simulating a fixed condition (Stewart et al. 2010).

The tests utilized one blast generator to launch, in free flight, a 0.38 m×0.37 m×4 cm (15×14.25×1.625 in.) impacting (flyer) mass consisting of a urethane pad mounted on an aluminum backing at the target. Each specimen, of a variety of materials and lengths, was impacted with various flyer mass and velocity combinations to achieve a range of impulses. The sheathing panels were instrumented to provide data to characterize the response of the panels that was suitable for developing and validating analytical models for the various configurations of panels. The instrumentation consisted of three types of measurements: (1) load measurements of the dynamic panel reactions, (2) strain measurements of the dynamic flexural strains in the panels, and (3) velocity measurements of the overall panel deflection histories. The response of the specimens were recorded with high-speed cameras. Using these results and other instrumentation, the deformation of the panel and forces transferred to the reaction wall from the uniform loading was recorded for various levels of simulated blast load up through formation of a midspan hinge as shown in Fig. 13.7.

The data collected was used to generate a resistance function for the sheathing material. Using a SDOF software, SBEDS (Protective Design Center 2006), the deformations from the tests were compared to the SDOF model derived from the experimental data. Figure 13.8 shows the agreement from one sheathing experiment for the maximum displacement of the system. Variations between the response in the experiment and the SDOF can be contributed to the quantification of fiction,

Fig. 13.7 Behavior of sheathing including midspan hinge development when subjected to increasing levels of simulated blast loads using hydraulic blast generator

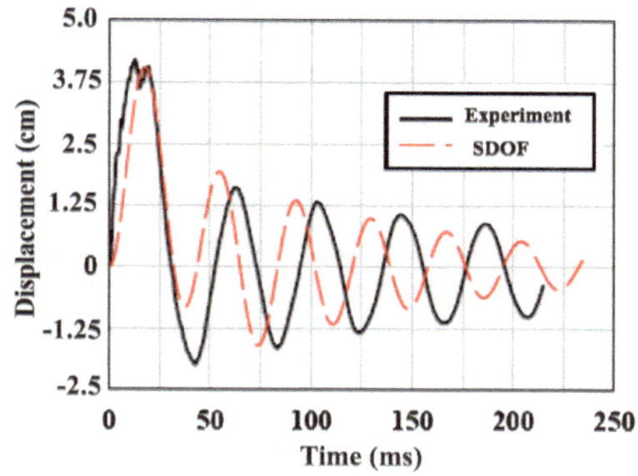

Fig. 13.8 Comparison of recorded panel deformation from single panel flexural experiment and derived SDOF model. Results show relatively good agreement and the ability to predict maximum deformation

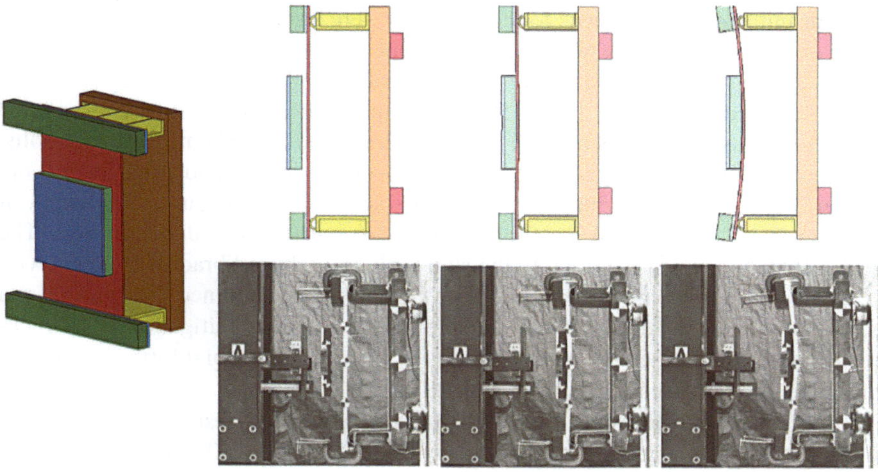

Fig. 13.9 Comparison of single panel flexural response from experiment and finite element model using LS-DYNA

the urethane behavior, and material models. The SDOF model was then used to calculate the dynamic reactions (Biggs and Biggs 1964) of the system to find the loads transferred into the stud system.

Additionally, the experiments were used to calibrate material and computational models for the sheathing for eventual use in system-level design. In this case, the sheathing setup was modeled with LS-DYNA using the K&C Concrete Model (Magallanes et al. 2010; Malvar et al. 2000) and compared to the experimental results through the duration of the test (Fig. 13.9). The LS-DYNA model is shown on the left and includes the specimen, backing plate, flyer plate, and urethane rotation system. Response of the panel experimentally and computationally is shown at three instances of time (before impact, at impact, and during the impact event) for comparison.

13.4.2 System Level Characterization for Design of Connections

The response of the sheathing and the corresponding dynamic reactions were used to design the curtain wall at the system level (stud, sheathing, and connection). As an initial estimate, the stud size, spacing, and thickness were optimized using the maximum load capacity of the sheathing using the results from the SDOF analysis and LS-DYNA models. In these calculations, the system was assumed to respond in a purely flexural mode with two simply supported end conditions. Ensuring that the connections behave in this manner in reality is critical to the overall performance of the wall system. Because of this, system-level experiments were conducted to design and optimize the connection details for the desired response.

Four initial design concepts for connections were tested using a setup with either three or four simultaneous blast generators depending upon wall height: connection angle, clip system, square bearing washer system, and track-washer system. The angle detail consisted of a 15.2×15.2×0.95 cm (6×6×.375 in.) steel angle bolted to the footer using anchors and 7.6 cm (3 in.) long, 1.6 cm (5/8 in.) diameter bolts. The square bearing washers were 10.2 cm (4 in.) in length and used the 1.6 cm (5/8 in.) diameter bolts with nut. The washer system was custom to the length of the size and is attached to the footer with a 1.6 cm (5/8 in.) diameter bolt. The clip system connected the track to the stud with an L-shaped bracket attached with screws. Throughout the test series, these details were combined, optimized, and adjusted. Not all details provided suitable results for blast mitigation. The track-washer system provided the best response for this application and further details on the connection is given in (Stewart et al. 2014).

The test setup, shown in Fig. 13.10 for one particular test, was designed to simulate a typical one-story steel stud wall. The 3.28 m (10 ft-6 in.) stud wall was placed on a footer and reacted against a concrete slab header held up by steel angles and steel tube supports. The top of the specimen was connected to the header using high-strength steel bolts. The angles were screwed into the specimen studs and rotations at the top of the wall were partially suppressed. The specimens consisted of 15.24×4.13 cm (6×1.625 in.) steel studs retrofitted with Sureboard sheathing panels in this specific example. On the outside, the panels attached to the steel studs using screws spaced at 10.16 cm (4 in.) along the edge and at 15.24 cm (6 in.) the field area. For the inside panel, 4.13 cm (1.625 in.) screws were used at spacing of 15.24 cm (6 in.) for attachment to the steel studs.

Fig. 13.10 Full-scale experiment of curtain wall system during which simulated blast loads were applied through simultaneous impacts with three hydraulic blast generators. The pretest setup is shown on the left and the behavior after 50 ms is shown on the right

The stud wall systems were loaded with three blast generators using the impacting masses attached to the piston rod, allowing for control of the duration and pulse shape (Freidenberg et al. July 2013). Figure 13.11 shows the progression of damage of the angle connection at various times after impact recorded with high-speed cameras. Displacements of the wall were measured with the high-speed camera footage and a tracking software and compared to various limit states for specific agency guidelines.

13.4.3 Live Explosive Validation Testing

Utilizing the connection designs and calibrated computational models from the blast generator experiments, the wall technology was optimized and finalized for integration into a design for a blast-resistant module (BRM). A BRM is a modular building which can provide office or storage space in an environment that is protected from specific levels of blast loading. Live field experiments were conducted as part of a validation effort for specific guidelines. The experiment

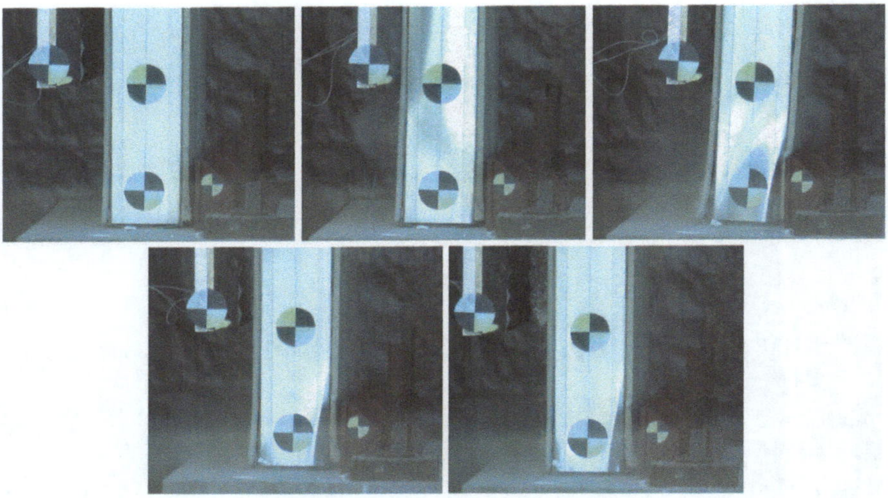

Fig. 13.11 Response of the wall bottom connection with angle backing. Progression of damage recorded by the camera shows the initial buckling of the stud web followed by a rebound response of the wall system

Fig. 13.12 Live explosive test specimens: blast-resistant module (*left*) and standard ISO container (*right*)

included an arena test of three BRMs all with slightly different designs and a standard ISO container (Fig. 13.12). The containers were organized in a ring formation around an enhanced 4,000 kg (9,000 lb) cylindrical ANFO charge (TNT equivalency of 3,600 kg (8,000 lbs)). All of the structures were located at a standoff distance of 46 m (150 ft) to provide a direct comparison of the responses. A photo of the test setup and blast event is given in Fig. 13.13, and an image sequence from the high-speed cameras showing the progression of the response of the BRM structure is given in Fig. 13.14

Pressure sensors were used to collect information regarding the incident and reflected pressures experienced by the structures during testing. Sensors were mounted to the front (directly loaded) walls of the BRMs at mid-height and the one-third locations along the span and at mid-height and midspan for the side walls.

Fig. 13.13 Field testing of stud wall system in blast-resistant module application in arena configuration with charge size of 4,000 kg (9,000 lbs) of ANFO

A single sensor was placed inside each structure as well to capture the peak internal pressure. To determine the free-field blast wave pressure at the specified standoff distance of 46 m (150 ft) as well as two additional distances of 53 m (175 ft) and 61 m (200 ft), free-standing sensors were placed at these locations along a single radial line extending from the charge. Average peak reflected pressure measured on the BRM in Fig. 13.12 was 0.46 MPa (66.5 psi) with corresponding impulse of 2.4 MPa-ms (345.9 psi-ms).

Posttest behavior of the BRM and the ISO container is given in Fig. 13.15 and shows the robustness of the design for blast loads of this magnitude. The main observable damages for the BRM included paint chipping and small punctures from projectiles on the external faces. No residual deflections were observed in any wall other than that which experienced reflected pressure, which had 1 cm (0.4 in) residual deformation. Further information regarding this test series can be found in (Durant 2015).

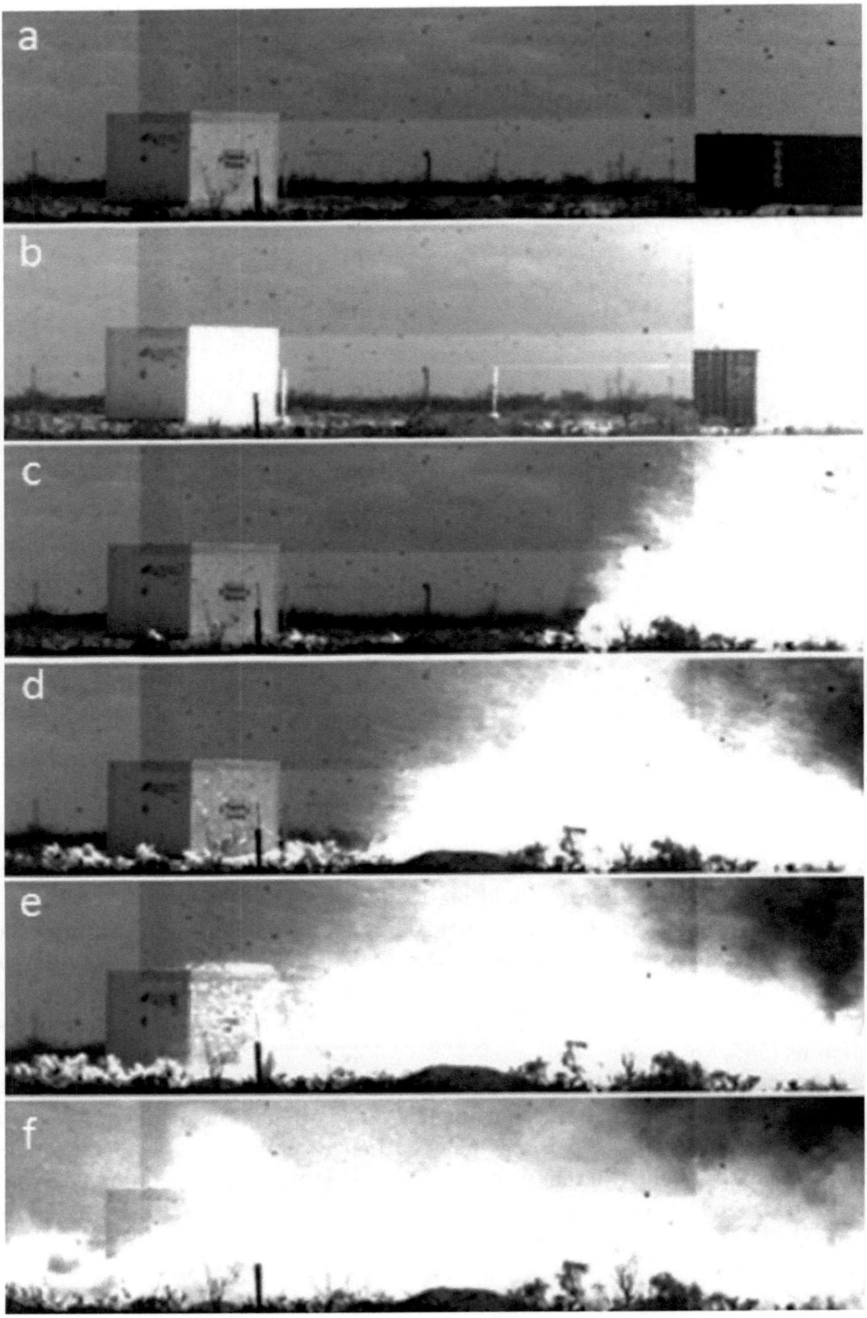

Fig. 13.14 (continued)

13.4.4 Predictive Tools

In order to practically implement the validated BRM units for other commercial and military applications, a rigorous set of analyses was conducted to determine the design capacity of the system given varying wall heights and blast loads using the data from the system-level tests and the live field experiments.

Because the wall involves a unique connection system at the header and footer, an accurate resistance and response cannot be developed using one of the generally prescribed equations consistent with a simply supported or fixed-fixed conditions. Thus, a custom resistance function for this particular system was developed and included the calibration of a system with rotational springs on both ends as shown in Fig. 13.16. In order to calibrate the rotational stiffness values to accurately represent the BRM wall header and footer connections, an iterative approach driven by the live field test data was utilized. The plot in Fig. 13.16 displays a relative example of the calibrated curve (for one scenario) compared with curves which would result from fixed-fixed or simple-simple boundary conditions for the same wall. As expected, the resistance lies in between a simple support and fixed connection.

With calibrated rotational stiffnesses for the boundary conditions, which is a characteristic of the connection properties and independent of the wall size, the only remaining variables are the wall height and stud spacing. For design purposes, a set of pressure-impulse (P-I) diagrams were developed with varying height and stud spacing for use with ASCE guidelines for petrochemical facilities (Durant 2015). The SDOF model can also be utilized to determine the capacity of various BRM wall systems to a range of blast loads for other design purposes.

13.5 Summary

The development of mitigating systems for blast loading from high explosives and vapor cloud explosions often requires the combination of various analysis and experimental strategies. In the case of the wall system presented in this chapter, the behavior of the sheathing was first characterized with experiments using an hydraulic blast generator and then represented with SDOF and computational models. The results were then used, along with another series of full-scale experiments, to design connection details at the system level. Finally, the system was

Fig. 13.14 Progression of the experimental explosive event and response of blast-resistant module from high-speed camera frames. Frame (**a**) shows the pretest state of the structure next to the ISO container. Frame (**b**) occurs at detonation and displays the flash from the ANFO charge being projected onto the BRM surface. The shock wave and fireball can be seen at the right of frame (**c**), and frame (**d**) provides a visual of the projectile impact on the BRM wall which preceded the shock wave. In frame (**e**), the structure is experiencing the reflected pressure from the shock wave on the front face, followed by the BRM being surrounded by pressure and dust in frame (**f**)

Fig. 13.15 Blast-resistant module and ISO container after live explosive testing with 4,000 kg of ANFO at 46 m. BRM experienced less than 1 cm (0.4 in) of permanent deformation

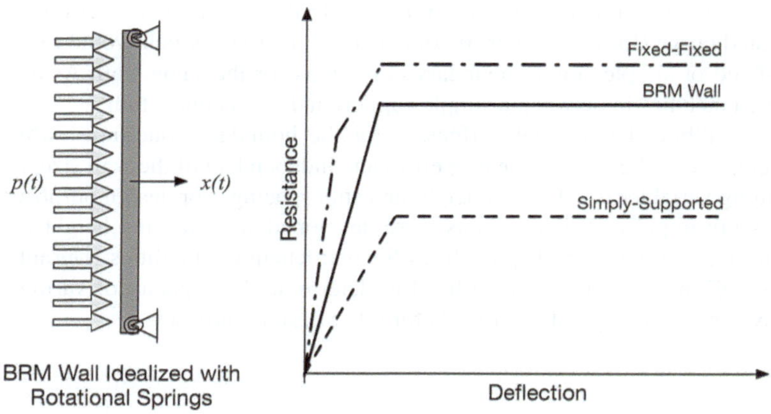

Fig. 13.16 BRM wall system was idealized as an SDOF system with resistance function determined from a beam with two rotational springs. The resistance function derived for the system lies in between that of a simple support and a fixed-fixed condition

implemented into a wall design for a blast-resistant module structure, which was validated through field experiments. The validated structure and the data collected were then used to develop simplified analysis tools for design for structures of various configurations.

Acknowledgements The authors would like to thank Mr. Ken Morrill, USG, and Sureboard for their contributions to the experimental programs included in this chapter.

References

ABAQUS User's Manual ©Hibbitt, Karlsson & Sorensen, Inc. 2000. USA.

Applied Research Associates. (2010). SDOF 3.0 user's guide. Applied Research Associates, Inc.

Baker, Q. A., Tang, M. J., Scheier, E. A., & Silva, G. J. (1996). Vapor cloud explosion analysis. *Process Safety Progress, 15*(2), 106–109.

Van den Berg, A. (1985). The multi-energy method: a framework for vapour cloud explosion blast prediction. *Journal of Hazardous Materials, 12*(1), 1–10.

Biggs, J. M., & Biggs, J. M. (1964). *Introduction to structural dynamics*. New York: McGraw-Hill College.

Birnbaum, N., Cowler, M., Itoh, M., Katayama, M., & Obata, H. (1987). Autodyn-an interactive non-linear dynamic analysis program for microcomputers through supercomputers. In: *Transactions of the 9th international conference on structural mechanics in reactor technology*, Vol. B.

Blair, B. E., & Duvall, W. I. (1954). *Evaluation of gages for measuring displacement, velocity, and acceleration of seismic pulses*. Washington, D.C.: Bureau of Mines.

Bounds, W. L. (2010). *Design of blast-resistant buildings in petrochemical facilities*. ASCE Publications, Reston, VA

Center, U. P. D. (2006). *User's Guide for the Single-Degree-of-Freedom Blast Effects Design Spreadsheets (SBEDS)*. Rep PDC-TR 06 2.

Center for Chemical Process Safety. (2012). *Guidelines for evaluating process plant buildings for external explosions, fires and toxic releasese*, 2nd edn. Hoboken, NJ: John Wiley and Sons, Inc.

Chopra, A. K. (1995). *Dynamics of structures*, vol 3. New Jersey: Prentice Hall.

Command, N. F. E. (2008). *UFC 3-340-02: Structures to Resist the Effects of Accidental Explosion*. Department of Defense

Cook, M. A. (1958). *The science of high explosives*, vol 139. RE Krieger Pub. Co., New York

Cooper, P. W. (1996). *Explosives engineering*. Vch Pub, United States

Cranz, K. J., Elder, J. D., & Roller, D. E. (1944) A translation of Cranz's Textbook of ballistics (Lehrbuch der Ballistik von G. Cranz). Technical Reports Section for Armor and Ordnance, National Defense Committee, Washington.

Devine, J. F. (1962). Vibration levels from multiple holes per delay quarry blasts. *Earthquake Notes, 33*, 32–39.

DiPaolo, B., & Woodson, S. (2006). An overview of research at erdc on steel stud exterior wall systems subjected to severe blast loads. In: *Structures congress 2006: structural engineering and public safety*

Durant, B. (2015). *Blast Resistant Module Development, Testing, and Analysis*. Tech. rep., PROTECTIVE TECHNOLOGIES, LLC

El-Dakhakhni, W., Mekky, W., & Rezaei, S. C. (2009). Validity of sdof models for analyzing two-way reinforced concrete panels under blast loading. *Journal of Performance of Constructed Facilities, 24*(311–325).

Forbes, D. J. (1999). Blast loading on petrochemical buildings. *Journal of Energy Engineering, 125*(3), 94–102.

Fowlers, G. R., Duvall, G., Asay, J., Bellamy, P., Feistmann, F., Grady, D., et al. (1970). Gas Gun for Impact Studies. *Review of Scientific Instruments, 41*, 984–996.

Freidenberg, A., Lee, C., Durant, B., Nesterenko, V. F., Stewart, L., & Hegemier, G. (2013). Characterization of the blast simulator elastomer material using a pseudo-elastic rubber model. *International Journal of Impact Engineering, 60*, 58–66.

Freidenberg, A., Aviram, A., Whisler, D., Kim, H., Stewart, L., & Hegemier, G. (July 2013). Demonstration of tailored impact to achieve blast-like loading. Submitted to *International Journal of Impact Engineering*, vol 71, pp. 97–105

Fried, L. E. (1994). *Cheetah 1.0 users manual*. Tech. rep. CA (United States): Lawrence Livermore National Lab.

Friedlander, F. G. (1946). The diffraction of sound pulses. i. diffraction by a semi-infinite plane. *Proceedings of the Royal Society of London Series A Mathematical and Physical Sciences, 186*(1006), 322–344.

Gram, M., Clark, A. J., Hegemier, G. A., & Seible, F. (2006). Laboratory simulation of blast loading on build and bridge structures. *WIT Transactions on the The Built Environment, 87*, 33–44.

Gugan, K. (1979). *Unconfined vapor cloud explosions*. Gulf Pub Co, Houston, TX

Hegemier, G., Seible, F., Arnett, K., Rodriguez-Nikl, T., Oesterle, M., Wolfson, J., et al. (2006). The ucsd blast simulator. In: *77th Shock and Vibration Symposium*

Hopkinson, B. (1914). A method of measuring the pressure produced in the detonation of high explosives or by the impact of bullets. *Philosophical Transactions of the Royal Society of London Series A, Containing Papers of a Mathematical or Physical Character*, pp. 437–456.

Hopkinson, B. (1915). *British ordnance board minutes 13565*. The National Archives, Kew, UK British ordinance boad minutes 13565.

Hugoniot, H. (1889). Sur la propagation du movement dans les corps et spécialement dans les gaz parfaits. *J Ecole Polytechnique, 58*, 1–125.

Hyde, D. W. (1991). *Conwep, conventional weapons effects program*. USA: US Army Engineer Waterways Experiment Station.

Kolsky, H. (1949). An investigation of the mechanical properties of materials at very high rates of loading. *Proceedings of the Physical Society Section B, 62*(11), 676.

Lee, J., & Moen, I. (1980). The mechans of transition from deflagration to detonation in vapor cloud explosions. *Progress in Energy and Combustion Science, 6*(4), 359–389.

Lekan, J., Gotti, D., Jenkins, A., Owens, J., & Johnston, M. (1996). *Users guide for the 2.2 second drop tower of the NASA Lewis Research Center.*

Li, Q., & Meng, H. (2002). Pressure-impulse diagram for blast loads based on dimensional analysis and single-degree-of-freedom model. *Journal of engineering mechanics, 128*(1), 87–92.

Löhner, R., Yang, C., Cebral, J., Soto, O., Camelli, F., Baum, J. D., et al. (2001). *Advances in feflo.* AIAA Paper 582

LSTC. (2007). *LS-DYNA Keyword User's Manual.*

Luccioni, B., Ambrosini, R., & Danesi, R. (2004). Analysis of building collapse under blast loads. *Engineering structures, 26*(1), 63–71.

Magallanes, J. M., Wu, Y., Malvar, L. J., & Crawford, J. E. (2010). Recent improvements to release iii of the k&c concrete model. In: *11th international LS-DYNA Users conference*, Livermore Software Technology Corporation Livermore, CA, pp 6–8.

Malvar, L., Crawford, J., & Morrill, K. (2000). *K&c concrete material model release iii–automated generation of material model input.* Karagozian and Case Structural Engineers, Technical Report TR-99-243.

Mays, G. C., & Smith, P. D. (1995). *Blast effects on buildings - design of buildings to optimize resistance to blast loading*. New York: Thomas Telford.

McGlaun, J., Thompson, S., & Elrick, M. (1990). Cth: a three-dimensional shock wave physics code. *International Journal of Impact Engineering, 10*, 351–360.

Nichols, A. (2007). *Users manual for ale3d: An arbitrary lagrange/eulerian 3d code system.* Lawrence Livermore National Laboratory

Oswald, C. (2008). Comparison of measured component responses to blast loads with single-degree-of-freedom analyses. In: *Proceedings of 20th international symposium on military aspects of blast and shock (MABS20).* Oslo, Norway

Rankine, W. J. W. (1870). On the Thermodynamic Theory of Waves of Finite Longitudinal Disturbance. *Philosophical Transaction of the Royal Society, 160*, 277–288.

Resler, E., Lin, S. C., & Kantrowitz, A. (1952). The production of high temperature gases in shock tubes. *Journal of applied physics, 23*(12), 1390–1399.

Rinehart, E. J., Henny, R. W., Thomsen, J. M., & Duray, J. P. (2010). *Dtra weapons effects testing: A thirty year perspective*. Tech. rep., DTIC Document.

Salim, H., & Townsend, P. (2004). Explosion-resistant steel stud wall system. In: *Structures 2004: Building on the Past, Securing the Future*. ASCE, Reston, VA

Schwer, L. (2010). A brief introduction to coupling load blast enhanced with multi-material ale: the best of both worlds for air blast simulation. In: *LS-DYNA Forum*, Bamberg.

Smith, P., & Herrington, J. (1994). *Blast and ballistic loading of structures*. Butterworth-Heinemann, Boca Raton, Florida.

Stewart, L., Morrill, K., & Natesaiyer, K. (2010). Development of high performance concrete panels for curtainwall systems. In: *Structures Congress 2012*, ASCE.

Stewart, L., Freidenberg, A., Rodriguez-Nikl, T., Oesterle, M., Wolfson, J., Durant, B., et al. (2014). Methodology and validation for blast and shock testing of structures using high-speed hydraulic actuators. *Engineering Structures, 70*, 168–180.

Stewart, L., Freidenberg, A., & Hegemier, G. A. (2014). Design and testing of steel stud system for blast mitigation. In: *Structures Under Shock and Impact 2014*.

Strehlow, R. A. (1973). Unconfined vapor-cloud explosions—an overview. In: *Symposium (International) on Combustion* (vol. 14, pp. 1189–1200). Elsevier, The Combustion Institute, Pittsburgh, PA

Tang, M., & Baker, Q. (1999). A new set of blast curves from vapor cloud explosion. *Process Safety Progress, 18*(4), 235–240.

Part V
Wind Hazards

Chapter 14
Wood-Frame Residential Buildings in Windstorms: Past Performance and New Directions

John W. van de Lindt and Thang N. Dao

Abstract Residential buildings in coastal areas are often at risk to hurricanes, which can result in both wind and storm surge damages, while tornadoes are one of the most devastating natural hazards that have occurred in all 50 states of the USA and can happen during any season of the year. This chapter focuses on summarizing some past studies on the performance of wood-frame residential buildings in recent major hurricanes and tornadoes. Damage data collected from hurricanes shows that in most hurricanes the damage to residential wood-frame buildings often comes from high winds, hurricane surge, flooding, and rainwater intrusion due to damage in the building envelope. Roof systems experienced extensive damage either directly from wind or due to failure of the flashing and coping. Hurricanes are often accompanied by heavy rain that results in substantial water intrusion through the breached area of the building, which in turn results in substantial financial loss to the structure and its contents. On the other hand, data collected from recent tornadoes in Tuscaloosa, Joplin, and Moore show that, for an EF-4 or EF-5 tornado, damage levels increase from the outer edges toward the centerline of a tornado track. Residential building damage in tornados is caused by high wind loading or debris impact, or both. A general procedure for performance-based wind engineering is proposed, and research needs for development of wood-frame performance-based wind engineering are also highlighted in this chapter.

J.W. van de Lindt (✉)
Department of Civil and Environmental Engineering, Center for Risk-Based Community Resilience Planning, Colorado State University, Fort Collins, CO, USA
e-mail: jwv@engr.colostate.edu

T.N. Dao
Department of Civil, Construction, and Environmental Engineering, University of Alabama, Tuscaloosa, AL, USA
e-mail: tndao@eng.ua.edu

P. Gardoni, J.M. LaFave (eds.), *Multi-hazard Approaches to Civil Infrastructure Engineering*, DOI 10.1007/978-3-319-29713-2_14

14.1 Introduction

14.1.1 Wood Building Performance in Hurricanes

Over 80 % of the total building stock in the USA, and well over 90 % of the residential stock in North America, are light-frame wood (wood-frame) construction. While wood-frame construction is the most prevalent type of building, it is also the most susceptible to wind damage. This was evident during the 2004 hurricane season, when four hurricanes made landfall in the USA and became even more publicized following Hurricane Katrina in 2005 (van de Lindt et al. 2007). Although several states along the US coastline have been struck by numerous hurricanes, a number of hurricanes making landfall affect every gulf coast and eastern seaboard state, as evidenced by Fig. 14.1.

Residential buildings in coastal areas are often at risk to hurricanes, which can result in both wind and storm surge damages. While the USA averaged $1.6 billion in normalized hurricane damage annually from 1950 to 1989, this figure rose dramatically between 1989 and 1995, to approximately $6 billion annually (Pielke and Pielke 1997). The average annual normalized hurricane loss in the USA was approximately $10 billion in recent years, with more than $150 billion in total damage in 2004 and 2005 (Pielke et al. 2008). In the past decade (2005–2015), the hurricanes that hit the gulf coast and eastern seaboard of the USA—Katrina, Rita, and Sandy—have been the most devastating, resulting in widespread damage

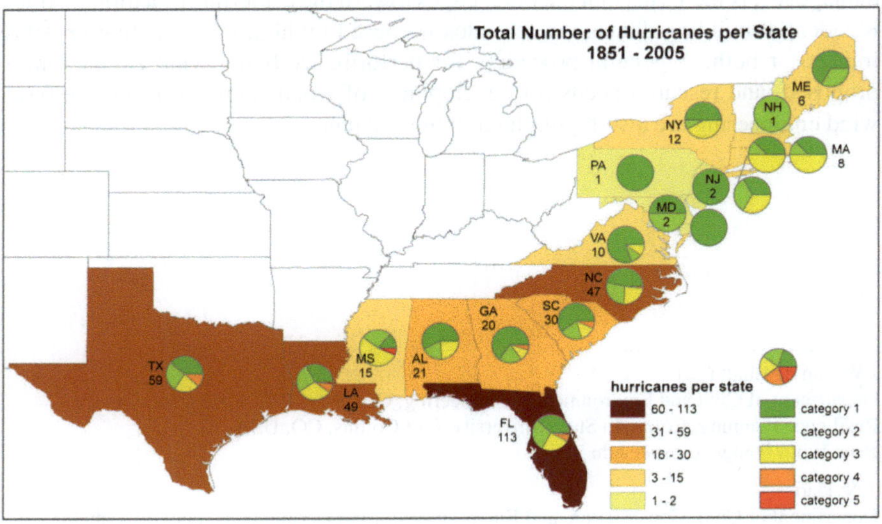

Fig. 14.1 Hurricanes making landfall in the USA (1851–2005) (excerpted from van de Lindt et al. 2007)

to residential buildings and infrastructure. Hurricane Katrina first made landfall near Buras, Louisiana, on August 29, 2005 (NIST 2006). The storm was classified at the upper end of Saffir–Simpson Category 3 by the National Hurricane Center (NHC), which estimated sustained maximum wind speeds of 125 mph, and storm surge heights up to 27 ft; less than 1 month later, Hurricane Rita hit land along the Texas–Louisiana border on September 24, 2005 (NIST 2006). This hurricane was also categorized as Category 3, with up to a 15 ft high storm surge when it made landfall. Those two hurricanes caused about 1400 fatalities and $70–$130 billion in damage, of which about $45–$65 billion was insured loss (NIST 2006). On October 22, 2012, a tropical depression formed in the southern Caribbean Sea off the coast of Nicaragua, which later became the Sandy hurricane. The hurricane then moved away from the Bahamas with winds of about 110 mph and made a turn to the northeast off the coast of Florida, having already caused about 70 or more fatalities in the Caribbean. On October 29, the hurricane made landfall at New Jersey as a huge storm, with winds covering almost a 1000 mile diameter and 11–12 ft of storm surge in some places. It affected more than 50 million people on the eastern seaboard and killed 73 people in the USA (FEMA 2015; National Geographic 2015).

Damage data collected from hurricanes (e.g., in van de Lindt et al. 2007) shows that in most hurricanes the damage to residential wood-frame buildings often comes from high winds, hurricane surge, flooding, and rainwater intrusion due to damage to the building envelope. Roof systems experienced extensive damage either directly from wind or due to failure of the flashing and coping. Roofing aggregate was responsible for the majority of wind-borne debris damage to windows (NIST 2006), which results in breach of the building envelope for structures hit with the debris. Hurricanes are often accompanied by heavy rain that results in substantial water intrusion through the breached area of the building, which in turn results in substantial financial loss to the structure and its contents (Dao and van de Lindt 2010). All of this type of damage was observed in hurricanes Katrina, Rita, and Sandy.

14.1.2 Wood Building Performance in Tornadoes

Tornadoes are found in all parts of the world, but the USA has by far the highest occurrence of tornadoes compared to any other countries. Tornadoes are one of the most devastating natural hazards that have occurred in all 50 states and can happen during any season of the year. The tornado record from 1950 to 2011, kept by the US National Oceanic and Atmospheric Administration (NOAA), documents 56,457 tornado events, of which 33,756 resulted in reported damage (Simmons et al. 2013). This means that, on average, there are about 925 documented US tornadoes annually, of which 553 cause reported damage. In particular, there were several years that saw major tornado outbreaks. In 1953, there were several outbreaks that resulted in 519 fatalities, over 500 injuries, and $32.5 billion (2011 GDP normalized US dollars) in property and structural damage. In 1965, there were 301 people killed and $40.0 billion (2011 GDP normalized US dollars) in reported damage from tornadoes. In 1974, 348 people died, 6500 people suffered injuries, and $26.0 billion (2011 GDP normalized US dollars) in damage was done from tornadoes (Simmons et al. 2013). In 2011, there were two major tornadoes that occurred in Tuscaloosa, AL, and Joplin, MO. These tornadoes resulted in 219 fatalities with over 13,000 homes destroyed and an estimated $5 billion in damage (Prevatt et al. 2012). The losses from tornado damage continue to increase due to an increase in population and the development of the related infrastructure.

On May 20, 2013, an EF-5 tornado hit Moore, Oklahoma, and caused $1.5–$2.0 billion in damage and 24 fatalities, including ten children. This tornado was approximately 1 mile wide based on ground reports, and the damage path was very similar to the path observed after a tornado 14 years earlier (on May 3, 1999). Several schools and many residential structures were damaged or destroyed. The performance of residential structures along and across a tornado path provides a unique and useful research opportunity to document the progression of failure of wood-frame homes when exposed to tornadic winds. Comparison of the damage allows for a qualitative (and/or quantitative) analysis of the performance of residential structures under different failure progressions. A reconnaissance team of engineers from around the USA was formed and traveled to Moore 5 days after the tornado to collect and assess residential structural damage in the aftermath of the tornado. Many team members had also participated in the reconnaissance team in the aftermath of the Tuscaloosa, AL, and Joplin, MO, tornadoes in 2011.

Data collected from recent tornadoes in Tuscaloosa, Joplin, and Moore show a consistent pattern of damage to residential structures. For an EF-4 or EF-5 tornado, damage levels increase from the outer edges toward the centerline of a tornado track. In general, the progression of residential structure failure often depends on wind speed, wind direction, wind-borne debris impact, and the capacity of building components. Residential building damage in tornados is caused by high wind loading or debris impact, or both. Failure of residential buildings in tornado winds is often due to:

- Lack of continuous sheathing and fiberboard sheathing, lack of gable-end sheathing, and/or garages without anchor bolts.
- Lack of garage-wall sheathing and limited use of plywood, with most of the sheathing being fiberglass or hardboard siding.
- Observations from recent tornadoes Tuscaloosa, AL, April 27, 2011, (van de Lindt et al. 2012); Joplin, MO, May 22, 2011 (Prevatt et al. 2012) and Moore, OK, May 20, 2013, (Graettinger et al. 2014; LaFave et al. 2016) show that the structures often did not have continuous load paths from the foundation to the roof.

When a house is hit by wind-borne debris, the internal pressure changes causing progressive damage of the structure. The failure progression depends on where the debris impacts occurred, the relative location of the house to the tornado track, and the orientation of the house with respect to the tornado path.

In order to document the damage due to tornadic winds, the assessment team collected damage data in a neighborhood at the western edge of Moore, as indicated by the circle on Fig. 14.2. This fairly new neighborhood was located on Kyle Dr. and SW 151st Street and built around 2006. The buildings in this neighborhood were believed to be less affected by wind-borne debris because the tornado traveled from west to east, and the area west of this neighborhood is flat open fields with only a few residential structures that were not damaged. Therefore, it was believed that the damage to the houses in this neighborhood mostly came from tornado wind loading rather than debris impact. The houses in this area had similar structural configuration, which made it easier to identify the wind field from the damage patterns as compared to other areas.

Figure 14.3a shows the damage to residential structures on SW 151st Street looking west. This figure shows houses on both sides of the street, and one can see that all homes on the left side still have garage doors and sustained little or no damage, while homes on the right side of the street lost garage doors that blew inward. Since all the garage doors had similar design and were installed on rails inside the door opening, the lateral reinforcement bars were placed inside the door (Fig. 14.3b), and these garage doors had higher capacity when experiencing negative pressure (outward load) as compared to lower capacity with positive pressure (inward load). This is because in the outward load case the load is transferred to the wall and rails through compression, while in the inward load case, the rails failed by bending into the garage (see Fig. 14.3b). With this reasoning, it can be seen from Fig. 14.3a that the residential houses in this picture experienced mainly one wind direction as the tornado passed this street, and the garage doors on the left were on the leeward walls (negative wind pressure), while those on the right side of the street were on the windward walls (positive pressure). This behavior can also be seen on streets where the direction of the street changes, and the garage doors change from leeward to windward as shown in Fig. 14.3c. This means that when the tornado passed this area, these houses mainly experienced horizontal winds (the failure pattern was similar to straight line winds seen during hurricanes). For those houses having garage doors on the windward side in straight wind, the garage doors often fail before roof sheathing panels.

Homes on Kyle Dr., a north-south street shown in the circled area on Fig. 14.1, indicated an increase in damage when located closer to the centerline of the tornado track. This damage pattern is similar to what was observed in the aftermath of the Tuscaloosa and Joplin tornadoes. Figure 14.4 shows the damage pattern of houses on Kyle Dr., in which the camera was pointed to the south looking away from the centerline of the tornado. It should be noted that the tornado vortex was counterclockwise, and this neighborhood was on the south side of the tornado centerline. Therefore, the houses on the south edge of the damage area experienced west-to-east winds, which is right to left in Fig. 14.4. In Fig. 14.4, the houses on the right (west) side near the middle of the street had the roof fail before the garage door failed. Failure of the roof happened on the leeward side, and the roof was intact on the windward side. The garage doors, on the leeward side, were not damaged on these homes. Similar damage was observed on the roofs of the houses on the left

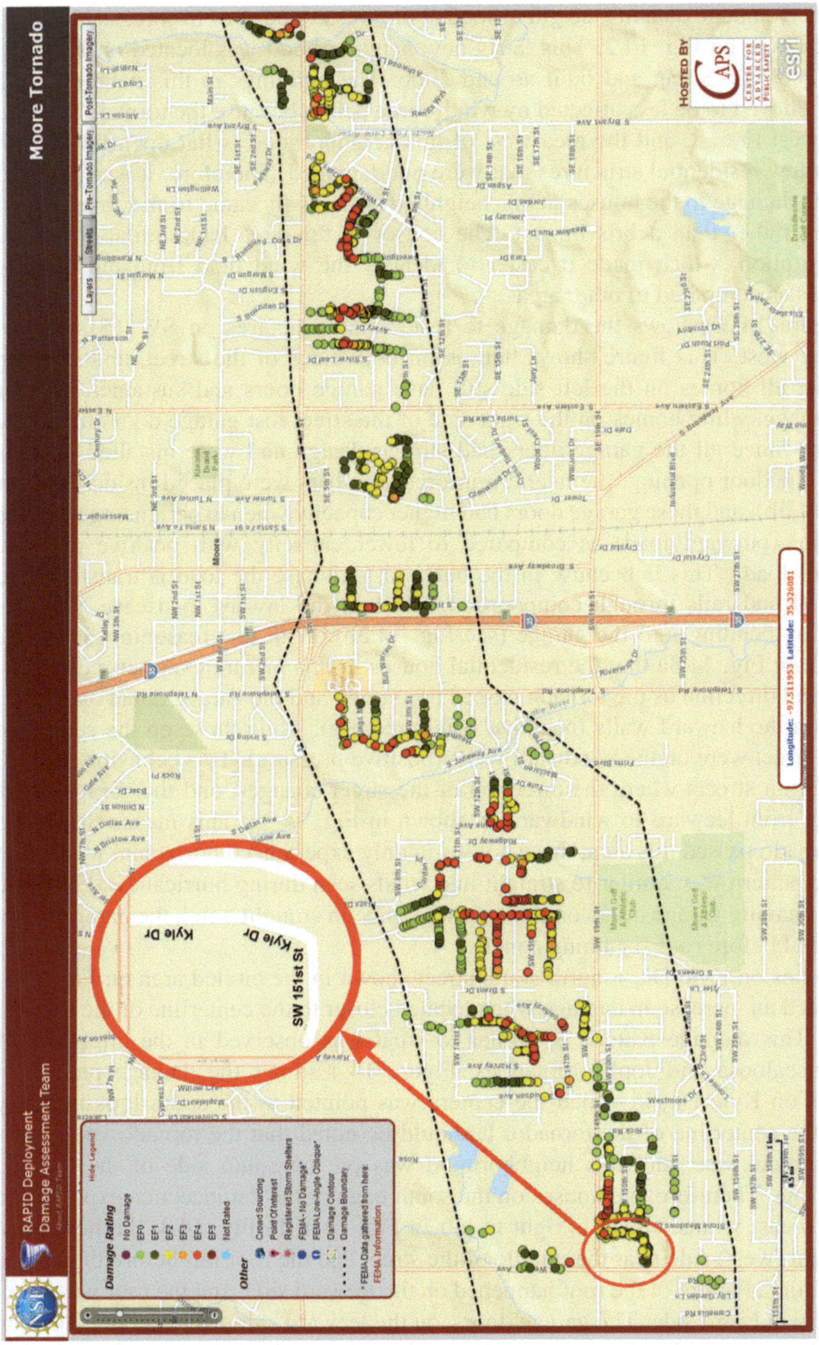

Fig. 14.2 Overview map of the Moore tornado path, with *circle* showing the neighborhood used for the detailed damage investigation (Dao et al. 2014)

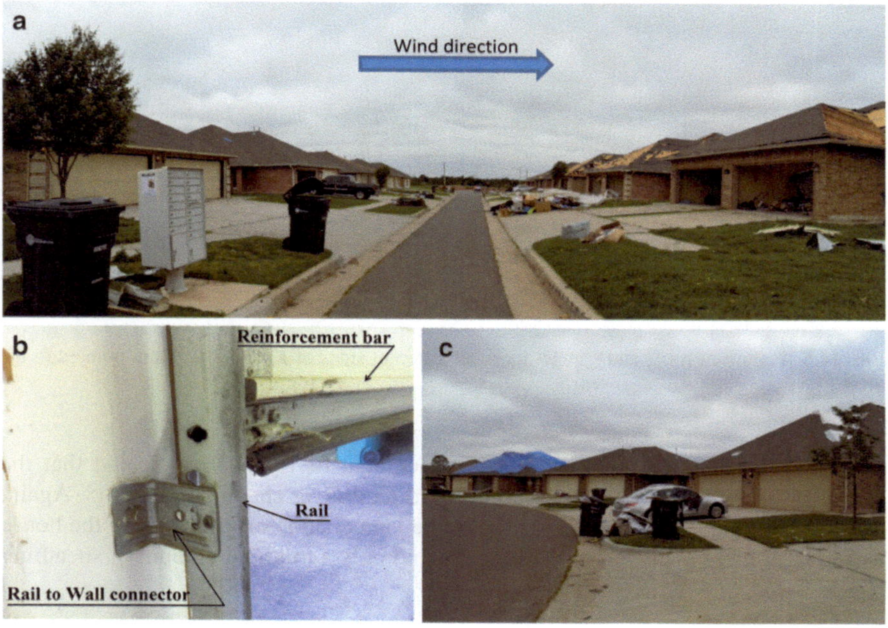

Fig. 14.3 Different garage door damage levels on the same street. (**a**) Leeward (*left*) and windward (*right*) sides of SW 151st Street. (**b**) Garage door supports and configuration. (**c**) Change of garage door damage level when home orientation changes from leeward to windward direction

Fig. 14.4 Damage pattern on Kyle Dr., Moore, Oklahoma (camera pointed to the south)

(east) side of the street (failure happened on the leeward roof, but the windward roof was intact), while the garage doors on the left side of the street all failed. This is because roof sheathing panels are weaker in uplift and stronger in compression due to nail withdrawal capacity. This failure pattern indicates that these houses experienced higher wind speed than at the edge of the damage area, and mainly in one direction of horizontal wind. It can also be seen in Fig. 14.4 that garage doors and roofs failed on both sides of Kyle Dr. to the north, where the photo was taken, and several walls were collapsed in the house on the northeast end of Kyle Dr. (far left in Fig. 14.4).

Fig. 14.5 Failure of roofs and garage doors on different sides of Kyle Dr. (camera pointed to the north)

A closer inspection near the northern segment of Kyle Dr. indicated that the homes in this area experienced wind in one direction, as shown in Fig. 14.5. Again, the house on the left had roof failure and the garage door survived, while the house on the right had the garage door blown in, and only a minimal amount of sheathing panels failed.

14.1.3 Current ASCE-7 Wind Load and History of Wind Engineering

Wind-engineering studies appear to be recorded first in 1759 when Smeaton (1759–1760) attached in a small-scale windmill model to the end of a rotating arm to do research on windmill sails. Since then, wind engineering has become an integral part of hydraulic and aerospace engineering, but the first US National Conference on Wind Engineering (Roshko 1970) only occurred less than 50 years ago. It took more than a century for wind engineering to emerge as a new engineering discipline because during the period there was a lack of a reliable methodology for quantification of wind characteristics and wind effects on structures. This barrier had been removed when the application of physical modeling to wind-engineering problems was invented by Cermak (1975) when he published his 1974 ASME Freeman Lecture "Application of Fluid Mechanics to Wind Engineering." The publication indicated the application of fluid mechanics in the development of boundary-layer wind tunnel (BLWT) which could be used to model wind loads on structures and wind flow in the boundary layer. This is the first time a reliable methodology for quantification of wind characteristics was introduced in wind engineering to study wind effects on structures. Since the invention of BLWT, the research on wind effects on structures came to a new era and led to many publications, codes, and standards; these includes wind load effects on bridges (e.g., Bienkiewicz et al. 1981; Cermak et al. 1981), buildings (e.g., Cermak and Peterka

1984; Kopp et al. 2008; ASCE-7 (American Society of Civil Engineers) 2010), and special structures (e.g., ASCE-7 2010).

Currently, the procedures for wind load calculation specified by ASCE-7 provide wind pressure and forces for the design of main wind force-resisting system (MWFRS) and for the design of components and cladding (C&C) for buildings and other structures. The procedures include the determination of wind velocity pressure, gust factor, wind directionality, and pressure or force coefficient. For the selection of pressure or force coefficient, it is often based on the results from research using BLWT. The wind load is then combined with other loads such as live load, dead load, earthquake load, snow load, etc., for use in structural design. Therefore the wind load is used in design for the purpose of life safety only, and the design procedure does not include other expectation of building performance. Also, the ASCE-7 wind load calculation procedures imply for general wind, and application to east coastal area where it was heavily affected by hurricane wind is accounted through assigning higher wind speed. And tornado wind loads are not considered in ASCE-7 wind load procedures.

14.2 Proposals for Performance-Based Wind Engineering

Performance-based design (PBD)has been defined many ways over the last decade with perhaps the most general definition being provided by Ellingwood (1998) as "an engineering approach that is based on (1) specific performance objectives and safety goals of building occupants, owners, and the public, (2) probabilistic or deterministic evaluation of hazards, and (3) quantitative evaluation of design alternatives against performance objectives; but does not prescribe specific technical solutions."

In the USA, performance-based design has been focused primarily on seismic, fire, and manufacturing engineering. PBD is, by and large, felt by most to be a system-level philosophy that allows inclusion of system-level behavior including the improvement in performance as a result of this assertion. However, in wind engineering most failures are understood to be at the component and subassembly level. A recent paper by Ellingwood et al. (2006) highlights the current status and future challenges for PBD for wood including performance-based wind engineering (PBWE). In that paper it was stated that guidelines for PBWE do not currently exist in the USA. It was also stated that extreme winds (with the exception of tornadic winds) are not viewed as a life safety issue in force-based design primarily because of the opportunity for prior warning, which is not true for earthquakes and fire. Thus, the parallel with these other hazards stops at the life safety performance expectation to some degree. Finally, it was articulated in Ellingwood et al. (2006) that models are needed which model both load and non-load-bearing walls as an integrated system.

14.2.1 Proposed Performance-Based Wind-Engineering Procedure for Hurricane Wind

The proposed PBWE procedure is an extension of the fragility studies outlined earlier (Rosowsky and Ellingwood 2002; Ellingwood et al. 2004). However, there are two distinct differences, namely, in the present study the focus is spread over four of the five performance descriptors listed in Table 14.1 and a system-level finite element model is used to more accurately address structural instability issues related to collapse. Returning to Table 14.1, five performance descriptors are proposed (van de Lindt et al. 2007). To date, only two of these have been addressed in previous studies: continued occupancy and life safety. Continued occupancy is assumed herein to correspond to loss of the first sheathing panel which is consistent with previous studies. Figure 14.6 shows a photograph of what, at first inspection, looks like a moderate gable opening during Hurricane Katrina. The owner was not able to remain in the structure following the hurricane, and according to the owner the

Table 14.1 Performance expectations and related model damage parameters for PBWE of wood

Performance expectation	Performance description	Model damage parameter	Study addressing issue
Occupant comfort	Little or no reduction in living/inhabitant comfort	Almost a durability issue; no damage or water entry limited to moisture, i.e., no pooling	Present study
Continued occupancy	Up to moderate reduction in comfort but no threat to safety or injury. Electrical, plumbing, and egress still present	Loss of first gable or roof sheathing panel	Lee and Rosowsky (2005); Ellingwood et al. (2004); present study
Life safety	Structural integrity is questionable; significant risk of serious injury might occur; safety normally provided is not present	Roof truss-to-wall connection failure; supporting column/post failure	Ellingwood et al. (2004); present study
Structural integrity	Visible signs of structural distress, i.e., permanent deformation and structure not safe	Collapse of roof; loss of lateral capacity	Present study
Manageable loss	Cost to repair structure is below a selected percentage of reconstruction/replacement value. This is dependent on numerous factors and is often the result of rainwater intrusion	Loss fragility based on the assembly of damaged components	Not addressed, but would likely require assembly-based vulnerability or other method

Fig. 14.6 Gable-end damage during Hurricane Katrina

2005 insurance estimate was equal to the 1998 purchase price of the home (van de Lindt et al. 2007). If this structure had not been insured for wind (and subsequent water) damage, this would most likely exceed what can be called "manageable loss" for most homeowners. Manageable loss is beyond the scope of the work presented herein, but can best be explained as the upper limit of the cost that a homeowner can (or is willing) to pay (whether borrowed or out of pocket) to be able to live in the structure comfortably. In Table 14.1 this is indicated as some percentage of the reconstruction/replacement value for repair. The concept of continued occupancy refers to the owner's ability to inhabit the dwelling following the event.

Life safety is perhaps the most difficult to define, but is summarized here as being a condition in which the safety normally afforded by a structure is no longer present. For wind damage, this can be characterized as failure of the roof-to-wall connection or supporting column/post failure. Figure 14.7 shows the collapse of a porch overhang as a result of poor (or no) anchorage during Hurricane Katrina. The life safety issue in this case arises from the fact that the joists frame back into the ceiling of the first level and failure then occurred within the living portion of the structure.

Two additional performance expectations that have not been explicitly addressed to date are as follows. Structural integrity, which can be summarized as the state at which the structure shows significant signs of distress, may include the collapse of the roof or the loss of lateral capacity either locally or globally. Although the general consensus is that complete loss of lateral capacity from wind load is rare, it is possible as evidenced by Fig. 14.8. This is a convenience store in

Fig. 14.7 Loss of a porch overhang due to lack of support post anchorage

Fig. 14.8 Collapse of a wood-frame (metal clad) building

Mississippi that was literally blown over in Hurricane Katrina. van de Lindt et al. (2007) describe this failure with the following sequence: The roof uplifted and there was a loss of roof sheathing. The front glass window "blew out" and the roof trusses collapsed. The trusses were tied to the walls with hurricane clips, but without roof sheathing they did not provide lateral stability for the trusses alone. The structural instability performance expectation includes life safety, meaning none of the performance expectations are necessarily mutually exclusive. For example, if the structure collapses, i.e., does not meet the expectation of structural integrity, clearly all of the other expectations were not met, albeit they may be tied to different hazard intensities. The performance expectation of life safety may not be met even when there is no local or global collapse.

The other performance expectation which has not been addressed is occupant comfort. This is intended to mean following the event since it is not anticipated that the homeowner would necessarily be present during a hurricane. In the present study, it is proposed to model this as water penetration at roof sheathing edges resulting in potential mold and other issues related to moisture. A detailed finite element model, which utilizes a new 6-DOF fastener model developed by Dao and van de Lindt (2008), is used to detect/model sheathing uplift and help develop fragilities as a function of edge uplift.

14.2.1.1 Performance Expectations

Consider Fig. 14.9 whose concept is adopted from performance-based seismic design. Current force-based design utilizes a single peak 3-second gust and designs with some level of safety or with both a load and resistance factor (e.g., ASCE 16 1996). In Fig. 14.9 the leftmost line corresponds most closely to current force-based design values. However, it is important to note that simply by defining multiple performance objectives the design philosophy is no longer the same. For this leftmost line, returning to the performance expectations and damage parameters in Table 14.1, and for a peak gust of 90 mph (145 km/h), an owner would expect no damage and no water intrusion. For a well-designed and well-constructed residential structure, this is typically the case provided wood is in a non-decaying state and fasteners are spaced appropriately. For the same leftmost line (squares in Fig. 14.9), one would expect to provide life safety at 170 mph (270 km/h) meaning no loss of truss-to-wall connections or supporting post failures. Although this is the performance expectation described here, the method described below is probabilistic, and thus there is always some probability of exceeding such an expectation, as examined by Ellingwood et al (2004). Therefore some level of exceedance probability must be selected, which for the present study is set at 50 % for illustrative purposes.

The concept of PBWE can be further explained by again returning to Fig. 14.9. Now, focusing on the rightmost line with triangles, one can see that no water intrusion or damage would be expected at 130 mph (210 km/h), life safety expected at 210 mph (335 km/h), and structural integrity expected at 250 mph (400 km/h).

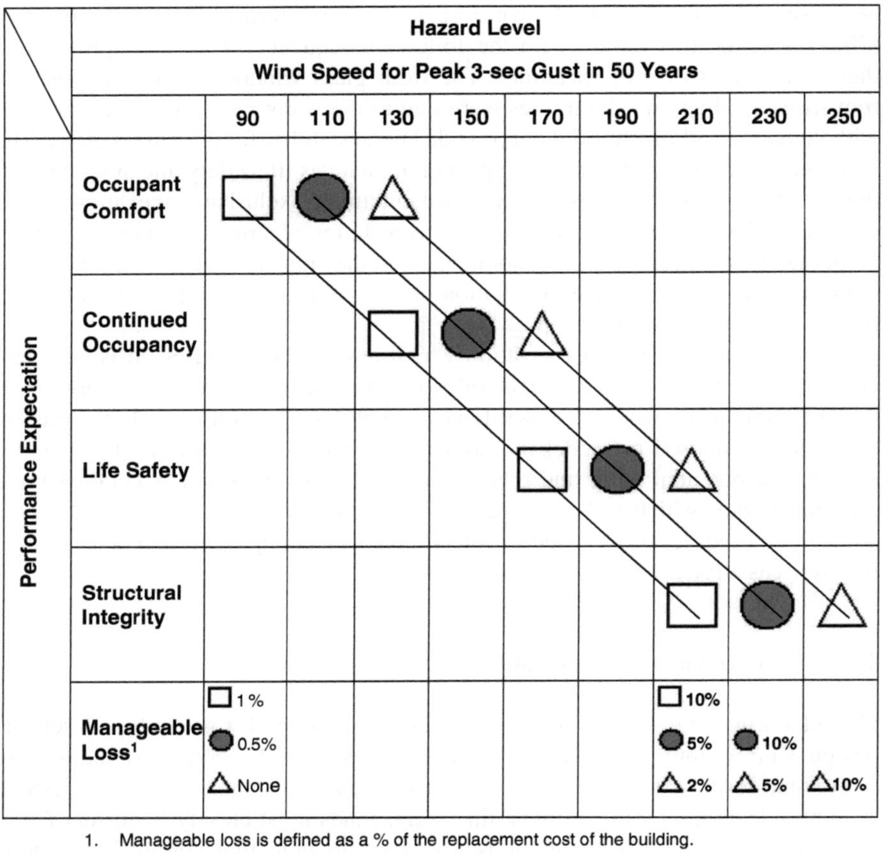

Fig. 14.9 Example of various levels of building performance as a function of hazard level. [1]Manageable loss is defined as a percentage of the replacement cost of the building

Of course, it should be noted that at wind speeds this high, debris acting as airborne missiles will ultimately have to be considered in performance-based design but is not here. The force exerted by the debris is understood, but unfortunately little beyond speculation is available for occurrence modeling since it is related to many things beyond the engineer's control, e.g., equipment left out in the open. Several studies have examined this concept with some recent work being completed by Lin and Vanmarcke (2008). Finally, recall that the pressure varies as the square of the wind velocity, so although the various performance expectations are linear when expressed as pressures and subsequently in terms of strength requirements, the force exhibited by these wind speeds increases substantially from occupant comfort to even continued occupancy.

Perhaps the most important aspect of PBWE will eventually be addressing manageable loss through modeling and detailed comparison of structural performance to estimated losses during high-wind events. For example, in Fig. 14.9 how would one ensure that at 210 mph the performance expectation level for the leftmost line (indicated by squares) only has losses not to exceed 10 % of the replacement value of the structure? Further, for the "triangle" performance level, these would not be expected to exceed 5 % at 210 mph. As mentioned earlier, this is not quantitatively addressed in this chapter. To accurately assess the damage in terms of dollars would require the full inclusion of damage due to wind-borne debris and a mechanism to assess volume and affect of rainwater entry. This would likely include an approach such as assembly-based vulnerability (Porter et al. 2001; Pei and van de Lindt 2009; Dao and van de Lindt 2012; van de Lindt and Dao 2012).

14.2.1.2 Additional Considerations

Articulating, or quantifying, the performance expectations of a peak 3-second gust in 50 years does not address other "failure" mechanisms that may occur during a hurricane as a result of the duration (sometimes in excess of 8 h). Figure 14.10 shows a photograph of a hurricane clip that lasted almost 4 h during Hurricane Katrina and finally failed (van de Lindt et al. 2007). Another factor is roof coverings and siding, which are not designed to carry wind load, but are envisioned to protect the structural components such as paneling from direct water exposure during storms, thus helping to maintain the integrity of the building envelope. In this study, the nonstructural

Fig. 14.10 Photo of missing fasteners in a hurricane clip. The roof lifted off after several hours of uplift pressures

siding and roof coverings are assumed to have been removed by the wind prior to the analysis performed on the wood components and assemblies. Finally, although considered beyond the scope of the current study, it is again stressed that it is imperative that PBWE ultimately considers projectiles and breakage of windows for an accurate assessment of risk caused by wind events.

14.2.2 Needs for PBWE of Wood-Frame Buildings

Many recent studies have focused on a framework for PBWE (e.g., van de Lindt and Dao 2009; Spence et al. 2015; McCullough and Kareem 2011; Griffis et al. 2012; Muthukumar et al. 2012; Unnikrishnan et al. 2012), but none have yet to fully develop and propose a methodology that leads close enough to code adoption. In most of the frameworks proposed, the fragility methodology was used to measure the performance levels either at the component or system levels of a building subjected to high wind. The construction of component or system fragility was often based on either reliability theory or Monte Carlo simulation for the loads and system resistance. A statistical approach is suitable for the frameworks with current available technologies in wind engineering and can be used for insurance policy decisions. However, for engineering practice it is likely that PBWE for wood-frame buildings will utilize a semi-prescriptive approach with certain design decisions achieving some level of statistical performance level.

To make PBWE for wood-frame buildings a viable option in engineering practice and effectively move beyond merely the applications of statistics and reliability theory, understanding building performance in high-wind events is essential. The data collected from high-wind events gives an overview of building performance for these, often vulnerable buildings to wind load, and the damage patterns in high-wind events allow forensic investigation into the failure mechanisms. However, the local wind speeds, quality of building construction, specific design of each individual building are often neglected in the collected data; in fact, they are often unknown once the damage has occurred. This is because the large amount of data needs to be collected in a short period before the sites are cleaned to begin the recovery process for a community. This is the reason most current wind load standards or codes are often based on experimental results from wind tunnel experiments using rigid-body models. Currently, wind load simulation is based on a wind tunnel test either for small- or large-scale structures. For small-scale wind tunnel tests, the wind load is often measured on a scaled solid model based on similitude laws in aerodynamics and fluid mechanics. A small model allows researchers to simulate the effects of terrain roughness and aerodynamic behavior around the tested structures for a relatively low cost compared with that for large-scale tests. But because the model is solid, the building performance under the wind load cannot be properly evaluated. For these reasons, several large facilities for wind load simulation of large- or full-scale structures have been recently built. Even though the development of large-scale wind load simulation helps to improve the understanding the performance of buildings under the load, it has some drawbacks:

(1) The construction of large-scale wind load simulation laboratories is expensive.
(2) The cost of experiments is often very high due to the cost of constructing large-scale structural models as well as the cost of operating the tests.
(3) It is difficult and sometimes impossible to model the effects of aerodynamic behavior around the prototype from adjacent structures in a large-scale wind load simulation laboratory; therefore the effects of wind turbulence due to the vortex in the built environment cannot be modeled properly, thereby making it difficult to evaluate wood-frame building performance in high-wind events with the effects from surrounding buildings in a community (including wind-borne debris impact).

14.3 Research Needs for Development of Wood-Frame PBWE

In order to develop PBWE for wood-frame building, the knowledge of how these types of buildings perform in high-wind events needs to be investigated further. This includes understanding of the wind loading and the building response in high wind, including both structural and nonstructural components such as the building envelope, finishing products, and building contents. As mentioned, atmospheric boundary-layer wind tunnels can be used to model the wind environment in the boundary layer properly, but cannot model the building performance accurately. This is because they do not account for the nonlinear behaviors of building materials and structures. To solve these drawbacks mentioned earlier, it will be necessary for the wind-engineering wood-frame building research community to develop a new testing method that includes both the wind environment and building behaviors in high wind.

In order to be able to estimate the performance of building structures within an extreme wind field, the mechanics behavior of the interaction between the high-wind field and structures needs to be explored numerically and then validated with these experimental studies. Although there has been a significant volume of studies focusing on numerical fluid–structure interaction (FSI) problems over the last several decades, a comprehensive study of such problems remains a challenge due to the strong nonlinearity of the problem and multidisciplinary approach required for a solution (Hou et al. 2012). For this reason, purely analytical solutions to the model equations for the interaction between high wind and structures are likely impossible to obtain. Recently, some limited laboratory studies (Haan et al. 2010) have focused on the pressure coefficients on a rigid structural model in which the interaction between the extreme wind field and structural model was not included. Thus, to investigate the fundamental physics involved in the complex interaction between fluids and structures, numerical methods should also be employed (Hou et al. 2012).

This may require an algorithm to model the interaction between the extreme wind field and structural dynamics.

After the numerical modeling and experimental tools are available, these can be applied into one of the PBWE frameworks that have been developed recently. This process may take several generations of PBWE before it can comprehensively be applied in engineering practice.

14.4 Closure

PBWE for wood-frame buildings is a relatively new topic in both research and design and owes its impetus to damaging hurricanes over the last 15 years and supercell tornado outbreaks across the Central and Southern USA. It is clear that PBWE will not take the same format as design for heavier and stiffer buildings made of steel frames and/or concrete walls/frames. Those types of buildings typically keep their structural system intact but lose windows, covering, and curtain walls, whereas wood-frame buildings suffer loss of components and cladding and their main wind force resisting system. So, while the underlying hybrid research required will be robust, it will be necessary to use a semi-prescriptive approach to truly implement PBWE in wood-frame design practice.

References

AF&PA/ASCE 16-95. (1996). *Standard for load and resistance factor design (LRFD) for engineered wood construction*. American Society of Civil Engineers, Reston, VA.

ASCE-7 (American Society of Civil Engineers), (2010). *Minimum design loads for buildings and other structures* (ASCE Standard 7-10). Reston, VA: American Society of Civil Engineers.

Bienkiewicz, B., Cermak, J. E., & Peterka, J. A. (1981, July). A new technique of modeling atmospheric turbulence for wind-tunnel tests of bridge models. In *Proceedings of the Fourth U.S. National Conference on Wind Engineering Research* (pp. 113–116). University of Washington, Seattle, WA.

Cermak, J. E. (1975, March). Applications of fluid mechanics to wind engineering—A Freeman Scholar lecture. *Transactions of the ASME, Journal of Fluid Engineering, I, 97*(1), 9–38.

Cermak, J. E., & Peterka, J. A. (1984, August). Wind-tunnel testing of highrise buildings. In *Proceedings of ASCE Specialty Conference Hurricane Alicia: One Year Later* (pp. 211–229). Galveston, TX.

Cermak, J. E., Bienkiewicz, B., & Peterka, J. A. (1981). Active Modeling of Turbulence for Wind-Tunnel Studies of Bridge Models, *Technical Report CER80-81JEC-BB-JAP43,* Fluid Dynamics and Diffusion Laboratory, Colorado State University, Fort Collins, Colo., Mar.

Dao, T. N., Graettinger, A., Alfano, C., Gupta, R., Haan, F. L., Prevatt, D., et al. (2014, April 3–5). Failure progression analysis of observed residential structural damage within a Tornado wind field. In *ASCE Structures Congress 2014*. Boston, MA.

Dao, T. N., & van de Lindt, J. W. (2008). New nonlinear roof sheathing fastener model for use in finite-element wind load applications. *ASCE Journal of Structural Engineering, 134*(10), 1668–1674.

Dao, T. N., & van de Lindt, J. W. (2010). Methodology for wind-driven rainwater intrusion fragilities for light-frame wood roof systems. *ASCE Journal of Structural Engineering, 136*(6), 700–706.

Dao, T. N., & van de Lindt, J. W. (2012). Loss analysis of woodframe buildings during hurricanes. I: Structure and hazard modeling. *ASCE Journal of Performance of Constructed Facilities, 26*(6), 729–738.

Ellingwood, B. (1998). Reliability-based performance concept for building construction. In *Structural Engineering World Wide 1998*. Paper T178-4. Elsevier (CD-ROM).

Ellingwood, B. R., Rosowsky, D. V., Li, Y., & Kim, J. H. (2004). Fragility assessment of light-frame wood construction subjected to wind and earthquake hazards. *ASCE Journal of Structural Engineering, 130*(12), 1921–1930.

Ellingwood, B. R., van de Lindt, J. W., Gromala, D., Rosowsky, D. V., Gupta, R., & Pryor, S. (2006). Performance-based engineering for light-frame wood construction in the United States: Status and challenges. In *Proceedings of the 2006 World Conference on Timber Engineering*. Portland, OR.

FEMA. (2015). Link. https://www.fema.gov/sandy-recovery-office. Accessed 17 July 2015.

Graettinger, A. J., Ramseyer, C., Freyne, S., Prevatt, D., Myers, L., Dao, T.N., Floyd, R., Holliday, L., Agdas, D., Haan, F.L., Richardson, J., Gupta, R., Emerson, R. & Standohar-Alfano, C. (2014). Tornado Damage Assessment in the Aftermath of the May 20th 2013 Moore Oklahoma Tornado, *Report to the National Science Foundation, 133* (2014), http://esridev.caps.ua.edu/MooreTornado/Images/MooreTornadoFinalReport.pdf

Griffis, L., Patel, V., Muthukumar, S., & Baldava, S. (2012) A Framework for Performance-Based Wind Engineering. *Advances in Hurricane Engineering,* pp. 1205–1216

Haan, F. L., Jr., Balaramudu, V. K., & Sakar, P. P. (2010). Tornado induced wind loads on a low-rise building. *ASCE Journal of Structural Engineering, 136*(1), 106–116.

Hou, G., Wang, J., & Layton, A. (2012). Numerical methods for fluid–structure interaction: A review. *Communications in Computational Physics, 12*(2), 337–377.

Kopp, G. A., Oh, J. H., & Inculet, D. R. (2008). Wind-induced internal pressures in houses. *ASCE Journal of Structural Engineering, 134*(7), 1129–1138.

LaFave, J. M., Gao, Z., Holder, D. E., Kuo, M. J., & Fahnestock, L. A. (2016). Commercial and residential building performance during the May 20, 2013, Tornado in Moore, Oklahoma. *ASCE Journal of Performance of Constructed Facilities, 30*(2), Article # 04014210.

Lee, K. H., & Rosowsky, D. V. (2005). Fragility assessment for roof sheathing failure in high wind regions. *Eng. Struct., 27*(6), 857–868.

Lin, N., & Vanmarcke, E. (2008). Windborne debris risk assessment. *Probabilistic Engineering Mechanics, 23*, 523–530.

McCullough, M. & Kareem, A. (2011). A Framework for Performance-Based Engineering in Multi-Hazard Coastal Environments. *ASCE Proceedings of the Structures Congress 2011,* pp. 1961–1972.

Muthukumar, S., Baldava, S., & Garber, J. (2012). Performance-Based Evaluation of an Existing Building Subjected to Wind Forces. *Advances in Hurricane Engineering,* pp. 1217–1228.

National Geographic. (2015). Link. http://voices.nationalgeographic.com/2012/11/02/a-timeline-of-hurricane-sandys-path-of-destruction/. Accessed 17 July 2015.

National Institute of Standards and Technology (NIST). (2006, June). *Performance of physical structures in Hurricane Katrina and Hurricane Rita: A reconnaissance report* (NIST Technical Note 1476). Gaithersburg, MD: National Institute of Standards and Technology.

Pei, S., & van de Lindt, J. W. (2009). Methodology for earthquake induced loss estimation: An application to woodframe buildings. *Structural Safety, 31*, 31–42.

Pielke, R. A., Gratz, J., Landsea, C. W., Collins, D., Suanders, M. A., & Musulin, R. (2008). Normalized hurricane damage in the United States: 1900–2005. *Natural Hazards Review, ASCE, 9*(1), 29–42.

Pielke, R. A. J., & Pielke, R. A. S. (1997). *Hurricanes: Their nature and impacts on society*. Chichester: Wiley.

Porter, K., Kiremidjian, A. S., & LeGrue, J. S. (2001). *Assembly-based vulnerability of buildings and its use in performance evaluation* (p. 23). Oakland, CA: Earthquake Engineering Research Institute.

Prevatt, D. O., van de Lindt, J. W., Back, E., Graettinger, A. J., Pei, S., Coulbourne, W., et al. (2012). Making the case for improved structural design: The Tornado outbreaks of 2011. *Leadership and Management in Engineering, 12*, 254–270.

Roshko, A. (Ed.). (1970). Wind loads on structures. In *Proceedings of the First U.S. National Conference on Wind Engineering*. California Institute of Technology, Pasadena, CA.

Rosowsky, D. V., & Ellingwood, B. R. (2002). Performance-based engineering of wood frame housing: A fragility analysis methodology. *Journal of Structural Engineering, 128*(1), 32–38.

Simmons, K. M., Sutter, D., & Pielke, R. (2013). Normalized tornado damage in the United States: 1950–2011. *Environmental Hazards, 12*(2), 132–147.

Smeaton, J. (1759–1760). An experimental investigation concerning the natural powers of water and wind. *Philosophical Transactions of the Royal Society of London, 51*, 100–174.

Spence, S. M. J., Bernardini, E., & Kareem, A. (2015). A first step towards a general methodology for the performance-based design of wind-excited structures. *ASCE Proceedings of the Structures Congress 2015*, pp. 1482–1493.

Unnikrishnan, V., Barbato, M., Petrini, F., & Ciampoli, M. (2012) Probabilistic Performance Based Risk Assessment Considering the Interaction of Wind and Windborne Debris Hazards. *Advances in Hurricane Engineering*, pp. 1183–1193.

van de Lindt, J. W., & Dao, T. N. (2009). Performance-based wind engineering for wood-frame buildings. *ASCE Journal of Structural Engineering, 135*(2), 169–177.

van de Lindt, J. W., & Dao, T. N. (2012). Loss analysis of woodframe buildings during hurricanes. II: Loss estimation. *ASCE Journal of Performance of Constructed Facilities, 26*(6), 739–747.

van de Lindt, J. W., Graettinger, A., Gupta, R., Skaggs, T., Pryor, S., & Fridley, K. (2007). Performance of woodframe structures during Hurricane Katrina. *ASCE Journal of Performance of Constructed Facilities, 21*(2), 108–116.

van de Lindt, J. W., Pei, S., Dao, T., Graettinger, A., Prevatt, D. O., Gupta, R., et al. (2012). Dual-objective-based Tornado design philosophy. *ASCE Journal of Structural Engineering, 139*(2), 251–263.

Chapter 15
An Engineering-Based Approach to Predict Tornado-Induced Damage

Xinlai Peng, David B. Roueche, David O. Prevatt, and Kurtis R. Gurley

Abstract The tornado risk assessment methodology currently used by both private and public agencies utilizes empirically derived loss models that rely on historical claims data for predicting future effects of tornadoes. The accuracy of these empirical models is dependent on many factors, including the quality and quantity of available historical data, accuracy of the tornado intensity models, and the universality of applying those empirical models from one region to another. A more rigorous approach may be the development of engineering-based damage assessment models, made applicable to construction in any region and to any tornado that varies in size and strength. This chapter presents a framework for an engineering-based tornado damage assessment (ETDA) for low-rise buildings. The model predicts damage on the most vulnerable sector of the built environment, nonengineered residential buildings. The model components include a translating tornado vortex model, a tornado-induced wind load calculation approach, a probabilistic wind-borne debris impact model, and a time-variant model for internal pressure changes within the structure. The time evolution of structural damage to a building is determined using successive time steps of component level wind loading vs. structural resistance as the tornado translates past the building. The output of this model is a percentage damage index for each component and the overall building damage ratio. The ETDA model is illustrated using four houses damaged in the 2011 Joplin, MO, tornado. Predicted damage using the ETDA model is in good agreement (within 15 %) of post-tornado damage observations reported by the third author.

15.1 Introduction

Tornadoes pose a great threat to the infrastructure, residential housing, and people in their paths. The average annual loss from tornadoes, over the period of 1949 through 2006, was nearly $1 billion dollars (Changnon 2009). In the United States, nearly

X. Peng (✉)
Forensic Engineer, Unified Investigations & Sciences, Fort Lauderdale, FL, USA
e-mail: xpeng@uis-usa.com

D.B. Roueche • D.O. Prevatt • K.R. Gurley
University of Florida, Gainesville, FL, USA
e-mail: david.roueche@ufl.edu; dprev@ce.ufl.edu; kgurl@ce.ufl.edu

P. Gardoni, J.M. LaFave (eds.), *Multi-hazard Approaches to Civil Infrastructure Engineering*, DOI 10.1007/978-3-319-29713-2_15

1200 tornadoes occur yearly, on average producing 90 fatalities and 1500 injuries (Simmons et al. 2012). Improved forecasting and warning systems have helped reduce injuries and loss of life, while economic losses are not mitigated by advanced warning. However, single tornado losses of $1 billion or more are becoming more frequent (Folger 2011; Graettinger et al. 2013), and the death tolls in recent violent tornadoes have also been high. Those recent events in Tuscaloosa, AL, Joplin, MO, and Moore, OK, have prompted a renewed interest to better understand the tornado and its interactions with man-made infrastructure, with the goal of reducing the impact of these events (Burgess et al. 2014; Dao et al. 2014; Graettinger et al. 2013; Prevatt et al. 2012).

To mitigate economic loss from tornadoes requires widespread creation of tornado-resilient communities. A major challenge in developing tornado-resilient housing is to demonstrate to the public both the need and benefits of effective retrofits, which will increase the likelihood that homeowners will strengthen their weak existing buildings and better prepare their community to recover from the impact of a tornado. Thus, a reliable and robust tornado damage assessment model is needed to provide a rational methodology for estimating damage in a "what if" tornado scenario and present the benefits of retrofitting. Such effort focusing on how tornado hazard risk information is presented significantly affects the public perception of risk, whereby presenting risk in terms of particular scenario outcomes (rather than simply discussing the probability of occurrence) is far more impactful on the general public and likely to stimulate positive behaviors (Bonstrom et al. 2012).

Tornadoes have complex three-dimensional wind fields as well as rapidly changing, temporal variations in wind velocity. The rotational airflow around the tornado vortex results in a large atmospheric pressure deficit that produces differential pressure on the building. Previous probabilistic wind models are mainly focused on the assessment of hurricane-induced damage (e.g., Gurley et al. 2005; Vickery et al. 2006a, b; Yau et al. 2010; Dao and van de Lindt 2011; van de Lindt and Dao 2011; Barbato et al. 2013; Grayson et al. 2013). Researchers have developed probabilistic models linking tornado wind speeds to the structural damage of buildings, such as van de Lindt et al. (2012) and Amini and van de Lindt (2013). These models quantified tornado-induced loads by scaling existing codified values in ASCE 7, but the validity of this approach has not been established. The insurance industry has relied upon empirical vulnerability models developed by fitting curves to tornado claims data (Walker 2011). However, empirical models are specific to the observed structures and regions, and applicability to other structures and locations may require modification. This necessitates creating a flexible, reliable analytical approach based upon an understanding of structural loads, strength, and load path.

This study presents an engineering-based tornado damage assessment (ETDA) framework that builds upon previous wind damage assessment methodologies by incorporating a translating vortex model to describe tornado wind field and atmospheric pressure variation, a probabilistic approach for generating wind-borne debris, and a time-variant internal pressure model for a low-rise building. The proposed methodology is illustrated and validated through the estimation of damage to four houses in the 2011 Joplin, MO, tornado. This approach is capable of estimating progressive structural damage as a tornado moves over a given structure or a portfolio of homes. It allows users to explicitly track the cumulative damage to buildings caused by pressure and wind-borne debris. It provides a means of developing fragility curves of residential buildings subjected to tornadoes, which will be discussed in a follow-up study.

Section 15.2 provides details of the ETDA framework, consisting of hazard simulation, load estimation, and capacity modeling. Section 15.3 presents an application of the model, illustrating the structural damage sustained by four houses in the 2011 Joplin tornado. The primary conclusions from this study, the known limitations of the current model, and its potential impacts are then given in Sect. 15.4.

15.2 Framework of Engineering-Based Tornado Damage Assessment (ETDA)

15.2.1 Overview

Figure 15.1 depicts a plan view of the position of a tornado with respect to a building. Two coordinate systems are defined in order to determine the direction of the resultant wind relative to the building's major axis. The building center is situated at the origin of the ground-fixed coordinate system (*X*'*Y*'-axes), whereas the tornado-fixed coordinate system (*XY*-axes) is attached to the tornado center, and its principal axis (*X*) direction aligns with the tornado translating direction.

The core of the ETDA framework is a progressive assessment of structural damage as a tornado translates past a building. It is necessary to establish a spatial window for the model that is sufficiently large to capture the initiation of damage. For this study, a window of $-4R_{\max}$ to $4R_{\max}$ was found to be adequate

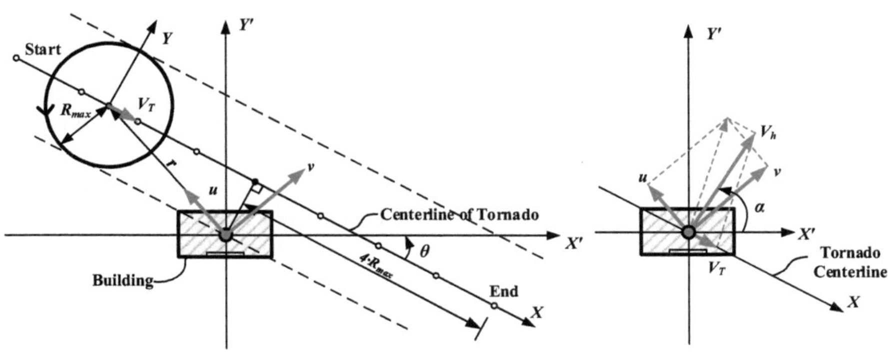

Fig. 15.1 *Left*: schematic of the progressive tornado-induced damage assessment and *right*: wind velocity vectors at the house center and definition of wind direction with respect to *X*'-axis

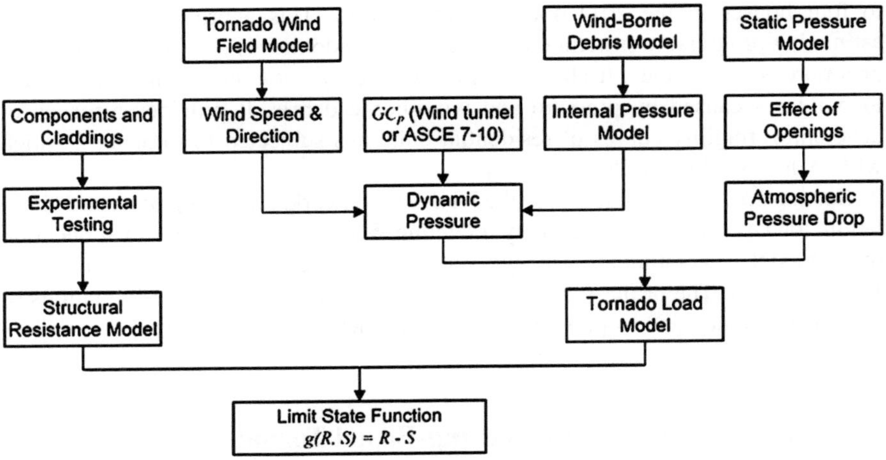

Fig. 15.2 Component level load-versus-resistance analysis for the ETDA model

to capture the full failure progression, where $R_{\max}$ is the radius to maximum wind speeds in the tornado vortex. Consequently, the step-by-step assessment process starts when the building is at a position of $X = 4R_{\max}$ and stops at a building position of $X = -4R_{\max}$. At each time step corresponding to a particular location, a component-based load-resistance analysis is conducted to determine which building components fail. Figure 15.2 summarizes the framework of the analysis at each time step. It consists of two parts: a module of simulating random tornado-induced loads (S) incorporating dynamic and static pressures (refer to Sect. 15.2.3 for details) and a simulation module of structural resistance (R) obtained either from in situ measurement or from laboratory testing. The limit-state function for a specific component is written as

$$g(R, S) = R - S \tag{15.1}$$

Component failure can be defined as the condition where $g(R, S) < 0$.

A flowchart of the ETDA framework is presented in Fig. 15.3. The preprocessing phase begins with users defining parameters regarding the house and the tornado. Structural dimensions, roof type, envelope covering, sheathing panel material, orientation, and distance of the home with respect to the tornado path are required in the definition of the house model. The information is used to assemble rectangular plan gable- or hip-roofed structures of one or multiple stories. User-defined tornado parameters include the maximum tangential and forward velocities and core radius.

Based upon the building information, damage indicator and tributary area matrices for building claddings and components are initialized. Prior to conducting progressive damage assessment, probabilistic capacity matrices for all building components and cladding, encompassing openings, wall cover, wall sheathing,

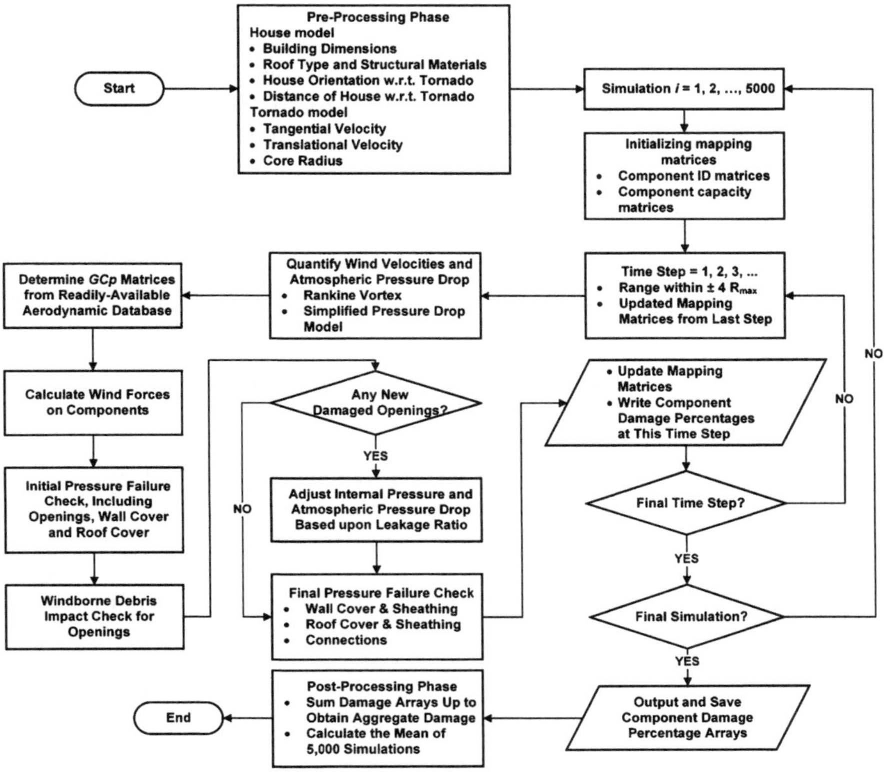

Fig. 15.3 Flowchart of the proposed tornado damage assessment model

roof cover, roof sheathing, and roof-to-wall connections, are sampled from their individual distributions, derived from laboratory testing and engineering judgment. Details of the capacity distributions are provided in Sect. 15.3.

Prior to conducting component-based damage assessment, the horizontal resultant wind velocity, wind angle, and pressure drop acting on the building need to be quantified. The tornado-induced velocity components in the x and y direction at the mean roof height and the pressure drop acting on the building are first determined from the Rankine vortex model at each time step. The resultant horizontal velocity vector is used with the GC_p matrices from a wind tunnel database to estimate the dynamic external pressures on the building. The effect of the pressure drop associated with the tornado vortex is superimposed with the dynamic pressures to describe the total wind loads on the building.

The damage assessment at each time step begins with determining the damage to glazing or openings in walls from wind pressures and wind-borne debris. Envelope breach alters internal pressure. The areas and external pressure coefficients of all breaches are used to determine the internal pressure coefficient (GC_{pi}). The modified pressures acting on the remaining building components are then compared against their capacities (sampled from probabilistic models) to evaluate the damage status. A series of checks are performed to determine if lost components will redistribute loading to adjacent components (e.g., loss of a roof-to-wall connection places additional load on adjacent connections). The integrated damage assessment ends with updating the mapping matrices of all components and compiling damage percentages of all components. These matrices are the initial conditions for analysis at the next time step. New wind-borne debris generated in the current location is also utilized at the next time step. The time-step analysis continues until the tornado center moves out of the scope of concern. The procedure is repeated for a sufficiently large number of simulations to account for the uncertainty of sampling and to obtain a robust estimate of the mean damage percentages of all components. Initially a numerical analysis was conducted to establish the sensitivity in experimental error relative to number of simulations.

A sensitivity analysis was conducted to evaluate the effects of number of simulations on the mean overall damage ratio. Damage ratios were developed for a series of simulations ranging from 100 to 100,000 (see Fig. 15.4). As expected, the standard error decreased as the number of simulations increased. Five thousand simulations were selected since the standard error of the mean was 0.5 % and each analysis run took a reasonably short computer processing time (2 min). This fast analysis time and small standard error compared very favorably with the benchmark run of 100,000 simulations that took nearly 300 min of computer time to complete. Thus 5000 simulations are used for each run in the study. The percentage of physical damage for all components is the outputs of the model that are used in post-processing.

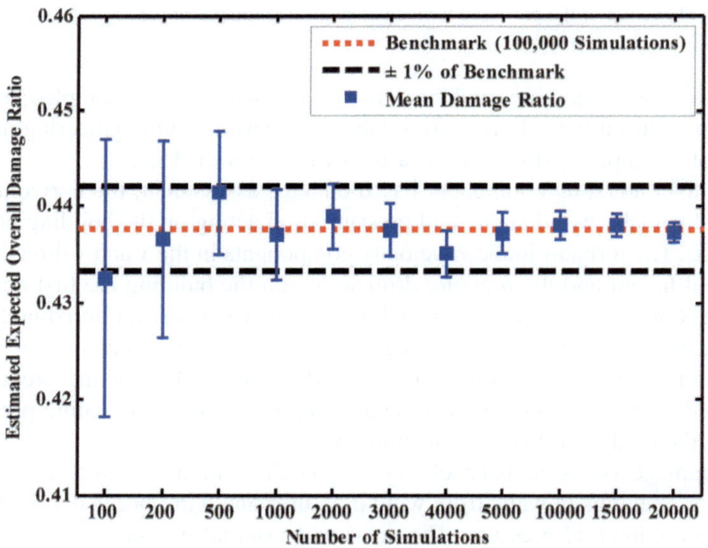

Fig. 15.4 Estimated mean damage ratio vs. number of simulations (whiskers are the 99 % confidence interval bounds of the estimate)

The following sections provide more detail on the engineering model of the tornado hazard, structural wind loading, and component capacity modules of the ETDA. Each of these modules is subject to further refinement as additional research informs future model development.

15.2.2 Tornado Hazard Model

15.2.2.1 Physical Vortex Model

Several physical vortex models have been developed to model and represent the tornado wind field, such as the viscous Burgers–Rott model and the Sullivan analytical vortex model (Wood and White 2011). The idealized theoretical Rankine vortex model is often employed to approximate the distribution of wind velocity footprint of a tornado. The distance from the house to the tornado center, r, is given by transforming the Cartesian coordinates (X, Y) to polar coordinates:

$$r = \sqrt{X^2 + Y^2} \tag{15.2}$$

The 3-s tangential velocity at the house center (origin of the coordinate system), $v(r)$, is expressed as a function of the distance from the house to the tornado center (Holland et al. 2006):

$$v(r) = \begin{cases} \frac{r \cdot V_{max}}{R_{max}}, & 0 \le r \le R_{max} \\ \frac{R_{max} \cdot V_{max}}{r}, & R_{max} \le r < \infty \end{cases} \tag{15.3}$$

in which V_{max} represents the maximum tangential 3-s gust wind velocity at the mean roof height, and R_{max} is the radius of tornado inner core, where the maximum tangential gust velocity occurs.

The relationship between radial and tangential velocities in tornadoes can vary significantly from tornado to tornado. Hoecker (1960) proposed an approximate relationship based on photogrammetric analysis of a tornado:

$$u(r) = \frac{v(r)}{2} \tag{15.4}$$

Substituting Eq. 15.3 into Eq. 15.4 yields the expression for radial velocity $u(r)$ as a function of r. Peterson (1992) suggested that tornado radial wind speed varied with the distance from tornado center in a same manner as tangential velocity, which is adopted for the empirical equation herein.

15.2.2.2 Wind Angle Relative to the Building Major Axis

The angle of the horizontal resultant wind velocity relative to the building major axis (X-axis) is critical to the determination of tornado-induced dynamic wind effects on the building. The tangential and radial wind velocities, estimated in accordance with the Rankine vortex model, and the predefined tornado forward speed add up to result in the horizontal resultant wind velocity V_h at the mean roof height. Two velocity components (V_X, V_Y) are denoted herein as the projections of V_h on the X- and Y-axes. In an effort to estimate wind angle relative to the building major axis (X‘-axis) in an efficient manner, a rotated coordinate system is introduced, as illustrated in Fig. 15.1. The new system is rotated counterclockwise by an angle, θ, from the initial coordinate system. The 2-D vector transformation equations are the following:

$$V_{X'} = V_X \cos\theta + V_Y \sin\theta \tag{15.5}$$

$$V_{Y'} = -V_X \sin\theta + V_Y \cos\theta \tag{15.6}$$

where $V_{X'}$ and $V_{Y'}$ are the coordinates or the components of the horizontal resultant wind velocity (V_h) after rotation. The purpose of using two arguments instead of one is to gather information on the signs of the vector in order to return the appropriate quadrant of the computed angle. The wind angle estimated will be employed to identify the GC_p matrices, which is detailed in Sect. 15.3.

15.2.2.3 Pressure Drop

According to the mass and momentum conservation principle, the Navier–Stokes equations for antisymmetric incompressible flow are given in cylindrical coordinates (Panton 2013):

$$\frac{1}{r}\cdot\frac{\partial(ru)}{\partial r} + \frac{\partial w}{\partial z} = 0 \tag{15.7}$$

$$\frac{\partial u}{\partial t} + u\cdot\frac{\partial u}{\partial r} + w\cdot\frac{\partial u}{\partial z} - \frac{v^2}{r} = -\frac{1}{\rho}\cdot\frac{\partial P}{\partial r} + \frac{\mu}{\rho}\cdot\left(\frac{\partial^2 u}{\partial r^2} + \frac{1}{r}\cdot\frac{\partial u}{\partial r} - \frac{u}{r^2} + \frac{\partial^2 u}{\partial z^2}\right) \tag{15.8}$$

$$\frac{\partial v}{\partial t} + u\cdot\frac{\partial v}{\partial r} + w\cdot\frac{\partial v}{\partial z} + \frac{uv}{r} = \frac{\mu}{\rho}\cdot\left(\frac{\partial^2 v}{\partial r^2} + \frac{1}{r}\cdot\frac{\partial v}{\partial r} - \frac{v}{r^2} + \frac{\partial^2 v}{\partial z^2}\right) \tag{15.9}$$

$$\frac{\partial w}{\partial t} + u\cdot\frac{\partial w}{\partial r} + w\cdot\frac{\partial w}{\partial z} = -\frac{1}{\rho}\cdot\frac{\partial P}{\partial z} + \frac{\mu}{\rho}\cdot\left(\frac{\partial^2 w}{\partial r^2} + \frac{1}{r}\cdot\frac{\partial w}{\partial r} + \frac{\partial^2 w}{\partial z^2}\right) + g \tag{15.10}$$

where μ is the viscosity coefficient; u, v, and w represent the radial, tangential, and vertical velocity components; ρ is the air density; g is the gravity acceleration; and P is denoted as the static atmospheric pressure, as a function of r. Some underlying assumptions are made to simplify. The tornado vortex is regarded as approximately axisymmetric and merely tangential and radial, implying that the vertical velocity is disregarded. Also, its viscous effect is negligible. Based upon these simple assumptions, the radial momentum equation, i.e., Eq. (15.8), reduces to

$$\frac{dP}{dr} = \frac{\partial P}{\partial r} = \frac{\rho v^2}{r} - \rho u \cdot \frac{\partial u}{\partial r} \tag{15.11}$$

where dP/dr is the atmospheric pressure change gradient at a distance of r with respect to the tornado center. This equation is the same as the cyclostrophic control equation (Simiu and Scanlan 1996). The static pressure $P(r)$ is expressed as a function of r. Substituting Eqs. 15.3 and 15.4 into Eq. 15.11 and integrating it from infinity to r yield the expression for a simplified pressure drop or static pressure model:

$$P(r) = \begin{cases} \rho \cdot \frac{V_{\max}^2}{2} \cdot \left(2 - \frac{3}{4} \cdot \frac{r^2}{R_{\max}^2}\right), & 0 \le r \le R_{\max} \\ \frac{5}{4} \cdot \frac{\rho V_{\max}^2}{2} \cdot \frac{R_{\max}^2}{r^2}, & R_{\max} \le r < \infty \end{cases} \tag{15.12}$$

For a completely enclosed building, the full effect of the static pressure from Eq. (15.12) is experienced as a pressure differential on the building, but for a building with openings or leakage, this effect diminishes. Kikitsu et al. (2010) explored the transient tornado-induced wind load on a low-rise building using a tornado simulator and suggested that the internal pressure markedly affected the overall uplift force on roofing systems. It was found that when structural leakage ratio is 0.04 %, approximately 80 % of the maximum pressure drop acts on structures. However, the mean static pressure resulting from pressure change decreases to 10 % of the maximum value in the case of the vented structures of 0.13 % leakage ratio. Interpolation is suggested when the leakage ratio falls between 0.04 % and 0.13 %. This is used in combination with Eq. (15.12) to estimate the static wind force acting on the building components and cladding.

15.2.3 Tornado-Induced Wind Loads

The total pressure acting on a component, P_T, consisting of the direct action of wind and the atmospheric pressure drop due to airflow rotation is given by (Simiu and Scanlan 1996)

$$P_T = q_h \cdot \left[GC_p - GC_{pi}\right] + P(r) \tag{15.13}$$

$$q_h = \frac{1}{2}\rho V_h^2 \tag{15.14}$$

where $P(r)$ is the pressure drop, obtained from Eq. (15.12); q_h is the velocity pressure at the eave height; and V_h represents 3-s gust wind velocity at the mean roof height, which is the vector sum of tangential, radial, and translational velocities. The GC_p values for straight-line winds are used herein for the evaluation of the dynamic pressures induced by tornadoes. The determination of the internal pressure coefficient GC_{pi} will be discussed in the next section.

15.2.3.1 Internal Wind Pressures

The internal pressure model proposed by Holmes (2001) is employed. For turbulent flow through an orifice, the following relationship between the flow rate, Q, and the pressure difference across the orifice, $p_e - p_i$, applies:

$$Q = kA \cdot \sqrt{\frac{2\left(p_e - p_i\right)}{\rho}} \tag{15.15}$$

where k is an orifice constant, typically 0.6, and A is the opening area.

The mean internal pressure for a building with multiple openings is derived by using Eq. (15.15) and applying mass conservation. The mass conservation relation can be written for a total of N openings in the envelope:

$$\sum_{1}^{N} \rho Q_j = 0 \tag{15.16}$$

Assuming quasi-steady and incompressible ideal flow, the air density ρ is regarded as constant. Substituting Eq. 15.15 into Eq. 15.17 yields an expression for the mean internal pressure:

$$\sum_{1}^{N} A_i \cdot \sqrt{\left|p_{e,j} - p_i\right|} = 0 \tag{15.17}$$

Figure 15.5 shows the plan view of a building with five openings in the envelope. Applying Eq. 15.17 to this situation:

$$A_1\sqrt{|p_{e,1} - p_i|} + A_{21}\sqrt{|p_{e,2} - p_i|} + A_3\sqrt{|p_{e,3} - p_i|} = A_4\sqrt{|p_{e,4} - p_i|} + A_5\sqrt{|p_{e,5} - p_i|} \tag{15.18}$$

In Eq. 15.18, the inflows through openings on the upwind and side walls are equal to the outflows through openings on the downwind walls. A desirable nonlinear

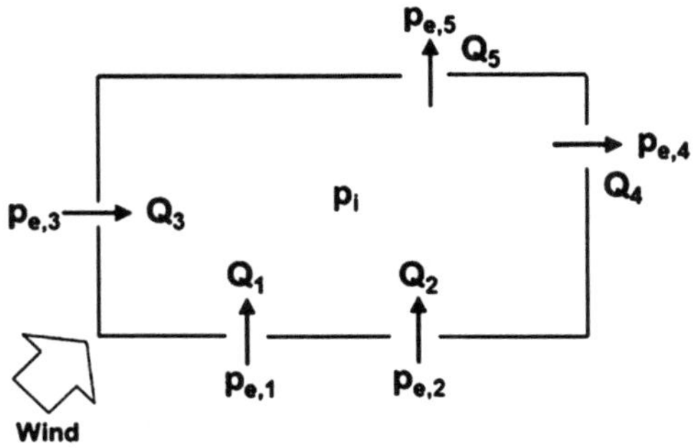

Fig. 15.5 Inflows and outflows for multiple openings (Holmes 2001)

equation in terms of internal pressure coefficient is built by normalizing two sides of the above equation by the reference velocity pressures. The internal pressure coefficients for partially enclosed buildings specified in ASCE 7 (2010) are plus and minus 0.55 and representative of the worst-case scenarios for the life of the structure. In consideration of this circumstance, the transient internal pressure coefficient of interest is obtained by searching a root of Eq. 15.18 between −0.55 and +0.55. These values represent the GC_{pi} values for a partially enclosed building in ASCE 7-10. Differences between internal pressures in buildings within straight-line wind and tornado flow are not well known, and therefore the ASCE 7-10 values are used as a reasonable estimate of the upper and lower bounds until more accurate information is made available.

15.2.3.2 Wind-Borne Debris Model

The failure of wall openings caused by wind-borne debris impacts is accounted for in this model. Gurley et al. (2005) developed a probabilistic debris risk model, using an exponential distribution, for the structural vulnerability analysis in the Florida Public Hurricane Loss model (FPHLM). Balderrama (2009) revisited that debris model and advanced the analysis methodology by employing state-of-the-art wind-borne debris research knowledge. His model is utilized in the present study, and a full description can be found in his thesis.

15.2.4 Probabilistic Wind Resistance Capacities

The load capacity of each building component is modeled as a random variable following a specific distribution with a prescribed mean and coefficient of variation. These values are determined from past laboratory testing available in the literature and can be updated as additional studies are performed or as new component types or materials are needed. Component load capacities for a typical wood-frame residential building are provided in Table 15.2. A portion of the resistance probabilistic models used for analysis are extracted from the engineering team report of FPHLM (Gurley et al. 2005). Additional models are taken from the published literature.

15.3 Applying the Engineering-Based Tornado Damage Assessment (ETDA) Approach

The methodology proposed in this study is capable of performing a scenario damage assessment, which is a process of analyzing specific tornado events by projecting their possible impact on a building or a portfolio. The ETDA model is demonstrated here by evaluating the performance of four houses damaged in the 2011 Joplin, MO, tornado.

15.3.1 2011 Joplin, Missouri (MO), Tornado

A tornado struck Joplin, MO, on May 22, 2011, rated EF-5 based upon the Enhanced Fujita (EF) tornado intensity scale (McDonald and Mehta 2006). According to the final report of the National Institute of Standards and Technology (NIST) (Kuligowski et al. 2014), the best estimated maximum wind speed is approximately 78.2 m/s (175 mph), and the average translational speed was estimated to be 13.4 m/s (30 mph). The tornado maintained a radius value of approximately 260 m (845 ft) for most of the path.

In the aftermath of the Joplin tornado, a team of engineering faculty and students and scientists, led by the third author, surveyed the damage with the objective of collecting perishable information about the performance of residential structures (Prevatt et al. 2011, 2012). The damage to over 1400 individual homes was documented using geo-tagged photographs and rated using the provisions of the EF scale. Four houses were selected from this dataset as case studies for the ETDA model. The criteria for selection were that the houses had (1) damage to the building envelope or roof structure only, (2) simple roof shapes, and (3) damage on all sides visible in photographs. The observed damage to four of these homes was compared to the output of the ETDA model as summarized in Sect. 15.3.4.

Figure 15.6 shows a snapshot of the wind field of the 2011 Joplin tornado as it translates along the *X*-axis from left to right. The tangential and radial velocities at each grid point in this plot are first estimated using Eqs. 15.3 and 15.4. Adding the tangential and radial velocity yields the wind field for a stationary tornado. To simulate a translating tornado, the forward motion is then added to the wind field, as shown in Fig. 15.6.

Radar estimates of wind speed profile and in situ estimates of pressure drop were not available for the Joplin tornado. To consolidate the applicability of the

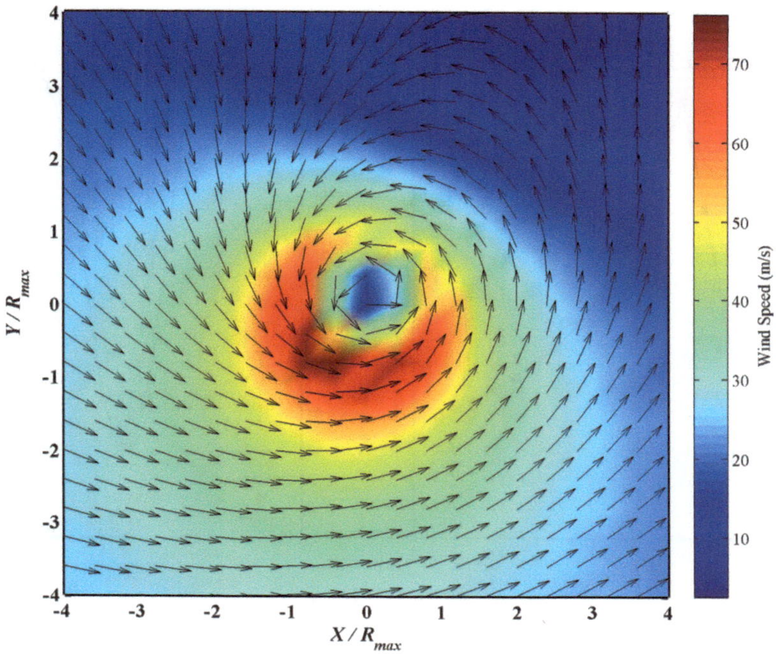

Fig. 15.6 Wind vector and isotachs representing a wind velocity field of 2011 Joplin, MO, tornado with a maximum wind speed of 78.2 m/s (175 mph), a forward speed of 13.4 m/s (30 mph), and an inner core radius of 260 m (845 ft)

tornado catastrophe model, the profiles generated using Eqs. 15.3 and 15.12 were compared against experimental results from a tornado simulator at the Iowa State University (Haan et al. 2010) and field measurements from an EF-4 tornado which struck Manchester, SD, on June 24, 2003 (Lee et al. 2004). Figure 15.7 depicts the variations of normalized tangential velocity and surface pressure drop with the distance of tornado center relative to the building. In the left graph, the abscissa denotes tangential velocity normalized by the maximum tangential velocity ($V_{\max}$), while the ordinate is representative of the ratio of the distance of the tornado center from the building (r) to the tornado core radius ($R_{\max}$). The red circles represent the data from the tornado simulator, whereas the analytical results obtained from the Rankine vortex model are in the black circle line. Comparison of the experimental and analytical results manifests a good agreement, implying that the Rankine vortex model is adequate to model the wind field of a tornado. The right plot presents the profile of static pressure drop on the ground. Results show that the analytical model basically agrees well with the in situ measurement, but slightly deviates from the experimental data. This disparity is most likely to be attributed to the relatively smaller swirl ratio generated in the simulator than obtained in a full-scale tornado.

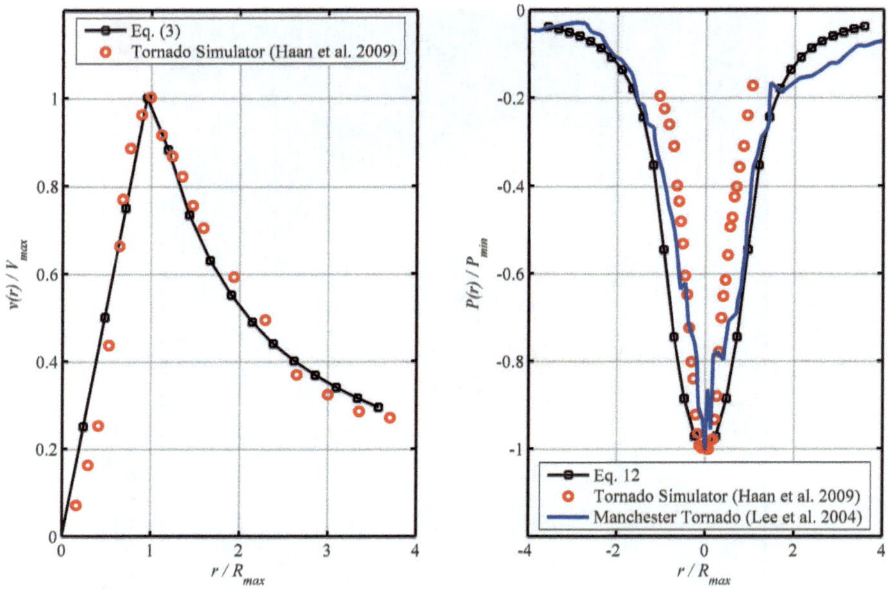

Fig. 15.7 *Left*: tangential velocity profile and *right*: pressure drop distribution

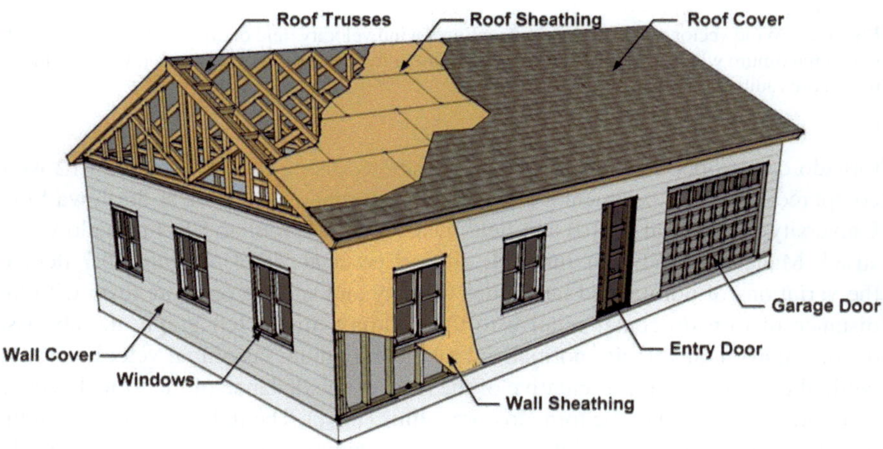

Fig. 15.8 House model used in study showing the overall layout, structure, and building openings

15.3.2 Resistance Capacity of Components

The house model used in this study is shown in Fig. 15.8 and is used to represent all four houses being considered from the Joplin tornado. The numbers of windows, entry doors, and garage doors for the four houses were counted from their damage survey pictures and are presented in Table 15.1, whereas Table 15.2 provides a

Table 15.1 Description of the prototype house models

Building	House 1	House 2	House 3	House 4
Roof type	Gable	Gable	Gable	Gable
Roof cover	Asphalt shingles	Asphalt shingles	Asphalt shingles	Asphalt shingles
Roof sheathing (2.4 × 1.2 m)	OSB	OSB	OSB	OSB
Garage door (3.7 × 2.4 m)	1	1	1	0
Windows (1.2 × 1.5 m)	12	14	13	16
Entry doors (1.2 × 2.4 m)	2	2	2	2
Exposure	Open country	Open country	Open country	Open country

Table 15.2 Summary of structural capacities

Building component	Mean/kPa (psf)	COV	Distribution type	Data source
Windows	2.4 (50)	0.2	Normal distribution	Gurley et al. (2005)
Entry doors	3.6 (75)	0.2	Normal distribution	Gurley et al. (2005)
Garage doors	2.5 (52)	0.2	Normal distribution	Shen (2013)
Roof cover (asphalt shingle)	2.9 (60)	0.2	Normal distribution	Gurley et al. (2005)
Roof and wall sheathing panel (8d 6″/12″)	4.4 (92)	0.15	Normal distribution	Henderson et al. (2013)
Wall cover (vinyl siding)	3.2 (66)	0.2	Normal distribution	Gurley et al. (2005)
Roof-to-wall connection (three 16d toe nails SP)	1.97 kN (442 lb)	0.38	Normal distribution	Shanmugam et al. (2009)

summary of the capacities of various assemblies. The capacity statistics of building components could be found in the previous studies, such as Gurley et al. (2005), Cope (2004), Shen (2013), Shanmugam et al. (2009), and Henderson et al. (2013). In the sampling scheme, the normal distribution is truncated at two standard deviations from the mean.

15.3.3 Wind Loads

In this study, the GC_p values for the roof and wall envelope are taken from the online database provided by the Wind Engineering Information Center at the Tokyo Polytechnic University (TPU). Wind tunnel testing was conducted on a 1/100-scale model in the atmospheric boundary-layer wind tunnel at the TPU. Test setup and data acquisition system are described in detail in Quan et al. (2007). Figure 15.9 shows the locations of pressure taps on the model tested, along with the layout of its roof and wall sheathing in the ETDA model. Wind directions are categorized as eight test cases, and the wind angle of the tornado relative to the X'-axis at each time step is rounded to the closest test case, as illustrated in Fig. 15.10. The data from the TPU database were analyzed employing the methodology proposed by Peng

Table 15.3 Four houses used for validation of the ETDA framework

House	Orientation w.r.t. tornado path (°)	Distance from tornado centerline (Y). Unit: m (ft)	Observed degree of damage	EF rating
No. 1	0	−417.6 (−1370)	6[a]	2
No. 2	90	−35.1 (−115)	6[a]	2
No. 3	0	304.8 (1000)	4[b]	1
No. 4	0	335.3 (1100)	4[b]	1

[a]DOD 6—"Large sections of roof structure removed; most walls remain standing" (McDonald and Mehta 2006)
[b]DOD 4—"Uplift of roof deck and loss of significant roof covering material (>20 %); collapse of chimney; garage doors collapse inward; failure of porch or carport" (McDonald and Mehta 2006)

et al. (2014) to estimate peak pressure coefficient at each tap, which was converted into GC_p based upon the method in St. Pierre et al. (2005).The GC_p value on a single sheathing panel was finally determined by arithmetically averaging the GC_p values at all taps within it. Figure 15.11 presents the GC_p values of all roof and wall sheathing panels for four wind directions. The GC_p values for roof covers, wall covers and sheathing, and openings were identified in the same manner.

15.3.4 Observed Damage vs. Estimated Damage

Examples are presented in this section to illustrate the assessment of the damage to four housesin the 2011 Joplin tornado and to provide a validation for the proposed methodology. No efforts were made to match the predicted and observed damages to the four houses. Summaries of the observed damages were developed from visual inspection of the damage photographs only, and then these results were compared to the output of the ETDA model. Figure 15.12 identifies the path of the tornado in Joplin, MO, and identifies four houses, including two houses on the right-hand side and two on the left-hand side of the tornado centerline. Their detailed location information is presented in Table 15.3 along with the summary of the damage, classified as Degrees of Damage (DOD) from the EF scale (McDonald and Mehta 2006). DODs are typically used to classify damage in tornadoes, and for one- and two-family residences, vary from 1 to 10, with 1 representing the threshold of visible damage and 10 representing all walls and roof completely destroyed and swept away. The contours in Fig. 15.12 represent wind speeds obtained using the Rankine vortex model in Sect. 15.2.2.2 converted to EF scale, while the dashed black line indicates the estimated tornado centerline as given in the NIST report (Kuligowski et al. 2014).

The overall damage ratio is defined as a sum of the physical damage percentages times the cost ratios for all subassemblies. The unit cost for each subassembly is presented in Table 15.4, which is mainly taken from Building News Inc. (2013).

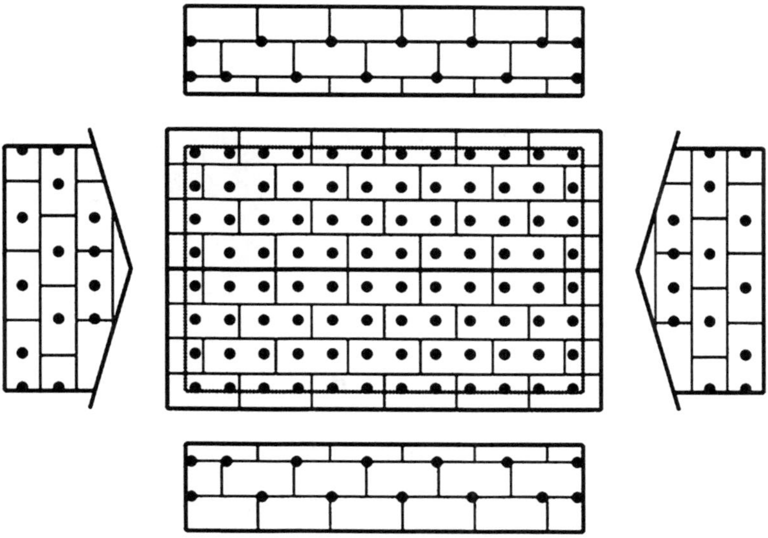

Fig. 15.9 Roof sheathing layout and tap location (after Quan et al. 2007)

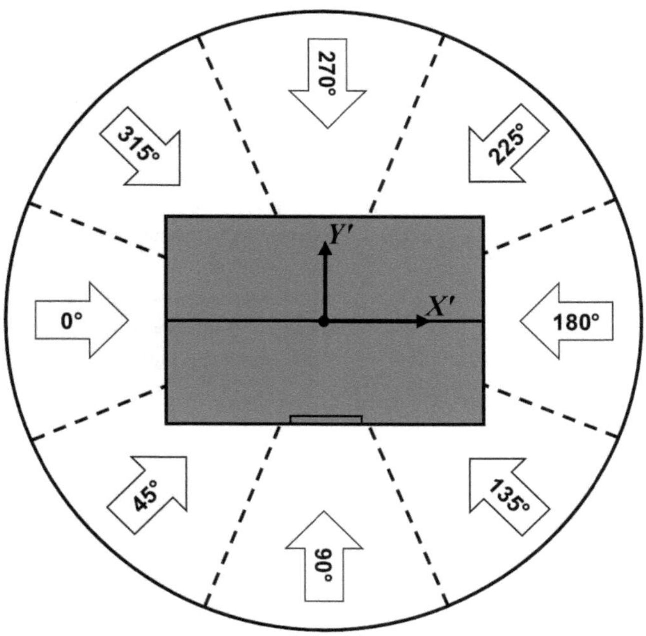

Fig. 15.10 Wind direction definition

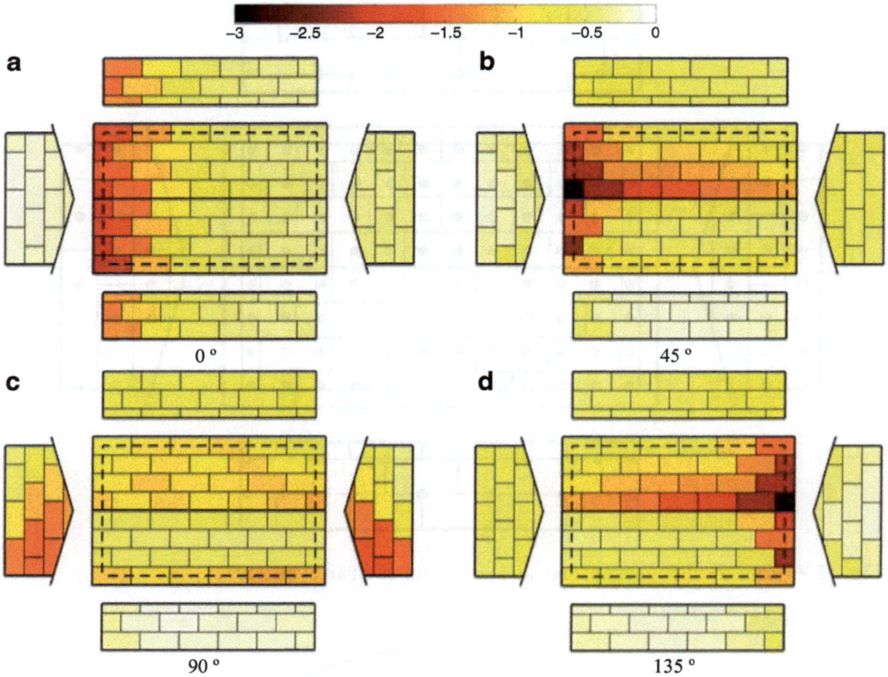

Fig. 15.11 GC_p values for building components and cladding (Quan et al. 2007)

Table 15.4 Unit costs for residential homes (Building News Inc. 2013; Gurley et al. 2005)

Item—residential homes	Unit	$/unit
Roof and wall sheathing	sf	1.50
Roof cover—asphalt shingle	sf	3.20
Exterior wall—frame	sf	6.54
Wood windows	ea	800.00
Wood exterior doors (one face)	ea	2110.00
Garage (roll-up doors)	ea	2250.00
Wall cover (vinyl siding)	sf	3.45

These values are multiplied by their individual quantities to determine repair or reconstruction cost of each subassembly. These costs add up to obtain overall building reconstruction cost. The weighted average of the subassembly damage percentages is the overall damage ratio of a house as a result of this tornado event.

Figure 15.13 illustrates the time histories of the expected horizontal resultant wind velocity and the variation of the corresponding wind direction at House 1 as the Joplin tornado translated. Many of the results in this study reflect the time-variant characteristics of the damage caused by the tornado. Rather than plotting these results (e.g., wind velocity, pressure drop, and expected damage percentage) against time, this study presents them as a function of distance from the center of

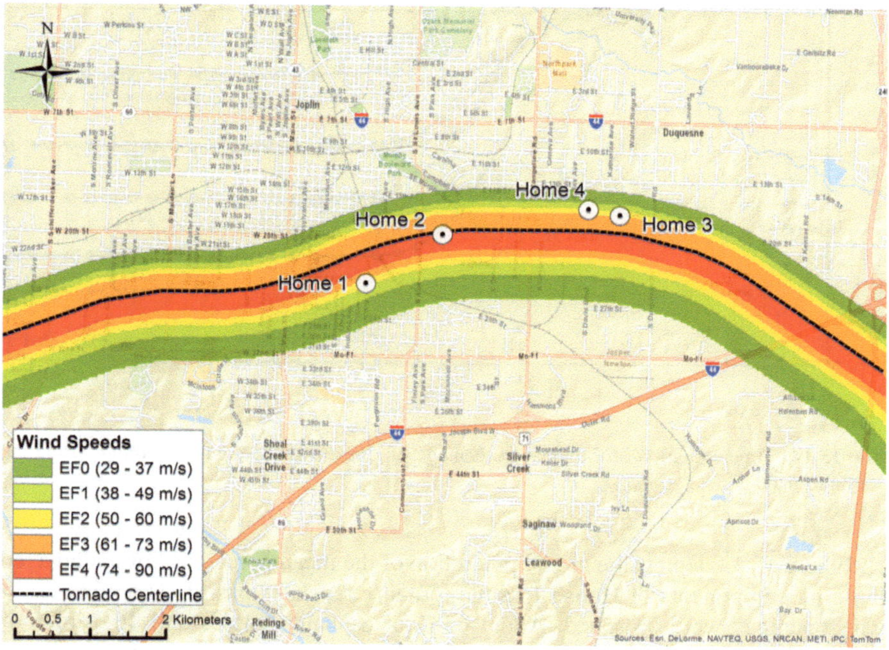

Fig. 15.12 Wind speed contours from the tornado wind field model. Also shown are locations of the four houses used for analysis in this study

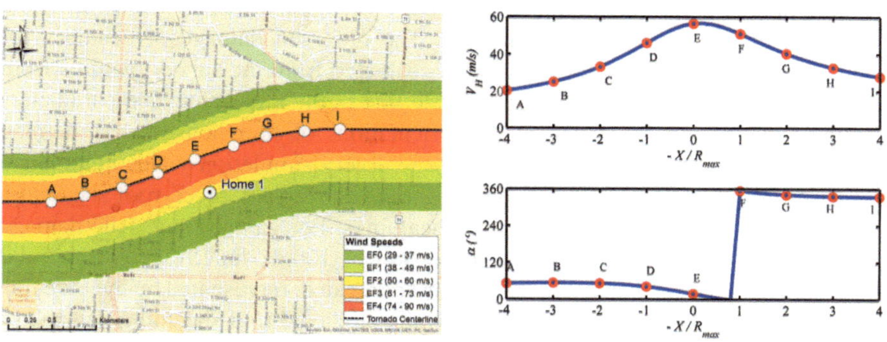

Fig. 15.13 Variation of mean horizontal velocity and wind angle with the tornado location for House 1

the tornado vortex to the building center in the X-axis direction. The complexity of transient tornado wind effects and rapid, time-variant direction change trend are reflected in the plot.

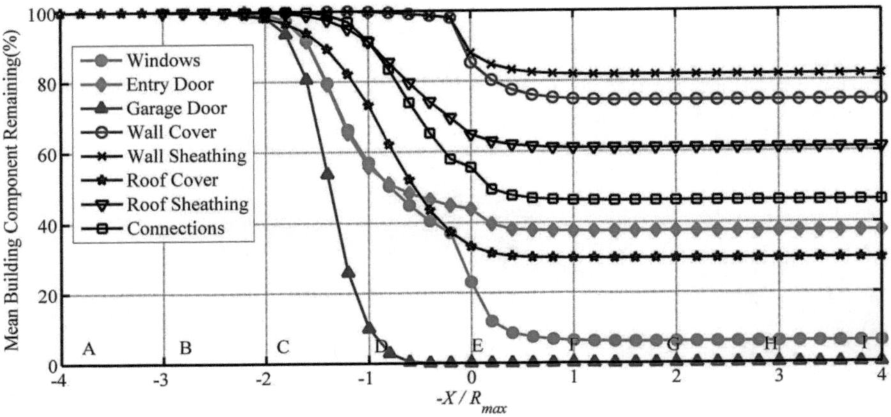

Fig. 15.14 Time evolution of the mean percentage of House 1 component remaining during the Joplin tornado

Figure 15.14 depicts the model prediction of the mean percentage of undamaged building components of House 1 as the tornado translates past it. The majority of wall openings were breached. About 65 % of asphalt shingles and 40 % of roof OSB sheathing panels were damaged, while most of the wall system remained undamaged.

Figures 15.15, 15.16, 15.17, and 15.18 show the comparison of observed damage to those predicted for the four houses. For each damaged house, the front view from post-disaster reconnaissance is provided. The observed damage percentages for their components and cladding were then evaluated from the survey pictures from various angles and presented in green circle line in the radar plot. The parameters for the tornado and the distances and orientations of four houses were defined and substituted into the ETDA methodology. Every simulation yields cumulative damage percentages for windows, doors, garage doors, roof cover, roof sheathing, roof connections, wall cover, and wall sheathings. For garage doors, the damage percentage is a binary value (0 %, did not fail or 100 %, failed). From the 5000 simulations, the mean damage percentage is aggregated, represented by the blue line in the radar plot. For example, in House 1, the predicted roof sheathing damage of 40 % indicates that for the 5000 simulations, on average, 40 % of the roof sheathing was damaged. Results show that the ETDA model provides a reliable and reasonable estimate for damage as compared with the observed damage.

An absolute error analysis was performed for the four houses to quantitatively evaluate the performance of the ETDA model, as illustrated in Fig. 15.19. The first eight groups represent the estimation errors for the subassemblies constituting a building. The absolute error for all subassemblies was below 25 %. The absolute error in the overall damage ratio is less than 15 % for all four houses, which is a good agreement with the observed damage for the limited dataset.

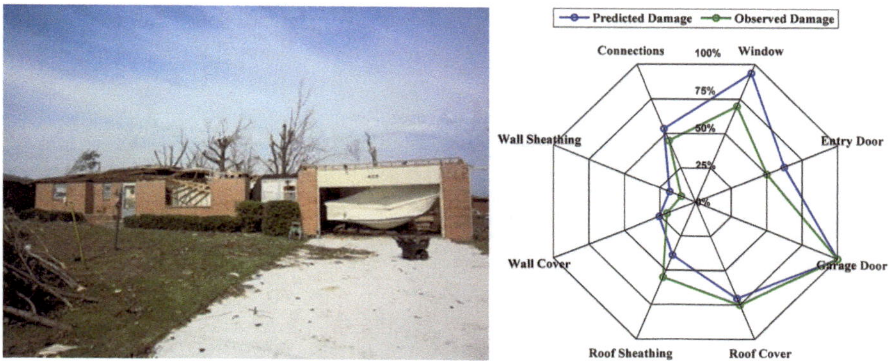

Fig. 15.15 *Left*: front view of House 1 and *right*: observed damage percent vs. estimated damage percent

Fig. 15.16 *Left*: front view of House 2 and *right*: observed damage percent vs. estimated damage percent

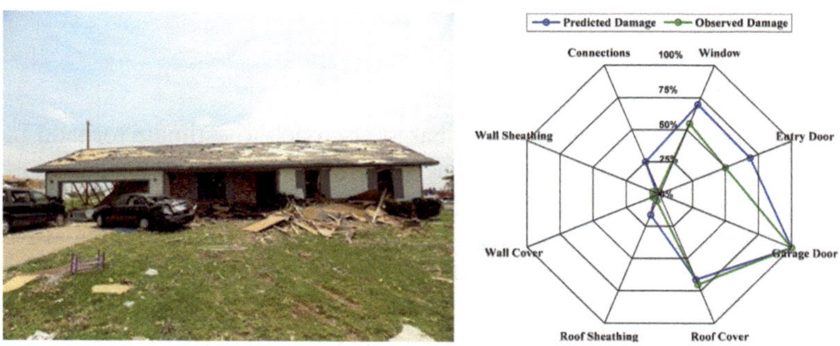

Fig. 15.17 *Left*: front view of House 3 and *right*: observed damage percent vs. estimated damage percent

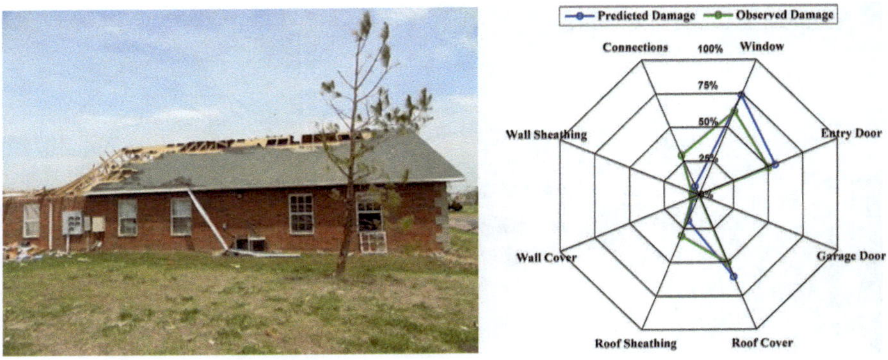

Fig. 15.18 *Left*: front view of House 4 and *right*: observed damage percent vs. estimated damage percent (no garage door for House 4)

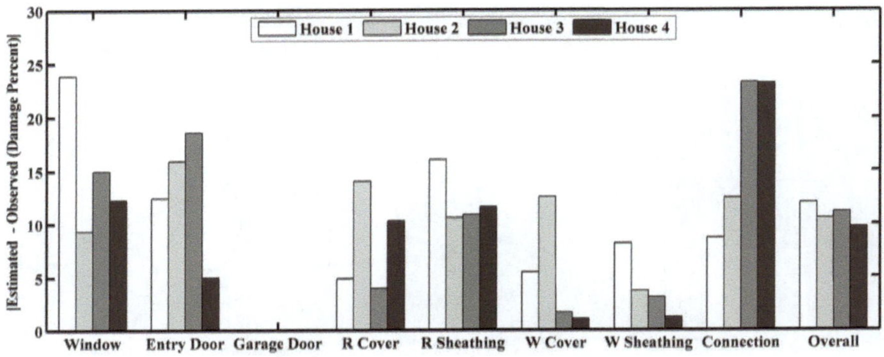

Fig. 15.19 Absolute value of differences between the estimated (5000 simulations) and observed damage percentages

15.4 Conclusions

This chapter proposed an engineering-based approach to estimate tornado-induced damage on a community. The engineering-based tornado damage assessment (ETDA) framework incorporates a rational physical vortex model representative of tornado velocity and its atmospheric pressure drop fields, an analytical tornado-induced wind load estimation approach, a probabilistic damage estimate from wind-borne debris methodology, and an internal pressure model. This framework was applied to predict building envelope damage sustained by four single-story family houses in the May 22, 2011, EF-5 tornado. The comparison demonstrated that the ETDA model reasonably predicts the structural performance of these houses. Results show that the difference between the expected and observed overall damage ratios for the four homes considered in this study is within 15 %. By extension, the ETDA model could be used to predict the structural performance of a larger portfolio of buildings. And since the tornado vortex model is typically applied for any size or intensity tornado, the ETDA could also be expected to accurately predict structural performance for a tornado of any size or intensity. Given the appropriate inputs for buildings within a town or community, the approach described herein can become applied as an engineering-based decision-making tool, capable of providing scenario analyses of various retrofit policies or impact of code changes on the future losses from tornadoes. The ETDA model is flexible to accommodate many tornadoes that vary in size, intensity, or the location of the tornado's central path. It is anticipated such research will be valuable in identifying highly vulnerable housing inventories and in preparation for response and recovery activities.

Further validation is still needed, however. Limitations still exist in the ETDA framework that increase the level of uncertainty in the results. As more is learned about tornado loading and the structural response of buildings, this uncertainty can be reduced. The physical tornado model can be updated as more full-scale data from tornadoes, and small-scale simulated tornado data, become available. The general catastrophe modeling approach is well established for estimating losses due to extreme winds in hurricanes, and so it is expected the results from tornado models will have similar reliabilities. The uncertainty related to wind velocity profiles in tornadoes could create significant variations. It is important to develop robust, absolute wind speed or pressure measurement devices that can monitor conditions on a structure during the tornado. The current tornado-induced load model combining wind pressures from straight-line winds, with the pressure drop within the tornado vortex and internal pressure equalization, needs to be validated.

On the loading and damage model, work is needed to confirm that the wind-borne debris model (borrowed again from straight-line wind flow) is appropriate for the tornado. Some field observations suggest this may not always be the case, which may limit its applicability to tornado loading.

The ETDA damage model is able to predict building envelope failures and failures to the roofing and roof structure only. It does not include failure capacities for wall racking, rigid body sliding, or overturning of the house. These latter damage states are not commonly observed in hurricanes and limited research is available. As research continues to expand our knowledge of tornadoes and tornado-structure interactions, those new findings can be incorporated into the ETDA framework as presented in this chapter.

Acknowledgments This work was supported by the National Science Foundation (NSF) under grant no. 1150975, NSF Graduate Research Fellowship under grant no. GMO2432, and a University of Florida Alumni Fellowship. Any opinions, findings, and conclusions are those of the authors and do not necessarily represent the views of the National Science Foundation. In addition, the authors would like to acknowledge Juan A. Balderrama from the Department of Civil and Coastal Engineering at the University of Florida for his assistance in developing the model.

References

Amini, M., & van de Lindt, J. (2013). Quantitative insight into rational tornado design wind speeds for residential wood-frame structures using fragility approach. *Journal of Structural Engineering*, 04014033.

ASCE 7. (2010). *Minimum design loads for buildings and other structures. ASCE 7-10*. Reston, VA: American Society of Civil Engineers: Structural Engineering Institute.

Balderrama, J. A. G. (2009). *Development of a hurricane loss projection model for commercial residential buildings*. Master, University of Florida, Gainesville, FL.

Barbato, M., Petrini, F., Unnikrishnan, V. U., & Ciampoli, M. (2013). Performance-based hurricane engineering (PBHE) framework. *Structural Safety, 45*, 24–35.

Bonstrom, H., Corotis, R. B., & Porter, K. (2012). Overcoming public and political challenges for natural hazard risk investment decisions. *IDRiM Journal, 2*(1), 26–48.

Building News Inc. (2013). *Sweets unit cost guide 2013*. The McGraw-Hill Companies, Inc. BNI Publication.

Burgess, D., Ortega, K., Stumpf, G., Garfield, G., Karstens, C., Meyer, T., et al. (2014). 20 May 2013 Moore, Oklahoma tornado: Damage survey and analysis. *Weather and Forecasting, 29*(5) 1229–1237.

Changnon, S. (2009). Tornado losses in the United States. *Natural Hazards Review, 10*(4), 145–150.

Cope, A. D. (2004). *Predicting the vulnerability of typical residential buildings to hurricane damage*. PhD, University of Florida, Gainesville, FL.

Dao, T., & van de Lindt, J. (2011). Loss analysis for wood frame buildings during hurricanes. I: Structure and hazard modeling. *Journal of Performance of Constructed Facilities, 26*(6), 729–738.

Dao, T., Graettinger, A., Alfano, C., Gupta, R., Haan, F., Prevatt, D., et al. (2014). Failure progression analysis of observed residential structural damage within a tornado wind field. In *Structures Congress 2014* (pp. 1448–1459). American Society of Civil Engineers.

Folger, P. (2011). *Severe thunderstorms and tornadoes in the United States*. DIANE Publishing, Washington, DC.

Graettinger, A. J., Ramseyer, C. C. E., Freyne, S., Prevatt, D. O., Myers, L., Dao, T., et al. (2013). *Tornado damage assessment in the aftermath of the May 20th 2013 Moore Oklahoma tornado*. Arlington, VA: National Science Foundation.

Grayson, M., Pang, W., & Schiff, S. (2013). Building envelope failure assessment framework for residential communities subjected to hurricanes. *Engineering Structures, 51*, 245–258.

Gurley, K., Pinelli, J. P., Subramanian, C., Cope, A., Zhang, L., Murphree, J., et al. (2005). *Florida public hurricane loss projection model engineering team final report*. International Hurricane Research Center, Florida International University, Miami, FL.

Haan, F., Balaramudu, V., & Sarkar, P. (2010). Tornado-induced wind loads on a low-rise building. *Journal of Structural Engineering, 136*(1), 106–116.

Henderson, D., Williams, C., Gavanski, E., & Kopp, G. A. (2013). Failure mechanisms of roof sheathing under fluctuating wind loads. *Journal of Wind Engineering and Industrial Aerodynamics, 114*, 27–37.

Hoecker, W. H., Jr. (1960). Wind speed and air flow patterns in the Dallas tornado of April 2, 1957. *Monthly Weather Review, 88*(5), 167–180.

Holland, A. P., Riordan, A. J., & Franklin, E. C. (2006). A simple model for simulating tornado damage in forests. *Journal of Applied Meteorology and Climatology, 45*(12), 1597–1611.

Holmes, J. D. (2001). *Wind loading of structures/John D. Holmes*. London and New York: Spon Press.

Kikitsu, H., Sarkar, P. P., & Haan, F. L. (2010). Experimental study on tornado-induced loads of low-rise buildings using a large tornado simulator. In *Proceedings of the Fifth US-Japan Workshop on Wind Engineering (UJNR)*, Chicago, IL.

Kuligowski, E. D., Lombardo, F. T., Phan, L. T., Levitan, M. L., & Jorgensen, D. P. (2014). *Final Report, National Institute of Standards and Technology (NIST) technical investigation of the May 22, 2011, tornado in Joplin, Missouri*. NIST NCSTAR-3, National Institute of Standards and Technology (NIST).

Lee, J. J., Samaras, T. P., & Young, C. R. (2004). Pressure measurements at the ground in an F-4 tornado. In Preprints, 22d Conf. on Severe Local Storms, Hyannis, MA, Amer. Meteor. Soc., CD-ROM, vol. 15.

McDonald, J. R., & Mehta, K. C. (2006). *A recommendation for an enhanced Fujita scale (EF-scale)*. Wind Science and Engineering Center, Texas Tech University. Lubbock, TX.

Panton, R. L. (2013). *Incompressible flow*. John Wiley & Sons., Hoboken, New Jersey.

Peng, X., Yang, L., Gavanski, E., Gurley, K., & Prevatt, D. (2014). A comparison of methods to estimate peak wind loads on buildings. *Journal of Wind Engineering and Industrial Aerodynamics, 126*, 11–23.

Peterson, R. E. (1992). Johannes Letzmann: A pioneer in the study of tornadoes. *Weather and Forecasting, 7*(1), 166–184.

Prevatt, D., van de Lindt, J., Back, E., Graettinger, A., Pei, S., Coulbourne, W., et al. (2012). Making the case for improved structural design: Tornado outbreaks of 2011. *Leadership and Management in Engineering, 12*(4), 254–270.

Prevatt, D., van de Lindt, J. W., Coulbourne, B., Pei, S., Graettinger, A., Gupta, R., et al. (2011). *The Joplin tornado of 2011 – Damage survey and case for tornado-resilient codes*. American Society of Civil Engineers, ASCE Publications, Reston, VA.

Quan, Y., Tamura, Y., Matsui, M., Cao, S., & Yoshida, A. (2007). TPU aerodynamic database for low-rise buildings. In *Proceedings of the 12th International Conference on Wind Engineering* (pp. 1615–1622). Cairns, Australia.

Shanmugam, B., Nielson, B. G., & Prevatt, D. O. (2009). Statistical and analytical models for roof components in existing light-framed wood structures. *Engineering Structures, 31*(11), 2607–2616.

Shen, Y. (2013). *Assessing the wind resistance of sectional door systems for facilities in hurricane-prone areas through full- and component-scale experimental methods and finite element analysis* (3584497). PhD, University of Florida, Gainesville, FL.

Simiu, E., & Scanlan, R. H. (1996). *Wind effects on structures: Fundamentals and applications to design*. New York: John Wiley.

Simmons, K. M., Sutter, D., & Pielke, R. (2012). Normalized tornado damage in the United States: 1950–2011. *Environmental Hazards, 12*(2), 132–147.

St. Pierre, L. M., Kopp, G. A., Surry, D., & Ho, T. C. E. (2005). The UWO contribution to the NIST aerodynamic database for wind loads on low buildings: Part 2. Comparison of data with wind load provisions. *Journal of Wind Engineering and Industrial Aerodynamics, 93*(1), 31–59.

van de Lindt, J., & Dao, T. (2011). Loss analysis for wood frame buildings during hurricanes. II: Loss estimation. *Journal of Performance of Constructed Facilities, 26*(6), 739–747.

van de Lindt, J., Pei, S., Dao, T., Graettinger, A., Prevatt, D., Gupta, R., et al. (2012). Dual-objective-based tornado design philosophy. *Journal of Structural Engineering, 139*(2), 251–263.

Vickery, P. J., Lin, J., Skerlj, P. F., Twisdale, L. A., Jr., & Huang, K. (2006a). HAZUS-MH hurricane model methodology. I: Hurricane hazard, terrain, and wind load modeling. *Natural Hazards Review, 7*(2), 82–93.

Vickery, P. J., Skerlj, P. F., Lin, J., Twisdale, L. A., Jr., Young, M. A., & Lavelle, F. M. (2006b). HAZUS-MH hurricane model methodology. II: Damage and loss estimation. *Natural Hazards Review, 7*(2), 94–103.

Walker, G. R. (2011). Modelling the vulnerability of buildings to wind: A review. *Canadian Journal of Civil Engineering, 38*(9), 1031–1039.

Wood, V. T., & White, L. W. (2011). A new parametric model of vortex tangential-wind profiles: Development, testing, and verification. *Journal of the Atmospheric Sciences, 68*(5), 990–1006.

Yau, S. C., Lin, N., & Vanmarcke, E. (2010). Hurricane damage and loss estimation using an integrated vulnerability model. *Natural Hazards Review, 12*(4), 184–189.

Chapter 16
Performance-Based Hurricane Engineering: A Multi-Hazard Approach

Vipin U. Unnikrishnan and Michele Barbato

Abstract Hurricanes are among the most costly natural hazards affecting communities worldwide. The landfall of a hurricane involves different hazard sources (i.e., wind, wind-borne debris, flood, and rain) that interact to generate the hazard scenario for a given structure. Thus, hurricanes can be viewed and must be analyzed as multi-hazard scenarios. In this chapter, a probabilistic Performance-Based Hurricane Engineering (PBHE) framework is used for the risk assessment of a residential structure subjected to hurricane hazard. The general multilayer Monte Carlo simulation (MCS) approach is specialized for the risk assessment of pre-engineered or non-engineered buildings. A case study of a hypothetical residential house subjected to the combined hazards of wind, wind-borne debris, flood, and rainfall is considered to illustrate the sequential procedure for the probabilistic risk assessment. The results obtained from the application example include the annual probability of exceedance of repair cost for the target residential building due to each hazard and their combined effects. These results highlight the importance of considering the interaction between different hazard sources.

16.1 Introduction

Structures located in coastal regions at tropical and subtropical latitudes are at high risk of suffering severe damages and losses from wind and surge hazards due to tropical storms. In the USA, the average economic loss due to hurricanes in the period 1900–2005 was about $10 billion (normalized to 2005 USD) per year (Pielke et al. 2008). As the population tends to concentrate on coastal regions and the number of residential buildings in hurricane-prone areas continues to rise, the societal vulnerability to hurricanes is increasing, with the prospect of even higher damages and losses in the future (Li and Ellingwood 2006). Most of the USA's densely populated Atlantic and Gulf Coast coastlines lie less than 10 ft above mean sea level (NOAA 2011) and are vulnerable to hurricane-induced surge. During Hurricane

V.U. Unnikrishnan • M. Barbato (✉)
Department of Civil and Environmental Engineering, Louisiana State University, Baton Rouge, LA, USA
e-mail: mbarbato@lsu.edu

P. Gardoni, J.M. LaFave (eds.), *Multi-hazard Approaches to Civil Infrastructure Engineering*, DOI 10.1007/978-3-319-29713-2_16

Katrina in 2005, hurricane-induced surge caused catastrophic damage to residential buildings and tragic loss of life (Eamon et al. 2007; van de Lindt et al. 2009).

Early studies on hurricane hazard assessment and mitigation focused on the damage/loss from individual hazards like wind (including water intrusion due to rainfall) or surge. Powell and Houston (1995) proposed a real-time damage assessment model based on a damage function relating various meteorological variables to the percentage of damage to the buildings. Thomalla et al. (2002) developed a storm surge and inundation model for the risk assessment of residential buildings. Discrete damage states were identified and assigned on the basis of inundation and component damage of the building. Li and Ellingwood (2006) developed a probabilistic risk assessment methodology to assess the performance and reliability of low-rise light-frame wood residential construction subjected to hurricane wind hazard.

More recently, the widespread losses observed in the recent hurricanes motivated researchers to consider the combined effects of hurricane wind and surge hazards. Phan et al. (2007) proposed a methodology for creating site-specific joint distributions of combined hurricane wind and surge for Tampa, Florida, using full-track hurricanes to compute the wind speed and the Sea, Lake, and Overland Surges from Hurricanes (SLOSH) model (Jelesnianski et al. 1992) to estimate surge heights. Lin and Vanmarcke (2010) developed an integrated vulnerability model to explicitly account for the correlation between wind-borne debris damage and wind pressure damage. This integrated vulnerability model was obtained by coupling a pressure damage model derived from the component-based model of the Florida Public Hurricane Loss Model (FPHLM, Gurley et al. (2005)) with the wind-borne debris risk model developed by Lin and Vanmarcke (2008). Friedland and Levitan (2011) developed a joint hurricane wind-surge damage scale based on a loss-consistent approach using Hazards United States Multi-hazards (HAZUS-MH) hurricane model damage and loss functions (FEMA 2012) and the US Army Corps of Engineers (USACE) flood depth loss functions (USACE 2000) for the assessment of damage from combined wind and flood events. Li et al. (2011) conducted a risk assessment for residential buildings by estimating the combined losses from hurricane wind, storm surge, and rainwater intrusion. The correlation between wind and surge was considered in their study by implementing a hurricane-induced surge model through regression analysis of historical data. Dao and van de Lindt (2011) presented a methodology based on the combination of existing wind tunnel data and rainwater intrusion model, for estimating the probability of rainwater intrusion into each room of typical wood-frame structures subjected to hurricanes.

Barbato et al. (2013) developed a PBHE framework and applied it for the risk assessment of residential buildings subjected to wind and wind-borne debris impact. They also observed that the interaction between different hazard sources can significantly affect the risk assessment and emphasized the need to consider the multi-hazard nature of hurricane events for accurate probabilistic loss analysis. Pei et al. (2014) developed joint hazard maps of combined hurricane wind and surge for Charleston, South Carolina. The surface wind speeds and surge heights from individual hurricanes were computed using the Georgiou's wind field model (Georgiou 1985) and the SLOSH model (Jelesnianski et al. 1992), respectively.

In this paper, the PBHE framework (Barbato et al. 2013) is adopted for the risk assessment of structural systems located in hurricane-prone regions. The interaction among the multiple hazard sources and its treatment within the PBHE framework are discussed, and the general multilayer MCS approach is specialized for the risk assessment of pre-/non-engineered buildings such as single-family residential buildings. A hypothetical case study is presented to illustrate the adopted methodology and the specialized multilayer MCS approach for loss analysis of residential buildings subject to hurricane hazard including all pertinent hazard sources (i.e., wind, wind-borne debris, surge, and rainfall).

16.2 Summary of PBHE Framework

The PBHE framework proposed in Barbato et al. (2013) disaggregates the performance assessment procedure for structures subject to hurricane hazard into elementary phases that are carried out in sequence. The structural risk within the PBHE framework is expressed by the probabilistic description of a decision variable, *DV*, which is defined as a measurable quantity that describes the cost and/or benefit for the owner, the users, and/or the society resulting from the structure under consideration. The fundamental relation for the PBHE framework is given by

$$G(DV) = \int\int\int\int\int G(DV|DM) \cdot f(DM|EDP) \cdot f(EDP|IM\ IP, SP) \cdot f(IP|IM, SP) \cdot f(IM) \cdot f(SP) \cdot \mathrm{d}DM \cdot \mathrm{d}EDP \cdot \mathrm{d}IP \cdot \mathrm{d}IM \cdot \mathrm{d}SP \tag{16.1}$$

where $G(\bullet)$ = complementary cumulative distribution function and $G(\bullet|\bullet)$ = conditional complementary cumulative distribution function, $f(\bullet)$ = probability density function and $f(\bullet|\bullet)$ = conditional probability density function, *IM* = vector of intensity measures (i.e., the parameters characterizing the environmental hazard), *SP* = vector of structural parameters (i.e., the parameters describing the relevant properties of the structural system and non-environmental actions), *IP* = vector of interaction parameters (i.e., the parameters describing the interaction phenomena between the environment and the structure), *EDP* = engineering demand parameter (i.e., a parameter describing the structural response for the performance evaluation), and *DM* = damage measure (i.e., a parameter describing the physical damage to the structure). In Eq. (16.1), *IM* and *SP* are assumed as uncorrelated and independent of *IP*, while *IP* is dependent on both *IM* and *SP*. By means of Eq. (16.1), the risk assessment is disaggregated into the following tasks: (1) hazard analysis, (2) structural characterization, (3) interaction analysis, (4) structural analysis, (5) damage analysis, and (6) loss analysis.

16.3 Multilayer Monte Carlo Simulation

Equation (16.1) can be solved using different techniques, e.g., closed-form analytical solutions (Shome and Cornell 1999; Jalayer and Cornell 2003; Zareian and Krawinkler 2007; Mackie et al. 2007), direct integration techniques (Bradley et al. 2009), and stochastic simulation techniques (Porter et al. 2001; Au and Beck 2003; Lee and Kiremidjian 2007). In PBHE, analytical solutions and direct integration techniques require the knowledge of the joint probability density function of the component losses, which is very difficult to obtain for real-world applications. Thus, in this study, the general multilayer MCS approach (Conte and Zhang 2007) is adopted and specialized to efficiently perform loss analysis for residential buildings subject to hurricane hazard. The result of the PBHE equation (Eq. (16.1)) is the annual loss curve, $G(DV)$, i.e., the complementary cumulative distribution function of the annual losses for the residential building under consideration due to hurricane events.

Figure 16.1 shows the flowchart of the general multilayer MCS technique applied to PBHE. Multilayer MCS takes into account all phases of the PBHE framework (namely, hazard analysis, structural characterization, interaction analysis, structural analysis, damage analysis, and loss analysis). Each of these analysis phases is performed in two steps: (1) a sample generation step of random parameters with known probability distributions, which are needed to describe the uncertainties in environmental actions, structural properties, interaction phenomena, analysis techniques, and cost estimates; and (2) an analysis step based on a deterministic model, which is used to model the propagation of uncertainties from input to output parameters of each analysis phase. It is noted here that the analysis steps are usually

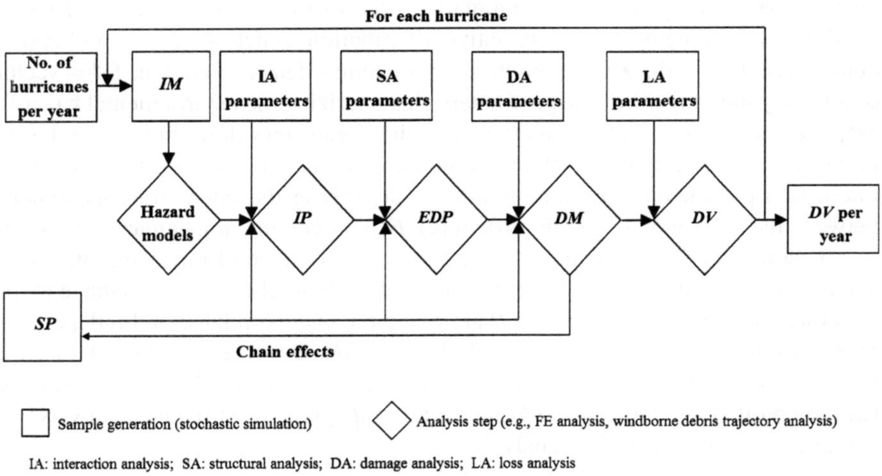

Fig. 16.1 General multilayer MCS approach for PBHE framework

more computationally intensive than the corresponding sample generation steps. Thus, it is useful to identify specific conditions under which one or more of the analysis steps can be avoided in order to reduce the computational cost of the multilayer MCS approach.

16.3.1 Specialized Multilayer MCS Approach for Pre-engineered and Non-engineered Buildings

Pre-engineered and non-engineered buildings, e.g., single-family residential buildings, are structures that are constructed by using design models with components consisting of products that are certified based on building code requirements (NAHB 2000). For these specific building typologies, component strength statistics are commonly available as functions of the environmental action intensity. Under these conditions, the damage analysis phase can be performed without requiring the statistical description of the structural response of the building. In fact, the probabilistic description of the strength for the building components subject to damage (i.e., windows, doors, walls, and roof) can be obtained from empirical relations available in the literature as a function of opportunely chosen *IP*. Thus, it is computationally convenient to eliminate the structural analysis phase from the multilayer MCS procedure. This simplification considerably reduces the computational cost of the multilayer MCS approach for probabilistic hurricane loss analysis of residential and other pre-engineered buildings. It is noted here that, for simple structures of risk category I and II (ASCE 2010), such as single-family residential buildings, simplified and computationally inexpensive models are often appropriate to perform the analysis steps required by the PBHE methodology.

16.3.2 Multi-Hazard Characterization of Hurricane Events

The multi-hazard nature of the phenomena related to hurricanes and their effects on the built environment can manifest itself in the following three different modalities (Barbato et al. 2013):

(1) Independent hazards, when different hazards affect the structure independently. For example, wind-borne debris and flood hazard can be considered as independent of each other because no mutual interaction between the two hazards has the effect of modifying the intensity of the corresponding actions. These hazards can occur individually or simultaneously.
(2) Interacting hazards, when the actions produced on a structure by different hazards are interdependent (e.g., wind and wind-borne debris hazards).
(3) Hazard chains, when the effects of some hazards modify sequentially the effects of other hazards. For example, the actions on a structure due to wind-borne debris can damage the structural envelope and increase the vulnerability of the subject structure to strong winds.

In the PBHE framework, the first two cases (i.e., independent and interacting hazards) are treated within the hazard analysis, by assuming proper interaction models between the hazards, e.g., by using a proper joint probability distribution function to describe the variability of the *IM* for different hazards as in Phan et al. (2007). The study of hazard chains requires modeling the structural system configuration and properties as a function of the level of structural damage caused by the different hazards. In particular, the presence of a hazard chain implies that the *SP* can change as a consequence of *DM* exceeding specified thresholds. Thus, structural characterization, interaction analysis, and structural analysis cannot be carried out without any information or assumption on the values of *DM*.

16.4 Case Study

The PBHE framework is illustrated here by considering a case study in which wind, wind-borne debris, flood, and rainfall hazards interact. This case study consists of a hypothetical residential development, located near the coast in Panama City, Florida, and composed of 25 identical concrete block gable roof structures (Fig. 16.2).

This building type was chosen to simplify the analysis, since the number of failure modes to be considered is smaller than for more common wood-frame buildings. The roof covers are considered as debris sources, whereas the windows

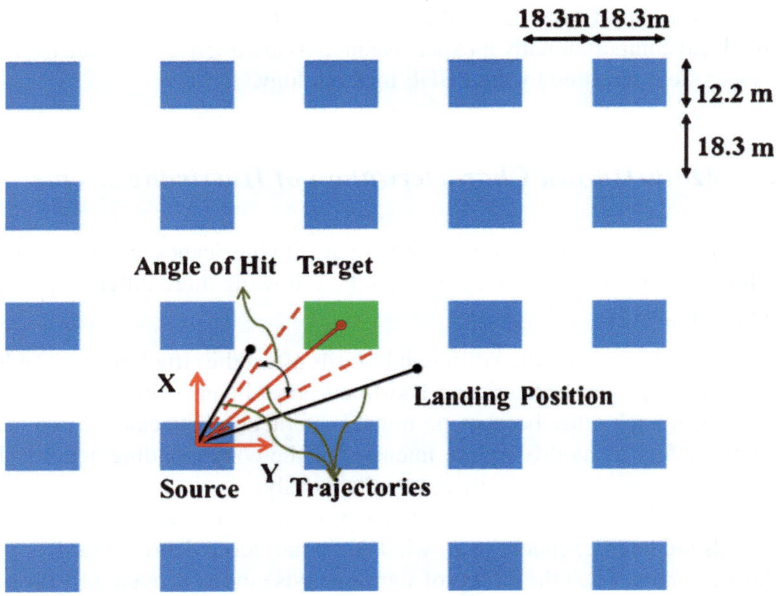

Fig. 16.2 Plan view of the residential development

and doors are considered as debris impact vulnerable components. The value of the target building is assumed to be $200,000 and the value of the content is assumed equal to $100,000. In this study, the cost associated with loss of usage is not considered.

16.4.1 Hazard Analysis

In this study, the 3-s wind speed (V) recorded at 10 m above the ground is considered as the *IM* for wind hazard. Among the different wind field models available in the literature (Batts et al. 1980; Peterka and Shahid 1998; Li and Ellingwood 2006), the Weibull distribution is adopted here to describe the hurricane wind speed variability. The two-parameter Weibull cumulative distribution function is given by

$$F(V) = 1 - \exp\left[-\left(\frac{V}{a}\right)^b\right]. \tag{16.2}$$

The two shape parameters a and b are site specific and are determined for 16 different wind directions by fitting to a Weibull distribution the hurricane wind speed records for the corresponding directions provided by the National Institute of Standards and Technology (NIST) (NIST 2005). The NIST wind speed records contain datasets of simulated 1-min hurricane wind speeds at 10 m above the ground in an open terrain near the coastline, for locations ranging from milepost 150 (near Port Isabel, Texas) to milepost 2850 (near Portland, Maine), spaced at 50 nautical mile intervals (92,600 m). Considering Panama City, Florida, as the location for the case study, the dataset corresponding to milepost 1000 is used for fitting the distribution.

The parameters needed to describe the wind-borne debris are (1) the relative distribution of different debris types, e.g., compact-type, rod-type, and sheet-type debris (Wills et al. 2002); (2) the physical properties of the debris, e.g., for sheet-type debris, M_{d} = mass per unit area of the debris and A_{d} = area of the single debris; (3) the resistance model for the debris sources (which contributes to determine the number of wind-borne debris generated by a given source under a specified wind speed); and (4) the trajectory model for the debris (which describes the debris flight path).

In residential developments, the wind-borne debris are predominantly sheet type, e.g., roof shingles and sheathing (Holmes 2010); hence, this paper focuses on sheet-type debris. The area and mass per unit area of debris are assumed to follow a uniform distribution defined in the range [0.108, 0.184] m^2 and [10.97, 14.97] $\mathrm{kg/m}^2$, respectively.

The debris generation model employed by the FPHLM (Gurley et al. 2005) is adopted in this study. This model is a component-based pressure-induced damage model, which provides the number of debris generated from each source house as a function of (1) the percentage of roof cover damage for a given 3-s gust wind speed and (2) the geometry of the house.

The debris trajectory model provides the landing position of the debris as identified by the random variables X = along-wind flight distance and Y = across-wind flight distance. These random variables are modeled using a two-dimensional Gaussian distribution (Lin and Vanmarcke 2008) described by the following parameters: μ_X = mean along-wind flight distance, $\mu_Y = 0$ m = mean across-wind flight distance, $\sigma_X = \sigma_Y = 0.35 \cdot \mu_X$ = standard deviation of the along-wind and across-wind flight distance, respectively. The parameter μ_X is computed as (Lin and Vanmarcke 2008)

$$\mu_X = \frac{2M_\mathrm{d}}{\rho_\mathrm{a}} \cdot \left[\frac{1}{2}C \cdot \left(K \cdot \tilde{T}\right)^2 + c_1 \cdot \left(K \cdot \tilde{T}\right)^3 + c_2 \cdot \left(K \cdot \tilde{T}\right)^4 + c_3 \cdot \left(K \cdot \tilde{T}\right)^5\right] \quad (16.3)$$

in which $\rho_\mathrm{a} = 1.225\ \mathrm{kg/m^3}$ = air density, $K = \frac{\rho_\mathrm{a} \cdot V^2}{2M_\mathrm{d} \cdot g}$ = Tachikawa number, $\tilde{T} = \frac{g \cdot T}{V}$ = normalized time, g = gravity constant, T = flight time in seconds, and C, c_1, c_2, and c_3 = nondimensional coefficients that depend on the shape of the debris. The flight time is assumed to follow a uniform distribution with range [1, 2.5] s. For the sheet-type debris considered in this study, $C = 0.91$, $c_1 = -0.148$, $c_2 = 0.024$, and $c_3 = -0.0014$.

In this study, a hurricane-induced surge model proposed by Irish et al. (2008) based on the regression analysis of historical data is used. The surge height (ζ) is considered as the intensity measure for flood hazard and is computed as (Irish et al. 2008)

$$\sqrt{\hat{\zeta}} = \left[\sqrt{\hat{R}_{\max}} \ \ 1\right] \cdot C\left(S_0\right) \cdot \begin{bmatrix} \widehat{\Delta p}^2 \\ \widehat{\Delta p} \\ 1 \end{bmatrix} \quad (16.4)$$

where $\hat{\zeta} = \frac{\zeta \cdot g}{V'^2}$, $\hat{R}_{\max} = \frac{R_{\max} \cdot g}{V'^2}$, $\widehat{\Delta p} = \frac{\Delta p}{p_{\mathrm{atmos}}}$, V' = peak 1-minute wind observed at a height of 10 m, $R_{\max}$ = radius of maximum wind (in km), Δp = central pressure deficit (in millibars), p_{atmos} = atmospheric pressure (in millibars), and $C\left(S_0\right) = \begin{pmatrix} -1.078 \times 10^{-1} & 3.996 \times 10^{-2} & 4.444 \times 10^{-4} \\ 3.974 \times 10^{0} & -1.093 \times 10^{0} & -1.653 \times 10^{-1} \end{pmatrix}$ = 2×3 curve fitting coefficient matrix (assuming an ocean slope of 1:5000). The radius of maximum wind ($R_{\max}$) is assumed to follow a normal distribution with a mean of 39.4 km and COV of 0.46, and the central pressure deficit (Δp) follows a lognormal distribution with mean of 70.4 mb and COV of 0.22.

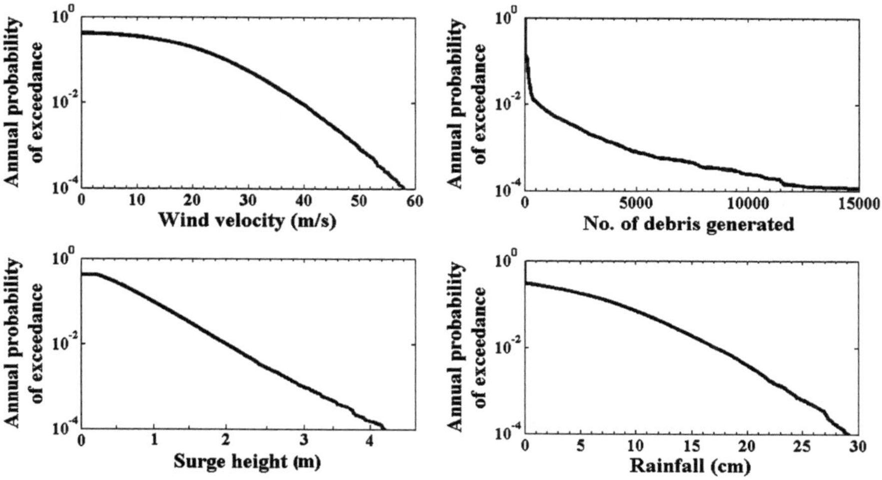

Fig. 16.3 Hazard curves for different hazard sources

In this study, the rainfall hazard model used in the FPHLM (Pita et al. 2012) is adopted to compute the impinging rainfall rate *(IRR)*, which is considered as the intensity measure. This model describes the *IRR* as a function of 3-s gust speed (V) and is given as

$$IRR = 0.84205V - 11.482 \quad (\text{units}: IRR = \text{cm},\ V = \text{m/s})\,. \tag{16.5}$$

The hazard curves for the different hazard sources are computed and plotted in a semilogarithmic scale (see Fig. 16.3).

16.4.2 Structural Characterization

The structural characterization phase provides the probabilistic description of the *SP*s. The *SP*s represent the geometrical and/or mechanical properties of the structure which influence the loading applied to the structure itself and, thus, the *IP*s. Geometrical properties can usually be treated as deterministic quantities, since they can be directly measured for existing structures or are characterized by a small variability. Table 16.1 shows the parameters corresponding to the target residential building (Gurley et al. 2006).

The position and dimension of the windows and doors of the target building are shown in Fig. 16.4.

The *SP*s considered in this case study also include (1) the wind pressure exposure factor (evaluated at h = height of the target building), K_h; (2) the external pressure coefficients, GC_{p}; and (3) the internal pressure coefficients, GC_{pi}. The

Table 16.1 Structural parameters of target building

Structural parameter	Dimension
Length	18.28 m (60 ft)
Width	12.19 m (40 ft)
Height of wall	3.05 m (10 ft)
Roof pitch	5/12
Eave overhang	0.61 m (2 ft)
Space between roof trusses	0.61 m (2 ft)
Roof sheathing panel dimension	2.44 m × 1.22 m (8 ft × 4 ft)

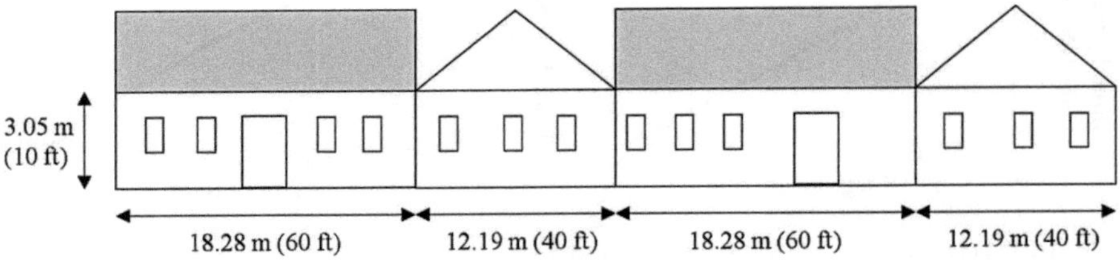

Fig. 16.4 Unfolded view of target building

pressure coefficients include the effects of the gust factor G and are different for different locations within the building (roof zones and windward/leeward/side walls) and/or different conditions of the envelope (enclosed or breached). The value of the topographic factor, K_{zt}, is assumed deterministically equal to one. The *SP* K_h is assumed as normally distributed with a mean value of 0.71 and a coefficient of variation (COV) of 0.19 (Lee and Rosowsky 2005). The statistical characterization of the external and internal pressure coefficients is given in Table 16.2 (Li and Ellingwood 2006).

16.4.3 Interaction Analysis

The choice of the *IP* vector is crucially dependent on the hazard sources, limit states, and performance levels of interest for both structural and nonstructural elements. In this study, the *IP* vector is selected to represent the effects of wind and windborne debris hazard on the different limit states of interest for low-rise residential buildings.

The interaction analysis for the wind hazard provides the statistical characterization of the wind pressure acting on the different components of the buildings, p_w. In this study, the wind pressure acting on the *j*th component of the building is computed as (ASCE 2010)

$$p_{w,j} = q_h \cdot \left(GC_{p,j} - GC_{pi,j}\right) \tag{16.6}$$

Table 16.2 Structural characterization of external and internal pressure coefficients

	Location/Condition	Mean	COV	Distribution
GC_p	Roof (zone 1)	−0.86	0.12	Normal
	Roof (zone 2)	−1.62	0.12	Normal
	Roof (zone 3)	−2.47	0.12	Normal
	Windward wall	0.95	0.12	Normal
	Leeward wall	−0.76	0.12	Normal
	Side wall	−1.05	0.12	Normal
GC_{pi}	Enclosed	0.15	0.33	Normal
	Breached	0.46	0.33	Normal

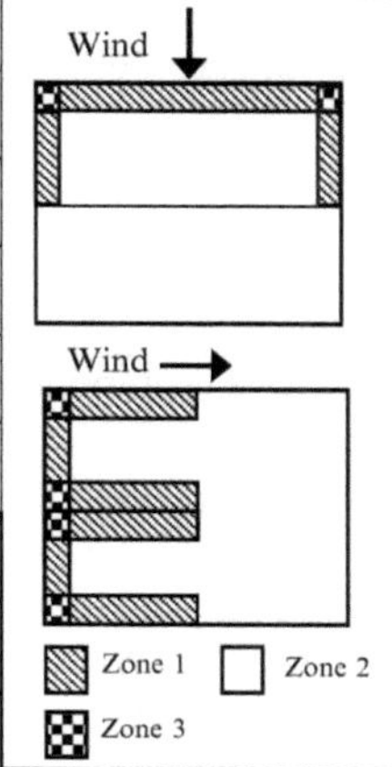

where the velocity pressure, q_h, evaluated at h, is given by

$$q_h = 0.613 \cdot K_h \cdot K_{zt} \cdot V^2. \tag{16.7}$$

The relevant *IP* components controlling the effects of wind-borne debris impact are (1) the number of impacting debris, n_d; (2) impact linear momentum, L_d; and (3) impact kinetic energy, KE_d. The impact linear momentum is well correlated with the damage to envelope components with a brittle behavior (e.g., glazing portions of doors and windows (Masters et al. 2010)), whereas the impact kinetic energy is better correlated with the damage to envelope components with a ductile behavior such as aluminum storm panels (Herbin and Barbato 2012; Alphonso and Barbato 2014). In this study, only envelope components with brittle behavior are considered.

The analysis step of the interaction analysis phase requires an impact model to evaluate n_d and L_d (Barbato et al. 2013). The debris impact model uses the debris flight path obtained from the trajectory model to check for any wind-borne debris impact with the target building. In the event of an impact, the horizontal component of the missile velocity and data relative to the missile size and mass (obtained from the debris generation model) are used to compute the impact linear momentum of the missile (i.e., the linear momentum corresponding to the wind-borne debris velocity component orthogonal to the impacted surface, conditional to the event of at least one impact on vulnerable components). The impact linear momentum is given by

$$L_d = M_d \cdot A_d \cdot u_d. \tag{16.8}$$

The debris horizontal velocity at impact, u_{d}, is a function of the wind velocity and the distance traveled by the debris (determined by its landing position), and is given by (Lin and Vanmarcke 2008)

$$u_{\mathrm{d}} = V \cdot \left[1 - \exp\left(-\sqrt{2 \cdot C \cdot K \cdot x}\right)\right] \tag{16.9}$$

in which $x = \frac{g \cdot X}{V^2}$ = dimensionless horizontal flight distance of the debris.

The *IP* component relevant to the flood hazard is the height of water due to the surge (h_{s}) which is calculated as the difference between the surge height (ζ) and the building ground elevation, which is assumed to be equal to 1 m in this study. The major *IP* for the rainfall hazard is the rainfall intrusion height (h_{r}), which can be computed as (Pita et al. 2012)

$$h_{\mathrm{r}} = \frac{IRR \cdot RAF}{A_{\mathrm{b}}} \cdot \left[\sum_j (d_j \cdot a_j) + a_0\right] \tag{16.10}$$

where RAF = rainfall admittance factor, d_j = percentage of damaged area for component j, a_j = area of component j, a_0 = area of preexisting openings in the building, and A_{b} = base area of the house.

The rainfall admittance factor accounts for the influence that building geometry exerts on the free-flow rain and measures the fraction of the rain that will actually fall on the building windward envelope (i.e., the impinging rain) (Pita et al. 2012). For low-rise buildings, the *RAF* ranges from 0.2 to 0.5 (Straube and Burnett 2000) and is assumed here to follow a uniform distribution.

16.4.4 Structural Analysis/Damage Analysis

In this study, the structural analysis phase is not performed explicitly, and the strength of vulnerable components is directly compared to the corresponding *IP*. This approach is computationally convenient and usually appropriate for non-engineered and pre-engineered structures. Following a procedure commonly used in performance-based earthquake engineering, the physical damage conditions are represented using a limit state function (*LSF*) for each damage limit state, i.e.,

$$LSF_j = DM_j - IP_j \tag{16.11}$$

in which DM_j correspond to the limit state capacity of component j for the given damage limit state. The limit states generally considered for residential buildings are (1) breaking of annealed glass windows/doors, (2) uplift of the roof sheathings,

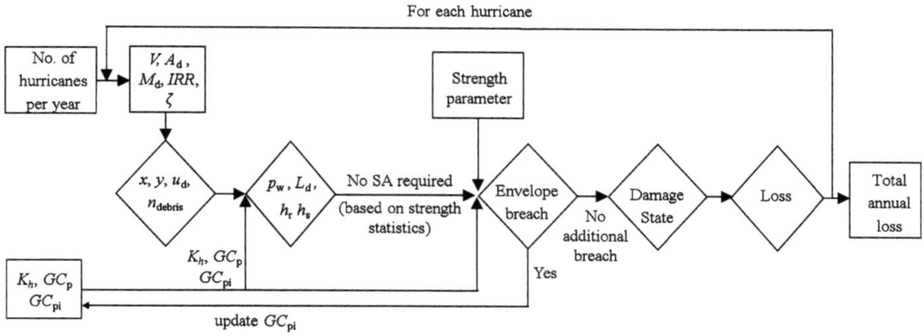

Fig. 16.5 Multilayer MCS approach for probabilistic hurricane loss estimation of residential buildings

Table 16.3 Statistics of the limit state capacity for different components

Component	Limit state	Mean	COV	Distribution
Roof cover	Separation or pull off	3.35 kN/m^2	0.19	Normal
Roof sheathing	Separation or pull off	6.20 kN/m^2	0.12	Lognormal
Door	Pressure failure	4.79 kN/m^2	0.20	Normal
Windows	Pressure failure	3.33 kN/m^2	0.20	Normal
Windows	Impact failure	4.72 kg-m/s	0.23	Lognormal
Roof-to-wall connections	Tensile failure	18.28 kN	0.20	Lognormal

(3) uplift of the roof covers, (4) roof truss failure, and (5) wall failure. The *IP*s are compared with the limit state capacity of different components of the building, and if the *IP*s assume values larger than the corresponding limit state capacity of the building component, the component is assumed to fail. In case of any breach in the building envelope, the iteration is repeated with updated *SP*s until no additional breach is observed (see Fig. 16.5). The statistics of the limit state capacity for different components of the building and their corresponding limit states are given in Table 16.3 (Datin et al. 2010; Gurley et al. 2005; Masters et al. 2010).

The damage states of the target building used in this case study are governed by the performance of the building envelope (damage state of the components) and are divided into five states, varying between 0 (no damage) and 4 (destruction) as shown in Table 16.4 (Vickery et al. 2006; Womble et al. 2006; Li et al. 2011). A rainfall intrusion limit state is used in conjunction with the other limit states for determining the damage state for the contents only. Thus, the damage state of the building is determined as the worst damage state among limit states (a) through (f), whereas the damage state for the content loss is determined as the worst damage state among limit states (a) through (g) (Table 16.4).

Table 16.4 Damage states for target building

Damage state	Qualitative damage description	Roof cover (a)	Window and door failures (b)	Roof deck (c)	Roof failure (d)	Wall failure (e)	Surge height (f)	Rainfall intrusion (g)
0	Very minor damage	≤2 %	No	No	No	No	None	>0 cm and ≤0.02 cm
1	Minor damage	>2 and ≤15 %	One opening failure	No	No	No	None	>0.02 cm and ≤0.25 cm
2	Moderate damage	>15 and ≤50 %	>1 and ≤ the larger of 20 % and 3	1–3 panel	No	No	>0.003 m and ≤0.61 m	>0.25 cm and ≤1.0 cm
3	Severe damage	>50 %	> the larger of 20 % and 3 and ≤50 %	>3 and ≤25 %	No	No	>0.61 m and ≤2.49 m	>1 cm and ≤2.5 cm
4	Destruction	–	>50 %	>25 %	Yes	Yes	>2.49 m	>2.5 cm

Table 16.5 Repair cost (% of building cost) and content loss (% of total content value) for different damage states

Damage state	Mean (%)	COV	Distribution
1	0.2	0.2	Lognormal
2	2	0.2	Lognormal
3	10	0.2	Lognormal
4	30	0.2	Lognormal
5	70	0.2	Lognormal

16.4.5 Loss Analysis

The *DV* in this case study is the repair cost of the building, and its annual probability of exceedance is calculated using the multilayer MCS (see Fig. 16.5). The number of hurricanes in each year is simulated according to a Poisson random occurrence model with annual occurrence rate obtained from the NIST database. For each generated hurricane, a peak wind speed, *V*, is generated randomly according to the Weibull distribution given by Eq. (16.2). For this value of *V*, the wind pressure is calculated using Eq. (16.6), the number of debris impacts is calculated by comparing the flight trajectory with the position of the target house, the surge height is calculated using Eq. (16.4), and the impinging rainfall rate by Eq. (16.5).

For each debris impact, the corresponding linear momentum is calculated using Eq. (16.8). The *IP*s are then compared with the limit state capacity of different components of the building, and if the *IP*s assume values larger than the corresponding limit state capacity of the building component, the component is assumed to fail. The building envelope is checked for any breach, in the event of which the internal pressure is modified. The remaining undamaged building components are checked for further damage due to the modified pressure. The amount of rainfall intrusion through the damaged components is calculated using Eq. (16.10). The damage state of the building is calculated based on the extent of the component damages, the surge height, and the rainfall intrusion (see Table 16.4). The repair cost and the content loss are then generated for the corresponding damaged state according to the probability distributions given in Table 16.5. In this study, it is assumed that the building is fully repaired after each hurricane. Research is ongoing to include the effects of downtime and reconstruction time, which depend on the extent of the damage and the time interval between consecutive hurricanes and are needed to estimate the cost associated with loss of use.

The single-year simulation described above is repeated a large number of times (e.g., in this example, 100,000 samples are used) to estimate the annual probability of exceedance (which coincides with the complementary cumulative distribution function of the *DV*) of the total loss. Figure 16.6 plots, in a semilogarithmic scale, the annual probabilities of exceedance of the loss for the target building for different hazard scenarios. The expected annual loss (EAL), which is defined as the average economic loss that occurs every year in the building (Lin et al. 2007) and is equal to

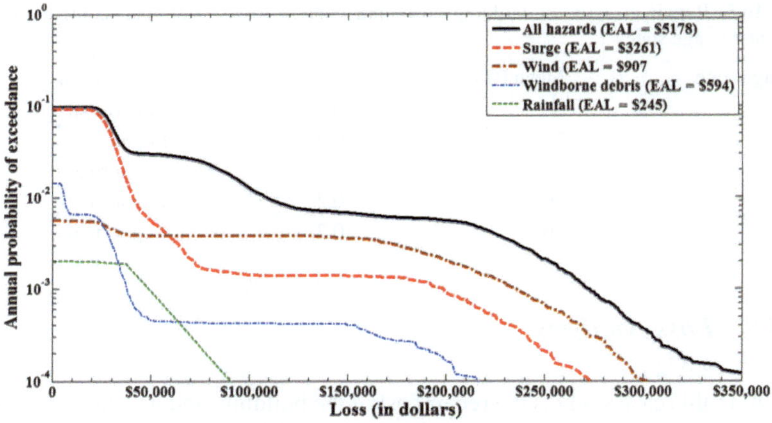

Fig. 16.6 Annual probability of loss exceedance for different hazards and their interaction

the area under each annual probability of exceedance curve, is also computed and shown in Fig. 16.6 for each of the different hazards and their interaction.

From the results presented in Fig. 16.6, it is observed that the loss due to surge hazard is predominant for repair costs lower than about $60,000, while the loss due to wind hazard is predominant for repair costs higher than about $60,000. This behavior can be explained by comparing the hazard curves for surge and wind hazard shown in Fig. 16.3 with the damage states corresponding to these two hazards. In particular, it is observed that storm surge values that can cause even significant damage to the structure have an annual probability of occurrence that is similar to wind speed values for which it is unlikely to have significant structural damage. However, the annual probability of occurrence for storm surge decreases significantly faster than the annual probability of occurrence of wind speed. Similarly, in comparison with the wind hazard, wind-borne debris hazard has a larger effect on loss for values lower than about $30,000, because the probability of damage to the windows due to wind-borne debris is lower than that due to wind pressure at lower wind speeds. It is also observed that the annual probability of loss exceeding the sum of the building cost and its content value is small but not negligible. This phenomenon is most likely due to (1) the assumption that the building is fully repaired after each hurricane and that more than one hurricane can take place in a single year, and (2) the assumption of lognormal distribution for the loss corresponding to a given damage state.

In terms of EAL, it is observed that, for the example considered in this study, the losses due to surge hazard are significantly higher than those due to other hazards. In addition, the EAL due to the interaction of all hazards is about 5 % higher than the sum of the EALs due to each individual hazard. This result suggests a moderate level of interaction among the different hazards for the case study considered here. The fact that the interaction is only moderate is most likely due to the predominance of the surge hazard effects on expected losses. However, the level of interaction among different hazards can be significant, particularly when the expected annual losses due to different interacting hazards considered independently are of similar magnitude (see, e.g., Barbato et al. 2013). Thus, in general, the multi-hazard nature of hurricane events must be taken into account for accurate probabilistic loss analyses.

16.5 Conclusion

In this chapter, the PBHE framework is applied to risk assessment of structures subjected to combined wind, wind-borne debris, flood, and rainfall hazards. Risk assessment analysis is disaggregated into the following basic probabilistic components: (1) hazard analysis, (2) structural characterization, (3) interaction analysis, (4) structural analysis, (5) damage analysis, and (6) loss analysis. In contrast to other existing performance-based engineering approaches, which consider explicitly only single hazards, the PBHE framework accounts for the multi-hazard nature of hurricane events by considering independent hazards, interacting hazards, and hazard chains. The general multilayer MCS approach is specialized for the risk assessment of pre-/non-engineered buildings such as single-family residential buildings.

The PBHE framework is illustrated through an example of the risk assessment analysis for a target building in a hypothetical residential development in Panama City, Florida. The annual probabilities of loss exceedance and the expected annual loss of the target building are computed for different individual hazards and their interaction. For the specific example considered in this paper, it is observed that the loss due to surge hazard is predominant for lower loss levels, whereas the loss due to wind hazard is predominant for higher loss levels. It is also observed that the interaction among different hazards is only moderate, because of the overall predominance of losses due to surge hazard when compared to the losses due to other hazards. However, in general, the multi-hazard nature of hurricane events needs to be taken into account for accurate probabilistic hurricane risk assessment, particularly when the losses due to different hazards are similar.

It is noteworthy that the presented probabilistic methodology differs from the HAZUS-MH approach because it is concerned with the design and/or retrofit of specific buildings and structures, whereas HAZUS-MH can be used to perform loss analysis for a region or for a hurricane event. Thus, the PBHE framework can be used for design and/or loss assessment of specific buildings and structures. This framework can also be used to compare the cost effectiveness of different storm mitigation strategies and can assist owners, insurers, designers, and policy makers in making informed decisions on design and retrofit of specific structures subject to hurricane hazard.

References

Alphonso, T. C., & Barbato, M. (2014). Experimental fragility curves for aluminum storm panels subject to windborne debris impact. *Journal of Wind Engineering and Industrial Aerodynamics, 134*, 44–55.

ASCE. (2010). *Minimum design loads for buildings and other structures*. Reston, VA: American Society of Civil Engineers.

Au, S., & Beck, J. (2003). Subset simulation and its application to seismic risk based on dynamic analysis. *Journal of Engineering Mechanics, 129*(8), 901–917.

Barbato, M., Petrini, F., Unnikrishnan, V. U., & Ciampoli, M. (2013). Performance-based hurricane engineering (PBHE) framework. *Structural Safety, 45*, 24–35.

Batts, M. E., Cordes, M. R., Russell, L. R., Shaver, J. R., & Simiu, E. (1980). *Hurricane wind speeds in the United States* (Report No. BSS-124). Washington, DC: National Bureau of Standards, U.S. Department of Commerce.

Bradley, B. A., Lee, D. S., Broughton, R., & Price, C. (2009). Efficient evaluation of performance-based earthquake engineering equations. *Structural Safety, 31*(1), 65–74.

Conte, J. P., & Zhang, Y. (2007, June 10–14). Performance based earthquake engineering: Application to an actual bridge-foundation-ground system. In *Proceedings of the 12th Italian National Conference on Earthquake Engineering*. Pisa, Italy.

Dao, T., & van de Lindt, J. (2011). Loss analysis for wood frame buildings during hurricanes. I: Structure and hazard modeling. *Journal of Performance of Constructed Facilities, 26*(6), 729–738.

Datin, P., Prevatt, D., & Pang, W. (2010). Wind-uplift capacity of residential wood roof-sheathing panels retrofitted with insulating foam adhesive. *Journal of Architectural Engineering, 17*(4), 144–154.

Eamon, C., Fitzpatrick, P., & Truax, D. (2007). Observations of structural damage caused by Hurricane Katrina on the Mississippi Gulf coast. *Journal of Performance of Constructed Facilities, 21*(2), 117–127.

FEMA. (2012). *HAZUS-MH 2.1 hurricane model technical manual*. Washington, DC: Federal Emergency Management Agency.

Friedland, C., & Levitan, M. (2011). Development of a loss-consistent wind and flood damage scale for residential buildings. In *Solutions to coastal disasters 2011* (pp. 666–677). American Society of Civil Engineers, Reston, Virginia.

Georgiou, P. N. (1985). *Design wind speeds in tropical cyclone-prone regions*. Ph.D. dissertation, Department of Civil Engineering, University of Western Ontario, London, Ontario.

Gurley, K., Pinelli, J. P., Subramanian, C., Cope, A., Zhang, L., & Murphree, J. (2005). Predicting the vulnerability of typical residential buildings to hurricane damage. In *Florida Public Hurricane Loss Projection Model (FPHLPM) engineering team final report* (Vol. II). Miami, FL: I.H.R. Center, Florida International University.

Gurley, K., Pinelli, J. P., Subramanian, C., Cope, A., Zhang, L., & Murphree, J. (2006). Development calibration and validation of vulnerability matrices of the Florida public hurricane loss projection model. In *Florida Public Hurricane Loss Projection Model (FPHLPM) engineering team final report* (Vol. III). Miami, FL: I.H.R. Center, Florida International University.

Herbin, A. H., & Barbato, M. (2012). Fragility curves for building envelope components subject to windborne debris impact. *Journal of Wind Engineering and Industrial Aerodynamics, 107–108*, 285–298.

Holmes, J. D. (2010). Windborne debris and damage risk models: A review. *Wind and Structures, 13*(2), 95–108.

Irish, J. L., Resio, D. T., & Ratcliff, J. J. (2008). The influence of storm size on hurricane surge. *Journal of Physical Oceanography, 38*(9), 2003–2013.

Jalayer, F., & Cornell, C. A. (2003). *A technical framework for probability-based demand and capacity factor design (DCFD) seismic formats* (PEER Report 2003/08). Berkeley, CA: Pacific Earthquake Engineering Center, University of California.

Jelesnianski, C. P., Chen, J., & Shaffer, W. A. (1992). *Slosh: Sea, lake, and overland surges from hurricanes* (NOAA Technical Report NWS 48). National Oceanic and Atmospheric Administration, U.S. Department of Commerce.

Lee, K. H., & Rosowsky, D. V. (2005). Fragility assessment for roof sheathing failure in high wind regions. *Engineering Structures, 27*(6), 857–868.

Lee, R., & Kiremidjian, A. S. (2007). Uncertainty and correlation for loss assessment of spatially distributed systems. *Earthquake Spectra, 23*(4), 753–770.

Li, Y., & Ellingwood, B. R. (2006). Hurricane damage to residential construction in the US: Importance of uncertainty modeling in risk assessment. *Engineering Structures, 28*(7), 1009–1018.

Li, Y., van de Lindt, J., Dao, T., Bjarnadottir, S., & Ahuja, A. (2011). Loss analysis for combined wind and surge in hurricanes. *Natural Hazards Review, 13*(1), 1–10.

Lin, N., Holmes, J., & Letchford, C. (2007). Trajectories of wind-borne debris in horizontal winds and applications to impact testing. *Journal of Structural Engineering, 133*(2), 274–282.

Lin, N., & Vanmarcke, E. (2008). Windborne debris risk assessment. *Probabilistic Engineering Mechanics, 23*(4), 523–530.

Lin, N., & Vanmarcke, E. (2010). Windborne debris risk analysis - Part I: Introduction and methodology. *Wind and Structures, 13*(2), 191–206.

Mackie, K. R., Wong, J.-M., & Stojadinovic, B. (2007). *Integrated probabilistic performance-based evaluation of benchmark reinforced concrete bridges* (PEER Report 2007/09). Berkeley, CA: Pacific Earthquake Engineering Research Center, University of California.

Masters, F. J., Gurley, K. R., Shah, N., & Fernandez, G. (2010). The vulnerability of residential window glass to lightweight windborne debris. *Engineering Structures, 32*(4), 911–921.

NAHB. (2000). A state-of-the-art review and application of engineering information for light-frame homes, apartments, and townhouses. In *Residential Structural Design Guide: 2000 Edition*. Washington, DC: U.S. Department of Housing and Urban Development Office of Policy Development and Research.

NIST. (2005). *Extreme wind speed data sets: Hurricane wind speeds*. http://www.itl.nist.gov/div898/winds/hurricane.htm.

NOAA. (2011). Tropical cyclone report: Hurricane Katrina 23-30 August 2005. http://www.nhc.noaa.gov/data/tcr/AL122005_Katrina.pdf (accessed April 18, 2016)

Pei, B., Pang, W., Testik, F., Ravichandran, N., & Liu, F. (2014). Mapping joint hurricane wind and surge hazards for Charleston, South Carolina. *Natural Hazards, 74*(2), 375–403.

Peterka, J., & Shahid, S. (1998). Design gust wind speeds in the United States. *Journal of Structural Engineering, 124*(2), 207–214.

Phan, L. T., Simiu, E., McInerney, M. A., Taylor, A. A., Glahn, B., & Powell, M. D. (2007). *Methodology for development of design criteria for joint hurricane wind speed and storm surge events: Proof of concept* (NIST Technical Note 1482). Gaithersburg: National Institute of Standards and Technology.

Pielke, R., Gratz, J., Landsea, C., Collins, D., Saunders, M., & Musulin, R. (2008). Normalized hurricane damage in the United States: 1900–2005. *Natural Hazards Review, 9*(1), 29–42.

Pita, G., Pinelli, J. P., Cocke, S., Gurley, K., Mitrani-Reiser, J., Weekes, J., et al. (2012). Assessment of hurricane-induced internal damage to low-rise buildings in the Florida public hurricane loss model. *Journal of Wind Engineering and Industrial Aerodynamics, 104–106*, 76–87.

Porter, K. A., Kiremidjian, A. S., & LeGrue, J. S. (2001). Assembly-based vulnerability of buildings and its use in performance evaluation. *Earthquake Spectra, 17*(2), 291–312.

Powell, M. D., & Houston, S. H. (1995, April 24–28). Real-time damage assessment in hurricanes. In *Proceedings of the 21st American Meteorological Society Conference on Hurricanes and Tropical Meteorology*. Miami, FL.

Shome, N., & Cornell, C. A. (1999). *Probabilistic seismic demand analysis of nonlinear structures* (Reliability of Marine Structures Program Report No. RMS-35). California: Department of Civil and Environmental Engineering, Stanford University.

Straube, J. F., & Burnett, E. F. P. (2000, September 18–21). Simplified prediction of driving rain deposition. In *Proceedings of the International Building Physics Conference*. Eindhoven.

Thomalla, F., Brown, J., Kelman, I., Möller, I., Spence, R., & Spencer, T. (2002). Towards an integrated approach for coastal flood impact assessment. In *Solutions to coastal disasters '02* (pp. 142–158). American Society of Civil Engineers.

USACE. (2000). *Generic depth-damage relationships*. Washington: U.S. Army Corps of Engineers.

van de Lindt, J., Li, Y., Bulleit, W., Gupta, R., & Morris, P. (2009). The next step for AF&PA/ASCE 16-95: Performance-based design of wood structures. *Journal of Structural Engineering, 135*(6), 611–618.

Vickery, P., Lin, J., Skerlj, P., Twisdale, L., & Huang, K. (2006). HAZUS-MH hurricane model methodology. I: Hurricane hazard, terrain, and wind load modeling. *Natural Hazards Review, 7*(2), 82–93.

Wills, J. A. B., Lee, B. E., & Wyatt, T. A. (2002). A model of wind-borne debris damage. *Journal of Wind Engineering and Industrial Aerodynamics, 90*(4–5), 555–565.

Womble, A. J., Ghosh, S., Adams, B., & Friedland, C. J. (2006). *Advanced damage detection for Hurricane Katrina: Integrating remote sensing and views™ field reconnaissance*. Buffalo, NY: MCEER.

Zareian, F., & Krawinkler, H. (2007). Assessment of probability of collapse and design for collapse safety. *Earthquake Engineering & Structural Dynamics, 36*(13), 1901–1914.

Chapter 17
Wall of Wind Research and Testing to Enhance Resilience of Civil Infrastructure to Hurricane Multi-Hazards

Arindam Gan Chowdhury, Mohammadtaghi Moravej, and Filmon Habte

Abstract The Wall of Wind (WOW) research facility at Florida International University (FIU) allows large- and full-scale testing of buildings and infrastructure, and is capable of simulating up to Category 5 hurricane wind speeds, making it the largest and most powerful university research facility of its kind. The WOW facility has made a significant impact on the mitigation of hurricane damage to civil infrastructure through extensive research conducted by the wind engineering team at FIU's International Hurricane Research Center (IHRC) and Department of Civil and Environmental Engineering (CEE). Since roofs are known to be the most vulnerable component of low-rise buildings, they have been the subject of many research projects conducted at the WOW. In this chapter, two large-scale studies performed on low-rise building roof coverings are reported and discussed. The first study discusses the results of full-scale wind testing on hip, ridge, and eave perimeter tiles and reveals how much valuable information can be assessed by full-scale modeling of building elements and details. In the second study, an experimental investigation conducted at WOW with the objective of evaluating the structural performance of standing seam metal roofs under high wind speeds is presented. Full-scale investigations, including wind-induced roof surface pressure measurements and roof panel deflection measurements, were conducted. The observed failure modes under realistic wind loading conditions were different from what is typically observed with standard uniform pressure testing methods. These experiments revealed new aspects of roof responses to high wind speeds and highlighted the importance of large- or full-scale modeling of buildings and structures that can incorporate realistic component and connection details and architectural features.

A. Gan Chowdhury (✉)
Department of Civil and Environmental Engineering, Faculty of Engineering,
Florida International University, Miami, FL, USA
e-mail: chowdhur@fiu.edu

M. Moravej • F. Habte
Department of Civil and Environmental Engineering, Florida International University,
Miami, FL, USA
e-mail: mmoravej@fiu.edu; fhabt003@fiu.edu

P. Gardoni, J.M. LaFave (eds.), *Multi-hazard Approaches to Civil Infrastructure Engineering*, DOI 10.1007/978-3-319-29713-2_17

17.1 Introduction

The Wall of Wind (WOW) at Florida International University (FIU) is the largest and most powerful university research facility of its kind and is capable of simulating Category 5 hurricanes—the highest rating on the Saffir-Simpson Hurricane Wind Scale. The WOW facility has made a significant impact on the mitigation of hurricane damage to civil infrastructure through extensive research conducted by the wind engineering team at FIU's International Hurricane Research Center and Department of Civil and Environmental Engineering. This document will focus on WOW interdisciplinary research and testing to enhance the built environment's resilience to hurricane multi-hazards, including wind, rain, and debris. Hurricane engineering research at FIU has confirmed the effectiveness of large- and full-scale *holistic* testing approaches in advancing the understanding of hurricane impacts on buildings and other infrastructure elements, such as traffic and electrical infrastructure elements. Holistic testing procedures entail testing of systems consisting of integrated assemblies of components, as opposed to individual component testing. The latter type of testing can be misleading because it misses the interaction effect of different parts (components) of a structure, which can often be decisive in the understanding of failure processes and progressive collapse. Full-scale testing also helps to achieve more realistic Reynolds and Strouhal numbers, which result in a more reliable application of model results to real design projects. A recent multi-scale experimental study at the WOW discussed various aspects of Reynolds number effects on a two bay bridge (Kargarmoakhar et al. 2015). Another capability which can greatly help to achieve a more realistic study of the wind-induced response is capturing the dynamic effects that cannot be correctly determined in small-scale models due to their higher rigidity. Studies on photovoltaic panels and comparisons of WOW full-scale to wind tunnel small-scale test results have clearly revealed the importance of accounting for those dynamic effects (Moravej et al. 2015).

WOW research has resulted in the development and validation of innovative damage mitigation products and techniques, including roof suction mitigation devices, rooftop equipment wind screens, and a nonintrusive roof-to-wall connection system. Moreover, results of full-scale experiments conducted at WOW were applied to improve Florida Building Code's wind load provisions on building roof mounted equipment for the State of Florida, including its high-velocity hurricane zones. The insurance industry views WOW testing as a development as revolutionary for wind engineering as crash testing—which led to airbags and other safety features—was for the automobile industry. The civil engineering community likens WOW to shake table testing, which has significantly contributed to the development of performance-based earthquake engineering. The new test-based data from the WOW facility will help create a sound scientific basis for developing risk- and performance-based design criteria embodied in code provisions and contribute to the attainment of a national objective: achieving more sustainable, hurricane-resilient, and energy-efficient communities.

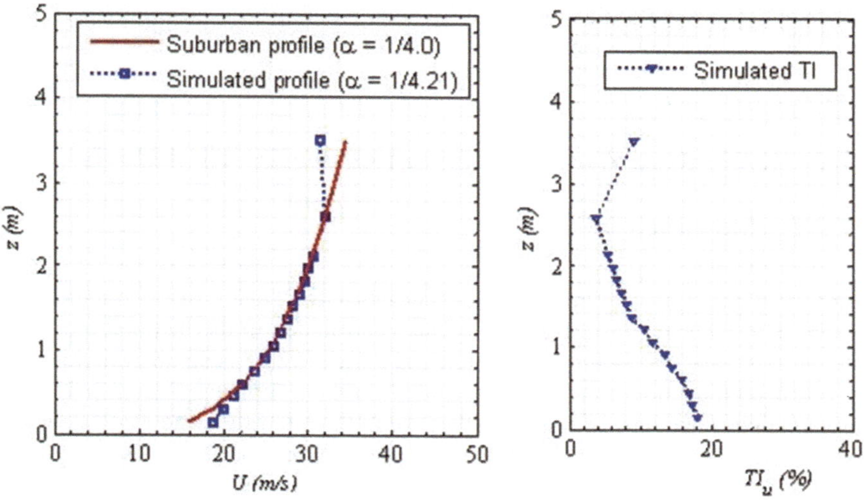

Fig. 17.1 Simulation of atmospheric boundary layer in the Wall of Wind. (a) Wind speed. (b) Turbulence intensity profile (Baheru et al. 2014a)

The WOW's 12 electric fans are arranged in a two-row by six-column pattern to produce a wind field of 6 m width and 4.25 m height, allowing aerodynamic testing of full-scale models of portions of small low-rise buildings. A contraction section is used downwind of the array of 12 fans to facilitate mixing and attaining a uniform flow field. The vertical flow directing vanes at the exit of the contraction zone guide the flow in the longitudinal direction. A 9.75 m long flow simulation box downwind of the contraction provides the required fetch length and confines the flow to develop the desired flow characteristics. Figure 17.1 shows a sample of mean wind speed and turbulence intensity profiles simulated by the WOW for suburban terrain. The profiles were generated using triangular-shape spires and floor roughness elements. The optimal shape and size of the spires and roughness elements were determined through a rigorous trial-and-error procedure based on wind profile measurements in the small-scale version of the 12-fan WOW (scale of 1:15). Based on the scaled model results, spires and roughness elements were constructed and installed in the full-scale 12-fan WOW.

Roofs are known to be the most vulnerable component of a building envelope. Suction generated due to wind flow separation coupled with positive internal pressures creates high dynamic pressures on the roof surface, which pose a significant threat to the structural integrity and functionality of roof structures. Damage to roof coverings is the prominent source of building performance problems during hurricanes. When the hurricane is accompanied by rain, water leaking through the failed roofing system can lead to major damage to interior contents, disrupt the functionality of critical and essential facilities, and even cause ceiling weakening and collapse that can result in injury to the occupants (FEMA 2005). In this document, two large-scale studies on low-rise building roof coverings performed at the WOW are reported and discussed.

17.2 Evaluating Wind Effects on Residential Tiled Roofs

17.2.1 Introduction

Roof tile damage usually initiates at the corners, eaves, rakes, or ridge regions and advances to other regions of the roof. This is attributed to the fact that hip, ridge, and eaves are located in flow separation and corner vortex generation regions, due to which high suction pressure is generated on their surfaces. Recent post-storm investigations have reported roof tile damage in the hip, ridge, and eave regions by hurricanes with wind speeds lower than the design wind speed (FEMA 2009). This implies that either current wind provisions are not accurate enough in representing peak uplift pressures in those roof regions or that conventional tile attachment methods are not as effective as expected under extreme conditions. Moreover, design pressures in current codes and standards are predominantly based on small-scale wind tunnel testing in which modeling of roof architectural features is difficult. Geometric scaling might also result in violations of Reynolds number similarity. The viscous stresses within high-frequency turbulent eddies are therefore larger than their full-scale counterparts and cause viscous dissipation of those eddies' energy (Simiu and Miyata 2006). This affects the local vortices at edges and corners in small-scale model testing, resulting in local pressures typically weaker than those at full scale (Simiu 2011).

Even though many efforts have been made to understand the nature of wind-induced uplift load on tiled roofs, most of the studies were limited to field and eave tiles (Kawair and Nishimura 2003; Robertson et al. 2007; Huang et al. 2009; Okada and Kikitsu 2005; Li et al. 2014; Hazelwood 1981; Tecle et al. 2013). Only a few research results are available on wind loading of hip and ridge tiles. Kawair and Nishimura (2003) investigated instantaneous uplift forces on tiled hip roofs using field measurements in natural wind. Their investigation pointed out that the windward tiles were exposed to high net uplift force due to the fact that external pressures and internal pressures developed in the cavity space between tiles and roof deck act in the same direction, while the leeward tiles were relieved due to the opposite directions of the two pressures (Kawair and Nishimura 2003). Similar conclusions were reached by Li et al. (2014) after performing full-scale experimental testing on roofs with high-profile tiles; the authors also concluded that accurate pressure measurements on roof tiles require modeling of the architectural details.

Very limited test-based data on roof tiles in eave, ridge, and hip-roof regions is available. Due to this scarcity of test-based data, available code and standard prescriptive attachment details for roof tiles were developed based on damage

Table 17.1 Roof tiles connection and attachment details (Baheru et al. 2014b)

Roof	Field tiles	Hip/ridge tiles
1	*Mechanically fastened:* The field tiles are firmly fastened to the roof deck using 2#8 screws per tiles	*Detail 1:* The hip/ridge tiles are attached to a wooden nailer board using 1#8 screw per tile. 20 gage hot-dipped zinc-galvanized steel strap spaced 61 cm o.c. was used to secure the wooden board to the roof deck as per RAS 118/119 requirement (Fig. 17.3)
2	*Adhesive:* The field tiles are attached to the roof deck using two-component double-patty-approved adhesive set system	*Detail 2:* The hip/ridge tiles are attached to a wooden nailer board set in a continuous bed of approved adhesive (Fig. 17.3)
3	*Adhesive:* The field tiles are attached to the roof deck using two-component single-patty-approved adhesive set system	*Detail 3:* The hip/ridge tiles are attached to a metal channel using approved adhesive (Fig. 17.3)

experienced during previous hurricanes (FRSA 2005). For example, hip and ridge attachment requirements of the Florida Building Code are enhanced based on damage experienced during 2004–05 hurricane seasons (FBC 2010). Therefore, in this study a full-scale model with realistic architectural features was tested under hurricane wind speeds of up to 67 m/s to gain a more practical and realistic understanding of the aerodynamics of wind-induced uplift pressures on tiles located in flow separation and cornering vortex regions. This helps to evaluate the performance of code prescriptive attachment details under severe wind pressures. Also, failure modes and mechanisms were identified to support the development of efficient mitigation techniques.

17.2.2 Methodology and Experimental Setup

The full-scale model used in testing was composed of a light-frame wood model base with plan dimensions of 2.70 m × 2.10 m and eave height of 2 m and three geometrically identical, interchangeable roofs with different tile attachment systems (Table 17.1). The roofs had a shape of half-cut hip roof with a pitch of 4:12. This shape was selected to represent hip, ridge, and eave/perimeter tiles. The overall plan dimensions of the roof were 4 m × 3 m with 0.30 m overhang on the gable-end side and 0.60 m overhang on the remaining three sides (Fig. 17.2). The structural roof decking was 19.1 mm thick plywood sheathing with a hot mop base sheet installed over its surface. High-profile concrete tiles were installed over the base sheet with standard accessory tile pieces used for the hip and ridge conditions. Tiles were installed by professional tilers in accordance with the Florida Building Code requirements.

Fig. 17.2 Tiled-roof light-frame test building model (Baheru et al. 2014a)

Full-scale experimental investigations were conducted in two phases. In the first phase, aerodynamic tests were conducted to collect high-resolution wind loading data on selected tiles at various wind directions. Pressure measurement tests were conducted at a mid-roof-height mean wind speed of 26.8 m/s and in nine wind directions. Test duration was 2 min. Pressure data were collected at 512 Hz sampling frequency and corrected for possible distortions caused by tubing length. Tile numberings and the wind directions tested are shown in Fig. 17.3. A total of 22 tiles (6 hip, 3 ridge, and 13 eave/perimeter tiles) were instrumented to capture the external pressure on the surface of the tiles and the internal pressure in the space between the tiles and the roof deck (Fig. 17.4). The locations of the instrumented tiles were carefully selected to capture the maximum loads and be representative of tiles in the flow separation and vortex generation regions. In the second phase of testing, destructive tests of tiled roofs were followed to scrutinize the performance of conventional tile attachment methods under high wind speeds. The different prescriptive roof tile attachment details used in the current study are described in Table 17.1. The eave tiles were attached to the substrate with *hurricane clips* as specified in RAS 118/119 (FBC 2010) for all the three roof types listed in Table 17.1. *Metal panel enclosure* was used to fill out the openings between the starting eave tiles and roof deck (Baheru et al. 2014b).

17.2.3 Results and Discussion

For each pressure tap, the time history of pressure coefficients $C_p(t)$ was calculated using Eq. (17.1):

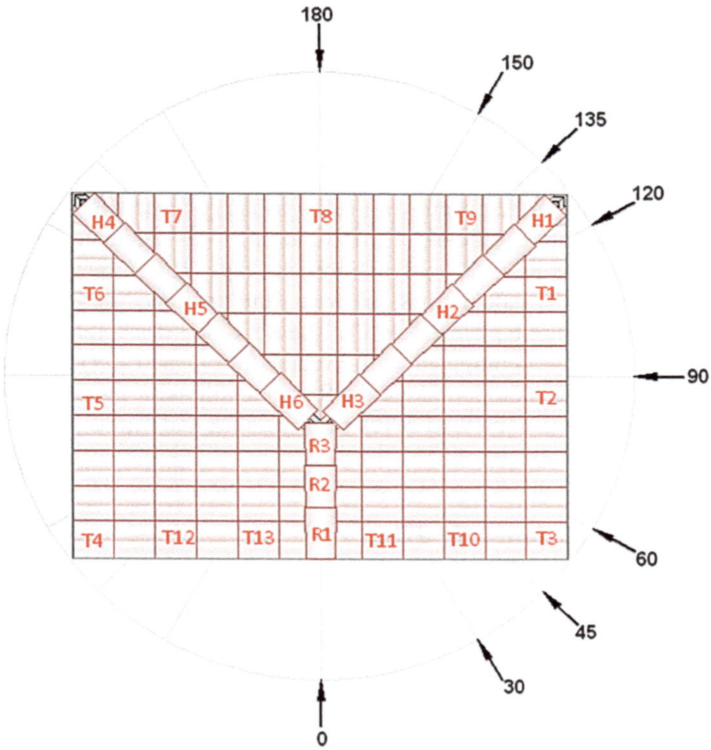

Fig. 17.3 Tile numbering and test wind directions (Baheru et al. 2014a)

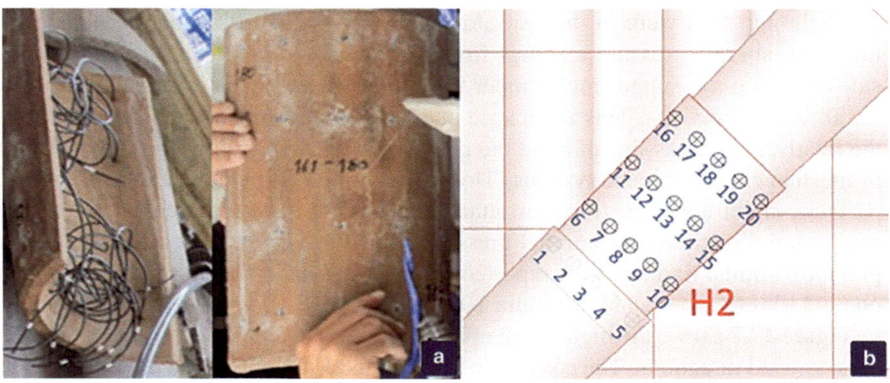

Fig. 17.4 Pressure taps layout. (**a**) Hip/ridge tile with installed pressure taps. (**b**) Numbering of pressure taps (Baheru et al. 2014a)

$$C_p(t) = \frac{p(t)}{\frac{1}{2}\rho V^2}, \quad (17.1)$$

where $p(t)$ represents time history of recorded pressure, V is the mean wind speed measured at model mid-roof height, and ρ is the density of air.

Each tile was instrumented with 20 external pressure taps, which were evenly distributed across its surface occupying equal tributary areas. Hence, an instantaneous external pressure coefficient C_{pe} for each tile was calculated by averaging the instantaneous C_p-s at its 20 external pressure taps. The instantaneous internal pressure coefficient C_{pi} calculations followed similar procedures but using only the two internal pressure taps for each tile. The net time-series pressure coefficient $C_{pnet}(t)$ for each tile was computed as the vector sum of its $C_{pe}(t)$ and $C_{pi}(t)$. The peak pressure coefficients reported are the 95 percentile peak pressure coefficients obtained using a statistical method developed by Sadek and Simiu (2002).

17.2.4 Wind Load on Hip Tiles

Figure 17.5 depicts the mean external pressure coefficients on the instrumented hip tiles as functions of wind direction. As illustrated in Fig. 17.5, the hip tiles experienced positive external pressure only when the wind direction was almost parallel to their hipline. Hip tiles H4, H5, and H6, which are located on the leeward hipline, exhibited the highest suction pressure at wind directions ranging from 90° to 180°. This indicates that for all the hip tiles, the critical external suction occurs when the flow direction is normal to their hipline.

Similarly, positive mean pressures were recorded underneath the hip tiles when the tiles were on the windward region, and negative pressures were experienced when the hip tiles were in the leeward region (Fig. 17.6). It should be noted that the mean internal pressure readings from the two pressure taps installed on either side of the hip-tile-supporting member were found to be almost equal in all hip tiles for all wind directions. This is because internal pressure equalization occurs due to the relatively free passage of air at the connection of tile-to-tile-supporting member in mechanical attachment systems. This development and equalization of internal pressure might not occur for tiles attached using adhesive-foam connections. For most wind directions, the internal pressure helped to alleviate the external suction pressure. Similar findings were reported in the past based on pressure measurements on field tiles (Kawair and Nishimura 2003; Li et al. 2014).

Figure 17.7 shows the highest net uplift peak pressure coefficients on hip tiles for various wind directions. The corner hip tiles H1 and H4 reported the largest net peak pressures from their respective hip lines, and the largest peak net negative pressure coefficient $C_{p,peak}$ was −1.45 on hip tile H4 at wind direction of 150°.

17.2.5 Wind Load on Ridge Tiles and Eave/Perimeter

Figure 17.8 shows the mean external pressure coefficients on the three ridge tiles as a function of wind direction. At wind direction of 0°, wind flow separation starts at tile R1, which causes high suction to be generated on its surface. This suction is much higher than that of ridge tiles R2 and R3. The ridge tiles experienced milder mean external suction as the wind approached the building from the hip-roof end (i.e., 90° to 180°) than when it approached the building from the gable end (i.e., 0° to 90°). This indicates a better aerodynamic performance of ridge tiles on hipped roofs with respect to those on gable roofs.

Eave tile T5 experienced negative mean external pressures in all wind directions, while positive pressure coefficients were measured on eave tile T9 for wind direction of 90° to 180°. Similar to hip tiles, the internal pressure measurements under each tile were almost equal, and their mean value was highly conforming with the mean external pressures. This resulted in alleviation of the positive and negative external pressures on the perimeter tiles.

Even though tile T9 is located near the corner hip tile H1, the largest mean external and largest peak net C_p measured were −0.6 and −0.85 respectively at wind direction of 60°, which are around 50 % less than the ones measured on H1 for the same wind direction. This shows the severity of the wind load on hip tiles.

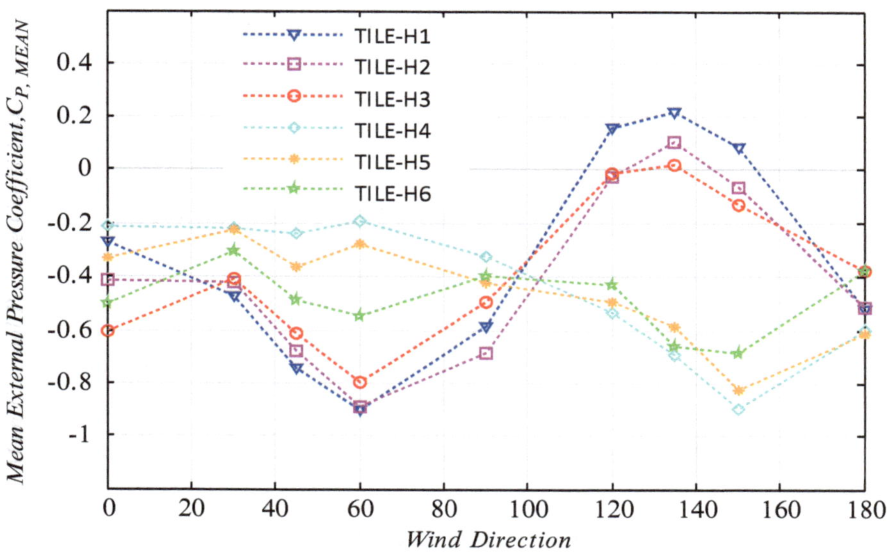

Fig. 17.5 Mean external pressure coefficients for hip tiles (Baheru et al. 2014a)

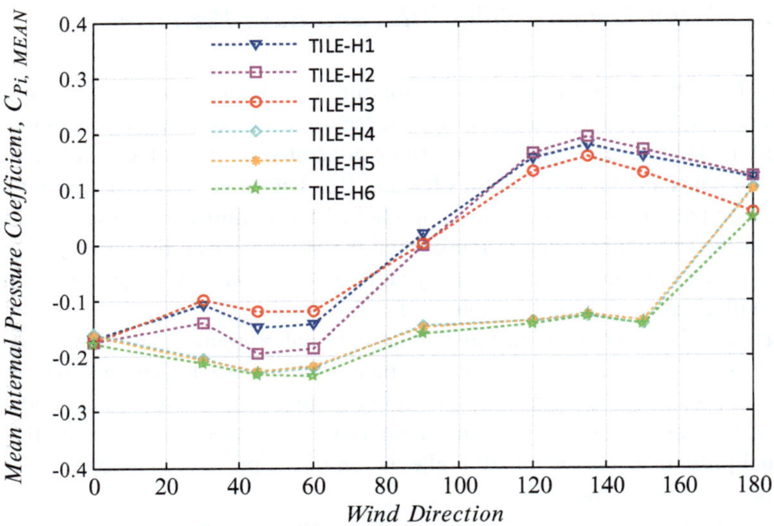

Fig. 17.6 Mean internal pressure coefficients for hip tiles (Baheru et al. 2014a)

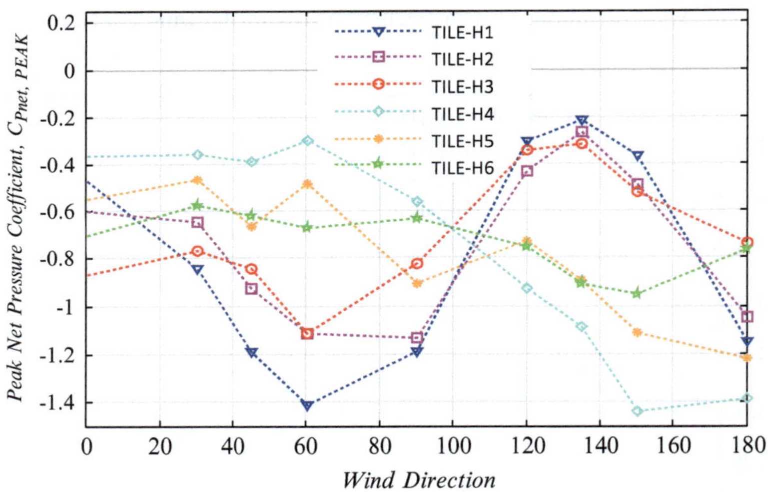

Fig. 17.7 Peak net pressure coefficients for hip tiles (Baheru et al. 2014a)

17.2.6 Performance of the Code Prescriptive Attachment Details

Destructive testing was used to examine the adequacy of three attachment methods endorsed by the code (Table 17.1) to withstand uplift pressures under simulated

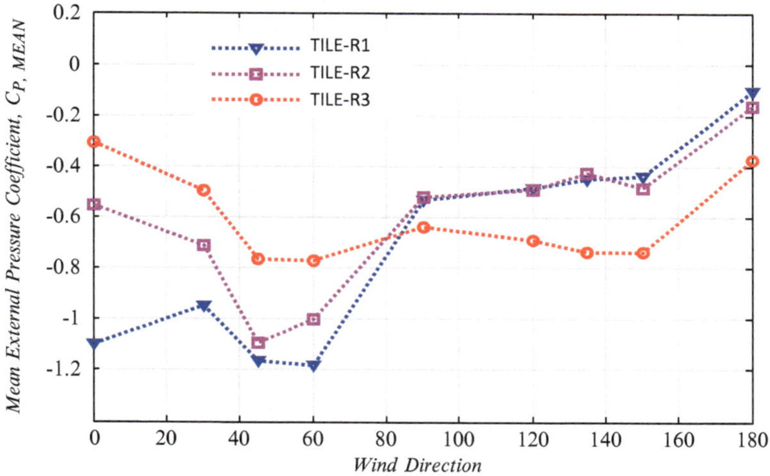

Fig. 17.8 Mean external pressure coefficients for *ridge* tiles (Baheru et al. 2014a)

wind speeds. The destructive tests were carried out for a total of 23 min duration at wind speeds ranging from 31 m/s to 67 m/s. From the aerodynamic pressure measurement tests, the wind direction producing maximum area-averaged net uplift force was determined to be the 60° wind direction. This direction was used in the destructive tests.

Figure 17.9a shows a failed eave tile. The failure occurred at a wind speed of 67 m/s during destructive testing of the mechanical attachment system. The failed eave tile was located near the windward hipline roof region (T9 in Fig. 17.3). At 60° wind direction, the net peak C_p value on tile T9 was −0.87, meaning the tile failure occurred around a suction of −2395 Pa. Pieces of tile T9 still remained attached to the roof deck after its failure; this shows that not only the strength of the tile-to-roof deck attachment system but also the strength of the individual tile was compromised.

The second type of attachment tested was the double-patty tile roof system, and no apparent failure was observed during the entire 23 min of high wind speed testing. During the destructive testing of the system, failure of the gable-end tile-to-roof deck attachment was observed on a tile located adjacent to tile T11 (Fig. 17.9b). Note that the gable-end tiles were connected to the roof deck system using two screws for all the three roof systems considered in this study. The failure could be attributed to the insufficiency of the gable-end tile-to-deck attachment in resisting the high suction on the gable-end tile itself and the adjacent tile T11. The significant overlap area of the two tiles will allow the transfer of the high suction wind load from tile T11 to the gable-end tile. This type of failure can easily be avoided by using additional adhesive attachment between the field and the gable-end tiles prior to applying cement mud filler. In addition to the failure of the gable-end tile-to-roof deck connection, cracks were observed in the cement paste filler used to fill the opening between the field tiles and hip, ridge, and gable-end tiles (Baheru et al. 2014a).

17.3 Performance of Standing Seam Metal Roofs Under Simulated Wind Loading: A Full-Scale Study

17.3.1 Introduction

Conventional methods of experimental testing used to evaluate the adequacy of metal roof panels in withstanding wind loads use defined positive/negative pressures that include neither the temporal nor the spatial pressure variations inherent in actual wind loading. This uniform pressure testing method might either set higher minimum design requirements for the entire system than necessary, or underestimate peak pressures at critical locations. Additionally, the modes of roof failures observed in uniform loading tests generally are not illustrative of the real failures which may happen in the field (Surry et al. 2005, 2007). This can result in a poor estimation of failure pressures or wind speeds. Hence, this method of using static uniform loading has been the object of continual debates and assessments by design engineers and researchers. Several research efforts have attempted to correlate results from uniform uplift pressure tests with those obtained under realistic wind loadings (Morrison and Kopp 2010; Sinno 2005; Farquhar et al. 2005).

In this study the performance of standing seam metal roofs under realistic wind load is evaluated by performing full-scale tests at the WOW. These included aerodynamic pressure measurement, deflection measurement, and rain-water intrusion. Furthermore, the pressure results were compared with the provisions of ASCE 7-10 (ASCE 2010) to evaluate the adequacy of design pressures obtained from the code.

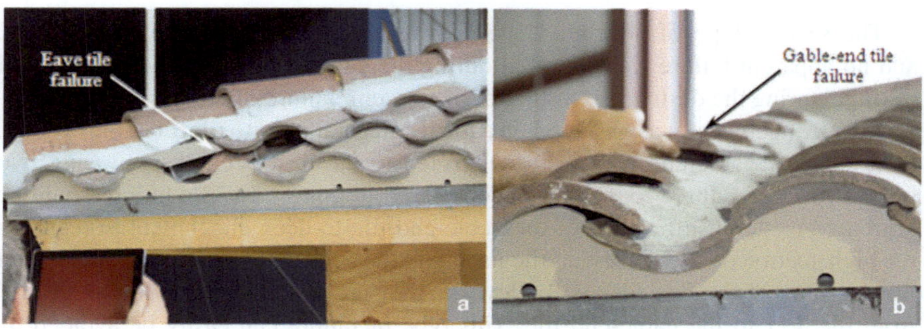

Fig. 17.9 Failure of tiles during high wind speed test. (**a**) Eave tile. (**b**) Gable-end tile (Baheru et al. 2014a)

17.3.2 Experimental Setup and Testing Protocols

The experimental study was performed in an open country wind field generated by the WOW. Vertical-leg and trapezoidal standing seam metal roofs were selected for testing (Fig. 17.10). The metal roofs were attached to a base structure designed to support interchangeable mono-sloped roofs with slope of 4.57°. The roof systems

Fig. 17.10 Standing seam metal roof mock-ups used in testing: (**a**) vertical-leg and (**b**) trapezoidal roof

were non-composite assemblies and were installed by professional installers in compliance with industry standards. The metal panels in both roofs consisted of 4.57 m long 24 gage metal panels attached to each other mechanically and using tape sealer, and each panel had three spans at 1.52 m o.c. The vertical-leg roof panels consisted of 460 mm wide flat pan sections and 51 mm tall ribs, while the trapezoidal roof panels had 76.2 mm tall ribs spaced at 610 mm. The vertical-leg roof had perimeter eave trims (attachments) installed on all four edges, but in the case of the trapezoidal roof, eave attachments were installed at the two sloping edges only. In the deflection measurement tests, to simulate the panel continuity at one end, the windward ends were screwed, but the leeward ends were free and had only one exposed fastener per panel rib. This fastener was installed through the vertical rib to make allowance for the factory lap cut in the panels that would not allow full panel/clip engagement.

A three-phase experimental investigation was conducted. In the first phase, aerodynamic pressure measurement tests were carried out on each roof at a low-eave height mean wind speed of 24.7 m/s for twelve wind directions. Each test had a duration of 1 min and data were collected at a sampling frequency of 512 Hz. In the second phase of full-scale testing, measurements of mid-panel and panel-rib deflections were carried out at different wind speeds and directions. Testing was performed at wind speeds ranging from 30.5 to 65.7 m/s and at the wind directions of 0°, 30°, 45°, 60°, 85°, 275°, and 315° (Fig. 17.11). Deflections were recorded using linear variable differential transformers (LVDTs) and Celesco SP2-type string pots. Data were collected at a sampling frequency of 100 Hz during a 1 min period. In the third phase, water intrusion tests were performed which will be discussed subsequently.

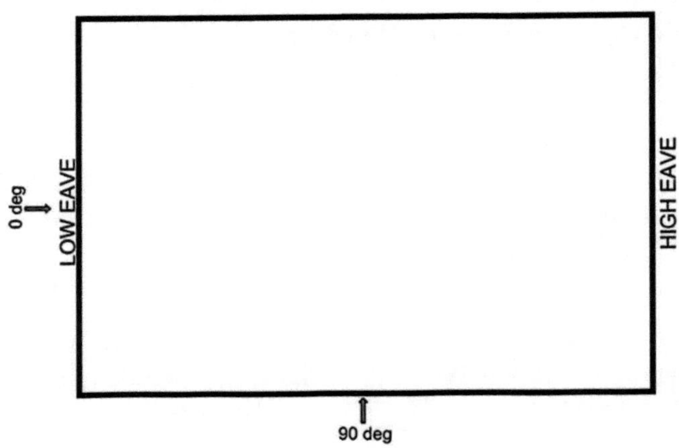

Fig. 17.11 Wind direction convention

17.3.3 Test Results and Discussion

17.3.3.1 Aerodynamic Pressure Results

The pressure results are presented as peak pressure coefficients in different roof regions. To compare the results with ASCE 7-10, the external pressure taps were grouped into different possible combinations of roof regions with surface area of 0.93 m^2 (which is the minimum components and cladding area specified in ASCE 7-10). The time series of measured pressures from pressure taps within each region were then spatially averaged. This produced time histories of pressures for different roof regions, from which the peaks can be computed. The peaks are computed using the method known as partial turbulence simulation (PTS) (Asghari Mooneghi 2014; Asghari Mooneghi et al. 2015). In this method, the WOW generated high-frequency turbulence and the missing low-frequency turbulence (which can be treated in a quasi-steady manner) are divided into distinct statistical processes, and their joint probability derived, from which a peak value with specified probability (95 % in this case) of not being exceeded in 1 h of full spectrum wind was selected. The peak pressure coefficients were calculated using Eq. (17.2).

$$C_{\mathrm{p\ peak}} = \frac{P_{\mathrm{peak}}}{\frac{1}{2}\rho V_{3\ \mathrm{sec}}^2}, \tag{17.2}$$

where P_{peak} is the estimated peak pressure and $V_{3\ \mathrm{sec}}$ is the three-second gust wind speed at model low-eave height.

In both standing seam roofs, relatively higher uplift pressures were recorded in the corner and edge roof regions when those regions were at the upwind position. This is an expected aerodynamic property of flat or near-flat roofs. Previous researches (Asghari Mooneghi et al. 2014; Mehta et al. 1992; Cochran and Cermak 1992; Irwin 2009) have shown that buildings experience highest suction due to the formation of conical vortices by oblique or cornering winds. The formation of the conical vortices creates strong suction forces on the roof surface below the vortices, which are highest at the front tip and decrease along the symmetry lines. Although the model geometry and the setup conditions were identical for both types of roofs, considerably higher pressures with larger nonuniformity were observed in all regions of the trapezoidal roof. This clearly demonstrated the significant effect of architectural features and connection details on the magnitude and spatial distribution of the wind-induced forces. Moreover, the results also reinforce the well-established notion that efficient aerodynamic design can play a critical role in mitigating roof uplift pressures.

According to ASCE 7-10, the experimental mono-sloped roof models are divided into three zones, namely, Zone 3′, Zone 3, and Zone 2′. ASCE 7-10 recommends using external GCp of −2.6, −1.8, and −1.6 for Zone 3′, Zone 3, and Zone 2′,

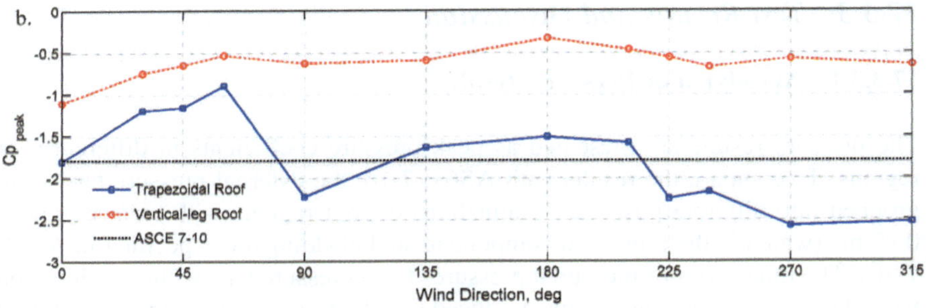

Fig. 17.12 Peak pressure coefficients in Zone 3

respectively. For the purpose of comparing the WOW C_p values with the ASCE value, a conversion factor (C_f) was calculated using Eq. (17.3)

$$C_f = \left(\frac{\frac{1}{2}\rho V_{h,3\ \mathrm{sec}}^2}{\frac{1}{2}\rho V_{10\ \mathrm{m},3\ \mathrm{sec}}^2 K_z K_{zt} K_d} \right), \tag{17.3}$$

where $V_{h,\,3\ \mathrm{sec}}$ is the 3-sec wind speed at the low-eave height, $V_{10\ \mathrm{m},\,3\ \mathrm{sec}}$ is the 3-sec wind speed at 10 m height, K_z is the velocity pressure exposure factor which in this case converts $V_{10\mathrm{m},3\,\mathrm{sec}}^2$ to $V_{h,3\,\mathrm{sec}}^2$, K_{zt} is the topographic factor, and K_d is a non-dimensional quantity smaller than unity that reflects the fact that the climatologically and aerodynamically or dynamically most unfavorable wind directions do not typically coincide (Habte et al. 2014). Assuming K_d to be 1.0 as directionality effects were not accounted for (Pierre et al. 2005), and using K_{zt} as 1.0, we get a conversion factor of 1.0.

Figure 17.12 illustrates the highest area-averaged peak suction pressure of all possible combinations of pressure taps that constitute a surface area of 0.93 m^2 inside roof zone 3 for both vertical-leg and trapezoidal roofs. Figure 17.12 also compares those maximum peak suctions with the ASCE 7-10 provisions and indicates that the ASCE 7-10 provisions are conservative for the vertical-leg roof; however, in the case of the trapezoidal standing seam roof, the ASCE 7-10 provisions were exceeded at some critical wind directions. Similar observations were also made for the other roof zones. This finding is in agreement with Morrison and Kopp (2010), which inferred that *GCp* values provided in ASCE 7-10 might be un-conservative when applied to standing seam metal roofs.

17.3.3.2 Wind-Induced Deflection Results

The mean and observed maximum net deflections were evaluated from time series of net deflections, $d_{net}(t)$, which were computed by subtracting the reference deflections measured at zero wind speed from the time series of recorded deflections.

During high wind loading, metal roofs experience vibratory deflections which are either upward or downward dominated, depending on the direction of the wind-induced forces on their surface. Most of the observed vibrations were upward dominated so all results discussed in this paper are for vertically upward deflections. The panel wind-induced vibrations were unsteady, nonuniform, and highly dependent on wind direction. An increase in wind speed was observed to increase not only the mean deflection but also the amplitude of the vibrations.

Relative to the mid-panels, lower deflections were observed along the panel ribs; this behavior is expected for standing seam roofs. The highest deflections were observed on the edge panels. In particular, as the wind direction approaches the direction normal to the panels' longer side, the deflection experienced by the free end side of the upwind edge panel becomes significantly higher than for the interior panels. During WOW deflection measurement tests, the windward side was fastened whereas the leeward side was free. Deflection measurements were taken at five different wind speeds and seven wind directions. The roof boundary conditions and the location of the deflection measurement devices were selected in accordance with the protocol normally followed in the ASTM E1592 tests. As a result, deflection measurement devices were not placed at the edge panels, which were later observed to experience the highest deflections.

Figures 17.13 and 17.14 show the variation of maximum deflections with wind direction for a wind speed of 56.8 m/s in the vertical-leg and trapezoidal roof, respectively. In the vertical-leg roof, LVDT 1 and LVDT 6 were used to record mid-panel deflections at the fixed low-eave end and free high-eave end, respectively, and SP1 was used to measure panel-rib deflections near the free high-eave end. Similarly, in the trapezoidal roof, LVDT 2 and LVDT 5 recorded mid-panel deflections near the fixed low-eave and free high-eave ends, respectively, while SP3 was used to measure panel-rib deflections near the free high-eave end. In all the deflection measurement devices depicted in Figs. 17.13 and 17.14, the critical wind directions were 85° and 275° (which are almost normal to the longer side

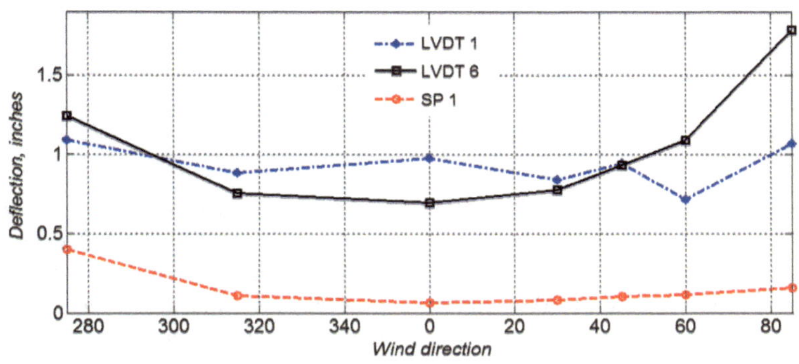

Fig. 17.13 Maximum deflection versus wind direction in vertical-leg roof, low-eave fixed (Chowdhury et al. 2014)

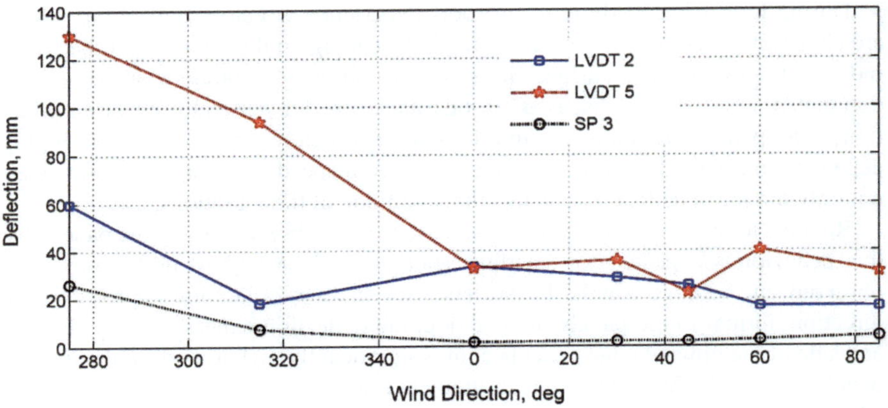

Fig. 17.14 Maximum deflection versus wind direction in trapezoidal roof (Chowdhury et al. 2014)

of the building). This shows that highest deflections occurred when the long side of the panels was normal to the approaching wind. The deflections experienced by the trapezoidal roof were significantly higher than for the vertical-leg roof. This is expected as the trapezoidal roofs were experiencing higher suction. Moreover, the trapezoidal roof was more flexible as the span between adjacent ribs in the trapezoidal roof was 152.4 mm longer than that of the vertical-leg roof. In both roofs the variability of deflection with wind direction was observed to be less at the location of the deflection measurement devices placed near the fixed end.

At low mid-panel deflections, the roof panel vibrations were observed to be mainly limited to the flat metal pan section. But as wind speed increases, the vibrations start creeping into the panel ribs, and at a sufficiently high wind speed, the flat pan section gets lifted entirely and the whole metal panel section starts vibrating from the ribs. At this stage, the rate at which deflection increases with increasing wind speed reduces significantly and the dominant frequency of vibration also changes. In a sense, at very high deflections, the metal panel assembly becomes increasingly stiffer until failure occurs.

17.3.3.3 Water Intrusion Test Results

The water intrusion tests were conducted with the objective of evaluating the performances of the two types of standing seam metal roofs under hurricane winds accompanied by simulated rain. The tests were carried out at a wind speed of 65.7 m/s, at a rain rate of 203.2 mm/h, and at various wind directions. The water intrusion tests were conducted for a duration of 2 min each. The wind-driven rain at the 12-fan WOW is generated using a plumbing system consisting of four vertical lines of spray nozzles attached to the supporting frame. The type and number of nozzles used is governed by the desired rain rate (Baheru et al. 2014a). A plastic bag was installed inside the model base under the roofs to collect water that might permeate through the roofs. In the vertical-leg roof, the open edge was sealed with tape and all metal panels were tested for water intrusion at high wind speed. In the trapezoidal roof, due to the occurrence of failure at two end panels (which is explained in the next section), water intrusion was investigated only in the interior panels, and the open edges were screwed to the purlins prior to testing. No case of water intrusion was observed in either standing seam roof for all angles tested (Chowdhury et al. 2014).

17.3.3.4 Observations of Permanent Deformations and Roof Failure

Wind-induced vibratory deflections exert tension stress on the metal panels, and cyclic loading at the clips and attachments, which can result in roof failure. If the metal panel of a standing seam roof blows off, it can become high-energy debris that can travel a considerable distance and cause injuries and/or damage to other properties (FEMA 2005). During the high wind speed experiments, two types of panel-rib/clip vibrations were observed to vibrate vertically and laterally. The vertical vibrations of the ribs/clips were created due to the unsteadiness of the uplift wind loading, while they were vibrating laterally as a result of the uneven and unsteady lateral forces exerted on the ribs/clips by adjacent panels. As a result of the uneven lateral forces, the clips were observed to rotate about their attachment points. It should be noted that neither the vertical nor the horizontal vibrations can be simulated in uniform uplift pressure tests. As a result, failure mechanisms observed during uniform static pressure tests might not be true representations of failure pressures or failure modes observed under real wind loading. Correct simulation of the actual failure mechanism is important because the component that fails first and its mode of failure determine the ultimate strength of the system (Dixon and Prevatt 2010).

In the vertical-leg roof, even though deflections of more than 110 mm were observed at the edge panels, no apparent permanent deformation or attachment failure was observed for the full range of wind speeds and directions tested. Similarly, in the trapezoidal roof, deflections of up to 190 mm were recorded without any visible roof failure. These show the ability of standing seam metal roofs to experience high elastic deflections. Clip failure was eventually encountered in the trapezoidal roof at low-eave height mean wind speed of 67 m/s. The main mode of failure was clip rupture and it occurred at the edge clips near the open end (Fig. 17.15). This failure is suspected to be due to fatigue generated by the vertical and lateral vibrations. It is important to mention that this type of failure might have been imposed by the high flutter experienced at the free end due to the fixed-free boundary condition used. This mode of failure is different from the clip slippage failure normally observed in uniform pressure tests. Once the first clip failure occurred, failure was observed to propagate along the open end as well as down the slope. Initiation of seam disengagement and panel-rib ruptures were also observed.

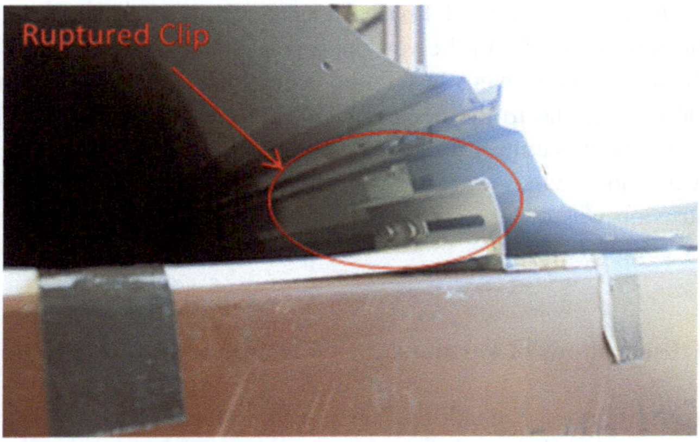

Fig. 17.15 Clip rupture at the free high-eave corner (Chowdhury et al. 2014)

17.4 Conclusions

Considering the multi-hazard nature of hurricanes due to high-velocity winds, wind-driven rain, and debris, an extensive study was performed on the behavior of building roofs, as they are the most vulnerable segments of buildings. In the first study presented in this chapter, full-scale wind load measurements on hip, ridge, and eave/perimeter tiles under simulated winds were conducted, and the performance of various types of code prescriptive tile attachment methods were evaluated. The following points summarize the major observations and conclusions:

- The highest net uplift pressures were recorded on the hip tiles when their hiplines were aligned normal to the direction of the approaching wind. From the ridge tiles, the gable-end tile experienced the highest suction, and the uplift pressure on ridge tiles was observed to be highly variable. Eave/perimeter tiles experienced relatively lower wind uplift load when compared with hip and ridge tiles.
- The internal pressure created in the cavity space between tiles and roof deck alleviated the effect of the external wind on the tiles. Internal pressure equalization occurred in hip and eave/perimeter tiles, but in the case of the ridge tiles, the internal pressure in the cavity space was uniform only when the wind direction was parallel to the ridge line.
- Failures of eave and gable-end tiles were observed during the destructive testing with mechanical and single-patty tile attachment, respectively. The failure of gable-end tile was due to the insufficiency of the attachment in resisting high suction wind pressure on the gable-end tile. The failure of eave tile despite the low pressure coefficient indicated that failure could be attributed to high vibration of the entire tile roof system under uplifting force. These failure results also highlighted the importance not only of tile-to-deck attachment adequacy but also of the strength of individual tiles and of the tiles' ability to act as a system.

Among all the three systems tested, the double-patty tile roof system performed best based on the angle of attack tested.

In the second study presented in this chapter, an experimental investigation was conducted with the objective of evaluating the structural performance of standing seam metal roofs under high wind speeds. Full-scale investigations, including wind-induced roof surface pressure measurements and roof panel deflection measurements, were conducted under realistic wind loading in the WOW facility at FIU. The following points summarize the major observations and conclusions of the research work:

- The magnitude of wind-induced pressures on the surface of the trapezoidal roof was significantly higher than on the vertical-leg roof, which showed that roof geometrical profile and eave details can considerably affect the nature of wind pressures on standing seam roofs. The comparison of WOW pressures with ASCE 7-10 provisions showed that the wind loading provisions in ASCE 7-10 can underestimate the wind loading on a mono-sloped trapezoidal standing seam roof.
- In both roofs, under the free-fixed boundary conditions tested, the highest deflections occurred in the free end of the edge panels when their longer side is situated normal to the approaching wind. The trapezoidal roofs experienced higher deflections due to their lesser metal panel rigidity. It was also observed that at high deflections metal panels become increasingly stiff and their fundamental frequency of vibration changes.
- In spite of high deflections, failure was not observed in the vertical-leg roof up to the highest wind speed tested. Failure did occur on the trapezoidal standing seam roof at a mean wind speed of 65.7 m/s at the open end of the edge panels. The main mode of failure was clip rupture, which is suspected to have been caused by vibrations experienced as a result of the turbulent nature of the wind loading. This mode of failure highlights the importance of the cyclic vibratory loads.

The results from these experimental studies reveal new aspects of building response, especially when subjected to high-speed winds, and highlight the importance of large- or full-scale modeling of buildings and structures, incorporating realistic details such as connection types and architectural features. This type of research at the Wall of Wind is important in promoting the progress of wind engineering science and technology; reducing life, property, and infrastructure losses; and helping prevent wind hazard events from becoming community disasters. The outcome will be a stronger, safer, and more economically and socially secure United States.

Acknowledgment The authors acknowledge the continuous support and expert advices of Dr. Peter Irwin during research carried out in FIU Wall of Wind and specifically thank him for his contributions to the projects discussed in this chapter. We also acknowledge the suggestions made by Dr. Emil Simiu to improve this chapter. The research was supported by funding from the National Science Foundation (Award # CMMI-1151003, CMMI-0923365) and the Florida Division of Emergency Management.

References

ASCE. (2010). *Minimum design loads for building and other structures*. American Society of Civil Engineers 7-10.

Asghari Mooneghi, M. (2014). *Experimental and analytical methodologies for predicting peak loads on building envelopes and roofing systems*. Florida Int'l University, FIU Electronic Theses and Dissertations. Paper 1846.

Asghari Mooneghi, M., Irwin, P., & Chowdhury, A. G. (2014). Large-scale testing on wind uplift of roof pavers. *Journal of Wind Engineering and Industrial Aerodynamics, 128*, 22–36.

Asghari Mooneghi, M., Irwin, P. A., & Chowdhury, A. G. (2015). *Partial turbulence simulation method for small structures*. Presented at the 14th International Conference on Wind Engineering, Porto Alegre, Brazil.

Baheru, T., Chowdhury, A., Bitsuamlak, G., Masters, F., & Tokay, A. (2014b). Simulation of wind-driven rain associated with tropical storms and hurricanes using the 12-fan Wall of Wind. *Building and Environment, 76*, 18–29.

Baheru, T., Habte, F., Moravej, M., & Gan Chowdhury, A. (2014a). *Full-scale testing to evaluate wind effects on residential tiled roofs*. Presented at the ICBEST, Aachen, Germany.

Chowdhury, A., Irwin, P., Habte, F., Mejia, A., 2014. *Full-Scale Testing to Evaluate Wind Effects on Residential Tiled Roofs*. A Report Submitted to the State of Florida Division of Emergency Management.

Cochran, L. S., & Cermak, J. E. (1992). Full-and model-scale cladding pressures on the Texas Tech University experimental building. *Journal of Wind Engineering and Industrial Aerodynamics, 43*, 1589–1600.

Dixon, C. R., & Prevatt, D. O. (2010). *What do we learn from wind uplift tests of roof systems?* In Structural Engineering Institute's 2010 Structures Congress Joint with the North American Steel Construction Conference, Orlando, FL.

Farquhar, S., Kopp, G., & Surry, D. (2005). Wind tunnel and uniform pressure tests of a standing seam metal roof model. *Journal of Structural Engineering, 131*, 650–659.

FBC. (2010). *Florida Building Code*. Tallahassee, FL: Florida Building Commission.

FEMA. (2005). *Summary Report on Building Performance, 2004 Hurricane Season*. Washington, DC: Federal Emergency Management Agency.

FEMA. (2009). *Performance of residential buildings (flood and wind), one- to two-family and multi-family*. Federal Emergency Management Agency.

FRSA, K. (2005). TRI-Concrete and Clay Roof Tile Installation Manual Fourth Edition (For Florida High Wind Applications, 150 mph). August.

Habte, F., Chowdhury, A. G., Yeo, D., & Simiu, E. (2014). Wind Directionality Factors for Non-hurricane and Hurricane-Prone Regions. *Journal of Structural Engineering,* 141(8), 04014208.

Hazelwood, R. A. (1981). The interaction of the two principal wind forces on roof tiles. *Journal of Wind Engineering and Industrial Aerodynamics, 8*, 39–48.

Huang, P., Mirmiran, A., Chowdhury, A. G., Abishdid, C., & Wang, T.-L. (2009). Performance of roof tiles under simulated hurricane impact. *Journal of Architectural Engineering, 15*, 26–34.

Irwin PA. (2009). *Wind engineering research needs, building codes and project specific studies*. 11th Americas Conf. on Wind Eng., san Juan, Puerto Rico.

Kargarmoakhar, R., Irwin, P. A., & Chowdhury, A. G. (2015). Reynolds number effects on twin box girder long span bridge aerodynamics. *Wind and Structures*, 20(2), 327.

Kawair, H., & Nishimura, H. (2003). Field measurement on wind force on roof tiles. In *Proceedings of the 11th International Conference on Wind Engineering* (pp. 599–606). Lubbock, TX: International Association for Wind Engineering and American Association for Wind Engineering Texas Tech University.

Li, R., Chowdhury, A., Bitsuamlak, G., and Gurley, K. (2014). Wind effects on roofs with high-profile tiles: experimental study. J. Archit. Eng., 10.1061/ (ASCE)AE.1943-5568.0000156, B4014002.

Mehta, K. C., Levitan, M. L., Iverson, R. E., & McDonald, J. R. (1992). Roof corner pressures measured in the field on a low building. *Journal of Wind Engineering and Industrial Aerodynamics, 41*, 181–192.

Moravej, M., Chowdhury, A. G., Irwin, P. A., Zisis, I., & Bitsuamlak, G. (2015). Dynamic effects of wind loading on photovoltaic systems. Presented at the 14th. International Conference on Wind Engineering (ICWE14), Porto Alegre, Brazil.

Morrison, M., & Kopp, G. (2010). Analysis of wind-induced clip loads on standing seam metal roofs. *Journal of Structural Engineering, 136*, 334–337.

Okada, H., & Kikitsu, H. (2005). Evaluation of wind resistance performance of clay tile roof based on survey of construction method and pulling-up tests. *Journal of Structural and Construction Engineering, 596*, 9–16.

Pierre, L. M. S., Kopp, G. A., Surry, D., & Ho, T. C. E. (2005). The UWO contribution to the NIST aerodynamic database for wind loads on low buildings: Part 2. Comparison of data with wind load provisions. *Journal of Wind Engineering and Industrial Aerodynamics, 93*, 31–59. doi:10.1016/j.jweia.2004.07.007.

Robertson, A. P., Hoxey, R. P., Rideout, N. M., & Freathy, P. (2007). Full-scale study of wind loads on roof tiles and felt underlay and comparisons with design data. *Wind and Structures, 10*, 495–510.

Sadek, F., & Simiu, E. (2002). Peak non-Gaussian wind effects for database-assisted low-rise building design. *Journal of Engineering Mechanics, 128*, 530–539.

Simiu, E. (2011). *Design of buildings for wind: A guide for ASCE 7-10 standard users and designers of special structures*. Wiley.

Simiu, E., & Miyata, T. (2006). *Design of buildings and bridges for wind: A practical guide for ASCE-7 standard users and designers of special structures*, John Wiley & Sons, USA.

Sinno, R. (2005). *Simulation of uplift loading on thin metal roofs (electromagnetic uplift testing)*. Mississippi State University.

Surry, D., Kopp, G. A., & Bartlett, F. M. (2005). Wind load testing of low buildings to failure at model and full scale. *Natural Hazards Review, 6*, 121–128.

Surry, D., Sinno, R. R., Nail, B., Ho, T. C., Farquhar, S., & Kopp, G. A. (2007). Structurally effective static wind loads for roof panels. *Journal of Structural Engineering, 133*, 871–885.

Tecle, A., Bitsuamlak, G., Suskawang, N., Chowdhury, A., & Fuez, S. (2013). Ridge and field tile aerodynamics for a low-rise building: A full-scale study. *Wind and Structures, 16*, 301–322.

Part VI
Geo-Hazards

Chapter 18
Accounting for Unknown Unknowns in Managing Multi-hazard Risks

Robert B. Gilbert, Mahdi Habibi, and Farrokh Nadim

Abstract A significant challenge in managing multi-hazard risks is accounting for the possibility of events beyond our range of experience. Classical statistical approaches are of limited value because there are no data to analyze. Judgment or subjective assessments are also of limited value because they are derived from within our range of experience. This chapter proposes a new framework, Decision Entropy Theory, to assess probabilities and manage risks for possibilities in the face of limited information. The theory postulates a starting point for assessing probabilities that reflect having no information in making a risk management decision. From this non-informative starting point, all available information (if any) can be incorporated through Bayes' theorem. From a practical perspective, this theory highlights the importance of considering how possibilities for natural hazards could impact the preferred alternatives for managing risks. It emphasizes the role for science and engineering to advance understanding about natural hazards and managing their risk. It ultimately underscores the importance of developing adaptable approaches to manage multi-hazard risks in the face of limited information.

18.1 Introduction

A significant challenge in managing multi-hazard risks is accounting for the possibility of events beyond our range of experience. A landslide in Oso, Washington, caused a debris runout that destroyed an entire community, including 43 lives, dammed a river creating a flood hazard upstream and then downstream when the dam was breached, and severed a transportation and utility corridor (GEER 2014). While this slope had failed multiple times in the past century, the debris never ran out far enough to impact the community until the event in 2014. The storm surge in Hurricane Katrina breached the levee system protecting neighborhoods below

R.B. Gilbert (✉) • M. Habibi
The University of Texas at Austin, Austin, TX, USA
e-mail: bob_gilbert@mail.utexas.edu

F. Nadim
Norwegian Geotechnical Institute, Oslo, Norway

P. Gardoni, J.M. LaFave (eds.), *Multi-hazard Approaches to Civil Infrastructure Engineering*, DOI 10.1007/978-3-319-29713-2_18

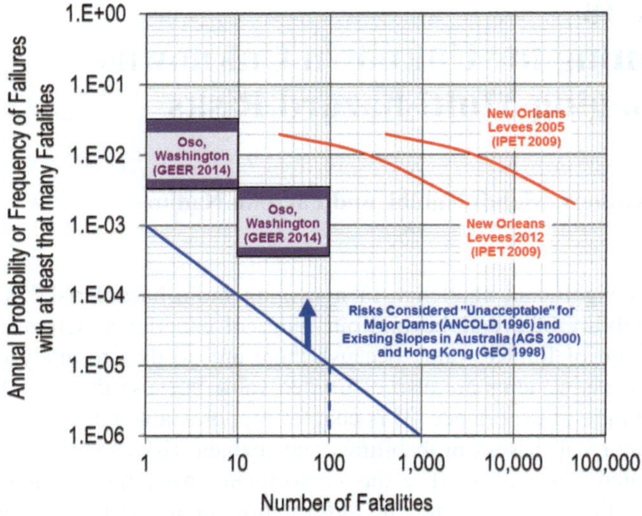

Fig. 18.1 Example comparisons of risks for natural hazards related to slopes and floods (AGS 2000; ANCOLD 1996; GEO 1998)

sea level, flooded the neighborhoods by connecting them to the ocean, and caused nearly 2000 deaths from drowning and exposure days after the storm. While storm surges had occurred in New Orleans before, the surge in Katrina was several meters higher than the maximum surge recorded previously in many locations (IPET 2009).

While rare and beyond our range of experience, these extreme hazards and multi-hazards are significant in terms of the consequences and the need to manage the associated risks (Fig. 18.1). Managing these risks requires first assessing probabilities of rare events (e.g., Liu and Nadim 2014; Nadim and Sparrevik 2013; Nadim 2011 and Lacasse and Nadim 2009). Probability assessments are typically based on historical data, observations, experience, and engineering judgment. However, historical data are sparse, and there is little or no experience with these extreme events because of their nature. In addition, models of combinations of highly complex events are inevitably simplified and uncertain. The lack of actual information with which to assess probabilities in these situations could lead to optimistic assessments that underrepresent risks or pessimistic assessments that overrepresent risks. It is therefore difficult to defend these assessments and rely on them to effectively manage the risks.

Taleb (2007) refers to possibilities beyond our experience as "black swan" events: "Before the discovery of Australia, people in the Old World were convinced that all swans were white, an unassailable belief as it seemed completely confirmed by empirical evidence… [The sighting of the first black swan] illustrates a severe limitation to our learning from observations or experience and the fragility of our knowledge. One single observation can invalidate a general statement derived from millennia of confirmatory sightings of millions of white swans." The United States Secretary of Defense, Donald Rumsfeld, infamously referred to possibilities beyond our experience as "unknown unknowns" (DOD 2002): "There are known knowns. These are things we know that we know. There are known unknowns. That is to say, there are things that we know we don't know. But there are also unknown unknowns. There are things we don't know we don't know." The challenge is to logically and defensibly account for "black swans" or "unknown unknowns" in managing risk.

This paper proposes a framework, Decision Entropy Theory, to assess probabilities and manage risks for possibilities beyond our experience. First, the challenge of assessing probabilities with limited data is described. Next, the mathematical basis for the proposed theory is presented and illustrated. Finally, practical insights for risk management are drawn from the theory.

18.2 The Challenge with Prior Probabilities

Probability theory is used to represent uncertainty. This theory is based on starting with a comprehensive set of all possible events, known as the sample space and denoted S. Probabilities for events are assessed based on available information via Bayes' theorem:

$$P\left(\text{Event } i\,|\text{Information} \cap S\right) = \frac{P\left(\text{Information }|\text{Event } i \cap S\right) P\left(\text{Event } i\,|S\right)}{\sum_{\text{all } j} P\left(\text{Information }|\text{Event } j \cap S\right) P\left(\text{Event } i\,|S\right)}$$

where $P\left(\text{Event } i\,|\text{Information} \cap S\right)$ is the probability for Event i given available information, which is the probability of interest and referred to as the posterior or updated probability; $P\left(\text{Information }|\text{Event } i \cap S\right)$ is the probability of obtaining the available information given that Event i occurs, which is referred to as the likelihood function; and $P(\text{Event } i|S)$ is the probability for Event i in the initial sample space, which is referred to as the prior or initial probability.

Consider the annual chance of a hazard, F. The available information is that this hazard has occurred x times in n years of experience. If we assume that occurrences follow a Bernoulli sequence,[1] then the likelihood function is given by the binomial distribution:

$$P\left(x \text{ occurrences in } n \text{ years }|F = f_i\right) = \left[\frac{n!}{x!\,(n - x!)}\right] f_i^x (1 - f_i)^{n-x}$$

If there are no occurrences in the experience, then the likelihood function does not distinguish between small chances of occurrence that are about an order of magnitude less than the inverse of the length of experience (Fig. 18.2).

[1] A Bernoulli sequence assumes that occurrences are independent from year to year and that the chance of occurrence each year is a constant.

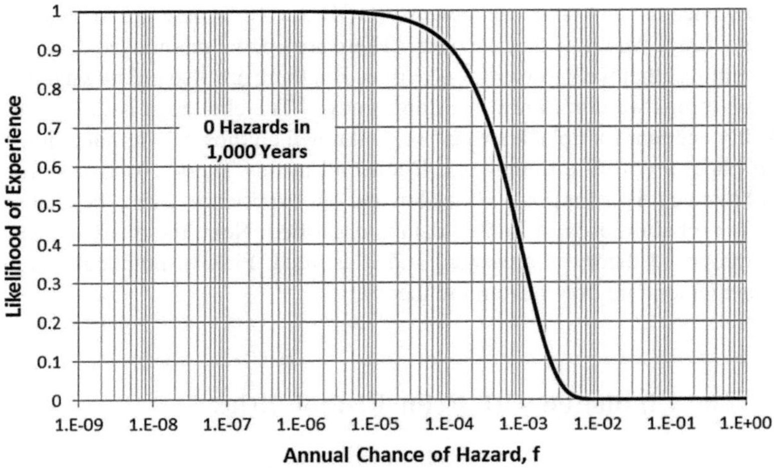

Fig. 18.2 Example likelihood function for no occurrences of a hazard in a period of experience versus the annual chance of occurrence

The challenge is the prior probability in Bayes' theorem, $P(F = f_i | S)$, because it is not conditioned or based on information. This prior probability represents "complete ignorance" in the words of Luce and Raiffa (1957) and is commonly referred to as a "non-informative prior" probability.[2] The non-informative prior sample space is important because it sets the stage for assessing or conditioning probabilities on any available information (whether subjective or objective). Consider a uniform prior distribution for the annual chance of occurrence between 0 and 1 (Fig. 18.3). The updated probability distribution for the annual chance of occurrence (Fig. 18.4) is the prior distribution filtered through the likelihood function; wherever the likelihood function is flat, such as for small chance of occurrence (Fig. 18.2), the updated distribution is entirely a reflection of the prior distribution.

The significance of the prior probability distribution is demonstrated using three different distributions: a uniform distribution on the annual chance of occurrence, a uniform distribution on the logarithm of the annual chance of occurrence,[3] and a uniform distribution on the return period or the inverse of the annual chance of occurrence[4] (Fig. 18.5). The updated probability distributions for the annual chance of occurrence are significantly different between the three possible prior

[2]The prior probability is non-informative because it does not depend on information: $P(S | \text{Information}) = P(S)$.

[3]Note that the logarithm of the annual chance of occurrence approaches negative infinity and the return period approaches positive infinity. A lower bound of 1×10^{-9} was used for the annual chance of occurrence (or an upper bound of 1×10^{9} on the return period). Since the likelihood function is flat approaching this lower bound, the choice of a lower bound will affect the results and underscores the significance of the shape of the prior probability distribution.

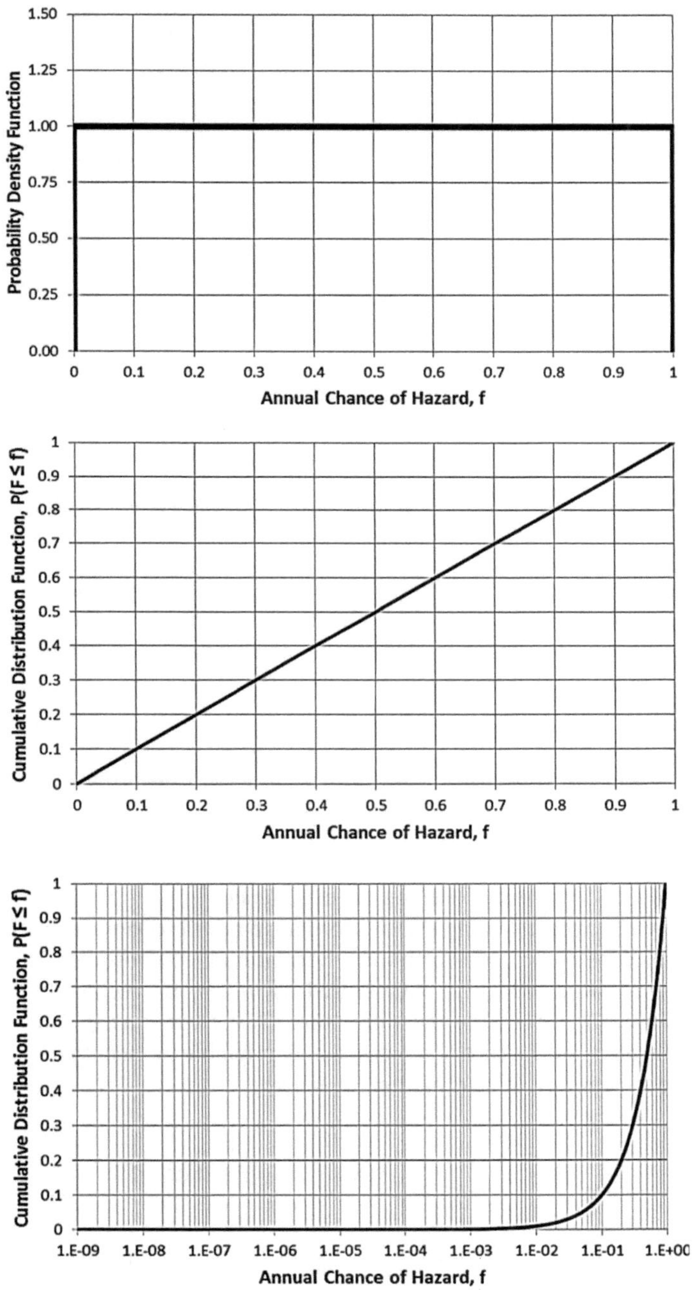

Fig. 18.3 A uniform prior probability distribution for the annual chance of occurrence. (Three different representations of the probability distribution are shown. Since we will typically be dealing with order-of-magnitude ranges for the chance of occurrence (Fig. 18.1), the representation with the cumulative distribution function on a logarithmic scale for the chance of occurrence will be used throughout this chapter.)

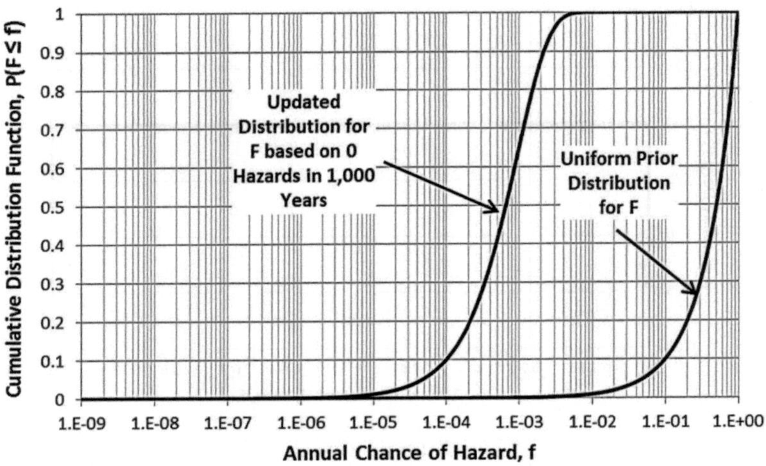

Fig. 18.4 Prior and updated probability distributions for the annual chance of occurrence

probability distributions (Fig. 18.6 and Table 18.1). The significance of the prior probability distribution for the annual chance of occurrence is further highlighted by considering the expected value for the updated probability distribution, which is the probability of an occurrence in 1 year (Fig. 18.7):

$$P\,(\text{Occcurrence in One Year}) = E(F) = \sum_{\text{all } f_i} f_i p_F\,(f_i|\,x \text{ occurrences in } n \text{ years})$$

where $p_F(f_i|x$ occurrences in n years) is the updated probability distribution for the annual chance of occurrence based on the experience.

The challenge of establishing a non-informative prior probability distribution has been the subject of theorists for centuries. Bernoulli (1738) postulated the principle of insufficient reason, which is paraphrased as: *If one is completely ignorant as to which state will occur, then the states should be treated as if they are equally probable.* Jaynes (1957, 1968) expressed the principle of insufficient reason as maximizing the entropy of information, H, where

$$H = \sum_{j=1}^{n \text{ States}} -\ln\left[P\,(\text{State } j\,|S)\right] P\,(\text{State } j\,|S)$$

The greatest "lack of information" is the set of prior probabilities that lead to the maximum entropy, which is obtained when these probabilities are equal for all states. In practice, this approach has been applied to a variety of problems (e.g., Tribus 1969; Box and Tiao 1973).

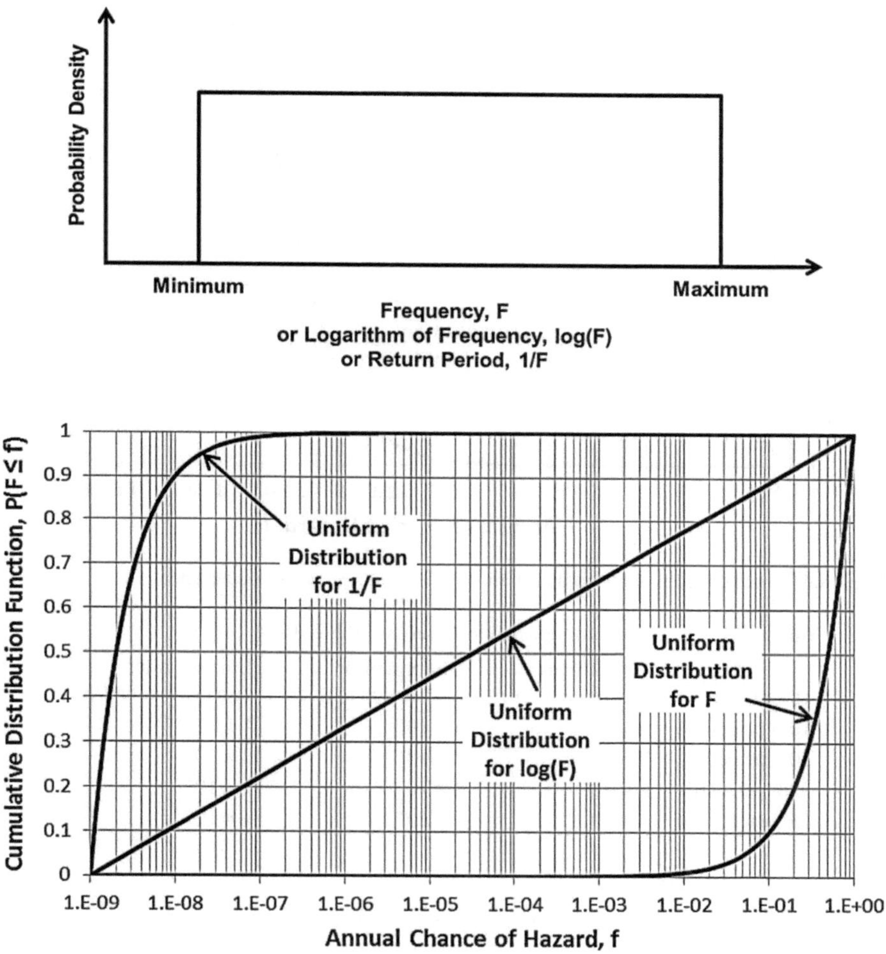

Fig. 18.5 Alternative prior probability distributions for annual chance of occurrence

While mathematically convenient, the principle of insufficient reason has consistently been criticized. Keynes (1921), who renamed it the principle of indifference, contends that this principle is ambiguous and can lead to paradoxical or contradictory conclusions. One of his examples was the specific volume of a substance, where the state of nature is bounded between 1 and 3 L^3/M. Based on the principle of insufficient reason, it would be equally probable for the specific volume to be between 1 and 2 or between 2 and 3 L^3/M. The density, the inverse of the specific volume, will be bounded between 1/3 and 1 M/L^3. In this case, the principle of insufficient reason suggests it would be equally probable that the density be between 1/3 and 2/3 M/L^3 or 2/3 and 1 M/L^3, meaning that it would be equally probable for the specific volume to be between 1 and 1.5 L^3/M or 1.5 and

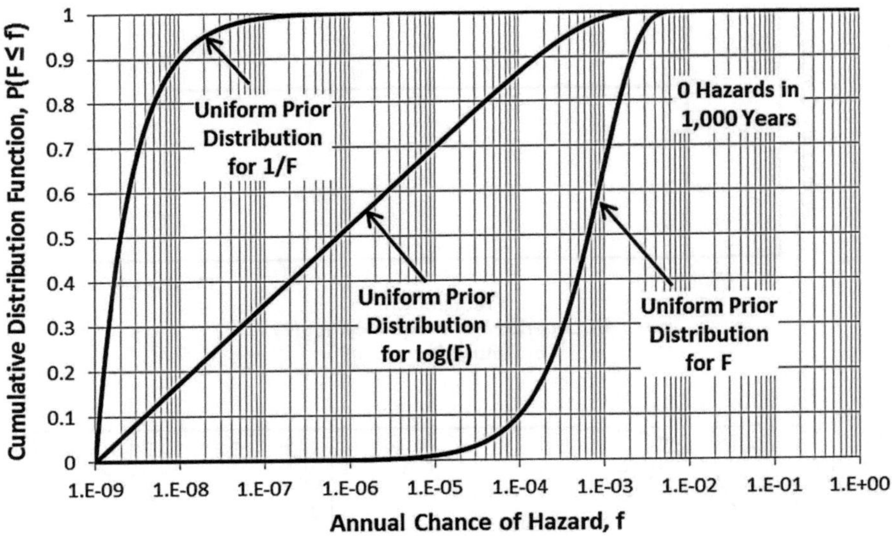

Fig. 18.6 Three different updated distributions for annual chance of occurrence

Table 18.1 Descriptors for updated distributions in Fig. 18.6

Descriptor	Uniform prior distribution for F	Uniform prior distribution for log(F)	Uniform prior distribution for $1/F$
10th percentile	1×10^{-4}	4×10^{-9}	1.1×10^{-9}
50th percentile	7×10^{-4}	8×10^{-7}	2×10^{-9}
90th percentile	2×10^{-3}	2×10^{-4}	1×10^{-8}
Expected value	1×10^{-3}	8×10^{-5}	1×10^{-8}

3 L^3/M. Therefore, these two perspectives give different prior probabilities for the same states of nature.

This difficulty of consistency is illustrated with the example of assessing the annual chance of occurrence for a hazard (Figs. 18.5, 18.6, and 18.7). There is no theoretical basis for defining the states of nature in terms of F, log(F), or $1/F$. One could justify log(F) on the basis that order-of-magnitude ranges are typically of interest for the chance of hazards (e.g., Fig. 18.1). One could justify $1/F$ on the basis that the hazards are typically represented by their return period.

Luce and Raiffa (1957) criticized the principle of insufficient reason in the context of decision making. They illustrated their criticism with an example decision problem where the consequences of different decision alternatives depend on uncertain states of nature. They showed they could affect which decision alternative was preferred by arbitrarily dividing states of nature into substates. Luce and Raiffa (1957) contended that this result is irrational; if there is "complete ignorance," then how the states of nature are labeled should not affect the preferred decision alternative.

Journel and Deutsch (1993) also criticized the principle of insufficient reason for producing irrational results. Their example is an oil reservoir in which the permeability of the reservoir varies spatially. They showed that maximizing the entropy of information in the input (the permeability field) generally minimizes entropy in the output of interest for making decisions (the water breakthrough time).

In summary, the challenge with assessing probabilities in practice is in establishing a non-informative prior probability distribution as a starting point. This non-informative prior probability distribution can significantly affect the assessed probabilities, particularly when there is limited information available. Existing approaches to establish non-informative prior probabilities based on the principle of insufficient reason do not provide a rational or consistent basis.

18.3 Decision Entropy Theory

We postulate a theory for establishing a non-informative sample space that characterizes the information in terms of making a decision between two alternatives. The greatest lack of information for the decision, i.e., the non-informative starting point, is defined by the following three principles:

1. An alternative compared to another alternative is equally probable to be preferred or not to be preferred.
2. The possible gains or losses for one alternative compared to another alternative are equally probable.
3. The possibilities of learning about the preference of one alternative compared to another with new information are equally probable.

The premise of this theory is that probabilities provide input to decision making (i.e., managing risk); therefore, non-informative probabilities are probabilities that do not inform the decision. The mathematical framework for implementing these principles is presented in Appendix 1.

The intent of this theory is to overcome the shortcomings associated with the principle of insufficient reason:

1. It is consistent: It always produces the same non-informative distribution of the possible outcomes of a decision no matter how those outcomes formulated in terms of states of nature. This non-informative distribution is what will affect the preferred decision alternative and the value of obtaining additional information about the decision.
2. It is rational: It follows the premise that the purpose of assessing probabilities is to support decision making. The non-informative distribution about the possible outcomes of a decision represents the maximum uncertainty in making a decision.

To illustrate this theory, consider a basic decision in risk management either eliminating a risk with mitigation or accepting the risk with an uncertain annual

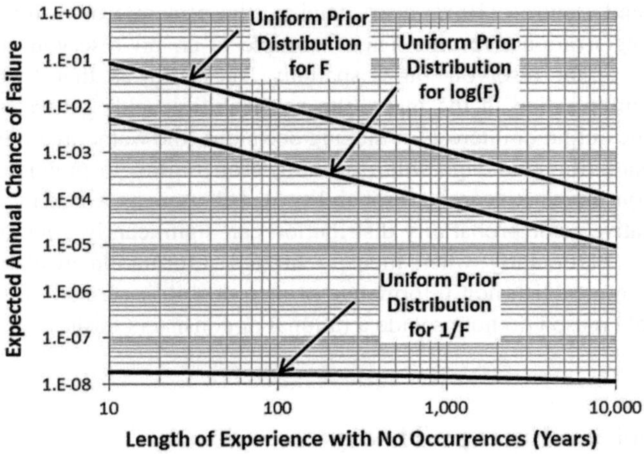

Fig. 18.7 Expected annual chance of failure of three different prior probability distributions for annual chance of occurrence

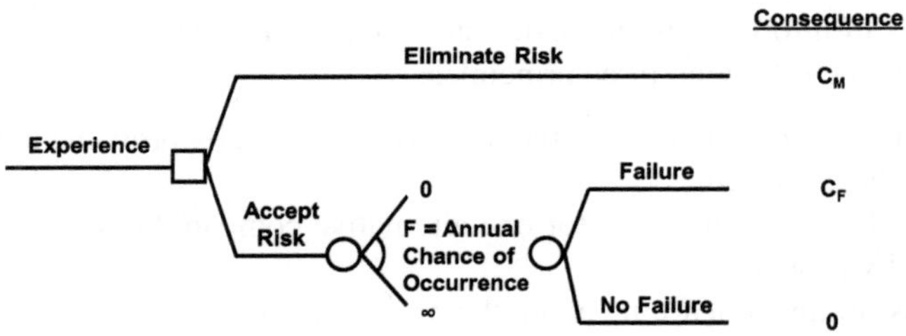

Fig. 18.8 Basic decision tree for risk management with uncertain chance of failure

chance of failure (Fig. 18.8). The alternative to accept the risk will be preferred when its expected annual consequence is smaller[4] than that for eliminating the risk:

$$\text{Prefer Accept Risk if } c_F \times F \leq c_M \text{ or } F \leq \frac{c_M}{c_F} = \frac{1}{c_F/c_M}$$

where c_M is the annual cost of eliminating the risk, c_F is the cost of a failure due to the occurrence of the hazard, and F is the annual chance of occurrence for the hazard. Therefore, the ratio $\frac{1}{c_F/c_M}$ is indicative of the threshold for "tolerable" risk in Fig. 18.1.

[4]Consequence will be considered here as a positive cost. The degree of preference for an outcome increases as the cost of that outcome decreases.

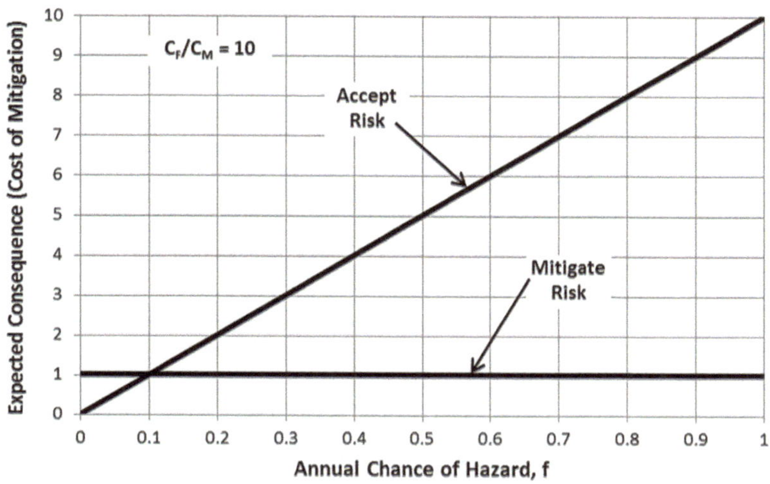

Fig. 18.9 Expected consequences versus annual chance of hazard for accepting risk and mitigating risk

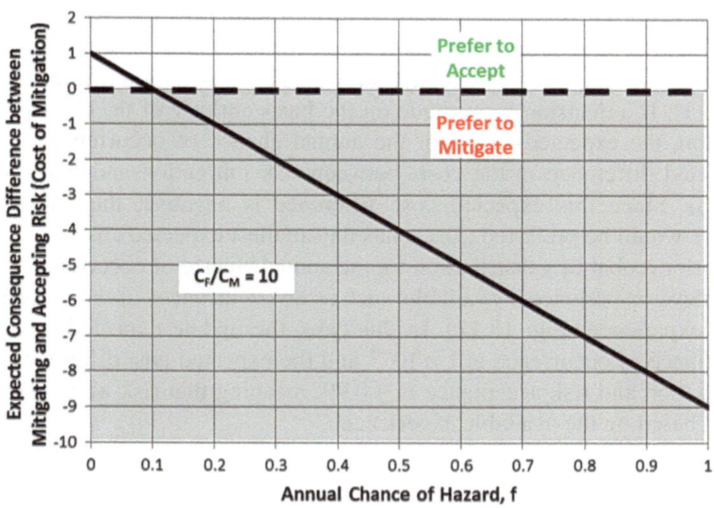

Fig. 18.10 Difference in expected consequences between mitigating risk and accepting risk versus annual chance of hazard

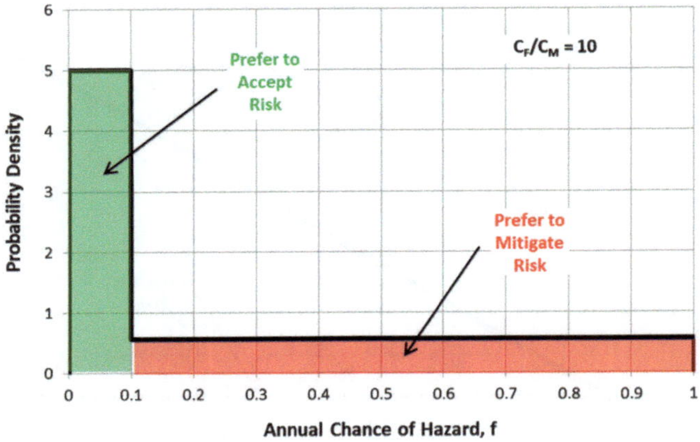

Fig. 18.11 Non-informative prior probability distribution for annual chance of hazard in decision between mitigating risk and accepting risk

The expected consequences for each alternative and the difference in expected consequences between alternatives are linear functions of the uncertain annual chance of occurrence for the hazard (Figs. 18.9 and 18.10). Subsequently, the non-informative prior probability distribution for the annual chance of occurrence based on the principle of Decision Entropy Theory is a bi-uniform distribution (Fig. 18.11). If a decision were made on the basis entirely of the prior probability distribution, the expected value for the annual chance of occurrence is 0.32 and the expected difference in the costs between risk mitigation and risk acceptance is $-2.2c_M$. Since this expected cost difference is negative, the risk mitigation alternative would be preferred (i.e., it has the smallest expected cost).

This prior probability distribution for the annual chance of occurrence is updated with whatever experience is available, such as no occurrences of the hazard in 1000 years of experience (Fig. 18.12). In this case, the update expected value for the annual chance of occurrence is 1×10^{-3} and the expected cost difference between risk mitigation and risk acceptance is $+0.99$, meaning that risk acceptance is now preferred based on the available experience.

18.4 Practical Insights for Risk Management from Decision Entropy Theory

The Decision Entropy Theory provides practical insights into managing risk in the face of uncertainty. The significance of uncertainty depends on its impact in risk management decisions. The value of additional information in a decision is

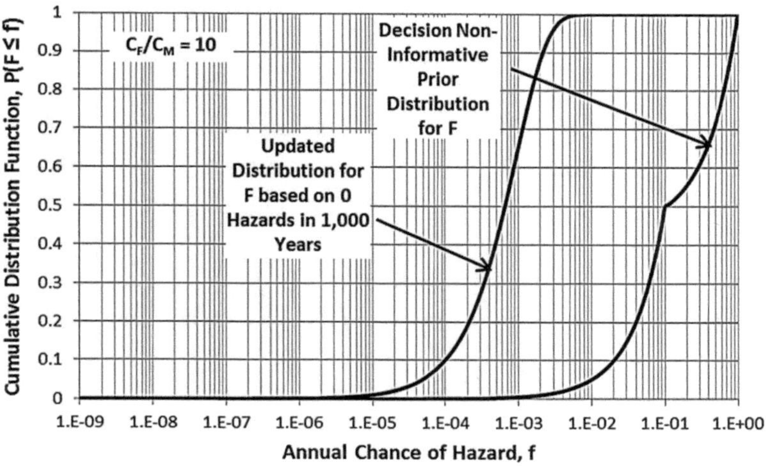

Fig. 18.12 Updated probability distribution for annual chance of hazard in decision between mitigating risk and accepting risk with no occurrences of hazard in 1000 years of experience

accentuated when a non-informative prior sample space is established. Finally, the possibility that available data are not relevant can have a significant impact on a risk management decision.

18.4.1 Significance of Uncertainty in Natural Hazards Depends on Risk Management Decisions

For the basic risk management decision (Fig. 18.8), the prior probability distribution for the annual chance of the hazard depends on the ratio of the cost of failure to the annual cost of mitigation, c_F/c_M. The larger the ratio, the smaller the threshold value for the annual chance of the hazard between preferring to accept versus mitigate the risk, $\frac{1}{c_F/c_M}$, and the more pronounced the left hand tail of the probability distribution for the annual chance of the hazard (Fig. 18.13). This manifestation of a "lack of information" at the start allows for the possibility that the actual chance of the hazard could be greater or smaller than the threshold (the decision point); note that the 50th percentile in the prior probability distribution is at $\frac{1}{c_F/c_M}$ (Fig. 18.13).

The result of starting open to the possibility of the decision going either way without any information is that subsequent information can be more influential in changing the decision (Fig. 18.14). To illustrate this point, consider the decision with $c_F/c_M = 1 \times 10^3$, meaning that the threshold value for the chance of the hazard is 1×10^{-3}. With the non-informative prior probability distribution, the preferred alternative is to mitigate the risk. However, an experience of no hazards occurring in a period of greater than 50 years is enough to change the preferred alternative to

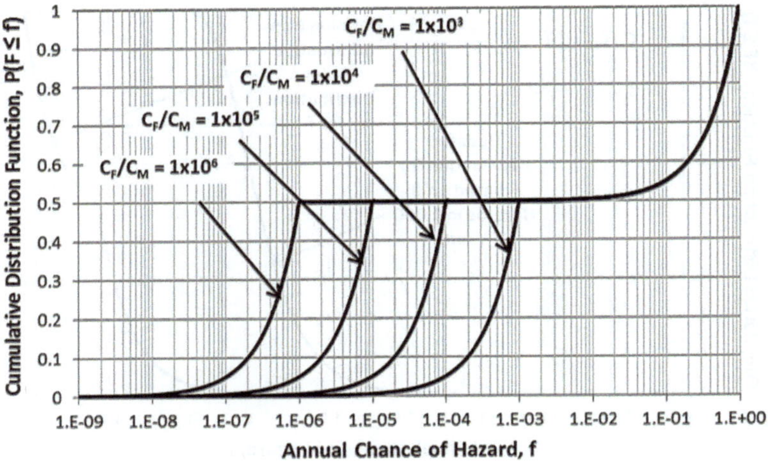

Fig. 18.13 Prior probability distributions for annual chance of hazard in decision between mitigating risk and accepting risk

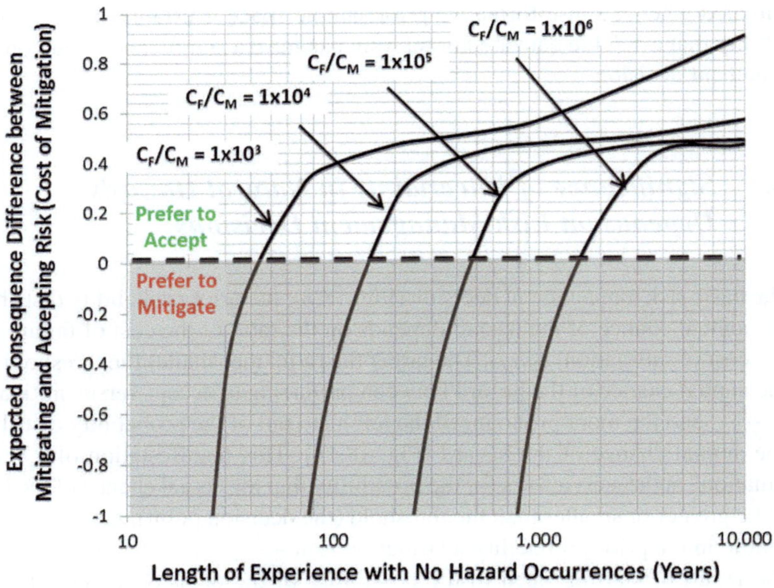

Fig. 18.14 Difference in expected consequences between mitigating risk and accepting risk versus length of experience with no hazard occurrences

accepting the risk (Fig. 18.14). For comparison, here is how other prior probability distributions would affect this decision:

- With a uniform prior probability distribution for the annual chance of the hazard, a period of greater than 1000 years with no hazards occurring would be required to change the preferred alternative to accepting the risk (the updated expected chance of occurrence is less than 1×10^{-3} for lengths of experience with no occurrences greater than 1000 years in Fig. 18.7).
- With a uniform prior probability distribution for the logarithm of the annual chance of the hazard, a period of greater than about years with no hazards occurring would be required to change the preferred alternative to accepting the risk (Fig. 18.7).
- With a uniform prior probability distribution for the return period or the inverse of the annual chance of the hazard, the preferred alternative would be accepting the risk without any information (Fig. 18.7).

Therefore, the greatest lack of information at the start (the non-informative prior probability distribution from Decision Entropy Theory) is not conservative or un-conservative in the context of risk management; it is intended to provide an unbiased starting point before incorporating any available information. In this particular example, the "black swan" that decision entropy is accommodating is the possibility that the annual chance of hazard may actually be "small" as opposed to assuming that it is "large" in the absence of information. While potentially counterintuitive (particularly for "conservative" engineers), this concept is of practical significance and used often implicitly in real-world decisions. We would essentially have never built a major dam (c_F/c_M of about 1×10^6 based on Fig. 18.1) if we needed to wait more than 1,000,000 years to assess the chance of extreme hazard occurrences. Decision entropy provides a rational and consistent basis to support making risk management decisions in the face of the inevitable lack of information about extreme hazard occurrences.

18.4.2 Value of Information Emphasized by a Non-informative Starting Point

The value of obtaining additional information (beyond the available experience) about the chance of hazard depends on how probable it is that additional information will change the risk management approach (Fig. 18.15). Quantitatively, the value of information is the maximum amount the decision maker would be willing to spend to pursue additional information. The value of information for the basic risk management decision is bounded by the cost required to eliminate the risk (c_M).

The value of perfect information about the annual chance of occurrence for the hazard depends both on the decision (i.e., c_F/c_M) and the available experience (Fig. 18.16). For relatively limited experience (small lengths of experience with no

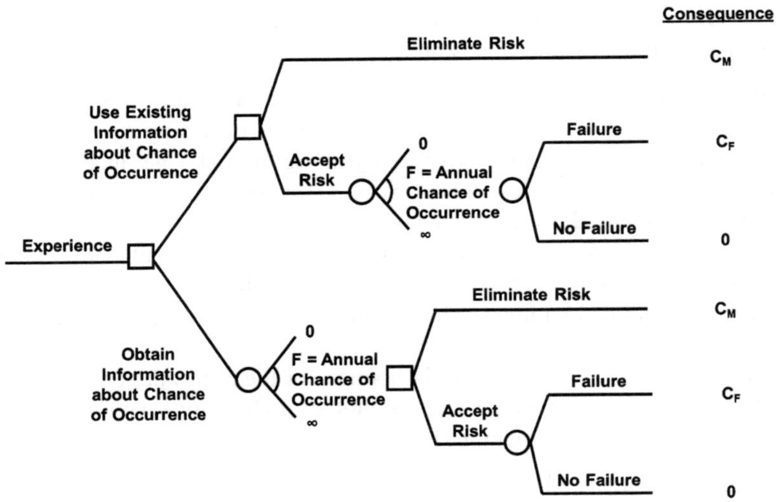

Fig. 18.15 Decision tree to assess value of information in managing risk

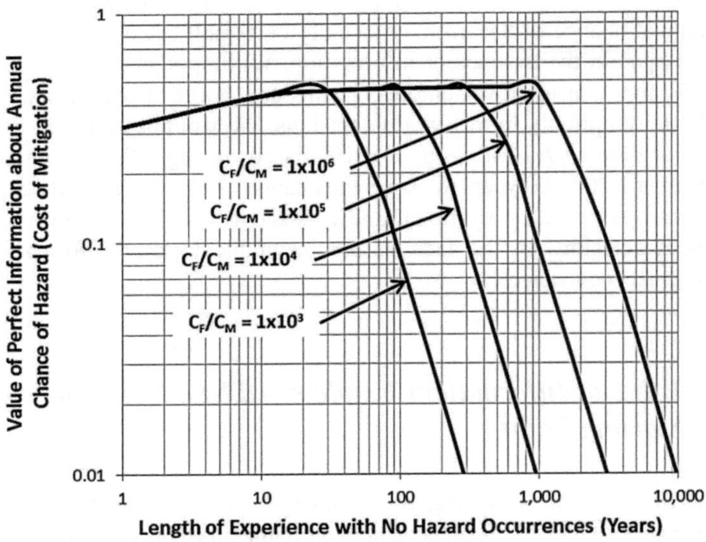

Fig. 18.16 Value of perfect information about annual chance of hazard versus length of available experience with no hazard occurrences

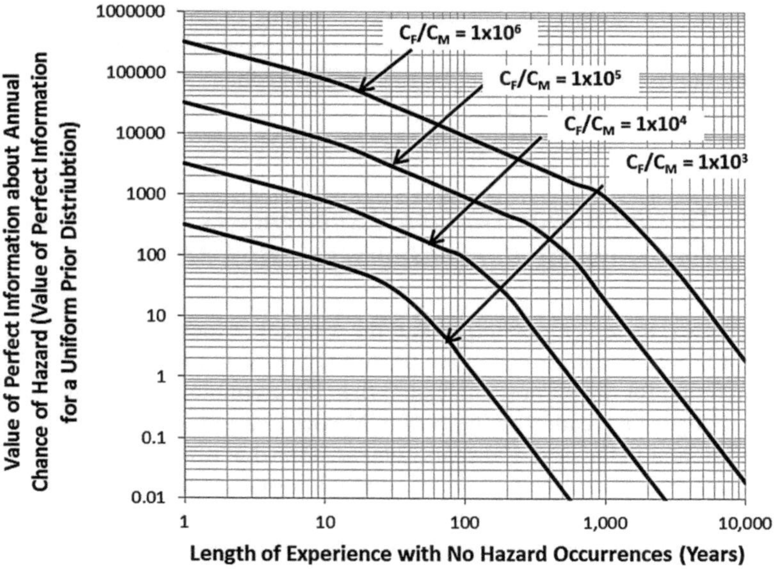

Fig. 18.17 Value of perfect information about annual chance of hazard relative to that obtained with a uniform prior probability distribution for the annual chance of hazard versus length of available experience with no hazard occurrences

hazard occurrences), the value of perfect information is between 25 and 50 % of the cost of risk mitigation. It is interesting that the value of perfect information increases initially with small amounts of experience (Fig. 18.16); this result reflects that potential to change this particular decision (i.e., mitigate the risk) based on additional information at first increases as length of experience with no occurrences increases. As the length of experience without an occurrence exceeds the threshold where the preferred alternative changes from risk mitigation to risk acceptance (Fig. 18.14), the value of perfect information decreases because it becomes less probable that new information beyond the available experience will change the risk management approach (Fig. 18.16). In practice, we will typically have experience bases less than 100 years, and the value of perfect information will be the greatest. Also, the value of perfect information is typically orders of magnitude greater than that obtained using a uniform prior probability distribution for the annual chance of the hazard (Fig. 18.17).

The value of obtaining additional information about the chance of hazard is of practical significance. The greater the value of information, the more important the role is of science and engineering to advance our understanding about natural hazards and managing their risks. For example, greater knowledge about the causes of long debris runouts from landslides could provide information about the chances of them occurring beyond simply waiting for a long record of experience. In addition, the greater the value of information, the greater the value of an adaptable approach to managing risk. For example, a means of mitigating the risk from debris runout that could readily be modified if additional information indicates that accepting the risk is preferred will be more effective than one that cannot easily be changed once it is implemented.

18.4.3 Possibility that Available Experience Is Not Relevant Can Have Significant Impact on Risk Management

In many cases, there is a possibility that the available experience about the annual chance of hazard occurrence may not be directly relevant to predicting what it will be in the future for the purpose managing risk. For example, the annual chance that the storm surge at a location on the Gulf of Mexico coast exceeds a particular height may be different in the next 100 years compared to what it was in the last 100 years due to changes in hurricane frequencies and intensities caused by climate changes. The limiting cases are that the experience is relevant or that it is not relevant.

The relevancy of the data affects the risk management decision because the relevancy is uncertain in making the decision (Fig. 18.18), where F_A is the annual chance of occurrence in the period of available experience, Data Set A, and F_B is the annual chance of occurrence in the future for purposes of managing risk, Data Set B.

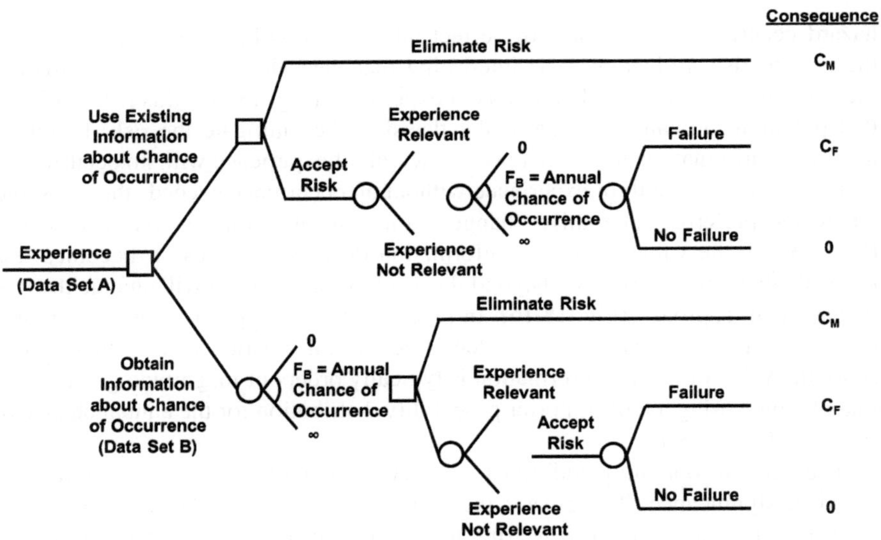

Fig. 18.18 Decision tree to assess impact of relevance of available experience in managing risk

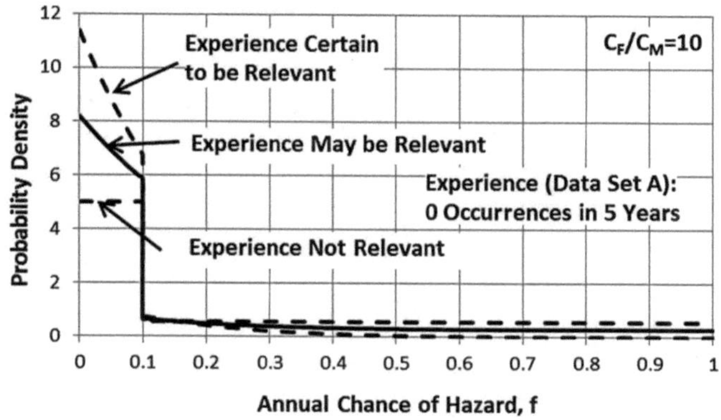

Fig. 18.19 Example prior probability distribution considering the possibility that the experience may be relevant to the risk management decision

The third principle of Decision Entropy Theory establishes the prior probability that the experience (Data Set A) is relevant. If the experience is relevant, then the greatest possible is learned from the experience. If the experience is not relevant, then the least possible (nothing) is learned from the experience. Therefore, the third principle establishes that the non-informative prior probability that the experience is relevant is equal to one-half (Appendix 2), and the prior probability distribution for the annual chance of the hazard in the decision (F_B) is in the middle of the two extremes where the experience is or is not relevant (Fig. 18.19)

To illustrate the significance of data relevancy, consider the value of obtaining additional information about the annual chance of occurrence in the future for purposes of managing risk, F_B (Fig. 18.18); see Appendix 2 for details. If the new data are not consistent with the experience, then the updated distribution reflects more weight on the new data versus the experience and the probability that the experience is relevant decreases (Fig. 18.20). Conversely, if the new data are consistent with the experience, then the updated distribution reflects the combined data and the probability that the experience is relevant increases (Fig. 18.20). Therefore, allowing for the possibility that experience may not be relevant allows for the preferred risk management decision to change more quickly with additional information.

The value of perfect information about the annual chance of occurrence for the purposes of making a risk management decision balances the two extremes where the experience is or is not relevant (Fig. 18.21). Keeping the possibility open that the experience may not be relevant can significantly increase the value of new information when the available experience is seemingly extensive (Fig. 18.21).

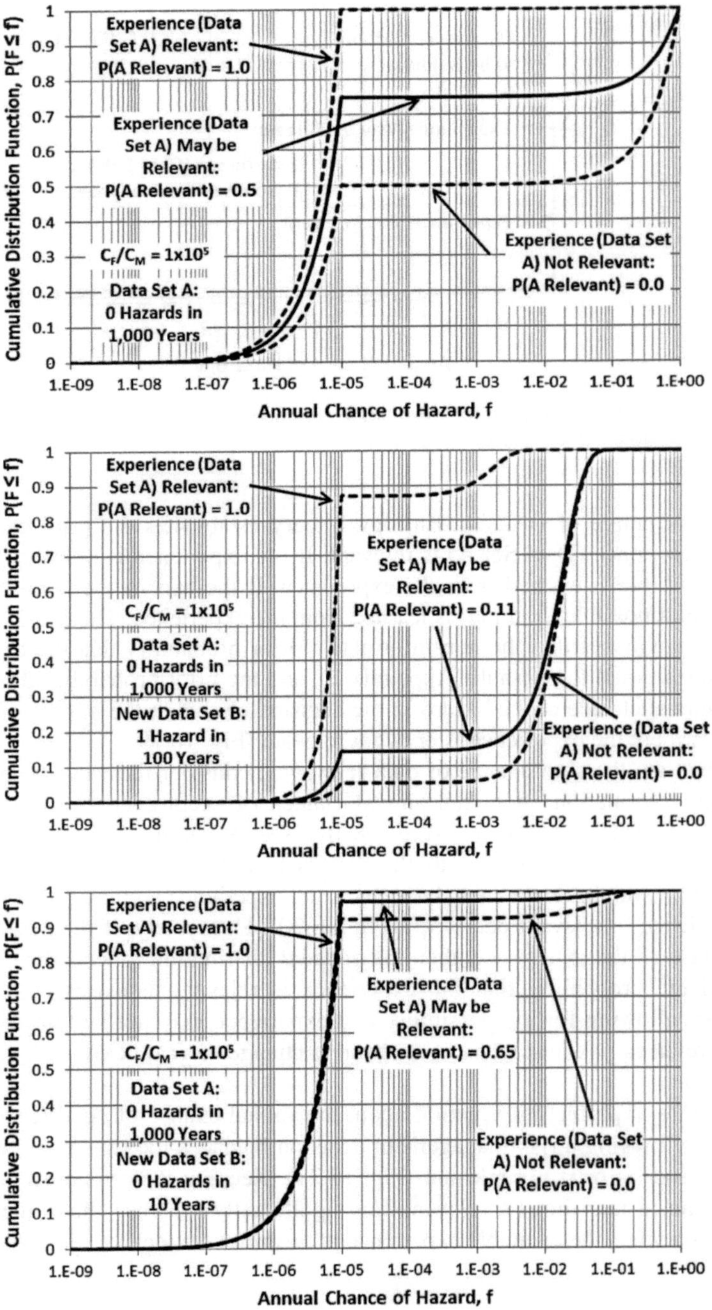

Fig. 18.20 Example prior (*top graph*) and updated probability distributions (*lower graphs*) considering possibility that experience may be relevant to the risk management decision

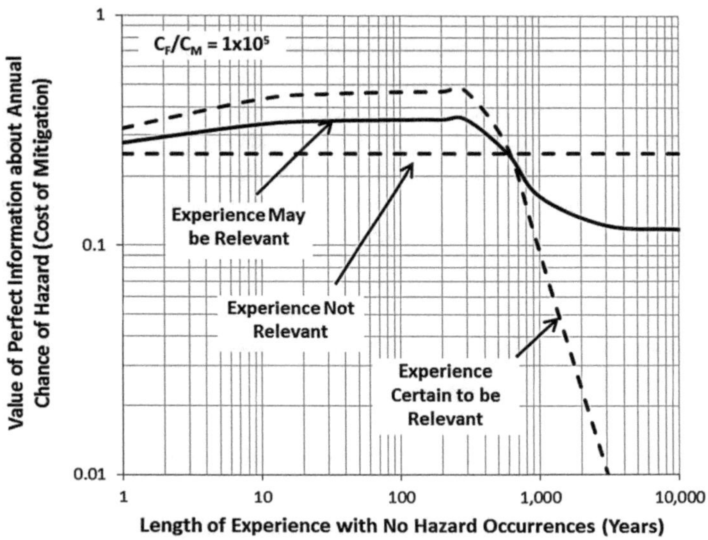

Fig. 18.21 Value of example updated probability distributions considering the possibility that the experience may not be relevant to the risk management decision

18.5 Summary and Conclusions

This chapter proposes a new framework, Decision Entropy Theory, to assess probabilities and manage risks for natural hazard possibilities that are beyond our experience. The theory postulates a starting point for assessing probabilities that reflect having no information in making a risk management decision between two alternatives:

1. An alternative compared to another alternative is equally probable to be preferred or not to be preferred.
2. The possible gains or losses for one alternative compared to another alternative are equally probable.
3. The possibilities of learning about the preference of one alternative compared to another with new information are equally probable.

From this non-informative starting point, all available information (if any) can be incorporated through Bayes' theorem. Decision Entropy Theory attempts to provide for consistency and rationality that is lacking in available approaches for assessing probabilities with limited information.

From a practical perspective, Decision Entropy Theory highlights the importance of considering how possibilities for natural hazards could impact the preferred alternatives for managing risks. A lack of information at the start means the greatest uncertainty in the preferred alternative, not the greatest uncertainty in the hazard. Limited information does not necessarily justify a conservative approach for managing risk; additional information could lead to either less expensive or more expensive means of risk management being preferred. The value of obtaining new information for managing risk is accentuated when limited information is available; therefore, the role science and engineering to advance our understanding about natural hazards and managing their risk is emphasized. Ultimately, this framework underscores the importance of developing adaptable approaches to manage multi-hazard risks in the face of limited information.

A.1 Appendix 1: Mathematical Formulation for Principles of Decision Entropy

The following appendix provides the mathematical formulation characterizing the entropy of a decision.

Principle Number 1 In a non-informative sample space, a selected alternative is equally probable to be or not to be preferred compared to another alternative.

Given that alternative A_i is selected and compared to A_j, the maximum lack of information for the decision corresponds to maximizing the relative entropy for the events that an alternative is and is not preferred:

$$\begin{aligned}&\text{Maximize } H_{\text{rel}}\left(\text{Preference Outcome}\middle|A_i \text{ Selected and Compared to } A_j\right)\\&= -P\left[A_i \text{ Preference to } A_j\right] \ln\left\{P\left[A_i \text{ Preference to } A_j\right]\right\}\\&- P\left[A_i\ \overline{\text{Preference}} \text{ to } A_j\right] \ln\left\{P\left[A_i\ \overline{\text{Preference}} \text{ to } A_j\right]\right\} - \ln(2)\end{aligned}$$

where H_{rel}(Preference Outcome| A_i Selected and Compared to A_j) is the relative entropy of the decision preference, $P[A_i$ Preferred to $A_j]$ is the probability that alternative A_i is preferred compared to alternative A_j, and $P\left[A_i\ \overline{\text{Preferred}}\ \text{ to } A_j\right]$ is the probability that alternative A_i is not preferred compared to alternative A_j or $1 - P\left[A_i \text{ Preferred to } A_j\right]$. If the preference for one alternative versus another is characterized by the difference in the utility values for each alternative, then the relative entropy of the decision preference can be expressed as follows:

$$\begin{aligned}&\text{Maximize } H_{\text{rel}}\left(\text{Preference Outcome}\middle|A_i \text{ Selected and Compared to } A_j\right)\\&= -P\left[u\left(A_i\right) - u\left(A_j\right) > 0\middle|A_i \text{ Compared to } A_j\right]\\&\times \ln\left\{P\left[u\left(A_i\right) - u\left(A_j\right) > 0\middle|A_i \text{ Compared to } A_j\right]\right\}\\&- \left\{1 - P\left[u\left(A_i\right) - u\left(A_j\right) > 0\middle|A_i \text{ Compared to } A_j\right]\right\}\\&\times \ln\left\{1 - P\left[u\left(A_i\right) - u\left(A_j\right) > 0\middle|A_i \text{ Compared to } A_j\right]\right\} - \ln(2)\end{aligned}$$

where $u(A_i)$ is the utility for alternative A_i.

The entropy is a measure of the frequencies or probabilities of possible outcomes divided between the two states: A_i Preferred to A_j and A_i $\overline{\text{Preferred}}$ to A_j. The natural logarithm of the entropy is used for mathematical convenience by invoking Stirling's approximation for factorials; note that maximizing the logarithm of the frequencies of possible outcomes is the same as maximizing the frequencies of possible outcomes since the logarithm is a one-to-one function of the argument. The relative entropy normalizes the entropy by the number of possible states (in this case two); the maximum value of the relative entropy is zero.

For two outcomes, A_i Preferred to A_j and A_i $\overline{\text{Preferred}}$ to A_j, the relative entropy H_{rel}(Preference Outcome| A_i Selected and Compared to A_j) is maximized in the ideal case where it is equally probable that one or the other alternative is preferred, or

$$\begin{aligned} &P\left[u(A_i) - u\left(A_j\right) > 0 \mid A_i \text{ Compared to } A_j\right] \\ &\quad = 1 - P\left[u(A_i) - u\left(A_j\right) > 0 \middle| A_i \text{ Compared to } A_j\right]. \end{aligned}$$

When $i = j$ (i.e., an alternative is compared to itself), the relative entropy becomes equal to its minimum possible value, $-\ln(2)$, because there is no uncertainty in the preference [$p\ln(p) \to 0$ as $p \to 0$ or $p \to 1$].

Principle Number 2 In a non-informative sample space, possible degrees of preference between a selected alternative and another alternative are equally probable.

To represent the maximum lack of information for the decision between two alternatives, maximize the relative entropy for the possible positive and negative differences in utility values between alternative A_i and A_j, $\Delta u_{i,j} = u(A_i) - u\left(A_j\right)$. The sample space where A_i is Selected and Compared to A_j is divided into two states, A_i Preferred to A_j and A_i $\overline{\text{Preferred}}$ to A_j, which are respectively subdivided into $n_{\Delta u_{i,j} \mid A_i \succ A_j}$ and $n_{\Delta u_{i,j} \mid A_i \preccurlyeq A_j}$ substates for each interval of $\Delta u_{i,j}$, where the operator $\succ$ means "preferred to" and the operator $\preccurlyeq$ means "not preferred to" or the complement of $\succ$. Maximizing the relative entropy for degree of preference is then given by the following:

$$\begin{aligned} &\text{Maximize } H_{\text{rel}}\left(\text{Preference Degrees}\middle| A_i \text{ Selected and Compared to } A_j\right) \\ &= H_{\text{rel}}\left(\text{Preference Degrees, } A_i \text{ Preferred to } A_j\right) \\ &+ H_{\text{rel}}\left(\text{Preference Degrees, } A_i\ \overline{\text{Preferred}} \text{ to } A_j\right) \end{aligned}$$

where

$$H_{\text{rel}}\left(\text{Preference Degrees},\ A_i \text{ Preferred to } A_j\right) = \sum_{n_{\Delta u_{i,j}\left|A_i \succ A_j\right.}} \begin{array}{l} -P\left(\Delta u_{i,j}\middle|A_i \text{ Preferred to } A_j\right) P\left(A_i \text{ Preferred to } A_j\right) \\ \times \ln\left[P\left(\Delta u_{i,j}\middle|A_i \text{ Preferred to } A_j\right) P\left(A_i \text{ Preferred to } A_j\right)\right] \end{array}$$

$$-\left(\tfrac{1}{2}\right)\ln\left(n_{\Delta u_{i,j}\left|A_i \succ A_j\right.} \times 2\right)$$

and

$$H_{\text{rel}}\left(\text{Preference Degrees},\ A_i\ \overline{\text{Preferred}} \text{ to } A_j\right) = \sum_{n_{\Delta u_{i,j}\left|A_i \preceq A_j\right.}} \begin{array}{l} -P\left(\Delta u_{i,j}\middle|A_i\ \overline{\text{Preferred}} \text{ to } A_j\right) P\left(A_i\ \overline{\text{Preferred}} \text{ to } A_j\right) \\ \times \ln\left[P\left(\Delta u_{i,j}\middle|A_i\ \overline{\text{Preferred}} \text{ to } A_j\right) P\left(A_i\ \overline{\text{Preferred}} \text{ to } A_j\right)\right] \end{array}$$

$$-\left(\tfrac{1}{2}\right)\ln\left(n_{\Delta u_{i,j}\left|A_i \preceq A_j\right.} \times 2\right)$$

The maximum value for H_{rel}(Preference Degrees| A_i Selected and Compared to A_j) is equal to zero, and it is achieved when $P\left(A_i \text{ Preferred to } A_j\right) = P\left(A_i\ \overline{\text{Preferred}} \text{ to } A_j\right) = 1/2$, all $P\left(\Delta u_{i,j}\middle|A_i \text{ Preferred to } A_j\right) = 1/n_{\Delta u_{i,j}\left|A_i \succ A_j\right.}$ and all $P\left(\Delta u_{i,j}\middle|A_i\ \overline{\text{Preferred}} \text{ to } A_j\right) = 1/n_{\Delta u_{i,j}\left|A_i \preccurlyeq A_j\right.}$. Note that the number of substates for intervals of $\Delta u_{i,j}$ in A_i Preferred to A_j or in A_i $\overline{\text{Preferred}}$ to A_j is not important in maximizing the relative entropy; the entropy is maximized when the possible intervals (however many there are) are as equally probable as possible.

Principle Number 3 In a non-informative sample space, possible expected degrees of information value for the preference between a selected alternative and another alternative are equally probable.

To represent the maximum lack of information for a value of information assessment for the decision between two alternatives, maximize the relative entropy for the possible positive and nonpositive expected changes with information in the differences in expected utility values between alternative A_i and A_j. These possible expected changes are termed the "information value" and denoted by $\Omega_{E_{k,l}}$:

$$\begin{aligned}\left(\Omega_{E_{k,l}}\right)_{i,j} &= E\left(\Delta u_{i,j}\left|E_k \cap S_{i,j}\right.\right) - E\left(\Delta u_{i,j}\left|E_l \cap S_{i,j}\right.\right) \\ &= E\left[u\left(A_i\right) - u\left(\overline{A_l}\right)\left|\mathrm{E}_k \cap S_{i,j}\right.\right] - E\left[u\left(A_i\right) - u\left(\overline{A_i}\right)\left|\mathrm{E}_l \cap S_{i,j}\right.\right]\end{aligned}$$

where E_k and E_l are two sets of possible information about the preference between A_i and A_j, $\Delta u_{\mathrm{i,j}} = u(A_i) - u(A_j)$. The sample space for $E_{k,l}$ is divided into two subsets, an expected positive information value (i.e., $\Omega_{E_{k,l}} > 0$) with $n_{\Omega_{E_{k,l}}>0}$ states and an expected nonpositive information value (i.e., $\Omega_{E_{k,l}} \leq 0$) with $n_{\Omega_{E_{k,l}}\leq 0}$ states. Maximizing the relative entropy for possible information values is then given by the following:

$$\text{Maximize } H_{\text{rel}}\left(\text{Information Value Degrees}\middle|\text{Information about } A_i \text{ Selected and Compared to } A_j\right) = H_{\text{rel}}\left(\text{Information Value Degrees, Information has Positive Value for } A_i \text{ Compared to } A_j\right) + H_{\text{rel}}\left(\text{Information Value Degrees, Information has Non-Positive Value for } A_i \text{ Compared } A_j\right)$$

where

$$H_{\text{rel}}\left(\text{Information Value Degrees, Information has Positive Value for } A_i \text{ Compared to } A_j\right) = \sum_{n_{\Omega_{E_{k,l}}>0}} \begin{array}{l} -P\left(\Omega_{E_{k,l}}\middle|\Omega_{E_{k,l}}>0\right)P\left(\Omega_{E_{k,l}}>0\right) \\ \times \ln\left[P\left(\Omega_{E_{k,l}}\middle|\Omega_{E_{k,l}}>0\right)P\left(\Omega_{E_{k,l}}>0\right)\right] \end{array} - \left(\frac{1}{2}\right)\ln\left(n_{\Omega_{E_{k,l}}>0}\times 2\right)$$

and

$$H_{\text{rel}}\left(\text{Information Value Degrees, Information has Non-Positive Value for } A_i \text{ Compared to } A_j\right) = \sum_{n_{\Omega_{E_{k,l}}\leq 0}} \begin{array}{l} -P\left(\Omega_{E_{k,l}}\middle|\Omega_{E_{k,l}}\leq 0\right)P\left(\Omega_{E_{k,l}}\leq 0\right) \\ \times \ln\left[P\left(\Omega_{E_{k,l}}\middle|\Omega_{E_{k,l}}\leq 0\right)P\left(\Omega_{E_{k,l}}\leq 0\right)\right] \end{array} - \left(\frac{1}{2}\right)\ln\left(n_{\Omega_{E_{k,l}}\leq 0}\times 2\right)$$

The maximum value for the relative entropy for possible information value is equal to zero and achieved when $P\left(\Omega_{E_{k,l}} > 0\right) = P\left(\Omega_{E_{k,l}} \leq 0\right) = 1/2$, all $P\left(\Omega_{E_{k,l}}\middle|\Omega_{E_{k,l}} > 0\right) = 1/\mathrm{n}_{\Omega_{E_{k,l}}>0}$ and all $P\left(\Omega_{E_{k,l}}\middle|\Omega_{E_{k,l}} \leq 0\right) = 1/\mathrm{n}_{\Omega_{E_{k,l}}\leq 0}$.

This principle is consistent with the first two principles where the alternative of obtaining information, $E_{E_{k,l}}$, is compared with the alternative of not obtaining information for a given preference comparison (i.e., $E_l = E_0 = \varnothing$): when the relative entropy of the information value is maximized, there is an equal probability that obtaining the information is preferred (i.e., has positive information value or $\Omega_{E_{k,l}} > 0$) and is not preferred (i.e., has nonpositive information value or $\Omega_{E_{k,l}} \leq 0$) and the possible positive and nonpositive degrees of information value are equally probable.

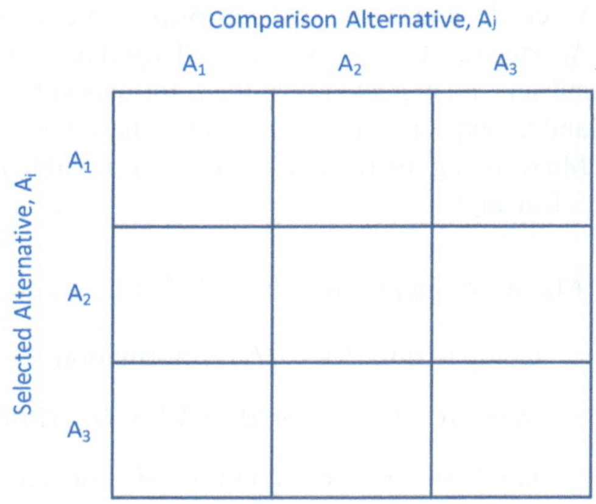

Fig. 18.22 Sample space for decision with three alternatives

A.1.1 Multiple Pairs of Alternatives

The principles of decision entropy establish a sample space for the comparison of any two decision alternatives, A_i and A_j. The sample space for the set of all possibilities of comparison for a given decision problem is denoted the decision sample space. In this sample space, each possible combination of an alternative that is selected, A_i, and an alternative that could have been selected, A_j, is equally probable (Fig. 18.22). For n_A alternatives, there are n_A^2 pairs of i, j and $P\left(A_i \text{ Selected and Compared to } A_j\right) = 1/n_A^2$ and $P\left(A_i \text{ Compared to } A_j \,|A_i \text{ Selected}\right) = 1/n_A$. The preferred decision alternative has the maximum expected degree of preference compared to all other alternatives:

$$E\left[u\left(A_i\right) - u\left(\overline{A_l}\right)|A_i \text{ Selected}\right] = \sum_{\text{all } j} \begin{array}{l} E\left(\Delta u_{i,j}\middle|A_i \text{ Selected and Compared to } A_j\right) \\ \times P\left(A_i \text{ Compared to } A_j \,|A_i \text{ Selected}\right) \end{array}$$

where

$$E\left(\Delta u_{i,j}\middle|A_i \text{ Selected and Compared to } A_j\right) = \sum_{\text{all } \Delta u_{i,j}} \Delta u_{i,j} P\left(\Delta u_{i,j}\middle|A_i \text{ Selected and Compared to } A_j\right)$$

The use of the expected degree of preference (or utility difference) as a measure of preference is consistent with utility theory analysis (e.g., Von Neumann and Morgenstern 1944; Hurwicz 1951; Savage 1951; Hodges and Lehmann 1952; Luce and Raiffa 1957; Raiffa and Schlaifer 1961; Benjamin and Cornell 1970; Ang and Tang 1984, etc.). In a conventional decision analysis, the sample space for utility values is not conditioned on a particular alternative being selected, A_i, meaning that the expected utility for a selected alternative can be calculated without considering the alternatives to which it is being compared. However, the absolute magnitude of expected utility is irrelevant[1]; its relevance depends on comparing it with the expected utility values for other alternatives. Therefore, it is the differences between utility values that are of interest.

Mathematically, comparing the expected degrees of preference, $E\left[u\left(A_i\right) - u\left(\overline{A_i}\right)\right.$ $\left.|A_i \text{ Selected}\right]$, is the same as comparing expected utility values in a conventional decision analysis. In a conventional analysis where the probabilities for utility values given that an alternative has been selected do not depend on the alternative to which it is being compared, the expected degree of preference for an alternative can be expressed as follows:

$$
\begin{aligned}
E\left[u\left(A_i\right) - u\left(\overline{A_i}\right)\middle|A_i \text{ Selected}\right] &= \sum_{\text{all } j} \begin{array}{l} E\left(\Delta u_{i,j}\middle|A_i \text{ Selected and Compared to } A_j\right) \\ \times P\left(A_i \text{ Compared to } A_j\middle|A_i \text{ Selected}\right) \end{array} \\
&= \sum_{\text{all } j} \begin{array}{l} \left[E\left(u_i\middle|A_i \text{ Selected}\right) - E\left(u_j\middle|A_j \text{ Selected}\right)\right] \\ \times P\left(A_i \text{ Compared to } A_j\middle|A_i \text{ Selected}\right) \end{array} \\
&= E\left(u_i\middle|A_i \text{ Selected}\right) - \sum_{j=1 \text{ to } n_A} E\left(u_j\middle|A_j \text{ Selected}\right) \times \left(\frac{1}{n_A}\right)
\end{aligned}
$$

Therefore, the expected degree of preference for an alternative in a conventional decision analysis is equal to the expected utility for that alternative minus a constant (the average expected utility for all alternatives). Therefore, the order of comparisons is the same whether the expected utility values, $E\left(u_i\middle|A_i \text{ Selected}\right)$, or the expected degree of preference values, $E\left[u\left(A_i\right) - u\left(\overline{A_i}\right)|A_i \text{ Selected}\right]$, are used in comparisons.

B.1 Appendix 2: Implementation of Bayes' Theorem with Bernoulli Sequence for Two Possibly Related Sets of Data

Define F_A as the annual chance of occurrence in the period of available experience, Data Set A, and F_B as the annual chance of occurrence in the future for purposes of managing risk, Data Set B (Fig. 18.18). If the annual chance of occurrence is the same in the past and the future, then the likelihood of obtaining the available data from a Bernoulli sequence (i.e., x_A occurrences in n_A years) for a particular value of f_{Bi} is given by:

$$
P\left(x_A \text{ in } n_A\middle|f_{Bi} = f_{Ai}\right) = \left\{\left[\frac{n_A!}{x_A!\left(n_A - x_A!\right)}\right] f_{Bi}^{x_A}\left(1 - f_{Bi}\right)^{n_A - x_A}\right\}
$$

[1]Utility values can be scaled arbitrarily with linear transformations.

Conversely, if the annual chance of occurrence is not the same in the past and the future, then the likelihood of obtaining the available data (i.e., x_A occurrences in n_A years) is given by:

$$P\left(x_A \text{ in } n_A \middle| f_{Bi} \neq f_{Ai}\right) = \sum_{\text{all } f_{Ai}} \left\{ \left[\frac{n_A!}{x_A!\,(n_A - x_A!)} \right] f_{Ai}^{x_A} (1 - f_{Ai})^{n_A - x_A} \right\} P\left(f_{Ai} | f_{Bi} \neq f_{Ai}\right)$$

where $P(f_{Ai})$ is the prior probability for the annual chance of occurrence in the experience. If the experience is relevant, then the prior probability for F_A is the same as that for F_B: $P\left(f_{Ai} | f_{Bi} = f_{Ai}\right) = P\left(f_{Bi}\right)$. Furthermore, $P\left(f_{Ai} | f_{Bi} \neq f_{Ai}\right) = P\left(f_{Ai} | f_{Bi} = f_{Ai}\right) = P\left(f_{Ai}\right) = P\left(f_{Bi}\right)$ since the probability for F_A does not depend on whether or not the experience is relevant (i.e., the prior probability of obtaining a particular set of data from Set A is the same whether or not Sets A and B are the same, and the probability of obtaining a particular set of data from Set A does not depend on the chance of occurrence in Set B if the two sets are different). Therefore,

$$P\left(x_A \text{ in } n_A \middle| f_{Bi} \neq f_{Ai}\right) = \sum_{\text{all } f_{Bi}} \left\{ \left[\frac{n_A!}{x_A!\,(n_A - x_A!)} \right] f_{Bi}^{x_A} (1 - f_{Bi})^{n_A - x_A} \right\} P\left(f_{Bi}\right)$$

meaning that the likelihood of the information from the experience is a constant with respect to f_{Bi} (i.e., the updated probability distribution will be same as the prior probability distribution for $F_B$if the data are not relevant). The composite likelihood function considering the possibility that the data may or may not be relevant is given by:

$$\begin{aligned} P\left(x_A \text{ in } n_A \middle| f_{Bi}\right) &= P\left(x_A \text{ in } n_A \middle| f_{Bi} = f_{Ai}\right) P\left(f_{Bi} = f_{Ai}\right) + P\left(x_A \text{ in } n_A \middle| f_{Bi} \neq f_{Ai}\right) P\left(f_{Bi} \neq f_{Ai}\right) \\ &= P\left(x_A \text{ in } n_A \middle| f_{Bi} = f_{Ai}\right) P\left(f_{Bi} = f_{Ai}\right) + P\left(x_A \text{ in } n_A \middle| f_{Bi} \neq f_{Ai}\right) \left[1 - P\left(f_{Bi} = f_{Ai}\right)\right] \end{aligned}$$

where $P\left(f_{Bi} = f_{Ai}\right)$ is the prior probability that the experience is relevant.

If the experience is relevant, then the greatest possible is learned from the experience. If the experience is not relevant, then the least possible (nothing) is learned from the experience. Therefore, the third principle of Decision Entropy Theory establishes that the probability that the experience is relevant is 0.5, or $P\left(f_{Bi} = f_{Ai}\right) = 0.5$.

If information could be obtained about the annual chance of hazard occurrence in the period of the decision (F_B) before making the decision (Fig. 18.18), then the likelihood of obtaining a particular set of data (i.e., x_A occurrences in n_A years and x_B occurrences in n_B years) is given by the following:

$$\begin{aligned} &P\left(x_A \text{ in } n_A \text{ and } x_B \text{ in } n_B | f_{Bi}\right) \\ &= P\left(x_A \text{ in } n_A \text{ and } x_B \text{ in } n_B \middle| f_{Bi} = f_{Ai}\right) P\left(f_{Bi} = f_{Ai}\right) \\ &+ P\left(x_A \text{ in } n_A \text{ and } x_B \text{ in } n_B \middle| f_{Bi} \neq f_{Ai}\right) \left[1 - P\left(f_{Bi} = f_{Ai}\right)\right] \end{aligned}$$

where

$$P\left(x_A \text{ in } n_A \text{ and } x_B \text{ in } n_B \middle| f_{Bi} = f_{Ai}\right) = \left\{\left[\frac{n_A!}{x_A!(n_A - x_A!)}\right] f_{Bi}^{x_A} (1 - f_{Bi})^{n_A - x_A}\right\} \times \left\{\left[\frac{n_B!}{x_B!(n_B - x_B!)}\right] f_{Bi}^{x_B} (1 - f_{Bi})^{n_B - x_B}\right\}$$

and

$$P\left(x_A \text{ in } n_A \text{ and } x_B \text{ in } n_B \middle| f_{Bi} \neq f_{Ai}\right) = \left\{\sum_{\text{all } f_{Bi}} \left\{\left[\frac{n_A!}{x_A!\,(n_A - x_A!)}\right] f_{Bi}^{x_A} (1 - f_{Bi})^{n_A - x_A}\right\} P(f_{Bi})\right\} \times \left\{\left[\frac{n_B!}{x_B!(n_B - x_B!)}\right] f_{Bi}^{x_B} (1 - f_{Bi})^{n_B - x_B}\right\}$$

Hence, the updated probability distribution for the annual chance of occurrence for the hazard in the risk management decision (F_B) is obtained from Bayes' theorem as follows:

$$P\left(F_B = f_{Bi} \middle| x_A \text{ in } n_A \text{ and } x_B \text{ in } n_B\right) = \frac{\left\{\begin{array}{l} P\left(x_A \text{ in } n_A \text{ and } x_B \text{ in } n_B \middle| f_{Bi} = f_{Ai}\right) P(f_{Bi} = f_{Ai}) \\ +P\left(x_A \text{ in } n_A \text{ and } x_B \text{ in } n_B \middle| f_{Bi} \neq f_{Ai}\right) [1 - P(f_{Bi} = f_{Ai})] \end{array}\right\}}{\sum_{\text{all } f_{Bj}} \left\{\begin{array}{l} P\left(x_A \text{ in } n_A \text{ and } x_B \text{ in } n_B \middle| f_{Bj} = f_{Aj}\right) P(f_{Bj} = f_{Aj}) \\ +P\left(x_A \text{ in } n_A \text{ and } x_B \text{ in } n_B \middle| f_{Bj} \neq f_{Aj}\right) [1 - P(f_{Bj} = f_{Ai})] \end{array}\right\}}$$

Likewise, the probability that the data from the experience (Data Set A) is relevant is updated with the data obtained from the period of the decision (Data Set B):

$$P\left(f_B = f_{Ai} \middle| x_A \text{ in } n_A \text{ and } x_B \textit{ in } n_B\right) = \frac{P\left(x_A \text{ in } n_A \text{ and } x_B \text{ in } n_B \middle| f_{Bi} = f_{Ai}\right) P(f_{Bi} = f_{Ai})}{\sum_{\text{all } f_{Bj}} \left\{\begin{array}{l} P\left(x_A \text{ in } n_A \text{ and } x_B \text{ in } n_B \middle| f_{Bj} = f_{Aj}\right) P(f_{Bj} = f_{Aj}) \\ +P\left(x_A \text{ in } n_A \text{ and } x_B \text{ in } n_B \middle| f_{Bj} \neq f_{Aj}\right) [1 - P(f_{Bj} = f_{Ai})] \end{array}\right\}}$$

References

AGS. (2000). Landslide risk management concepts and guidelines. *Australian Geomechanics, 35*(1), 49–92.

ANCOLD. (1996). *Commentary on ANCOLD guidelines on risk assessment*. Sydney, Australia: Australian National Committee on Large Dams.

Ang, A. A-S. & Tang, W. H., (1984). Probability Concepts in Engineering Planning and Design, Volume II - Decision, Risk and Reliability, John Wiley & Sons, New York.

Bernoulli, D. (1738). Specimen Theoriae Novae de Mensura Sortis. *Commentarii Academiae Scientiarum imperialis Petropolitanae, Tomus, V*, 175–192 [Trans. (1954) Exposition of a new theory on the measurement of risk. *Econometrica, 22*(1), 23–36].

Benjamin, J. R., & Cornell, C. A. (1970). *Probability, statistics, and decision for civil engineers*. New York, NY: McGraw-Hill.

Box, G. E. P., & Tiao, G. C. (1973). *Bayesian inference in statistical analysis*. Reading, MA: Addison-Wesley.

DOD. (2002). News Transcript, Presenter: Secretary of Defense Donald H. Rumsfeld, Washington, DC: United States Department of Defense.

GEO. (1998). *Landslides and Boulder falls from natural terrain: Interim risk guidelines*. Geotechnical Engineering Office Report 75, Government of Hong Kong.

GEER. (2014). The 22 March 2014 Oso Landslide, Snohomish County, Washington. Contributing Authors: J. R. Keaton, J. Wartman, S. Anderson, J. Benoit, J. deLaChapelle, R. Gilbert, & D. R. Montgomery. Geotechnical Extreme Event Reconnaissance, National Science Foundation.

Hodges, J. L., Jr., & Lehmann, E. L. (1952). The uses of previous experience in reaching statistical decisions. *Annals of Mathematical Statistics, 23*, 396–407.

Hurwicz, L. (1951). Some specification problems and applications to econometric models (abstract). *Econometrica, 19*, 343–344.

IPET. *Performance evaluation of the New Orleans and Southeast Louisiana Hurricane protection system*. (2009). Final Report, Interagency Performance Evaluation Task Force, U.S. Army Corps of Engineers.

Jaynes, E. T. (1957). Information theory and statistical mechanics. *Physical Review, 106*(4), 620–630.

Jaynes, E. T. (1968). Prior probabilities. *IEEE Transactions on System Science and Cybernetics, 4*(3), 227–241.

Journel, A. G., & Deutsch, C. V. (1993). Entropy and spatial disorder. *Mathematical Geology, 25*(3), 329–355.

Keynes, J. M. (1921). *A treatise on probability*. London: The MacMillan and Company Limited.

Lacasse, S., & Nadim, F. (2009). Landslide risk assessment and mitigation strategy, Chapter 3. In *Landslides – disaster risk reduction*. Berlin: Springer-Verlag.

Liu, Z. Q., & Nadim, F. (2014). A three-level framework for multi-risk assessment. In *Geotechnical safety and risk IV* (pp. 493–498). London: Taylor & Francis Group.

Luce, R. D., & Raiffa, H. (1957). *Games and decisions*. New York, NY: Wiley.

Nadim, F. (2011, October 24-26). Risk assessment for earthquake-induced submarine slides. In *5th International Symposium on Submarine Mass Movements and their Consequences, ISSMMTC*. Kyoto, Japan: Springer.

Nadim, F., & Sparrevik, M. (2013, June 17-19). Managing unlikely risks posed by natural hazards to critical infrastructure. In *22nd SRA Europe Conference*. Trondheim, Norway.

Savage, L. J. (1951). The theory of statistical decision. *Journal of the American Statistical Association, 46*(253), 55–67.

Raiffa, H., & Schlaifer, R. (1961). *Applied statistical decision theory*. Boston, MA: Harvard University Graduate School of Business Administration.

Taleb, N. N. (2007). *The black swan: The impact of the highly improbable*. New York, NY: Random House, Inc.

Tribus, M. (1969). *Rational descriptions, decisions and designs*. New York, NY: Pergamon Press.

Von Neumann, J., & Morgenstern, O. (1944). *Theory of games and economic behavior* (3rd ed.). Princeton, NJ: Princeton University Press.

Chapter 19
Bayesian Risk Assessment of a Tsunamigenic Rockslide at Åknes

Zenon Medina-Cetina, Unni M. Eidsvig, Vidar Kveldsvik, Sylfest Glimsdal, Carl B. Harbitz, and Frode Sandersen

Abstract This chapter introduces a comparison between two methods for estimating the risk of a tsunamigenic rockslide at Åknes, Norway. The first method follows a classical approach based on best estimates of the risk factors (i.e., hazard, vulnerability, and elements at risk). The second method follows a more recent approach based on Bayesian networks, which introduces the notion of causal effects and defines full probability distributions for each risk factor. The Bayesian approach is thought to be more powerful in terms of number and quality of inferences. It allows for conducting diagnosis and prognosis risk assessment analyses and it traces the influence of new evidence as it becomes available, either from experimental observations, model predictions, informed expert beliefs, a combination of them, or even "interventions" in the model to reproduce optimal decision-making processes (e.g., by introducing the stochastic model of an early warning system). Both methods illustrate the interaction of multiple natural threats when implemented in the Storfjord area where the Åknes rockslide is located. Results generated from the proposed methods are based on available evidence; however a key component on both approaches is the evidence assimilation from the experts who provided technical information, but also their beliefs in terms of probability measures (i.e., informed expert's beliefs). The use of informed expert's beliefs introduced a unique approach for incorporating fine engineering judgment into risk measures in a systematic manner. Results obtained from each method showed significant qualitative differences in terms of inference capabilities, but in terms of the expected risk estimates, their orders of magnitude were relatively similar, which validated the state of risk at the Åknes rockslide.

Z. Medina-Cetina (✉)
Zachry Department of Civil Engineering, Texas A&M University, CVLB 808 S, College Station, TX 77843-3136, USA
e-mail: zenon@tamu.edu

U.M. Eidsvig • V. Kveldsvik • S. Glimsdal • C.B. Harbitz • F. Sandersen
Norwegian Geotechnical Institute NGI, Sognsveien 72, 0855 Oslo, Norway

P. Gardoni, J.M. LaFave (eds.), *Multi-hazard Approaches to Civil Infrastructure Engineering*, DOI 10.1007/978-3-319-29713-2_19

19.1 Introduction

The International Centre for Geohazards (ICG) at the Norwegian Geotechnical Institute (NGI) organized a 3-day workshop on the risk associated to the threat of the Åknes tsunamigenic rockslide (located in the Stranda municipality in Norway) in October of 2007. The objective of the NGI workshop was to identify the Åknes' hazard (probability of the coupling of multiple threats), the region vulnerability (probability of impacting certain elements at risk conditioned on specific coupling of threats' intensities), and the elements at risk in Storfjorden (human life and material damage).

This work follows a reference paper stemmed from the NGI workshop and published by the same authors of this paper, where a "classical" risk assessment approach was initially introduced (Eidsvig et al. 2011). However, for the easiness of the comparison analysis with the proposed Bayesian risk assessment approach presented in this paper, a summary of the reference paper is being reproduced here. The proposed new method based on the use of Bayesian networks was at the time considered of exploratory nature. This paper aims at introducing a more robust, systematic, and causal probabilistic approach for quantifying risk, particularly suitable for multi- and varying hazard, vulnerability, and consequence conditions, which makes it depart significantly from the classical approach.

A number of publications on the Åknes rockslide followed since the publication of the authors' reference paper, including a 3D geological conceptual model of the (Jaboyedoff et al. 2011), a study on the effectiveness on risk communication (Rød et al. 2011), an analysis of the meteorological effects on seasonal rockslide displacements (Grøneng et al. 2011), and simulations on run-up impacts due to the expected tsunami effects (Harbitz et al. 2014). However, none of them introduced improvements on risk assessment methodologies at Åknes. Also, these publications further helped to convey a sound message on the relevance of a potential rockslide at Åknes, particularly for the communities likely to be impacted and for the decision-makers involved in its monitoring, civil protection, and those coordinating government, commercial, and industrial stakeholders. The major contribution of all research at Åknes as of today is a set of unique technological and engineering resources implemented to protect the surrounding populations via an early warning system (EWS) (Blikra 2006).

It is worth mentioning that the approach followed in the workshop was the construction of event trees based on opinions of some of the engineers, scientists, and stakeholders involved in the Åknes/Tafjord project (Blikra et al. 2006). Results obtained from the use of these event trees included the identification of potential triggering events, rockslide failure mechanisms, likely tsunami generation, propagation and onshore impacts, and a first overview of the consequences and required countermeasures for some of the Åknes neighbor villages (Eidsvig et al. 2008).

The ICG/NGI workshop generated a clear picture of the current status on the study of the Åknes tsunamigenic rockslide threat. It offered an assessment on the progress of the risk-related evidence collection, including field monitoring data, numerical modeling of the threats (rockslide and tsunami), elements at risk in the surrounding area (i.e., demographics), and particularly important, the opinions of experts and civil authorities regarding the state of preparedness for the eventual implementation of countermeasures via evacuations.

Results from the workshop showed the need for establishing mechanisms for assessing the risk of dependent threats such as the rockslide-tsunami into an efficient decision-making framework. This meant to formally include the uncertainty influence of each participating event, the uncertainty associated with potential scenarios during the operation of an EWS, and the uncertainty associated to the definition of optimal threshold warning levels (Medina et al. 2008). By having a probabilistic template that could replicate the complexity of the decision-making process involved in the Åknes, early warning system would lead to concrete answers: When should we issue a warning? How long should we wait to issue the next warning level? Is the effectiveness of the early warning the same through the different seasons of the year or through different data flows (i.e., what if some sensors stop working)?

This work addresses recent efforts on the risk quantification in some of the communities surrounding the Åkness rockslide. Four control locations of interest are considered in the Storfjorden: Hellesylt, Geiranger, Tafjord, and Eidsdal. Risk quantification in this area was carried out by the participation of a group of NGI experts, who collected information including historic rockslide evidence, numerical analyses of the physical mechanisms associated to the coupling of the rockslide and tsunami threats, a catalog of potential consequences of a tsunamigenic rockslide, and historic evidence of tsunamis' vulnerability.

A comparative analysis of results showing preliminary risk estimates in Storfjorden based on two risk quantification methodologies are presented: following *classical* approach (**C**) and a *Bayesian network* approach (**BN**). Both approaches fulfill the same definition of risk introduced by UNDRO (1979), where $Risk = Hazard \times Vulnerability \times Losses$. The **C** approach is based on expected values of the risk factors and calculates the hazard as one unique event irrespective of the coupling dependency of the threats, whereas the **BN** approach takes into account not only the threats dependencies but the dependencies between all events contributing to the state risk at a given time.

19.2 Risk Methodology Assessment

19.2.1 Risk Definition

This work defines *Risk* as a "measure of the probability and severity of an adverse effect to life, health, property, or the environment" (Nadim 2004). The proposed quantitative risk definition and the one used extensively in engineering applications is (UNDRO 1979)

$$R = H \times V \times L, \tag{19.1}$$

in which R, H, V, and L refer to *risk*, *hazard*, *vulnerability*, and *losses* or "elements at risk," respectively. In this context, *hazard* is defined as "the probability that a particular threat occurs within a given period of time." Physical *vulnerability* is "the degree of expected loss in an element or system in relation to a specific hazard," ranging from zero (no loss expected) to unity (total loss expected). And *loss* is defined as "any negative consequence, financial, or otherwise." The unit associated with loss is typically life's losses or monetary value.

19.2.2 Uncertainty Assessment

Traditional approaches for estimating risk-related events follow the tree-type decision-making principles based on monotonic Boolean logic (NUREG-0492 1981). Event trees allow for outlining potential scenarios in a discrete manner, where each branch of the tree represents a mutually exclusive risk outcome. More recent risk assessment methodologies include BN (Einstein and Sousa 2007; Medina-Cetina and Nadim 2008), which allows to represent multiple events based on causal relationships such as the coupling of multi-hazard processes (e.g., tsunamigenic rockslide) and to incorporate different information sources at the time (i.e., monitoring data, theoretical modeling, and informed experts' beliefs). In addition, BNs can index real-time risk "states" of complex and time-sensitive systems for the issuing (or not) of early warning systems (Medina-Cetina and Nadim 2008), find optimal threshold curves according to the "states" of the risk-related events (Medina-Cetina et al. 2008), and help to refine the risk assessment by disaggregating the events into sub-events (e.g., the effect of potential triggering factors such as rainfall, seismic activity, snowfall, or the effectiveness of the early warning and countermeasures on the consequences).

In either case, once risk-related events are identified, it follows to quantify their contributing uncertainties. This work introduces a first notion of risk in Storfjorden by using functions relating the risk components given in Eq. (19.1) and adding uncertainty bounds by the experts whenever possible. This will be referred to as the *classic* approach **C**, which simply relates the risk components through prescribed functions based on historic data or theoretical modeling. A schematic representation of this approach is presented in Fig. 19.1.

The **BN** approach on the other hand builds on the same information gathered for the assessment of the risk factors as used in the **C** approach and uses the same proposed ranges for the sake of comparison of results. What makes the **BN** approach unique is that different groups of independent experts at NGI working on

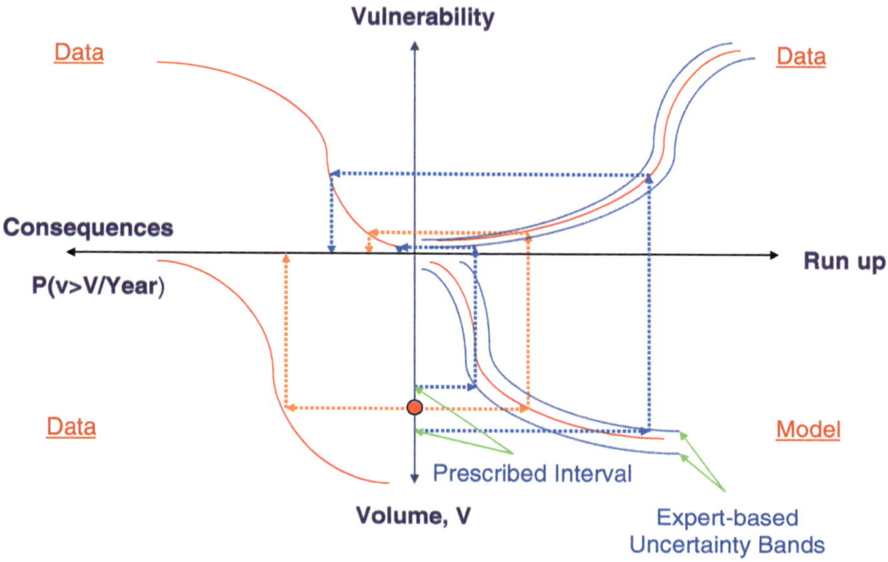

Fig. 19.1 Schematic representation for the risk quantification of the Åknes' tsunamigenic rock-slide (Eidsvig et al. 2011)

the modeling of rockslides, tsunami, and risk quantification were asked to propose informed belief measures for defining probability distributions to be assigned to the risk factors (Eq. 19.1) via a Markov conditioning network scheme depicting cause-effect relationships, resulting on an improved representation of the likelihood of occurrence of each of the risk factors taking place in the risk assessment process. Some of these beliefs were a result of the informed experts' judgment built after running numerical models, assessing field information, and relaying in previous experience. In this way, the **BN** incorporated these beliefs with the purpose of introducing probabilistic and causal dependencies into the risk estimates. This approach illustrated the relevance in simulating the uncertainty propagation as new evidence becomes available, making possible to update the "state of risk" on real time as new information flows into the Bayesian network. This analysis is known as "prognosis." In addition, the **BN** approach allows for defining likely scenarios that would yield a given state of risk. A schematic representation of the proposed **BN** model to reproduce the state of risk of the Åknes tsunamigenic rockslide is presented in Fig. 19.2.

A detailed discussion on the collection of evidence for integrating quantitative risk estimates is discussed in Sect. 19.3, followed by the application of the **C** and **BN** methodologies to the Åknes case study in Sect. 19.4 and the concluding remarks in Sect. 19.5.

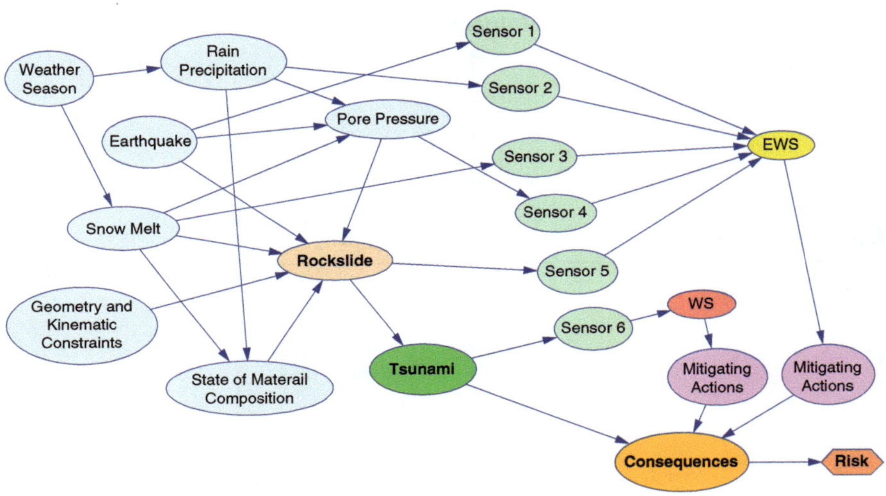

Fig. 19.2 Bayesian network for risk quantification of a tsunamigenic rockslide including the effect of an EWS (Medina-Cetina and Nadim 2008)

19.3 Evidence Collection of Risk Components

19.3.1 Hazard

The Åknes threat is not the rockslide itself but its tsunamigenic potential. This means that the hazard is comprised of two parts (Eidsvig et al. 2008):

- The probability of a rockslide at Åknes
- The probability of generating a tsunami caused by the Åknes rockslide

In the following sections, it is described how each of these parts is related when estimating the expected risk.

19.3.1.1 Rockslides

Use of Historic Data

To set the baseline of the state of risk, it is required to assess the Åknes hazard using historic data showing evidence found on similar threats. Hungr et al. (1999) examined the magnitude (in terms of volume) and frequency of rockfalls and rockslides along the main transportation corridors of Southwestern British Columbia in Canada. Following their approach on the use of an empirical power law, it is possible to represent the relationship between frequency and rockslide magnitude in Storfjorden. Blikra et al. (2005) evaluated the rockslide hazard in Storfjorden

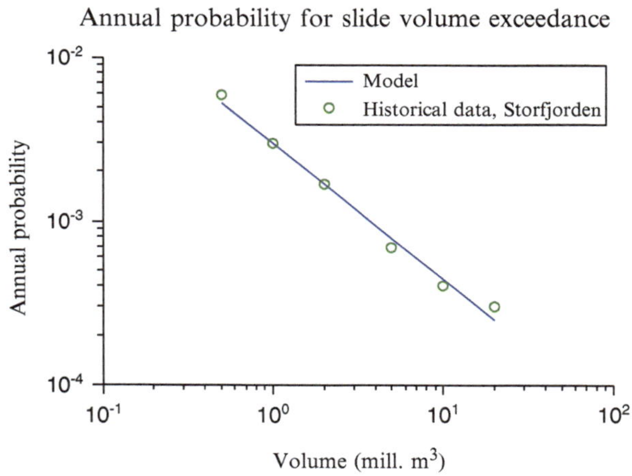

Fig. 19.3 Rock volume exceedance probability for Storfjorden (Eidsvig et al. 2008)

Table 19.1 Annual failure frequency for rockslide volume intervals, calculated from Eq. (19.2)

Volume intervals (Mm^3)	Annual failure probability
[0.5, 2.0]	0.0036
[2.0, 4.0]	0.00074
[4.0, 7.0]	0.00035
[7.0, 12.0]	0.00022
[12.0, 20.0]	0.00013
[20.0, 35.0]	0.00009
>35	0.00016

by investigating the spatial and temporal pattern of previous rockslide events. He mapped rock avalanche deposits in this area by swath bathymetry combined with seismic reflection data, which allowed for estimating the corresponding rockslide volumes and time of occurrence. An estimate of the probability (frequency estimate) that a certain rock volume is exceeded as deduced from data representing the last 10,000 years is plotted in Fig. 19.3.

This figure indicates the rock volume probability function by the use of a power regression model. Consequently, the failure probability for several intervals of rockslide volumes can be calculated using the relation:

$$P_f\,(V_1 < V < V_2) = G\,(V_1) - G\,(V_2)\,, \tag{19.2}$$

where V_1 and V_2 are rock volumes given in Mm^3, and P_f $(V_1 < V < V_2)$ is the probability of failure of a rock volume between V_1 and V_2. In this way, the failure probability can be calculated for any interval of rockslide volumes. For instance, selected volume intervals and their corresponding probability are shown in Table 19.1.

According to Blikra et al. (2006), the values shown in Fig. 19.3 should be considered a lower bound of the rockslide probability. This is because there is

Fig. 19.4 Åknes rock slope. The *white lines* indicate the contour of the potentially unstable area as derived through various investigations (slightly modified after Derron et al. 2005). The length of the "top scarp"/*upper crack* is about 800 m. The length along the dip direction of the slope inside the assumed unstable area is about 1100 m. *U*, *M*, *L* Upper, middle, and lower borehole sites (Eidsvig et al. 2011)

certainty on the continuous motion of the Åknes rockslide, meaning that it is imminent a displacement that could result in the generation of a tsunami. This is not the case at other sites in Storfjorden, where data reflects the probability of triggering a rockslide in Åknes equally for the whole Storfjord area. Therefore, estimates included in Fig. 19.3 and Table 19.1 approximate the lower bound of rockslide probability for the Åknes threat.

Use of Site-Dependent Data

Recent studies by Kveldsvik et al. (2008) show that about 650,000 m^2 of the Åknes rockslide is potentially unstable (Fig. 19.4). Further analysis of on-site geological mechanisms derived by Ganerød et al. (2008) and displacements measured at the slope surface using discontinuous deformation analysis (DDA) (Shi and Goodman 1985; Shi 1988) helped to generate a block segmentation model (Fig. 19.5).

Additionally, there is displacement data available that can be divided in three data sets:

(a) The 1961–1983 photogrammetric data set
(b) The 1983–2004 photogrammetric data set and
(c) The 2004–2006 data set derived from measurements by total station, GPS, and rod extensometers

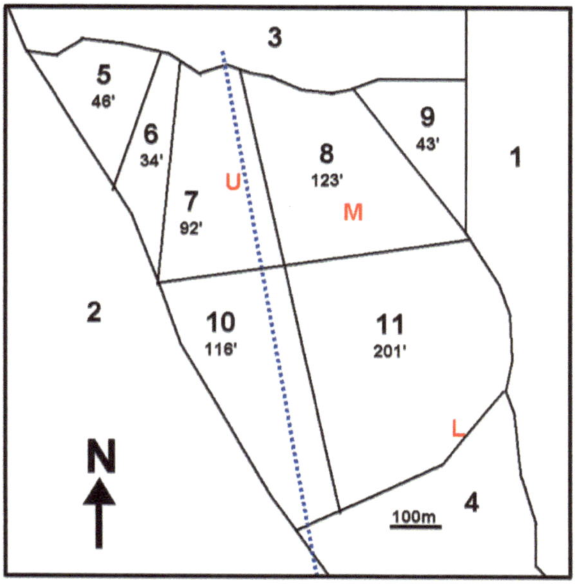

Fig. 19.5 Block model of the Åknes rock slope. *Black numbers*: block numbers and areas in 1000 m^2 for each assumed unstable block. *Red letters*: location of the upper, middle, and lower boreholes. *Blue broken line*: profile for numerical modeling. Modified after Kveldsvik et al. (2008)

This historic evidence shows that the highest displacement rates have been measured in blocks 5–7 in the upper half of the assumed unstable part of the slope, showing a horizontal component of about 7 cm/year since 1983 and about 13 cm/year from 1961 to 1983. Somewhat larger displacement rates are indicated in blocks 5 and 6 compared to block 7. Block 8 has been shown to move at a smaller rate than block 7. Block 9 appears to move little or not at all. In the lower half, most of block 11 has moved insignificantly from 2004 to 2006, and measuring points in the photogrammetric data sets are too few for drawing any conclusions for the period before 2004. Measurements do not exist for block 10. There is evidence that rockslides occurred on the western flank of block 10 in 1940 and in 1960 or 1961. Another rockslide event is also known to have occurred on the upper western flank between 1850 and 1900 (Kveldsvik et al. 2007).

Estimates of Potential Sliding Volumes

Based on the block model described above (Fig. 19.5) and using the field information about depth to the basal sliding surface, it is possible to estimate potential rockslide volumes. For instance, Fig. 19.5 indicates the location of a vertical cross section passing through the blocks that are likely to slide. The resulting slide is shown in Fig. 19.6 (Kveldsvik 2008). In this way, following the NGI informed expert's beliefs founded in the assimilation of field data and modeling of the rockslide mechanisms, and after investigating likely sliding rock volumes and their corresponding probabilities of occurrence in the area, it was concluded that

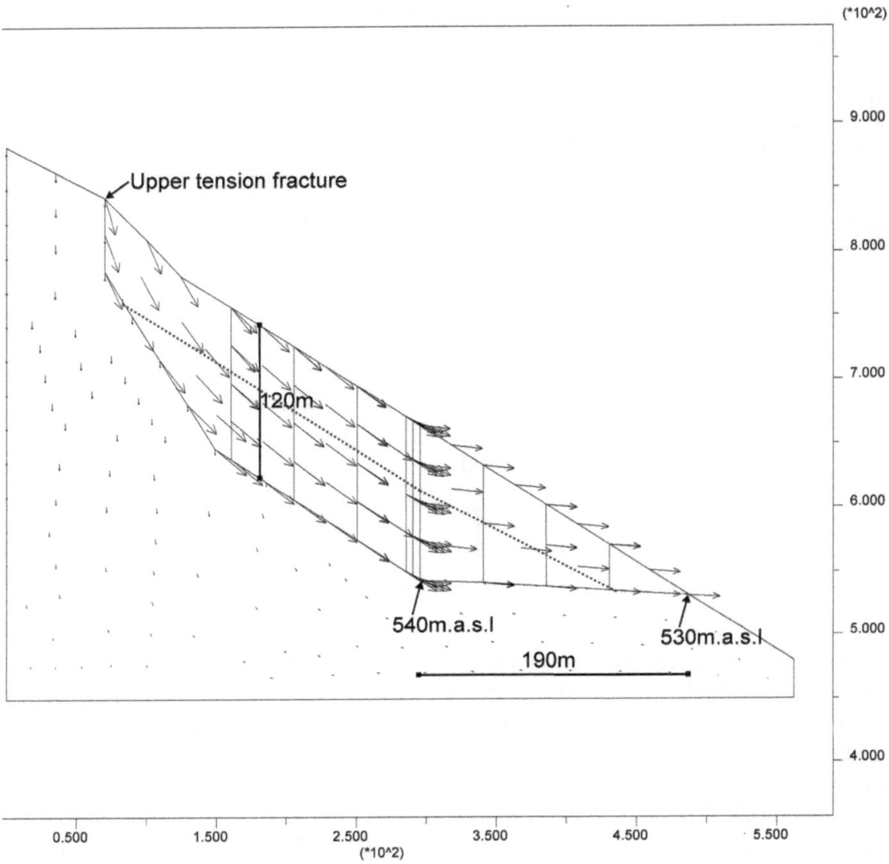

Fig. 19.6 Distinct element model of the upper half of the Åknes rock slope (block 7, Fig. 19.4). The lower fracture daylights at the boundary between blocks 7 and 10 (Fig. 19.5). 540 m.a.s.l. corresponds to sliding at 120 m depth by the upper borehole. The total area of the sliding blocks is 35,710 m^2. The *arrows* show computed displacement vectors in the model. The *blue broken line* shows the groundwater table (Kveldsvik 2008)

the shape of the probability curve showed in Fig. 19.5 was the same at Åknes, but leveled up, representing higher order of magnitudes. The site-specific curve representing this effect as compared with the one obtained using historic data is presented in Fig. 19.7.

Tsunami

Due to limited historic evidence of tsunamis in Storfjorden, the best effort at the moment for estimating its hazard measures is by numerical simulation. The

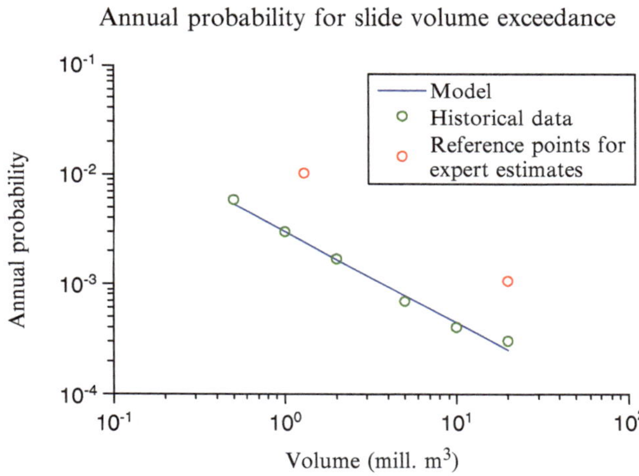

Fig. 19.7 Rock volume exceedance probability for Storfjorden (from historical data, *green dots*) and for Åkneset (estimates from expert's beliefs, *red dots*)

tsunami modeling component for the Åknes risk assessment is based on a simplified linear shallow water model which is applied for both the tsunami generation and propagation as described by Harbitz and Pedersen (1992). In this model, the landslide is simplified and described as a flexible box with a prescribed velocity progression. The shape of the box is defined by the physical extensions of the slide and is given as input to the numerical model (length, width, and height). The box is rounded to avoid numerical noise due to sharp edges, and the landslide propagation follows a straight line. The landslide progression is determined by the impact velocity and the run-out distance. Further details on the landslide representation can be found in Harbitz (1992) and Løvholt et al. (2005).

Additionally, according to the NGI informed experts' beliefs, the tsunami modeling uncertainties (e.g., landslide parameters among others) are within reasonable limits as to offer realistic tsunami propagation results. Verification of these calculations can be found in Harbitz et al. (1993) where the tsunami model is calibrated against the 1934 Tafjord tsunami. In this particular case, the predicted run-up heights showed good agreement with the measured run-up heights along the fjord system.

Tsunami Model Prediction in Storfjorden

The rockslide parameters considered in the present work are described in Table 19.2. The parameters are determined in the following way: After deciding on reference volumes for the different rockslide scenarios, the tsunami run-out length is determined by using a historical database for the relation between height, run-out, and

Table 19.2 Rockslide scenarios

Volume (Mm^3)	Slide dimensions (m)			Impact velocity (m/s)	Run-out distance (m)
	Length (L)	Width (W)	Height (H)		
2	400	200	20	50	800
4	500	200	30	55	900
6	500	200	45	60	1000
8	500	270	45	60	1000
12	800	200	60	60	1050
23	500	440	80	65	1200
35	800	440	80	40	1250

Table 19.3 Estimated run-up at selected locations. The run-up is given in meters and is assumed to be at 80th percentile

Location/rockslide volume	2 Mm^3	4 Mm^3	6 Mm^3	8 Mm^3	12 Mm^3	23 Mm^3	35 Mm^3
Hellesylt	3.5	6	9	13	13	36	36
Geiranger	2.5	4.5	8	11	11	32	34
Eidsdal	0.4	0.8	1.6	2	2	7	7
Tafjord	1	2	3	4	4	13	14

volume (NGI 2008). The impact velocities are then found by running a numerical block slide model (PCM) for different friction coefficients and tuning to match the run-out lengths determined above. The simulations with the PCM model are performed without taking the effect of water into account, since the run-out distance of large rockslides are only slightly affected by the ambient water. The largest scenario starts closer to the fjord and will not achieve the same high impact velocity as the ones starting higher up.

The run-up is determined by using the surface elevation of the highest wave of the leading part of the wave train. The surface elevation is measured at a single point (time series) outside each location before the wave starts to amplify due to shoaling. Based on run-up factors of nonlinear waves found in laboratory experiments (Pedersen and Gjevik 1983), the run-up is determined by multiplying the measured surface elevation with these factors. The chosen factor depends on the bathymetric slope and is varying from two (vertical wall) to five (extreme cases with gentle slopes). The bathymetry and topography applied in the tsunami modeling are provided by NGU (2007).

Estimates of run-up heights for the prescribed rockslide volumes are presented in Table 19.3. Also, to illustrate a potential tsunami scenario, Fig. 19.8 shows the estimated maximum surface elevation in Storfjorden (during the computational time) along with the corresponding run-ups in the selected locations for a 35 Mm^3 rockslide. Results correspond to the 80th percentile estimates.

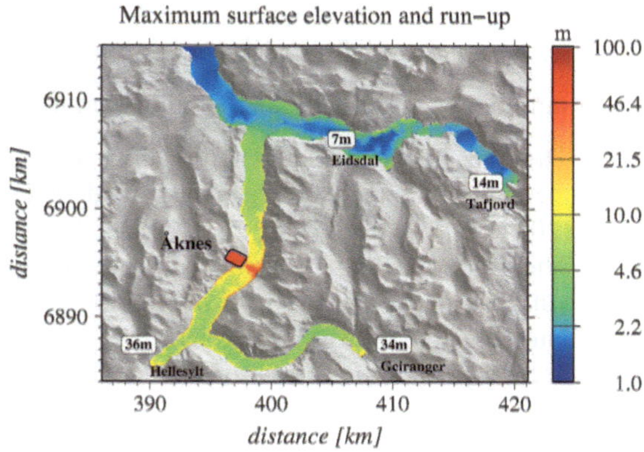

Fig. 19.8 Maximum surface elevation (in meters, logarithmic scale) during the computational time for the 35 Mm^3 scenario. The estimated run-up height at Hellesylt, Geiranger, Eidsdal, and Tafjord is given in the *white boxes*. Results correspond to the 80th percentile

Run-up heights for other volumes as those defined in Table 19.3 are computed by linear interpolation between the calculated run-up of the closest volumes. The estimated run-up value for one volume is considered to be at the 80th percentile, while half of the run-up is considered at the 20th percentile.

19.3.2 Vulnerability

Vulnerability is defined herein as the fraction of people in the affected areas losing their lives in case of a tsunamigenic rockslide. The quantitative vulnerability models for people are based on historic data from statistics of previous tsunami disasters. This is obtained in two steps: (1) a review of life's losses caused by earlier tsunamis and (2) a formulation of a continuous model relating inundation height with life's losses.

Vulnerability is therefore related to specific tsunami threats. This consideration is due to both the intensity of the tsunami and the susceptibility of the vulnerable elements, which in this case are the people living in the exposed areas.

The present vulnerability study is developed in the following way:

- General assumptions and limitations:
  - Only vulnerability for people is considered, i.e., material damage is omitted.
  - The vulnerability considerations assume that no warning has been issued.
- Assumptions and limitations regarding tsunami intensity:
  - The inundation height is assumed to be the most important parameter for tsunami intensity and vulnerability.
  - The effect of other physical tsunami parameters (as flow velocity) on vulnerability is omitted. The reason for this is that the quantification will be based

on published data, and most data about loss and destructions is given together with tsunami heights or inundation only.

- Assumptions and limitations regarding people's susceptibility to the tsunami:
  - The effect of the built and natural environment on the vulnerability is omitted. Details about the latter can be found in Dominey-Howes and Papathoma (2007).
  - The importance of the vertical placing of the people is included, since vulnerability is a function of inundation height, and higher elevations are less inundated than lower elevations.
  - Escape possibilities have not been considered.

19.3.2.1 Use of Historic Data

To make the vulnerability quantification method more general and applicable for wider land topographies and population patterns, the inundation height is proposed as the parameter of reference. Data has been collected from different sources containing fatality rates associated to inundation heights from tsunamis in Western Norway (Jørstad 1968; Tinti et al. 1999; Furseth 2006; and Furseth personal communication 2007), from the December 2004 Indian Ocean tsunami (Rosetto et al. 2007 and EEFIT 2006), and from the July 2006 Java tsunami (Reese et al. 2007). Below, an empirical model is introduced based on this data set.

19.3.2.2 Vulnerability Model

This section introduces a simplified model of vulnerability as a function of the inundation height. The shape of the vulnerability curve is suggested to be formed such that:

- Below a certain threshold in inundation height, the tsunami causes no harm.
- Above a certain limit, approximately every exposed person is fatally impacted by the tsunami.

Based on these assumptions, an S-shaped function on the general form

$$S = \frac{1}{1 + ke^{-\lambda H}}, \tag{19.3}$$

is proposed, where S is the vulnerability, k and λ are constants, and H represents the inundation height. Notice that this is a continuous function that mathematically never equals zero, for which this should be used with caution for the smallest inundation heights (i.e., not for inundation heights less than 1 m). Based on the historic data discussed above, three vulnerability curves are proposed: a best fit to the data

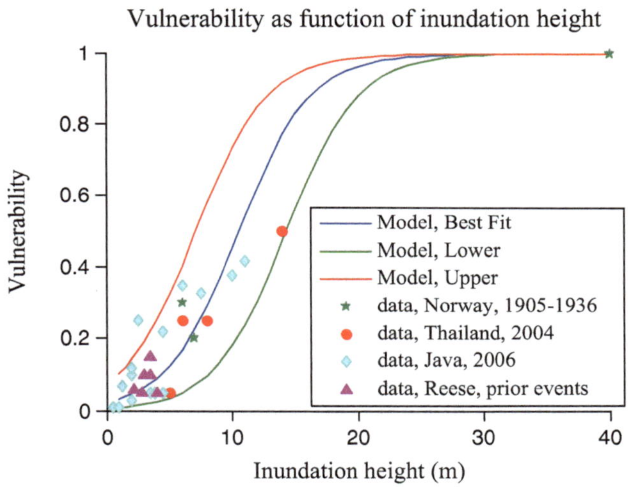

Fig. 19.9 Empirical quantification of vulnerability as a function of inundation height

and an upper and a lower vulnerability estimate based on the data uncertainty. These curves together with the empirical data set are plotted in Fig. 19.9.

Since vulnerability is a function of inundation height, the vulnerability varies within each location. To obtain one value for the vulnerability for each site, the vulnerability is averaged in space throughout the village. The averaging takes into account the proportion of people located at each elevation. The resulting vulnerability is a weighted average vulnerability. For the vulnerability calculations, the following assumptions and simplifications are considered:

- The inundation height at one location equals the difference between run-up and elevation height at the location. This simplification implies an assumption that the water above the inundated land is horizontal.
- In the spatial averaging procedure, the weighting of the vulnerability is performed according to the percentage distribution of the elements at risk (here people) at different elevation levels (i.e., at elevation 0 m, 5 m, 10 m, 15 m, 20 m, etc.). In this study, the percentage distribution was estimated roughly by looking into maps on housing pattern and by the location placing of roads and typical public places. In this way, the localization of the elements at risk is taken into consideration in the vulnerability estimate, and the number of elements at risk within the inundation limit of the largest scenario could be used as elements of risk for all scenarios.

The sequence of the proposed vulnerability assessment procedure is described in Table 19.4. An example of the type of maps used in the vulnerability calculation as applied for Hellesylt is presented in Fig. 19.10.

Table 19.4 Example of the vulnerability assessment assuming a run-up $R = 11$ m and that 20 % of the people are located at 0 m, 30 % located at 5 m, and 50 % located at 10 m. S is the vulnerability function

Elevation, E	$E = 0$ m	$E = 5$ m	$E = 10$ m
Fraction of people at elevation E	0.2	0.3	0.5
Inundation height, I	$I(0) = R = 11$m	$I(5) = R - 5$ m $= 6$ m	$I(10) = R - 10$ m $= 1$ m
Contribution to vulnerability	$0.2{*}S(11)$	$0.3{*}S(6)$	$0.5{*}S(1)$
Resulting vulnerability	$0.2{*}S(11) + 0.3{*}S(6) + 0.5{*}S(1)$		

The vulnerability assessment discussed above considers the following sources of uncertainty (Eidsvig et al. 2011):

- Uncertainty in run-up for a given volume (i.e., a given rockslide volume could result in a range of run-up values)
- Uncertainty in vulnerability for a given run-up (i.e., a given run-up could result in a range of vulnerability values)

Where the uncertainty in run-up for a given volume is due to uncertainty in (a) the geometry of the rockslide (the uncertainty is estimated from numerical experiments with different configurations), (b) the velocity of the rockslide (the uncertainty is quantified by analyses of the effect on run-up values from varying velocities and dynamical analyses estimating the probable velocity range for the rockslide), and (c) the numerical model and run-up calculations (the uncertainty is quantified by numerical and experimental data and judgment).

Finally, the uncertainty in vulnerability for a given run-up value is observed to be due to (a) the uncertainty in the inundation height estimation and spatial vulnerability averaging procedure; (b) the simplification with inundation height as the only parameter in the vulnerability model; (c) the spread of the empirical data, the limited amount of empirical data, and the uncertainty in the empirical data; and (d) the interpretation of vulnerability.

19.3.3 Consequences

The quantification of elements at risk (people's lives) is based on the number of people present in the exposed areas averaged over time. The exposed area is defined as the area within the inundation limit of the largest rockslide scenario. Averaged estimates of elements at risk for the locations of interest are described in Table 19.5 where it is assumed that the inhabitants and tourists are present 50 % of the time in the exposed areas. The tourist season is assumed to last two months with one fourth of the maximum tourist number present in the exposed area. The numbers are averaged through the year and through day and night.

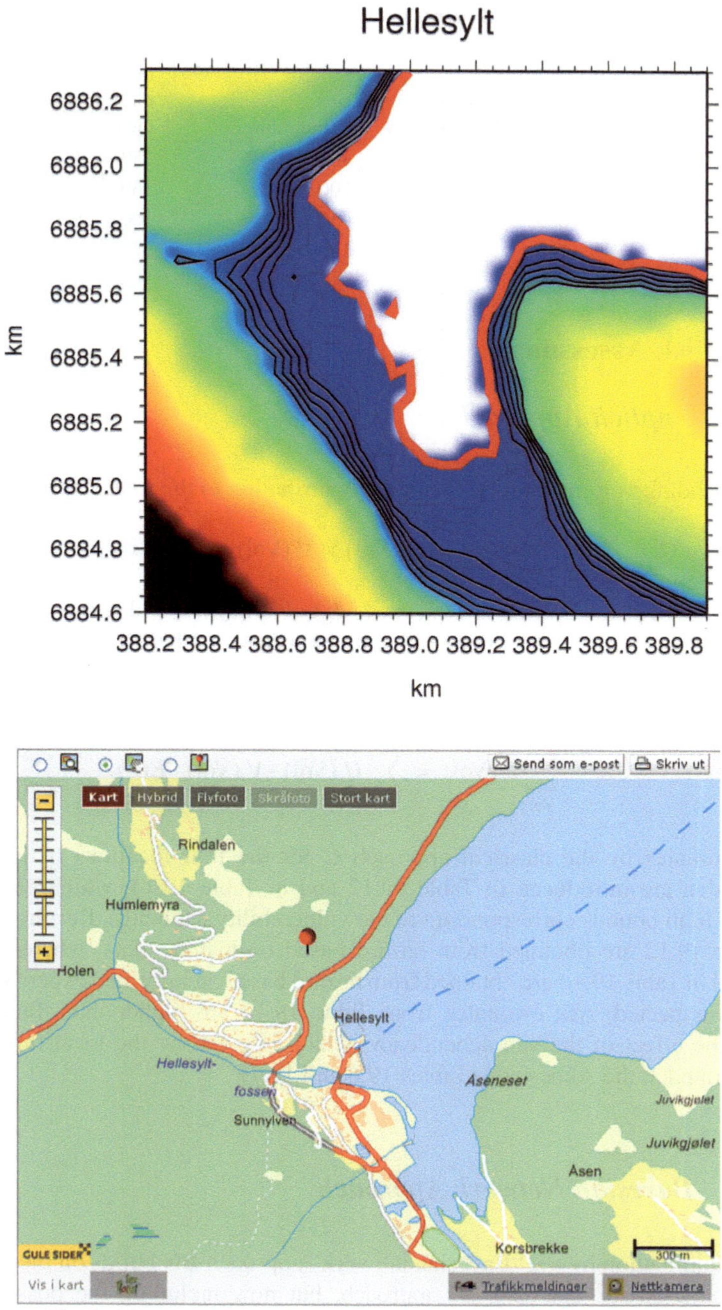

Fig. 19.10 Maps from Hellesylt used for the vulnerability assessment. *Upper figure*: contour map, *black lines* 10 m contour. The contour map is produced from topography provided by NGU (2007). *Lower figure*: map of the area (Gulesider 2008)

Table 19.5 A cutting from table of people in the exposed areas, Blikra et al. (2006)

Community	Tourists (maximum number in high season)	Population	Assumed average through the year and through day/night for the exposed areas
Hellesylt	5000	300	250
Geiranger	15,000	100	360
Tafjord	600	50	100
Eidsdal	1500	160	270

19.4 Risk Assessment

19.4.1 Implicit Approach

For one sliding volume interval (Vol), the contribution to the total risk is given by

$$R\,(\mathrm{Vol}) = H\,(\mathrm{Vol}) \times V\,(\mathrm{Vol}) \times L, \tag{19.4}$$

where *R*, *H*, *V*, and *L* refer to *risk*, *hazard*, *vulnerability*, and *losses* or elements at risk, respectively. Risk is defined as the expected number of fatalities per year averaged over several hundreds of years. The value of risk for the whole range of volume intervals is found by summing the risk over all volume intervals:

$$R = \sum_{\mathrm{Vol}} R\,(\mathrm{Vol}) = \sum_{\mathrm{Vol}} H\,(\mathrm{Vol}) \cdot V\,(\mathrm{Vol}) \cdot E(L), \tag{19.5}$$

Risk estimates by the classical approach ***C*** for the four locations of interest in Storfjorden are introduced in Table 19.12 and in Table 19.14, which present the low and high bounds corresponding to the vulnerability estimates. Results included in Table 19.12 are obtained from hazard based on historic data, whereas results included in Table 19.14 are obtained from hazard based on informed expert's beliefs. As it is expected, risk estimates from Table 19.14 are higher than Table 19.12 due to the effect of the site dependency, which is given by the local information concentrated at the Åknes site (Tables 19.6 and 19.7).

19.4.2 Bayesian Network Approach

Risk quantification by the "Bayesian network" approach **BN** builds on the evidence collection used in "classical" **C** approach, but now including the possibility of carrying out the uncertainty present at each event in what is defined as the risk

Table 19.6 Contribution to risk from each volume interval and the aggregated risk for all volumes, using failure probabilities from historic data (Fig. 19.3), assuming no warning has been issued

Location	Hellesylt		Geiranger		Tafjord		Eidsdal	
Risk (fatalities per year) for each volume interval (in Mm^3)	Low	Worst case	Low	Worst case	Low	Worst case	Low	Worst case
R([0.5,2])	0	0.065	0	0.036	0	0.0025	0	0.001
R([2,4])	0.0015	0.019	0.0011	0.013	0	0.0007	0	0.002
R([4,7])	0.0009	0.025	0.0006	0.030	0.00007	0.0005	0	0.002
R([7,12])	0.0012	0.027	0.0012	0.029	0.00004	0.0004	0.00024	0.002
R([12,20])	0.0036	0.032	0.0023	0.044	0.00007	0.0017	0.00018	0.003
R([20,35])	0.009	0.022	0.010	0.032	0.00006	0.0031	0.00019	0.003
R(>35)	0.031	0.040	0.036	0.058	0.00040	0.0096	0.00048	0.005
R(all volumes)	0.05	0.23	0.05	0.24	0.0006	0.02	0.001	0.02

Table 19.7 Contribution to risk from each volume interval and the aggregated risk for all volumes using probabilities from informed expert's beliefs (Fig. 19.7), assuming no warning has been issued

Location	Hellesylt		Geiranger		Tafjord		Eidsdal	
Risk (fatalities per year) for each volume interval (in Mm^3)	Low	Worst case	Low	Worst case	Low	Worst case	Low	Worst case
R([0.5,2])	0	0.28	0	0.16	0	0.011	0	0.004
R([2,4])	0.006	0.08	0.005	0.06	0	0.003	0	0.009
R([4,7])	0.004	0.11	0.003	0.13	0.0003	0.002	0	0.009
R([7,12])	0.005	0.11	0.005	0.10	0.0002	0.002	0.001	0.007
R([12,20])	0.016	0.14	0.010	0.20	0.0003	0.007	0.001	0.013
R([20,35])	0.038	0.10	0.045	0.14	0.0003	0.013	0.001	0.012
R(>35)	0.133	0.17	0.154	0.24	0.0017	0.041	0.0020	0.020
R(all volumes)	0.20	1.00	0.22	1.04	0.003	0.08	0.005	0.07

process. This is possible by generating probability distributions relating the causal events defined in a BN including the risk factors.

This approach relies on the use of informed experts' beliefs founded in the areas of rockslides, tsunamis, and risk assessment. For this purpose, a group of NGI experts collaborated together to translate their empirical and theoretical knowledge on the Åknes slide into uncertainty measures, being at the time the best known effort to quantify systematical risk in the Storfjord area.

19.4.2.1 Bayesian Paradigm

The Bayesian paradigm fits well the needs for integrating multivariable, multilevel, objective and subjective information sources, and even spatiotemporal referenced risk measures (Medina-Cetina 2006). Its most elemental definition is

$$\pi\left(\text{hypothesis}\middle|\text{evidence}\right) = \frac{f\left(\text{evidence}\middle|\text{hypothesis}\right)\pi\left(\text{hypothesis}\right)}{p\left(\text{evidence}\right)}$$
$$= \pi\left(\boldsymbol{\theta}\middle|\mathbf{d}_{\text{obs}}\right) = \frac{f\left(\mathbf{d}_{\text{obs}}\middle|\boldsymbol{\theta}\right)\pi\left(\boldsymbol{\theta}\right)}{\int f\left(\mathbf{d}_{\text{obs}}\middle|\boldsymbol{\theta}\right)\pi\left(\boldsymbol{\theta}\right)d\boldsymbol{\theta}} \propto \beta f\left(\mathbf{d}_{\text{obs}}\middle|\boldsymbol{\theta}\right)\pi(\boldsymbol{\theta}), \tag{19.6}$$

where the prior $\pi(\boldsymbol{\theta})$ represents the a priori state of information or probability of the evidence, associated to a set of parameters $\boldsymbol{\theta}$ (e.g., parameters of a model simulating a threat); the likelihood $f\left(\mathbf{d}_{\text{obs}}\middle|\boldsymbol{\theta}\right)$ represents the a priori state of information associated to the potential of the evidence or parameters $\boldsymbol{\theta}$ to reproduce the observations $\mathbf{d}_{\text{obs}}$, assumed to be certain. The posterior $\pi\left(\boldsymbol{\theta}\middle|\mathbf{d}_{\text{obs}}\right)$ is thus the joint probability function between the a priori states of information associated to both the prior and the likelihood. Similarly to the risk definition, this formulation permits the addition of other dependencies in the form of hyper-parameters (e.g., parameters defining the prior).

19.4.2.2 Risk Definition

The Bayesian network approach follows the probabilistic definition of risk (Einstein and Sousa 2007):

$$E[R] = P[T] \cdot P\left[\mathbf{C}\middle|T\right] \cdot u(\mathbf{C}), \tag{19.7}$$

where $P[T]$ is the *hazard* or probability of occurrence of a possible threat T; $u(\mathbf{C})$ is the *loss* or *utility* of a set of consequences $\mathbf{C}$, all certain to happen; and $P\left[\mathbf{C}\middle|T\right]$ is called *vulnerability*, representing the probability of loss or utility conditioned to certain level of intensity T. This definition can be expanded for including the dependency of its components relative to other events, as in the case of coupling of events leading to a main hazard (e.g., tsunamigenic rockslide).

19.4.2.3 Bayesian Networks

Medina-Cetina and Nadim (2008) introduced the use of Bayesian artificial intelligence for quantifying risk of a tsunamigenic rockslide. Figure 19.2 actually presents the proposed probability template or BN for representing this problem, when an early warning system is in place. A **BN** is a graphical model for reasoning under uncertainty where the nodes represent variables either continuous or discrete, and arcs represent influences between them, which aim at depicting causal relations (Jensen 2001). Causal analysis focuses on the dynamics of the evidence as it is generated (Pearl 2003). Inferences through causal analysis as in BNs cope with uncertainty with limited or multiple evidence, as opposed to Boolean inference which follows a monotonic reasoning. BNs in this way aim at predicting the effects of interventions, sudden changes, identifying causes of current events, and, most importantly, updating inferences based on new evidence.

Modeling BN requires the Markov's assumption: arcs present in the network are the only ones that represent direct dependencies. The process of introducing evidence into the network (at any node) is called probability propagation, inference, or belief updating. These can be diagnosis (from symptoms to cause), predictive or prognosis (from the causes to the symptoms), or a mixture of them called intercausal.

In order to illustrate the applicability of **BN** approach in the Åknes case, a simplified version of the BN presented in Fig. 19.2 is now introduced in Fig. 19.11. In this figure, the "rockslide" node adds up all the potential triggering factors and the probability of the rockslide displacement itself in one node. The "rockslide" has the potential to generate a "tsunami" and produce "consequences" (economical and/or life's losses). In the same BN, the influence of the "season" (tourist or non-tourist) is also considered. Evidence from each node was collected as described in Sect. 19.3.

It is worth mentioning that the simplified model does not diminish the importance of making inferences as long as it accurately represents the causal dependencies between the key events considered in the risk process. The same applies if thinking

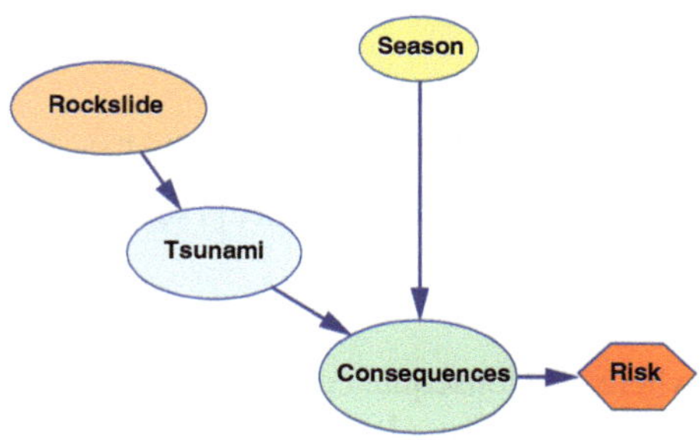

Fig. 19.11 Simplified Bayesian network for the risk quantification of the Åknes' tsunamigenic rockslide

on expanding the network to more detailed participating events. For the simplified model, all events associated to the nodes defined in Fig. 19.11 are assumed to include the corresponding expanding branches.

19.4.2.4 Risk Assessment

After the topology of the network is designed, it is required to define the probability distributions associated to each node, reflecting the available evidence, which is given in the form of probability tables (PTs) for the "parent" nodes such as "rockslide" and "season" and conditional probability tables (CPTs) for the "children" nodes such as "tsunami" and "consequence." The states of probability defined for each node are discrete to facilitate the transmission of the informed experts' beliefs on the "expected ranges" for each variable contained in each node. In this work probability distributions for the "rockslide," "tsunami," and "consequences" nodes were given by the corresponding NGI's experts.

Once the collected evidence is set in the form of PTs and CPTs, it is possible to simulate the information transferring from node to node and quantify risk. The forward propagation passes the uncertainty from the "rockslide" event up to the "risk" utility node. This requires to assessing each of the nodes' "beliefs" in between. The transferring of information by computing the corresponding states of probability once the node "rockslide" is instantiated is given as follows:

$$P\,(\text{Rockslide}) = \text{Bel}\,(\text{Rockslide})\,. \tag{19.8}$$

Then, using the principle of total probability

$$P\left(\text{Tsunami}\middle|\text{Rockslide}\right) \leftarrow \text{Bel}\,(\text{Tsunami}) = \sum_i P\left(\text{Tsunami}\middle|\text{Rockslide}_i\right)\,\text{Bel}\,(\text{Rockslide}_i), \tag{19.9}$$

$$P\,(\text{Season}) = \text{Bel}\,(\text{Season})\,, \tag{19.10}$$

$$P\left(\text{Consequence}\middle|\text{Tsunami, Season}\right) \leftarrow \text{Bel}\,(\text{Consequence}) = \sum_i\sum_j P\left(\text{Consequence}\middle|\text{Tsunami}_i, \text{Season}_j\right)\text{Bel}\,(\text{Tsunami}_i)\,\text{Bel}\left(\text{Season}_j\right), \tag{19.11}$$

where $P(\text{Consequence}|\text{Tsunami, Season})$ reflects the definition of vulnerability according to Eq. (19.7). And finally, the expected "risk" given by

$$E\left[\text{Risk}\middle|C, EM\right] = \sum_i U\left(O_i\middle|\text{Consequence}_i\right)\text{Bel}\,(\text{Consequence}_i), \tag{19.12}$$

where $U(O_i|\text{Consequence}_i)$ is the expected utility outcome O_{ij} conditioned on Consequence$_i$ for the combination i. Notice that Eq. (19.11) fulfills the definition of risk given previously by Eq. (19.7).

19.4.2.5 Risk Assessment in Storfjorden

Following the BN design given in Fig. 19.11, PTs for "rockslide" and "season" are assumed to be constant for the four sites considered in this study. CPTs for "tsunami" and "consequences" change according to the location of interest. For the PT of "rockslide," two sources are considered: the hazard obtained from historic data in Storfjorden and the hazard estimated from an informed expert's belief founded in field observations and theoretical modeling of the rockslide mechanisms at Åknes (Table 19.8).

The "season" node, although certain by nature, is given in the form of a random event. This is introduced to illustrate the ability of the BN to forward the information content into the risk index, but also to corroborate the ability of the BN to back-propagate the uncertainty for prescribed risk scenarios. The "season" PT is presented in Table 19.9. Notice that this event reflects the uncertainty of the tourist season.

The remaining nodes (children of "rockslide" and "season") are site dependent, since the probability of having consequences for specific threat intensities (vulnerability) changes from site to site due to topographic and population differences (Tables 19.10 and 19.11, respectively). CPTs of "tsunami" and "consequences" for Hellesylt are given below. Both tables were provided by NGI's experts. CPTs for Geiranger, Tafjord, and Eidsdal were also estimated and can be found in the NGI Report 20061032 (2008).

Table 19.8 Disaggregated rockslide hazard "R" according to historic data and expert's beliefs (on the Åknes site-specific data)

Volume Mm^3	Historic P(R)	Expert P(R)
No	0.9941	0.9269
0–0.5	0.00061	0.05
0.5–2	0.0036	0.0158
2–4	0.00074	0.0032
4–7	0.00035	0.0015
7–12	0.00022	0.0009
12–20	0.00013	0.0006
20–30	9.00E-05	4.00E-04
>35	0.00016	0.0007

Table 19.9 "Season" probability table

Season	P(S)
Tourist	0.25
Non-tourist	0.75

Table 19.10 "Tsunami" conditional probability table CPT for Hellesylt

P(T/R)		Volume, $\times 10^6$ m^3								
		No	0–0.5	0.5–2	2–4	4–7	7–12	12–20	20–35	>35
Run-up, m	No	1	0	0	0	0	0	0	0	0
	0–1	0	1	0.4	0.1	0.1	0	0	0	0
	1–5	0	0	0.45	0.65	0.3	0.2	0.1	0	0
	5–10	0	0	0.15	0.25	0.4	0.3	0.15	0.1	0
	10–15	0	0	0	0	0.2	0.4	0.15	0.15	0.05
	15–20	0	0	0	0	0	0.1	0.2	0.15	0.15
	20–25	0	0	0	0	0	0	0.15	0.15	0.2
	25–30	0	0	0	0	0	0	0.15	0.2	0.2
	>30	0	0	0	0	0	0	0.1	0.25	0.4

Table 19.11 "Consequence" CPT for Hellesylt

P(C|RU, R)		Consequences, loss of lives							
Run-up, m	Season	0	1–3	3–10	10–30	30–60	60–100	100–300	>300
No	TS	1	0	0	0	0	0	0	0
	NTS	1	0	0	0	0	0	0	0
0–1	TS	0.99	0.01	0	0	0	0	0	0
	NTS	0.999	0.001	0	0	0	0	0	0
1–5	TS	0.1	0.1	0.3	0.3	0.15	0.05	0	0
	NTS	0.25	0.4	0.25	0.1	0	0	0	0
5–10	TS	0	0	0.1	0.2	0.3	0.2	0.15	0.05
	NTS	0	0.2	0.3	0.3	0.15	0.05	0	0
10–15	TS	0	0	0	0	0.1	0.4	0.4	0.1
	NTS	0	0	0.1	0.3	0.4	0.2	0	0
15–20	TS	0	0	0	0	0	0	0.3	0.7
	NTS	0	0	0	0.1	0.4	0.4	0.1	0
20–25	TS	0	0	0	0	0	0	0.1	0.9
	NTS	0	0	0	0	0	0.2	0.8	0
25–30	TS	0	0	0	0	0	0	0	1
	NTS	0	0	0	0	0	0	1	0
>30	TS	0	0	0	0	0	0	0	1
	NTS	0	0	0	0	0	0	1	0

With the PTs and CPTs of each site defined, it follows to estimate the risk index by using Eqs. (19.8)–(19.12). Calculations of the uncertainty propagation were obtained using the program GeNIe (2007) developed by the Decision Systems Laboratory at the University of Pittsburgh. The expected risk index for each site is presented in Fig. 19.12 for both hazard estimates accounting for (a) the historic data and (b) the informed expert's beliefs on the threat imposed by the Åknes rockslide

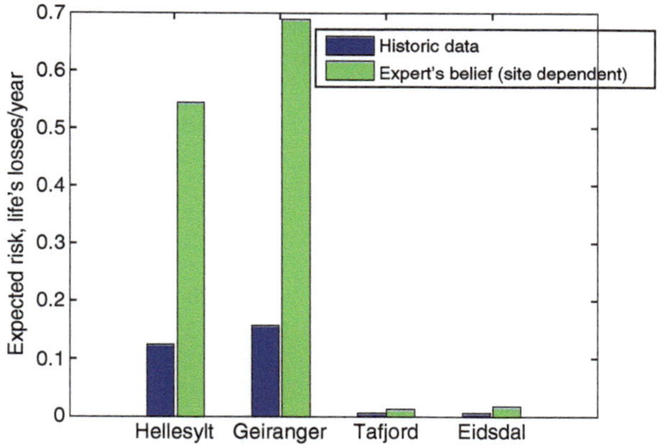

Fig. 19.12 Expected risk at the locations of interest based on historic data and informed expert's beliefs (site dependent)

Table 19.12 Expected risk (life losses) in Storfjorden based on historic data and expert's beliefs (site dependent)

Site	Season	Storfjorden	Åknes
Hellesylt	Undefined	0.12	0.54
	NTS	0.07	0.32
	TS	0.28	1.21
Geiranger	Undefined	0.16	0.69
	NTS	0.02	0.07
	TS	0.58	2.54
Tafjord	Undefined	0.00	0.01
	NTS	0.00	0.01
	TS	0.01	0.03
Eidsdal	Undefined	0.00	0.02
	NTS	0.00	0.01
	TS	0.01	0.03

(site-dependent hazard). Results from the **BN** approach (expected values) lie within the risk bounds obtained using the **C** approach, showing consistency on the use of the proposed risk assessment methodologies.

Two conditions of interest are the cases when evidence on the "season" is introduced, i.e., estimating the risk index for the non-tourist season (NTS) and for the tourist season (TS). Table 19.12 presents the computed expected risk values for the undefined season condition and for the NTS and TS conditions. This table shows that once evidence is introduced into the "season" node, a significant change in the risk is observed, particularly in Hellesylt and Geiranger. In Storfjorden, the lowest risk is expected during the NTS, the highest risk during the TS. Although these results look simplistic, they illustrate the capability of the BN to disaggregate influencing events into the risk index, with the possibility of refining the network branches to other events.

Table 19.13 Conditional risk (life's losses) on prescribed rockslide volumes (Mm^3) in Storfjorden

Volume	Hellesylt	Geiranger	Tafjord	Eidsdal
0	0	0	0	0
0–0.5	0	0	0	0
0.5–2	8	4	0	0
2–4	13	13	0	0
4–7	28	38	1	1
7–12	51	45	1	2
12–20	131	207	2	5
20–35	166	307	5	6
>35	201	350	7	12

An important benefit of the BN is the easiness in which evidence can be forwarded by updating the risk index. This is the case of prescribing potential scenarios of rockslide volumes and estimating the resulting risk, an exercise that can be interpreted as a pre-reliability analysis or simply belief updating, once that a specific rockslide volume is assumed to fall into the fjord. For this condition, Table 19.13 presents the expected risk associated to known evidence on the rockslide volumes. In this case, the order of magnitude for the risk is significantly higher to the overall risk estimates given before in Table 19.12, since the rockslide volume is assumed to be not uncertain anymore. Again, the higher risk is found on Hellesylt, showing consistency with the physical mechanism of the tsunami propagation described in the CPTs, given by the NGI's tsunami experts.

Similar to the belief updating, and probably the most important feature of **BN** approach, is the capacity to back-propagate evidence throughout the network as a way to generate a diagnosis based on current evidence. This is the case when evidence is introduced in the "consequence" node, assuming that the expected risk is known. Table 19.14 introduces the diagnosis analysis for Hellesylt, showing a natural redistribution of probabilities according to the risk state.

For instance, with a prescribed expected risk zero, the diagnosis corresponds to the ideal probability state of the "rockslide" with more favorable hazard condition than the one given by the informed expert's belief. On the other hand, if the maximum risk is prescribed, the "rockslide" hazard shows the most likely distribution needed to cause the most catastrophic condition. Diagnosis results for all prescribed risk estimates should be carefully observed during the operation of an EWS. In between the extremes discussed above, smooth changes on the "rockslide" and also the "tsunami" probability distributions take place, which clearly illustrate the nonlinear dependencies between them.

A final proof of logical passing of information through the BN is the convergence of the "season" node to the less and most critical risk conditions. This means that for a prescribed risk zero, the initial probability conditions are retrieved (i.e., NT $\sim$ 0.75, TS $\sim\sim$ 0.25), and as the risk is increased, this condition converges to the most critical, which is being at the touristic season (i.e., NT $\sim$ 0.00, TS $\sim$ 1.00).

Table 19.14 Diagnosis analysis for Hellesylt after introducing evidence in the "consequences" node

Volume, Mm^3	0	0.94	0.00	0.00	0.00	0.00	0.00	0.00	0.00
	0–0.5	0.05	0.04	0.00	0.00	0.00	0.00	0.00	0.00
	0.5–2	0.01	0.67	0.66	0.60	0.46	0.35	0.09	0.08
	2–4	0.00	0.20	0.20	0.19	0.15	0.11	0.03	0.03
	4–7	0.00	0.06	0.08	0.11	0.15	0.16	0.05	0.04
	7–12	0.00	0.02	0.04	0.07	0.13	0.17	0.06	0.07
	12–20	0.00	0.01	0.01	0.02	0.05	0.09	0.19	0.21
	20–35	0.00	0.00	0.00	0.01	0.03	0.05	0.18	0.18
	>35	0.00	0.00	0.00	0.01	0.03	0.07	0.41	0.40
Run-up, m	No	0.94	0.00	0.00	0.00	0.00	0.00	0.00	0.00
	0–1	0.06	0.05	0.00	0.00	0.00	0.00	0.00	0.00
	1–5	0.00	0.80	0.70	0.52	0.24	0.14	0.00	0.00
	5–10	0.00	0.16	0.28	0.40	0.51	0.43	0.15	0.13
	10–15	0.00	0.00	0.02	0.07	0.18	0.25	0.08	0.05
	15–20	0.00	0.00	0.00	0.01	0.07	0.13	0.05	0.17
	20–25	0.00	0.00	0.00	0.00	0.00	0.05	0.17	0.17
	25–30	0.00	0.00	0.00	0.00	0.00	0.00	0.22	0.20
	>30	0.00	0.00	0.00	0.00	0.00	0.00	0.32	0.28
Season	Tourist	0.25	0.10	0.23	0.33	0.46	0.49	0.27	1.00
	Non Tourist	0.75	0.90	0.77	0.67	0.54	0.51	0.73	0.00
Consequences, life's losses	0	E							
	1–3		E						
	3–10			E					
	10–30				E				
	30–60					E			
	60–100						E		
	100–300							E	
	300–800								E
Risk		0	2	6.5	15	45	80	150	450

19.5 Conclusions

This chapter integrates a first effort on the risk assessment of the Aknes' tsunamigenic rockslide in Storfjorden (Stranda municipality, Norway). Two risk assessment methodologies were considered: an implicit approach **C**, based on empirical relations between the risk components, and a **BN** approach based on causal relationships between the events taking place in the risk process. Expected risk from both methods showed similar values. However, some advantages on the use of the **BN** over the **C** approach were illustrated by giving a better understanding of the risk process through forward (prognosis) and back propagation (diagnosis) of the available evidence, which would improve significantly the decision-making process during the operation of the EWS.

The risk assessment considered two hazard conditions: one based on historical data and another based on informed expert's beliefs (site dependent). Overall risk measures from both methods showed a significantly higher risk when using the informed expert's beliefs due to the influence of the site dependency. Also, results from the risk analyses showed that Hellesylt and Geiranger had the highest risk estimates, while Eidsdal and Tafjord had relatively lower values.

Acknowledgments The elaboration of this chapter was funded by the ICG project "Risk and Vulnerability"; by the EU project IRASMOS, Integral Risk Management of Extremely Rapid Mass Movements (contract no. 018412); and by the EU Project TRANSFER, Tsunami Risk and Strategies for the European Region (contact no. 037058).

References

Blikra, L. H., Longva, O., Harbitz, C., & Løvholt, F. (2005). Quantification of rock-avalanche and tsunami hazard in Storfjorden, Western Norway. In K. Senneset, K. Flaate, & J. O. Larsen (Eds.), *Landslides and avalanches ICFL 2005 Norway* (pp. 57–64). London: Taylor & Francis Group.

Blikra, L. H., Anda, E., Høst, J., & Longva, O. *Åknes/Tafjord prosjektet –Sannsynlighet og risiko knyttet til fjellskred og flodbølger fra Åknes og Heggurdaksla* (p. 20). (2006). NGU Report 2006.039. www.aknes-tafjord.no

Derron, M.-H., Blikra, L. H., & Jaboyedoff, M. (2005). High resolution digital elevation model analysis for landslide hazard assessment (Åkerneset, Norway). In K. Senneset, K. Flaate, & J. O. Larsen (Eds.), *Landslide and avalanches ICFL 2005 Norway*. London: Taylor & Francis Group.

Dominey-Howes, D., & Papathoma, M. (2007). Validating a tsunami vulnerability assessment model (the PTVA model) using field data from the 2004 Indian Ocean tsunami. *Natural Hazards, 40*, 113–136.

EEFIT. (2006). The Indian Ocean tsunami of 26 December 2004: Mission findings in Sri Lanka & Thailand. In A. Pomonis, T. Rosetto, N. Peiris, S. M. Wilkinson, D. Del Re, R. Koo, S. Gallocher (Eds.), *Earthquake engineering field investigation team (EEFIT) report*. The Institution of Structural Engineers, London, UK. http://www.eefit.org.uk

Eidsvig, U., Kveldsvik, V., & Lacasse, S. (2008). Evaluation of Åknes rock slide hazard using event trees. In *Proceedings of the American Rock Mechanics Association*, San Francisco CA.

Eidsvig, U. M., Medina-Cetina, Z., Kveldsvik, V., Glimsdal, S., Harbitz, C. B., & Sandersen, F. (2011). Risk assessment of a tsunamigenic rockslide at Akne's. *Natural Hazards, 56*, 529–545.

Furseth, A. (2006). Skredulykker i Norge. Tun Forlag AS 2006, 1. opplag.

Ganerød, G. V., Grøneng, G., Rønning, J. S., Dalsegg, E., Elvebakk, H., Tønnesen, J. F., et al. (2008). Geological model of the Åknes rock slide, Western Norway. *Engineering Geology*, in print.

GeNie. (2007). *Graphical network Interface GeNie*. Decision Systems Laboratory, Department of Information Science and Telecommunications and the Intelligent Systems Program at the University of Pittsburgh, USA.

Einstein, H. H., & Sousa, R. (2007). Warning systems for natural threats. *Georisk, 1*(1), 3–20.

Grøneng, G., Christiansen, H. H., Nilsen, B., & Blikra, L. H. (2011). Meteorological effects on seasonal displacements of the Åknes rockslide, Western Norway. *Landslides, 1*, 1–15.

Gulesider Map of Hellesylt Norway. http://www.gulesider.no/kart/map.c?q=hellesylt+&imgt=MAP&id=a_10000281166 [Accessed August 2008].
Harbitz, C. B. (1992). Model simulations of tsunamis generated by the Storegga Slide. *Marine Geology, 105*, 1–21.
Harbitz, C. B., & Pedersen, G. (1992). *Model theory and analytical solutions for large water waves due to landslides*, vol. 4. Oslo, Norway: Department of Mathematics, University of Oslo, Preprint series.
Harbitz, C. B., Glimsdal, S., Løvholt, F., Kveldsvikb, V., Pedersen, G. K., & Jensen, A. (2014). Rockslide tsunamis in complex fjords: From an unstable rock slope at Åkerneset to tsunami risk in Western Norway. *Coastal Engineering, 88*, 101–122.
Hungr, O., Evans, S. G., & Hazzard, J. (1999). Magnitude and frequency of rock falls and rock slides along the main transportation corridors of Southwestern Columbia. *Canadian Geotechnical Journal, 36*(2), 224–238.
Jaboyedoff, M., Oppikofer, T., Derron, M.-H., Bilkra, L. H., Bohme, M., & Saintot, A. (2011). Complex landslide behaviour and structural control: A three-dimensional conceptual model of Aknes rockslide, Norway Slope Tectonics. *Geological Society, London, Special Publications, 351*, 147–161.
Jensen, F. V. (2001). *Bayesian networks and decision graphs*. New York, NY: Springer.
Jørstad, F. (1968). Waves generated by landslides in Norwegian fjords and lakes, Publication 79 (pp. 13–31). Oslo, Norway: Norwegian Geotechnical Institute.
Kveldsvik, V., Einstein, H. H., Nilsen, B., & Blikra, L. H. (2008) Numerical analysis of the 650,000m^2 Åknes rock slope based on measured displacements and geotechnical data. *Rock Mechanics Rock Engineering* (Submitted).
Kveldsvik, V., Einstein, H. H., Nadim, F., & Nilsen, B. (2007). Alternative approaches for analyses of a 100,000 m^3 rock slide based on Barton-Bandis shear strength criterion. *Landslides*, http://ejournals.ebsco.com/direct.asp?ArticleID=414B898789CF80825C7D, 10.1007/s10346-007-0096-x.
Kveldsvik, V. (2008). *Static and dynamic stability analyses of the 800m high Åknes rock slope, western Norway*. PhD dissertation, Department of Geology and Mineral Resources Engineering, Norwegian University of Science and Technology.
Løvholt, F., Harbitz, C. B., & Haugen, K. B. (2005). A parametric study of tsunamis generated by submarine slides in the Ormen Lange/Storegga area off Western Norway: Marine and petroleum. *Geology, 22*, 219–231.
Medina-Cetina, Z. (2006). *Probabilistic calibration of a soil model*. PhD Thesis, The Johns Hopkins University, Baltimore, MD.
Medina-Cetina Z., & Nadim F. (2008). Stochastic design of an early warning system. In T. Glade (Ed.) *Georisk, assessment and management of risk for engineered systems and geohazards*. Special Volume on Early Warning Systems.
Medina-Cetina, Z., Feinberg, J., & Nadim, F. (2008). Optimal threshold definition of early warning systems. In: *Proceedings of the European Geosciences Union*. Vienna.
Mei, C. C. (1989). *The applied dynamics of ocean surface waves. Advanced series on ocean engineering* (Vol. 1, p. 740). London: World Scientific.
Nadim, F. (2004). *Risk and vulnerability analysis for geohazards. Glossary of risk assessment terms*. ICG Report 2004-2-1, Oslo, Norway. http://www.engmath.dal.ca/tc32/2004Glossary_Draft1.pdf.
NGI. *Mapping of rock slides into reservoirs and lakes; Norwegian lakes exposed to tsunami generating rock slides*. (2008). Norwegian Geotechnical Institute (NGI), report 20061327–2.
NGU. (2007). Norwegian Geological Survey: bathymetry and topology for the Storfjorden area.
NUREG-0492. (1981). Fault tree handbook. US Nuclear Regulatory Commission. http://www.nrc.gov/reading-rm/doc-collections/nuregs/staff/sr0492/sr0492.pdf [Accessed August 2008].
Pearl, J. (2003). Statistics and causal inference: A review. *Test Journal, 12*(2), 281–345.
Pedersen, G., & Gjevik, B. (1983). Run-up of solitary waves. *Journal of Fluid Mechanics, 135*, 283–299.

Reese, S., Cousins, W. J., Power, W. L., Palmer, N. G., Tejakusuma, I. G., & Nugrahadi, S. (2007). Tsunami vulnerability of buildings and people is South Java – Field observations after the July 2006 Java tsunami. *Natural Hazards and Earth System Sciences, 7*, 573–589.

Rød, S. K., Botan, C., & Holen, A. (2011). Communicating risk to parents and those living in areas with a disaster history. *Public Relations Review, 4*, 354–359.

Rossetto, T., Peiris, N., Pomonis, A., Wilkinson, S. M., Del Re, D., Koo, R., et al. (2007). The Indian Ocean tsunami of December 26, 2004: Observations in Sri Lanka and Thailand. *Natural Hazards, 42*, 105–124.

Shi, G. H., & Goodman, R. E. (1985). Two dimensional discontinuous deformation analysis. *International Journal for Numerical and Analytical Methods in Geomechanics, 9*, 541–556.

Shi, G. H. (1988). *Discontinuous deformation analysis: A new numerical model for the statics and dynamics of block systems*. PhD dissertation, Department of Civil Engineering, University of California, Berkeley.

Tinti, S., Marami, A., Baptista, M. A., Harbitz, C., & Izquierdo, A. (1999). The unified European catalogue of tsunamis; A GITEC experience. In *International Conference on Tsunamies*, 26th – 28th May 1998, Paris, France, pp. 84–99. Paris, France: Commisariat à l'Energie Atomique (CEA).

United Nations UN. (2006). Global Survey in Early Warning Systems. International Strategy for Disaster Reduction ISDR. Available from: http://www.unisdr.org/ppew/info-resources/ewc3/Global-Survey-of-Early-Warning-Systems.pdf [Accessed August 2008].

UNDRO (United Nations Disaster Relief Organization). (1979). *Natural disasters and vulnerability analysis*. Geneva, Switzerland: UNDRO.

Chapter 20
Rock Moisture Dynamics, Preferential Flow, and the Stability of Hillside Slopes

Ronaldo I. Borja, Jinhyun Choo, and Joshua A. White

Abstract This chapter investigates the relevant hydrologic and geotechnical processes triggering failure of steep hillside slopes under rainfall infiltration. Despite decades of extensive study, the fundamental controls responsible for this commonly observed slope failure mechanism are yet to be quantified. The work focuses on the triggering mechanisms of slope failure induced by rainfall events and highlights the multiphysical nature of the problem. In hillside slopes, fluid supply from the rain and fluid input from the fractures of an underlying bedrock create moisture dynamics that could undermine the stability of slopes. The impact of such dynamics is difficult to predict, let alone quantify. In this chapter, the influence of rainfall input into the slope surface and the accompanying rock moisture dynamics are investigated using a hydromechanical model that couples the interaction between fluid flow and solid deformation. Both single-porosity and double-porosity formulations are employed, the latter formulation pertaining to the case where the solid matrix exhibits two dominant porosity scales. Nonlinear finite element simulations of the failure of hypothetical hillside slopes similar in configuration to the two well-documented test slopes, the CB1 and Ruedlingen test slopes, reveal the impacts of slope/bedrock topography, rainfall history, rock moisture dynamics, and preferential flow pattern on the failure of hillside slopes.

20.1 Introduction

Hydrologically driven slope instability threatens lives and property worldwide. On January 4, 1982, a 0.22 m storm superimposed on approximately 0.6 m of pre-storm seasonal rainfall triggered thousands of landslides in the central Coast Ranges of California. The landslides that resulted from this storm led to 24 fatalities and

R.I. Borja (✉) • J. Choo
Department of Civil and Environmental Engineering, Stanford University, Stanford, CA 94305, USA
e-mail: borja@stanford.edu

J.A. White
Atmospheric, Earth, and Energy Division, Lawrence Livermore National Laboratory, Livermore, CA 94550, USA

P. Gardoni, J.M. LaFave (eds.), *Multi-hazard Approaches to Civil Infrastructure Engineering*, DOI 10.1007/978-3-319-29713-2_20

millions of dollars in property damage (Brown III et al. 1984; Smith and Hart 1982). One of the observed failures was a debris flow that occurred in the 1200 block of Oddstad Boulevard in the city of Pacifica (Shlemon et al. 1987), which extended 230 m down a slope of approximately 21°. This event was particularly devastating in that three children lost their lives.

Other notable international examples of hydrologically driven slope failure within the last thirty years include the following events:

- Mameyes, Puerto Rico, 1985: 0.56 m of rainfall within a 24-h period (with rates as high as 70 mm/h) triggered debris flows resulting in 129 deaths (Jibson 1992).
- Rio Limon, Venezuela, 1987: 0.174 m of rainfall in less than 5 h triggered numerous shallow landslides and debris flows resulting in 210 deaths (Schuster et al. 2002).
- Rio de Janeiro Petropolis, Brazil, 1988: intense rainfall triggered landslides that resulted in 320 deaths (Nieto and Barany 1988; Ogura and Filho 1991).
- Antofagasta, Chile, 1991: rainfall rates as great as 60 mm/h during a 3-h period triggered landslides that resulted in 101 deaths (Van Sint Jan and Talloni 1993).
- Vargas, Venezuela, 1999: heavy rainfall exceeding 0.9 m over a three-day period, with daily values greater than the 1,000 year return period (Martinez 2000), triggered thousands of landslides and resulted in an estimated 30,000 deaths (USAID 2000).
- Guinsaugon, Philippines, 2006: heavy rainfall triggered massive landslides burying an elementary school that had 246 students and seven teachers (Lagmay et al. 2006).
- Colorado, United States, 2013: nearly continuous rainfall caused widespread landslides and flooding in the northern Colorado Front Range, resulting in three deaths (Coe et al. 2014).

Despite decades of extensive slope stability model development, the fundamental controls connecting the hydrologic and geotechnical processes triggering slope failure are still not well quantified. This is evident from the La Conchita landslide of January 11, 2005, in southern California that occurred without warning. This lack of understanding is a direct result of the simplified physics in current models, with the omission of the effect of variable density and partial saturation from slope stability calculations (Borja and Andrade 2006; Borja et al. 2013). Furthermore, there exist mechanisms of deformation that cannot be explained by a hydromechanical formulation based on only one dominant porosity scale. In slopes made up of aggregated soils, hydromechanical responses occur at two porosity scales, requiring a so-called double-porosity formulation (Barenblatt et al. 1960; Borja and Koliji 2009; Choo et al. 2015; Choo and Borja 2015; Gerke and van Genuchten 1993; Warren and Root 1963).

This chapter aims to develop and test a physics-based slope deformation model that couples the solid deformation with fluid flow processes in variably saturated soils with either one or two dominant porosity scales. The chapter also aims to assess the capability of the coupled model to predict stresses and deformation necessary to trigger slope failure in variably saturated soils. Apart from the full coupling of solid deformation with fluid flow processes, the exchange of water between the sliding earth mass and underlying bedrock is also quantified to allow a better understanding of the effects of bedrock fractures on the spatial and temporal variations of degree of saturation, effective stress, and deformation pattern within the variably saturated slope.

To facilitate an efficient solution, recently developed stabilized low-order finite element approximation schemes employing equal orders of interpolation for the solid displacement and pore pressure fields are utilized. Such stabilized formulation is critical for numerically solving the spatial pore pressure distributions at the two porosity scales accurately and efficiently. The coupled model is used to simulate two hillside slope failures similar in geometric configuration to the CB1 and Ruedlingen test slopes. The highly instrumented CB1 slope failed as a large debris flow in November 1996, while the Ruedlingen slope also failed as a debris flow in March 2009, thus providing insight into the mechanisms of failure of two highly instrumented test slopes.

20.2 Hydromechanical Model

A thermodynamically consistent measure of effective Cauchy stress $\bar{\boldsymbol{\sigma}}$ for unsaturated porous material with double porosity is given by the expression (Borja and Koliji 2009)

$$\bar{\boldsymbol{\sigma}} = \boldsymbol{\sigma} + B\bar{p}\mathbf{1}\,, \tag{20.1}$$

where $\boldsymbol{\sigma}$ is the total Cauchy stress tensor, B is the Biot coefficient, $\mathbf{1}$ is the Kronecker delta tensor, and

$$\bar{p} = \sum_{i=M,m} \psi^i \sum_{\alpha=\ell,g} S^{\alpha i} p_{\alpha i}\,, \tag{20.2}$$

is the overall mean pore pressure. In the expression for $\bar{p}$, α denotes either liquid (ℓ) or gas (g), i pertains to either the macropore (M) or micropore (m) scales, $p_{\alpha i}$ is the intrinsic α-fluid pressure for porosity scale i, $S^{\alpha i}$ is the degree of saturation for fluid α in porosity scale i, and ψ^i is the pore fraction for porosity scale i. We recall the closure condition

$$\sum_{i=M,m} \psi^i = 1\,, \qquad \sum_{\alpha=\ell,g} S^{\alpha i} = 1\,. \tag{20.3}$$

For single-porosity materials, the relevant expression is given in Borja (2004, 2006).

The balance of linear momentum for a two-scale porous medium ignoring inertia loads is obtained from Eqs. (20.12)–(20.14) of Borja and Koliji (2009) as

$$\mathrm{div}(\boldsymbol{\sigma}) + \rho \boldsymbol{g} = \sum_{i=M,m} \sum_{\alpha=\ell,g} c^{\alpha i} \boldsymbol{v}_{\alpha i} = \bar{\boldsymbol{c}}\,, \tag{20.4}$$

where ρ is the total mass density of the mixture, $\boldsymbol{g}$ is the gravity acceleration vector, $\boldsymbol{v}_{\alpha i}$ are the corresponding velocities, and $c^{\alpha i}$ are the mass exchange terms satisfying the closure condition

$$\sum_{i=M,m} \sum_{\alpha=\ell,g} c^{\alpha i} = 0\,. \tag{20.5}$$

In some cases, the following stronger condition holds

$$\sum_{i=M,m} c^{\alpha i} = 0 \qquad \alpha = \ell, g\,, \tag{20.6}$$

implying that mass transfer between the submixtures is limited to fluids of the same type, see Passman et al. (1984).

If we assume passive air condition, in which the pore air pressures are zero at the two porosity scales, then the overall mean pore pressure reduces to the form

$$\bar{p} = \psi S^m p_m + (1-\psi) S^M p_M\,, \tag{20.7}$$

where S^m and S^M are the degrees of saturation in the micropores and macropores, respectively; p_m and p_M are the corresponding intrinsic pore water pressures; and ψ is the void (or pore) fraction for the micropores. The pore air pressures are generally very close to the atmospheric pressure at shallow depths, such as what can be expected of a shallow landslide, so the assumption of passive air condition is reasonable for this case. Single-porosity formulation can readily be recovered by setting either $\psi = 0$ or $\psi = 1$.

The balance of mass for the pore water is given by a rate equation of the form

$$\frac{\phi^i}{K_w}(\dot{p}_i + \tilde{\boldsymbol{v}}_i \cdot \nabla p_i) + \psi^i B \mathrm{div}(\boldsymbol{v}) + \mathrm{div}(\boldsymbol{q}_i) = c^i\,, \quad i = M, m\,, \tag{20.8}$$

where the superimposed dot denotes a material time derivative following the solid motion, ϕ^i is the volume fraction for porosity scale i, K_w is the bulk modulus of the pore water, $\boldsymbol{v}$ is the solid velocity, and $\boldsymbol{q}_i = \phi^i \tilde{\boldsymbol{v}}_i$ is the relative discharge (Darcy) velocity of the pore fluid at porosity scale i. Without loss of generality, hereafter we assume that the water in the pores is incompressible and neglects the first term in Eq. (20.8).

The finite element equations over domain Ω are obtained from the following variational equations (letting $\boldsymbol{\eta}$, ω_M, and ω_m denote the weighting functions for solid displacements, macropore pressure, and micropore pressure, respectively):

1. Balance of linear momentum:

$$\int_{\Omega} \boldsymbol{\eta} \cdot \left(\operatorname{div}(\boldsymbol{\sigma}) + \rho \boldsymbol{g} - \bar{\boldsymbol{c}}\right) \mathrm{d}V = 0\,. \tag{20.9}$$

2. Balance of mass in the macropores, with $B = 1$:

$$\int_{\Omega} \omega_M \left((1-\psi)\operatorname{div}(\boldsymbol{v}) + \operatorname{div}(\boldsymbol{q}_m) + c^m\right) \mathrm{d}V = 0\,. \tag{20.10}$$

3. Balance of mass in the micropores, with $B = 1$:

$$\int_{\Omega} \omega_m \left(\psi \operatorname{div}(\boldsymbol{v}) + \operatorname{div}(\boldsymbol{q}_M) - c^m\right) \mathrm{d}V = 0\,. \tag{20.11}$$

One can use integration by parts and the divergence theorem to obtain alternative forms of the variational equations. Substituting the constitutive equation for solid, Darcy's law for fluid flow, and the relevant water retention curves for the macropores and micropores leads to a $\boldsymbol{u}/p_M/p_m$ formulation, with $\boldsymbol{u}$ = displacement of the solid matrix, and (p_M, p_m) = macropore and micropore pressures, respectively, see Choo et al. (2015), Choo and Borja (2015) for details.

We employ mixed finite elements leading to the matrix form of the finite element equations with a 3×3 submatrix block structure of the coefficient matrix representing the positions of the solid matrix displacement vector $\boldsymbol{d}$ and nodal pore pressure vectors $\boldsymbol{p}_M$ and $\boldsymbol{p}_m$. In general, the double-porosity formulation increases the size of the matrix problem significantly. Thus, it is desirable to use equal-order (linear) interpolations for the displacement and pore pressure degrees of freedom to alleviate computational difficulty. However, in the fully saturated range, equal-order interpolation leads to spurious pore pressure oscillations associated with LBB instability in the undrained limit (Brezzi and Bathe 1990; White and Borja 2008). These oscillations can be alleviated and, in some cases, completely eliminated, by stabilization of the low-order elements. The stabilization is done on the 2×2 submatrix block of the tangent operator associated with the two pore pressure degrees of freedom, i.e.,

$$\begin{bmatrix} A & B_1 & C_1 \\ B_2 & D & E_1 \\ C_2 & E_2 & F \end{bmatrix} \rightarrow \begin{bmatrix} A & B_1 & C_1 \\ B_2 & \bar{D} & \bar{E}_1 \\ C_2 & \bar{E}_2 & \bar{F} \end{bmatrix}, \tag{20.12}$$

where the submatrices $\bar{D}$, $\bar{E}_1$, $\bar{E}_2$, and $\bar{F}$ include stabilization terms.

As noted in Choo et al. (2015), the required stabilization for the problem at hand is treated as a twofold saddle point problem to a circumvent pore pressure oscillations in the macropore and micropore regions. The inf-sup condition in the discrete space for full saturation (i.e., $S^M = S^m = 1$) takes the form

$$\sup_{\boldsymbol{v}^h \in \mathcal{S}_u^h} \frac{\int_\Omega \bar{p}^h \nabla \cdot \boldsymbol{v}^h \, \mathrm{d}\Omega}{\|\boldsymbol{v}^h\|_1} \geq C(\|(1-\psi)p_M^h\|_0 + \|\psi p_m^h\|_0) \geq C\|\bar{p}^h\|_0 \tag{20.13}$$

for all $p_M^h \in \mathcal{S}_p^h$ and $p_m^h \in \mathcal{S}_p^h$, with $C > 0$ independent of h. Thus, we see that the pressure p in a single-constraint problem is now replaced with the mean pore pressure $\bar{p}$ weighted according to the pore fractions (cf. White and Borja 2008). To this end, the polynomial pressure projection (PPP) technique advocated by Bochev and Dohrmann (2006), Bochev et al. (2006), and used for single-constraint problem by White and Borja (2008) in the context of coupled solid deformation-fluid flow, is now used to stabilize the problem with two pore pressure constraints, see Choo and Borja (2015) for further discussions of this new stabilization technique.

20.3 Constitutive Laws

A number of constitutive laws must be introduced to close the boundary value problem for poromechanics problems with two dominant porosity scales. These constitutive laws are motivated by continuum principles of thermodynamics and, more specifically, by the expression for the rate of change of internal energy developed by Borja and Koliji (2009).

Restricting to infinitesimal deformation, the first constitutive law relates the effective Cauchy stress tensor $\bar{\boldsymbol{\sigma}}$ with the deformation of the solid matrix and takes the rate form

$$\dot{\bar{\boldsymbol{\sigma}}} = \mathbb{C} : \dot{\boldsymbol{\epsilon}}\,, \tag{20.14}$$

where $\mathbb{C}$ is a fourth-order tangent tensor that is obtained in this work from an elastoplastic constitutive law, and $\boldsymbol{\epsilon}$ is the infinitesimal strain tensor. A similar expression may be developed for finite deformation poromechanics.

The second constitutive law relates the relative flow vector with volume fraction and pore water pressure at each porosity scale. Noting that the flow vector in porosity scale i is $\boldsymbol{q}_{iw} = \phi^{iw}\tilde{\boldsymbol{v}}_i$, where ϕ^{iw} is the volume fraction of water in porosity scale i, Darcy's law is postulated to be valid for the two porosity scales:

$$\boldsymbol{q}_{iw} = -k_{ri}\frac{k_i}{\mu_w} \cdot (\nabla p_i - \rho_w \boldsymbol{g})\,, \quad i = M, m\,, \tag{20.15}$$

where k_{ri} and k_i are the relative and intrinsic permeabilities at porosity scale i, respectively, and μ_w is the dynamic viscosity of water. The relative permeability depends on the local degree of saturation. Note that porosity may also vary with finite deformation (see Song and Borja 2014a,b), although it is considered constant in the present formulation.

The third constitutive law relates the local matric suction s_i with the local degree of saturation S^i for a given volume fraction. In this work the van Genuchten equation (Van Genuchten 1980) is used at each porosity scale, which is written as

$$S^i(s_i) = S_1^i + (S_2^i - S_1^i)\left[1 + (s_i/s_{\alpha i})^{n_i}\right]^{-m_i}, \quad i = M, m, \tag{20.16}$$

where S_1^i and S_2^i are the residual water saturation and maximum water saturation at porosity scale i, respectively; $s_{\alpha i}$ are scaling suctions; and n_i and m_i are parameters that determine the shape of the water retention curves. Note that the fitting parameter m_i is related to the parameter n_i according to the relation

$$m_i = 1 - 1/n_i\,. \tag{20.17}$$

The relative permeabilities of the water phase in the macropores and micropores are given, respectively, by

$$k_{r_i} = \theta^{1/2}\left[1 - (1 - \theta_i^{1/m_i})^{m_i}\right]^2, \quad \theta_i = \frac{S^i - S_1^i}{S_2^i - S_1^i}, \quad i = M, m\,. \tag{20.18}$$

The fourth constitutive law involves mass transfer terms c^i and relates the diffusive mass transfer term to the pressure difference at the two porosity scales. A vast number of studies have advanced constitutive relations of this form (see e.g., Warren and Root 1963, Dykhuizen 1990, Gerke and van Genuchten 1993). Among them, a first-order mass transfer equation of the following form is common:

$$c^M = \frac{\bar{\alpha}}{\mu_w}(p_m - p_M)\,, \quad c^m = \frac{\bar{\alpha}}{\mu_w}(p_M - p_m)\,, \tag{20.19}$$

where $\bar{\alpha}$ is a dimensionless parameter that depends on the characteristics of the interface between the macropores and micropores, such as permeability, spacing, and shape. Several specific forms of the parameter $\bar{\alpha}$ have been suggested based on theoretical and experimental results (e.g., Warren and Root 1963, Dykhuizen 1990, Gerke and van Genuchten 1993). This study employs an equation proposed in Gerke and van Genuchten (1993), given by

$$\bar{\alpha} = \bar{k}\frac{\beta}{a^2}\gamma\,, \tag{20.20}$$

where $\bar{k}$ is the interface permeability, a is the characteristic length of the macropore spacing, β is a dimensionless coefficient that accounts for macropore geometry, and γ is a dimensionless scaling coefficient.

The final constitutive law inferred from the rate of change of internal energy relates the mean pore pressure difference with the macropore volume fraction (Borja and Koliji 2009). For the infinitesimal theory where the volume fractions may not be changing significantly, it is reasonable to ignore this particular constitutive relation.

20.4 The CB1 and Ruedlingen Slopes

Validation of a mechanistic model for rainfall-induced landslide triggering requires extensive full-scale field testing under a variety of geometric, geologic, and hydrologic conditions. Ideally, mechanistic models of slope failure must be validated against test slopes that failed, since very little insight into failure can be inferred from stable slopes. This requires that test slopes be steep enough and/or that the rainfall be sufficiently intense to insure that the test slope would fail during testing. Combined with the need for an extensive instrumentation, designing a slope for failure could imply that some of the expensive instruments might be damaged or lost from the ensuing mass movement.

Two well-documented test slopes that experienced failure are the CB1 slope in Coos Bay, Oregon, United States, and the Ruedlingen slope near the bank of the Rhine River in northern Switzerland. The CB1 slope was a hydrologic catchment (Anderson et al. 1997; Ebel et al. 2007; Loague and VanderKwaak 2001; Montgomery et al. 1997; Torres et al. 1998) and was not designed to validate any slope failure model. However, intense rains of November 1996 caused the slope to fail as a large debris flow, leaving behind one of the most comprehensive hydrologic response data sets in existence for a steep, deforested catchment that experienced slope failure. The Ruedlingen slope, on the other hand, was a geotechnical test slope designed to investigate slope failure phenomena. In March 2009, the slope failed as a debris flow after 15 h of artificial rain (Askarinejad et al. 2012).

The CB1 and Ruedlingen slopes have similar geology, with the soil layer composed mainly of sand matrix over a highly fractured sedimentary bedrock. However, the genesis of the soil layers and the clay contents for the two slopes are somewhat different. Whereas the Ruedlingen soil is a sediment with significant clay fractions that provide plastic characteristics to the soil, the CB1 soil is a nonplastic colluvium with the soil particles bonded solely by natural vegetation. Both slopes have inclinations on the order of 43° (average for CB1, maximum for Ruedlingen), but the soil thicknesses are quite different: the Ruedlingen slope has a variable soil depth ranging from 0.5 m to more than 5.0 m, while CB1 has a more uniform thickness of approximately 1.5 to 2.0 m.

Comparing the soil mechanical properties of the two slopes, the similarities are restricted to the effective friction angle and cohesion. For the Ruedlingen soil, the effective friction angle is around 31° and the cohesion is approximately 15 kPa; for the CB1 soil, the effective friction angle is close to 33°, while the apparent cohesion (developed primarily from root cohesion) ranges from zero to 10 kPa. Comparing their hydraulic properties and hydrogeology, the similarities are very pronounced: both slopes have soil saturated hydraulic conductivity in the range of 10^{-5} to 10^{-4} m/s. Furthermore, the bedrocks for both slopes are substantially fractured and conducive of water paths. As a fluid source, bedrock fractures could induce fluid flow that further destabilizes the slope.

Considering the comparison described above, the deformation properties of the two soils can be expected to be somewhat different. Whereas the Ruedlingen soil is expected to behave like a normally to over-consolidated clay due to significant plastic content, the CB1 soil is expected to behave like a compacted sand. In any case, since the two slopes are so steep, the stress points at any point in either slope are expected to lie on the dilatant side of the critical state line. Thus, constitutive models of either the Mohr-Coulomb or Drucker-Prager type, without a compression cap, would be sufficient for these two slopes, since the stress ratio (deviatoric stress to mean normal stress ratio) is not expected to engage the compression cap anywhere within the slope.

The next section is devoted to the numerical simulation of failure triggering mechanisms in slopes with similar geometry, geology, and geotechnical properties to CB1 and Ruedlingen. The simulations are restricted to failure triggering; post-failure studies, including the actual debris flow events, are not included in the simulations.

20.5 Numerical Simulations

The numerical simulations presented in this section consider two slope geometries: the first is similar to the CB1 catchment, while the second is a hypothetical slope modified from the Ruedlingen geometry. In both slopes the soil is underlain by a bedrock located at shallow depths.

The first slope example demonstrates the continuum modeling of a hillside slope failure similar to CB1 for which a pronounced scarp zone ensued following a large debris flow. The effect of fluid input from a fractured bedrock is also briefly discussed in this example. The second slope example investigates how single- and double-porosity modeling of flow in a hillside slope could impact the ensuing hydromechanical responses. This investigation is motivated by the fact that soil deposits in hillside slopes often manifest two dominant porosity scales, which has inspired a number of studies employing double-porosity concepts (see, e.g., Stadler et al. 2012, Klaus et al. 2013, Laine-Kaulio et al. 2014 and references therein). It must be emphasized that neither simulation is aimed at capturing the exact triggering mechanisms at CB1 or Ruedlingen, but rather, they are simply presented to provide a general picture of how failure might occur in hillside slopes with geometry, geology, and geotechnical properties similar to those found in natural slopes.

The methodology for continuum modeling of landslide triggering has been developed extensively and described in the authors' previous publications, see Borja and White (2010b), Borja et al. (2012a,b). The readers are simply referred to these references for further details of the modeling technique.

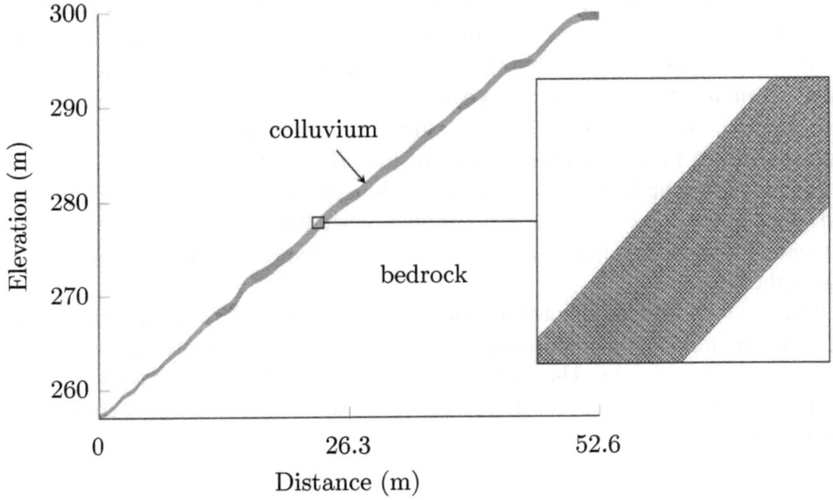

Fig. 20.1 Geometry and mesh for Slope #1

20.5.1 Slope #1: Bedrock Moisture Dynamics

The slope considered in this study has a geometry similar to that of CB1. Assuming a rigid bedrock, only the soil domain was discretized into 128,832 four-node quadrilateral mixed elements deforming in plane strain, as shown in Fig. 20.1. The finite elements utilize equal-order interpolation for the displacement and pore pressure degrees of freedom; however, some numerical stabilization for single-porosity problems is required for these elements to circumvent the so-called LBB instability under fully saturated, undrained loading; see White and Borja (2008).

Rainfall was applied on the surface of the slope as a flux infiltrating into the face of the slope. Two simulations were conducted, the first in which a no-flow condition was imposed on the slope-bedrock interface (impermeable bedrock), and the second in which the bedrock was assumed to be a fluid source (fractured bedrock). The displacements were fixed at the soil-bedrock interface so that sliding of the soil could only take place in the form of localized plastic shearing on the soil elements adjacent to the bedrock. Landslide is "triggered" once the yield zone emerges on the slope face.

For purposes of modeling the solid deformation response, a linear-elastic perfectly plastic Mohr-Coulomb plasticity model was used. As mentioned in the previous section, a compression cap would not be necessary since the stress points are expected to lie on the dilatant side of the critical state line. Table 20.1 summarizes the material parameters used in this example.

Table 20.2 summarizes the material parameters for fluid flow and water retention characteristics. Even though hysteresis in the water retention behavior is expected to influence the hydromechanical response of a slope subjected to cyclic wetting

Table 20.1 Solid deformation parameters for Slope #1

Parameter	Value
Bulk Modulus, K (MPa)	50
Poisson's Ratio, ν	0.25
Cohesion, c (kPa)	4
Friction angle, ϕ (deg)	40
Dilation angle, ψ (deg)	25

Table 20.2 Hydrologic parameters for Slope #1

Parameter	Value
Permeability, k (m^2)	3.4×10^{-11}
Dynamic viscosity, μ_w (kPa·s)	10^{-6}
Residual saturation, S_1	0.32
Maximum saturation, S_2	1.0
Scaling suction, s_a (kPa)	0.4
van Genuchten exponent, n	3.0

and drying (Ebel et al. 2010), only the wetting curve was considered in the present simulations. The parameters shown in Table 20.2 are similar to those reported for the soil deposit in CB1.

The initial stress field was generated by applying a static gravity load with solid grain density $\rho_s = 2.2$ t/m^3, water density $\rho_w = 1.0$ t/m^3, and porosity 0.5. After the initialization phase, the plastic strains were stored at the Gauss quadrature points and the nodal displacements were reset to zero. An initial suction of 1.5 kPa was assigned for the entire slope to reflect on-and-off precipitation before the main rainfall event. This is a simpler way of establishing the initial saturation condition compared to the modeling approach reported in previous studies (Borja and White 2010a; Borja et al. 2012a). Results of these previous studies revealed that landslide would be triggered only when the soil was completely saturated, which justifies the simpler approach of establishing the initial saturation condition used in the present simulations.

The main rainfall event was simulated by prescribing a uniform fluid flux of 30 mm/h until a yield zone emerges on the slope face. During this time, the slope face could become a seepage face when the phreatic surface reaches the slope face. To accommodate this condition, the pore water pressure was monitored on the slope surface at each time step. When the pore water pressure on the slope face became positive, the flow boundary condition was converted from that of a prescribed flux (natural boundary condition) to that of zero pressure (essential boundary condition).

Figure 20.2 shows the localized deformation pattern within the slope after 9.82 h of continuous rain. A plastic zone emerged at a point on the slope face (see inset) that coincides with the observed position of the scarp zone for CB1, see Borja et al. (2012a,b). It must be noted that the slope model shown in Fig. 20.2 was derived from the local topography of CB1; the results shown in this figure thus suggest that failure of the slope is sensitive to the local topography of the slope as well as that of the soil-bedrock interface.

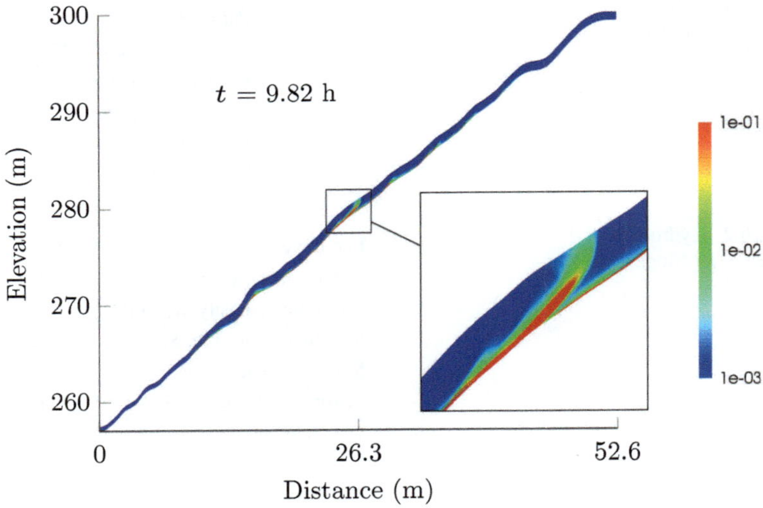

Fig. 20.2 Equivalent plastic strain (in percent) at 9.82 h for Slope #1

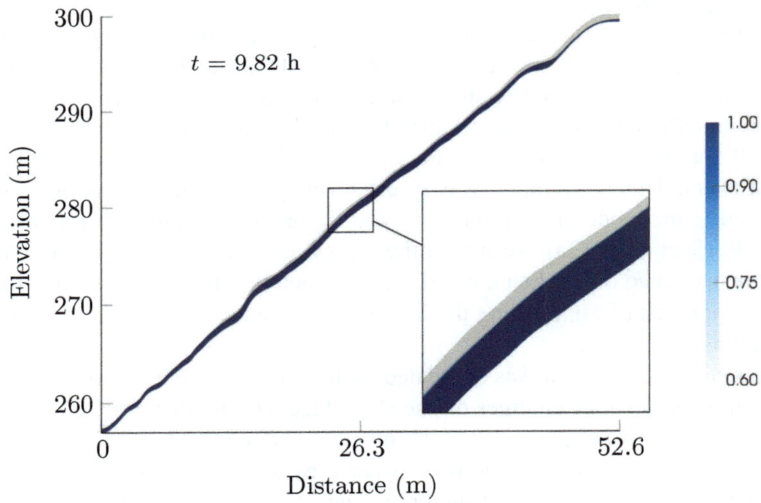

Fig. 20.3 Degree of saturation at 9.82 h for Slope #1

Figures 20.3 and 20.4 show, respectively, contours of degree of saturation and pore water pressure in the slope at $t = 9.82$ h. The contours indicate that loss of suction from increased saturation led to the development of the localized plastic zone. Localized plastic deformations formed in much of the high-pressure regions, notably near the soil-bedrock interface.

It has been observed from the field that the bedrock at CB1 contains high-density fractures that could be a source of fluid for the slope. Experimental results

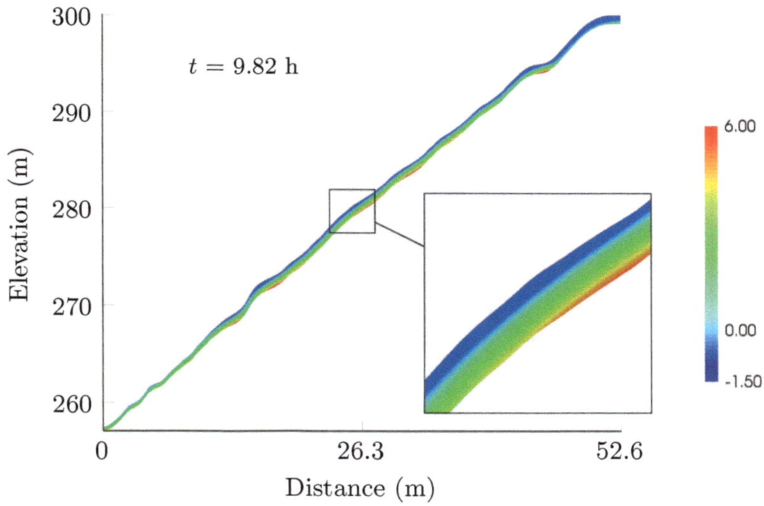

Fig. 20.4 Pore water pressure (in kPa) at 9.82 h for Slope #1

conducted for the catchment suggest that fluid could migrate through the soil-bedrock interface (Anderson et al. 1997). Determining the spatial distribution of fracture density is a challenging task; nevertheless, this aspect cannot be ignored in the mechanistic modeling of hillside slope failure. To gain some insight into the role of the bedrock fractures as a source of fluid, the analysis for Slope #1 was repeated but with an added input flux from the bedrock equal to 10 % of the applied rainfall.

With the additional flux applied from the bedrock, Fig. 20.5 shows that the localized plastic zone emerged at $t = 8.52$ h, about an hour earlier compared to the case with an impervious bedrock. Plastic deformation is also more intense than in the previous case; however, the location of the plastic zone remains essentially the same. Other fluid input scenarios from the bedrock also have been tested (not reported here), yielding essentially the same conclusion, i.e., that the surface and bedrock topography strongly influence the location of the plastic zone, but fluid input from the bedrock impacts the timing of slope failure. Not investigated in this chapter is the impact of rainfall history, which has been shown to be equally significant in determining the mechanism of failure, see Borja and White (2010b), Borja et al. (2012a,b).

20.5.2 Slope #2: Double-Porosity Simulation

The second example involves a modified version of the Ruedlingen slope and investigates the impact of double porosity on triggering slope failure. Figure 20.6 shows the finite element mesh for the slope domain consisting of 2,492 quadrilateral

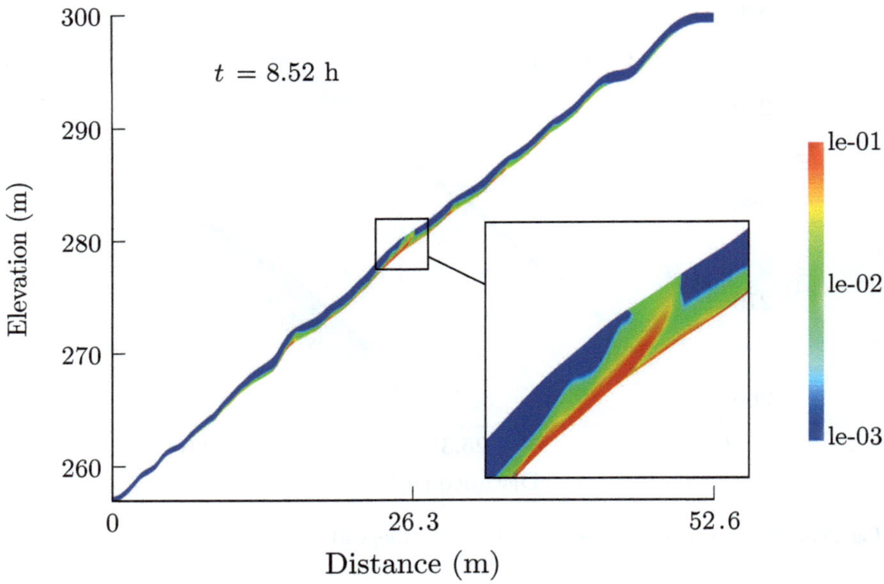

Fig. 20.5 Fractured bedrock simulation: equivalent plastic strain (in percent) at 8.52 h for Slope #1

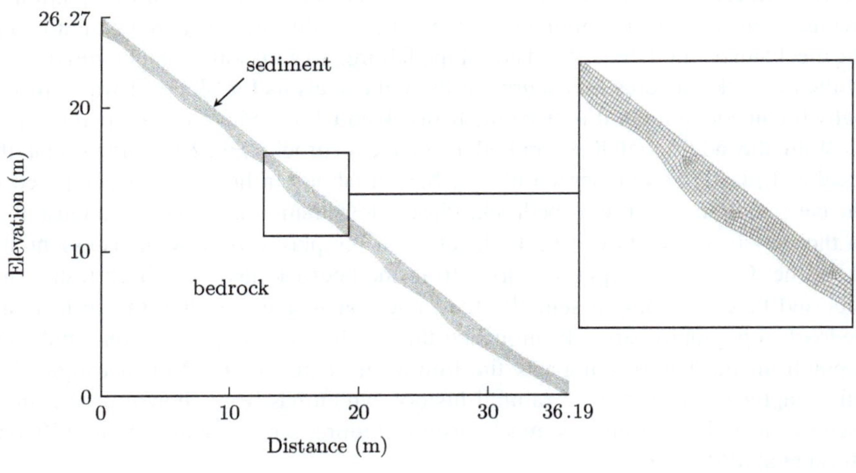

Fig. 20.6 Geometry and mesh for Slope #2

mixed elements. In double-porosity simulation, each node contains displacement and two pressure degrees of freedom for the macropores and micropores. The method proposed by Choo and Borja (2015) was used for numerical stabilization in the double-porosity range.

Table 20.3 Solid deformation parameters for Slope #2

Parameter	Value
Bulk modulus, K (MPa)	20
Poisson's ratio, ν	0.35
Cohesion, c (kPa)	4
Friction angle, ϕ (deg)	35
Dilation angle, ψ (deg)	25

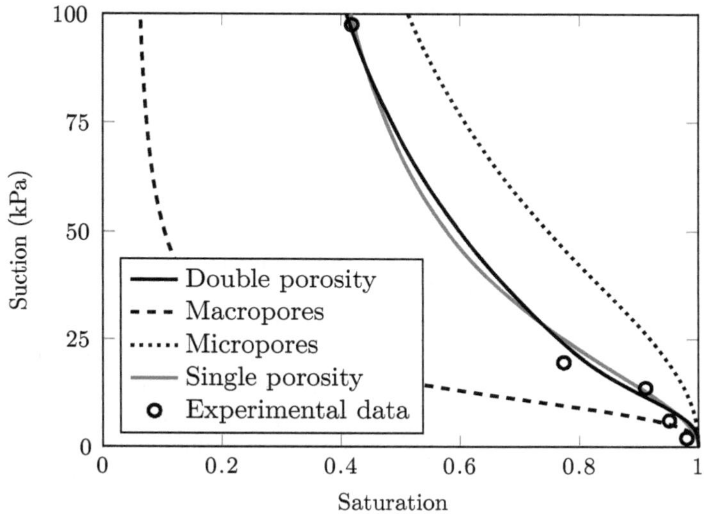

Fig. 20.7 Water retention curves for single- and double-porosity modeling of Slope #2. Data were calibrated from an aggregated soil (Carminati et al. 2007)

Solid and fluid densities of $\rho_s = 2.6$ t/m^3 and $\rho_w = 1.0$ t/m^3, respectively, and an overall porosity of 0.39 have been assumed in the simulations. The porosity value was derived from an aggregated soil (Carminati et al. 2007). The Mohr-Coulomb model was used for the solid matrix, with material parameters given in Table 20.3. For double-porosity modeling, the overall porosity was assumed to be partitioned into 0.09 for the macropores and 0.3 for the micropores. The overall permeability for single porosity k and the macro-/micro-permeabilities for double porosity k_M and k_m were calculated to satisfy the equation $\phi k = \phi^M k_M + \phi^m k_m$, as in Mehrabian and Abousleiman (2015). Water retention characteristics of the two porosity scales were also set to be comparable to the overall water retention characteristics of the single-porosity medium. Using the experimental data for an aggregated soil reported in Carminati et al. (2007), an overall van Genuchten curve for single-porosity medium and local (macro/micro) curves for double-porosity media were calculated according to the procedure described by Choo et al. (2015).

Figure 20.7 shows the water retention curves used for the single and double-porosity simulations. In this figure, the black curve is the superposition of the

Table 20.4 Hydrologic parameters for Slope #2

Parameter	Value		
	Single porosity	Double porosity	
		Macro	Micro
Permeability, k (m^2)	10^{-11}	4.33×10^{-11}	4.33×10^{-15}
Dynamic viscosity, μ_w (kPa·s)	10^{-6}	10^{-6}	10^{-6}
Residual saturation, S_1	0.2	0.05	0.2
Maximum saturation, S_2	1.0	1.0	1.0
Scaling suction, s_α (kPa)	25	12	50
van Genuchten exponent, n	1.92	3.0	2.2

local curves (dashed) and is comparable to the gray curve used for the single-porosity simulation. Table 20.4 summarizes the parameters used for hydrologic characterization of Slope #2. The diffusive mass transfer in the double-porosity simulation was modeled with the following parameters: $a = 0.01$ m, $\beta = 11$, $\gamma = 0.4$, and $\bar{k} = 0.01 \times \min(k_{rM}k_M, k_{rm}k_m)$.

The boundary and initial conditions were specified as in the previous example. An initial uniform suction of 20 kPa was assumed throughout the slope. Rainfall was applied as a flux of 50 mm/h on the slope surface, assuming an impermeable bedrock. For double-porosity simulation, two separate boundary conditions for the macropores and micropores were prescribed. Given a much higher permeability of the macropores, rainwater was assumed to infiltrate into the slope only through the macropores. Saturation of the micropores is possible only through diffusive mass transfer, consistent with the assumptions of Gerke and van Genuchten (1993). Note that due to the high contrast in water retention characteristics between the macropores and micropores, the macropores were assumed to be initially much less saturated, and most of the pore water was assumed to be stored in the micropores before the rain. This is a typical condition in the field.

Figure 20.8 shows contours of plastic strains developed in the sediment after 127 min of rain. Similar to the previous example, a localized plastic shear zone emerged on the upper part of the slope. Figures 20.9 and 20.10 show the degree of saturation and pore water pressure, respectively, at this same time instant. The location of the localized plastic shear zone correlates well with where the sediment is fully saturated, and the pore water pressures are positive, suggesting once again that the primary trigger of the localized shear zone is the loss of suction in the soil. In contrast to Slope #1 where the saturation progressed uphill from bottom to top, the saturation for Slope #2 propagated in the opposite direction, i.e., from top to bottom of the slope.

Next, the hydromechanical behavior of the same slope is investigated using the double-porosity formulation. In this case the zone of intense deformation emerged at a much later time, at 178 min. Figure 20.11 portrays the contour of plastic strain

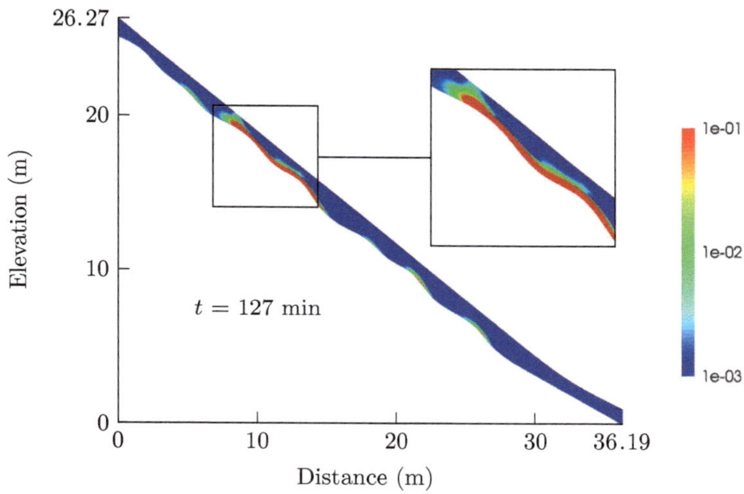

Fig. 20.8 Single-porosity simulation on Slope #2: equivalent plastic strain (in percent) at 127 min

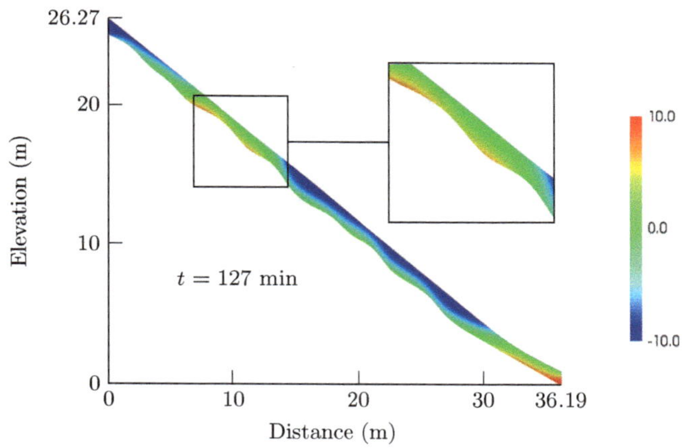

Fig. 20.9 Single-porosity simulation on Slope #2: pore water pressure (in kPa) at 127

at this time instant, revealing that the pattern of plastic strain development is quite different from the single-porosity simulation. Note that the intense plastic shear zone is located much lower than that determined from the single-porosity simulation shown in Fig. 20.8.

To explain why the location of the plastic zone changed with the double-porosity formulation for the same slope, Figs. 20.12 and 20.13 present the contours of overall degree of saturation and mean pore water pressure at $t = 178$ min, respectively. Note that the effective stress is determined by the mean pore water pressure $\bar{p}$, so the plastic yield zone is also dependent on $\bar{p}$. The results shown in these figures suggest

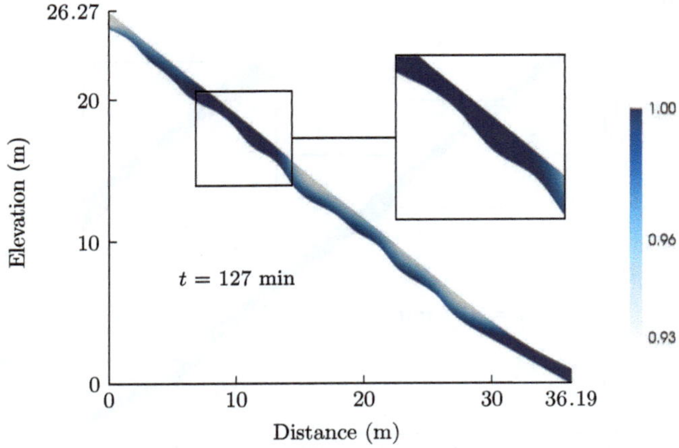

Fig. 20.10 Single-porosity simulation on Slope #2: degree of saturation at 127 min

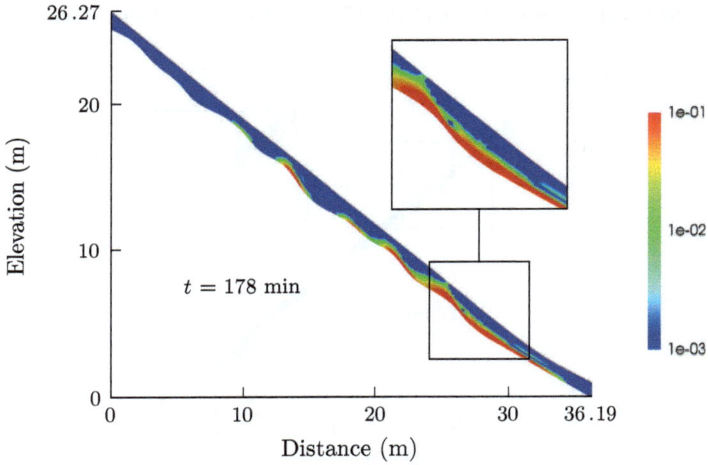

Fig. 20.11 Double-porosity simulation on Slope #2: equivalent plastic strain (in percent) at 178 min

that the saturation zone developed much more slowly than in the single-porosity simulation, even if the overall water retention curves are about the same, since the permeability of the macropores was high enough to allow the rainwater to flow more freely underneath the slope. This phenomenon is called preferential flow, which can only be captured with the double-porosity formulation. This example thus highlights the importance of an explicit treatment of the two porosity scales and the capture of the preferential flow pattern for an accurate prediction of landslide triggering.

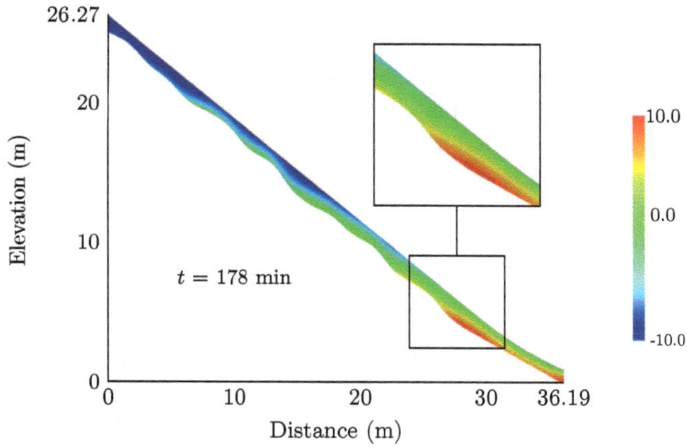

Fig. 20.12 Double-porosity simulation on Slope #2: mean pore water pressure (in kPa) at 178 min

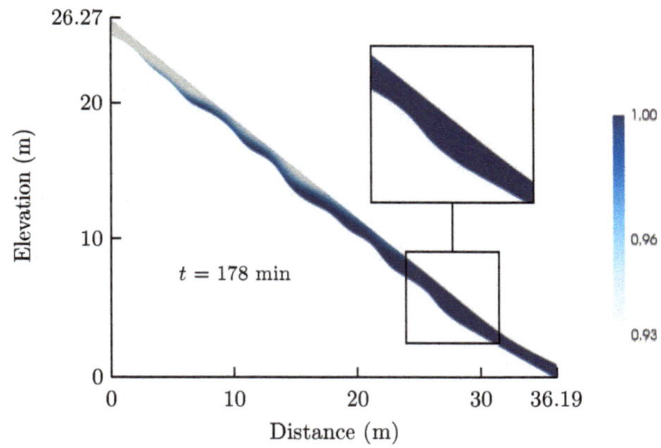

Fig. 20.13 Double-porosity simulation on Slope #2: overall degree of saturation in the colluvium at 178 min

20.6 Closure

Landslide triggering during rainfall events is influenced by a number of factors, including the local slope and bedrock topography, the hydromechanical properties of soil and bedrock, and rainfall history, among others. In addition, natural vegetation could influence the geotechnical properties of the soil in the field, making the laboratory testing of pristine soil materials of little value when used for predicting in situ landslides in the presence of natural vegetation. The mechanistic approach presented in this chapter is useful for investigating solid-fluid interaction in hillside slopes subjected to rainfall events. It allows quantification of the fundamental controls responsible for commonly observed slope failure mechanisms. When used with accurate data, the mechanistic model can also be used to predict the timing of failure of hillside slopes due to rainfall events.

Of interest to the computational modelers is the subject of double porosity, which has been treated explicitly in this work and implemented within the mixed finite element framework. An explicit treatment of the two porosity scales is important for capturing the preferential flow pattern in double-porosity media such as aggregated soils. This flow pattern is unique to double-porosity media and could ultimately determine whether or not a given hillside slope would fail.

Acknowledgements This material is based upon the work supported by the US Department of Energy, Office of Science, Office of Basic Energy Sciences, Geosciences Research Program, under Award Number DE-FG02-03ER15454, by the US National Science Foundation, under Award Number CMS-1462231. The second author acknowledges financial supports provided by the Fulbright Program and the John A. Blume Earthquake Engineering Center.

References

Anderson, S. P., Dietrich, W. E., Montgomery, D. R., Torres, R., Conrad, M. E., & Loague, K. (1997). Subsurface flow paths in a steep, unchanneled catchment. *Water Resources Research, 33*, 2637–2653.

Askarinejad, A., Casini, F., Bischof, P., Beck, A., & Springman, S. M. (2012). Rainfall induced instabilities: A field experiment on a silty slope in northern Switzerland. *Italian Geotechnical Journal, 46*, 50–71.

Barenblatt, G. I., Zheltov, I. u. P., & Kochina, I. N. (1960). Basic concepts in the theory of seepage of homogeneous liquids in fissured rocks. *Journal of Applied Mathematics and Mechanics, 24*(5), 1286–1303.

Bochev, P. B., & Dohrmann, C. R. (2006). A computational study of stabilized, low-order C0 finite element approximations of Darcy equations. *Computational Mechanics, 38*(4–5), 323–333.

Bochev, P. B., Dohrmann, C. R., & Gunzburger, M. D. (2006). Stabilization of low-order mixed finite elements for the Stokes equations. *SIAM Journal on Numerical Analysis, 44*(1), 82–101.

Borja, R. I. (2004). Cam-Clay plasticity. Part V: A mathematical framework for three-phase deformation and strain localization analyses of partially saturated porous media. *Computer Methods in Applied Mechanics and Engineering, 193*, 5301–5338.

Borja, R. I. (2006). On the mechanical energy and effective stress in saturated and unsaturated porous continua. *International Journal of Solids and Structures, 43*, 1764–1786.

Borja, R. I., & Andrade, J. E. (2006). Critical state plasticity. Part VI: Meso-scale finite element simulation of strain localization in discrete granular materials. *Computer Methods in Applied Mechanics and Engineering, 195*, 5115–5140.

Borja, R. I., & Koliji, A. (2009). On the effective stress in unsaturated porous continua with double porosity. *Journal of the Mechanics and Physics of Solids, 57*(8), 1182–1193.

Borja, R. I., & White, J. A. (2010). Conservation laws for coupled hydromechanical processes in unsaturated porous media: Theory and implementation. In L. Laloui (Ed.), *Mechanics of Unsaturated Geomaterials* (pp. 185–208). Wiley-ISTE

Borja, R. I., & White, J. A. (2010). Continuum deformation and stability analyses of a steep hillside slope under rainfall infiltration. *Acta Geotechnica, 5*(1), 1–14.

Borja, R. I., Liu, X., & White, J. A. (2012). Multiphysics hillslope processes triggering landslides. *Acta Geotechnica, 7*(4), 261–269.

Borja, R. I., White, J. A., Liu, X., & Wu, W. (2012). Factor of safety in a partially saturated slope inferred from hydro-mechanical continuum modeling. *International Journal for Numerical and Analytical Methods in Geomechanics, 36*(2), 236–248.

Borja, R. I., Song, X., & Wu, W. (2013). Critical state plasticity. Part VII: Triggering a shear band in variably saturated porous media. *Computer Methods in Applied Mechanics and Engineering, 261–262*, 66–82.

Brezzi, F., & Bathe, K. J. (1990). A discourse on the stability conditions for mixed finite element formulations. *Computer Methods in Applied Mechanics and Engineering, 82*, 27–57.

Brown III, W. M., Sitar, N., Saarinen, T. F., & Blair, M. L. (1984). *Debris flows, landslides, and floods in the San Francisco Bay region, January 1982: Overview of summary of a conference held at Stanford University*, August 23–26, 1982. National Research Council and USGS.

Carminati, A., Kaestner, A., Ippisch, O., Koliji, A., Lehmann, O., Hassanein, R., et al. (2007). Water flow between soil aggregates. *Transport in Porous Media, 68*, 219–236.

Choo, J., White, J., & Borja, R. (2016). Hydromechanical Modeling of Unsaturated Flow in Double Porosity Media. *International Journal of Geomechanics*. doi: 10.1061/(ASCE)GM.1943-5622.0000558, D4016002.

Choo, J., & Borja, R. I. (2015). Stabilized mixed finite elements for deformable porous media with double porosity. *Computer Methods in Applied Mechanics and Engineering, 293*, 131–154.

Coe, J. A., Kean, J. W., Godt, J. W., Baum, R. L., & Jones, E. S. (2014). New insights into debris-flow hazards from an extraordinary event in the Colorado Front Range. *GSA Today, 24*(10). doi: 10.1130/GSATG214A.1.

Dykhuizen, R. C. (1990). A new coupling term for dual-porosity models. *Water Resources Research, 26*(2), 351–356.

Ebel, B. A., Loague, K., Dietrich, W. E., Montgomery, D. R., Torres, R., Anderson, S. P., & Giambelluca, T. W. (2007). Near-surface hydrologic response for a steep, unchanneled catchment near Coos Bay, Oregon: 1. Sprinkling experiments. *American Journal of Science, 307*, 678–708.

Ebel, B. A., Loague, K., & Borja, R. I. (2010). The impacts of hysteresis on variably saturated hydrologic response and slope failure. *Environmental Earth Sciences, 61*, 1215–1225.

Gerke, H. H., & van Genuchten, M. T. (1993). A dual-porosity model for simulating the preferential movement of water and solutes in structured porous media. *Water Resources Research, 29*(2), 305–319

Gerke, H. H., & van Genuchten, M. T. (1993). Evaluation of a first-order water transfer term for variably saturated dual-porosity flow models. *Water Resources Research, 29*(4), 1225–1238

Jibson, R. W. (1992). The Mameyes, Puerto Rico, landslide disaster of October, 7, 1985. In J. E. Slosson, A. G. Keene, & J. A. Johnson (Eds.), *Landslides/Landslide Mitigation* (pp. 37–54). Reviews in Engineering Geology, Boulder. Colorado: Geological Society of America.

Klaus, J., Zehe, E., Elsner, M., Külls, C., & McDonnell, J. J. (2013). Macropore flow of old water revisited: experimental insights from a tile-drained hillslope. *Hydrology and Earth System Sciences, 17*, 103–118.

Laine-Kaulio, H., Backnäs, S., Karvonen, T., Koivusalo, H., & McDonnell, J. J. (2014). Lateral subsurface stormflow and solute transport in a forested hillslope: A combined measurement and modeling approach. *Water Resources Research, 50*, 8159–8178.

Lagmay, A. M. A., Ong, J. B. T., Fernandez, D. F. D., Lapus, M. R., Rodolfo, R. S., Tengonciang, A. M. P., et al. (2006). Scientists investigate recent Philippine landslide. *Eos Transactions, AGU 87*(12), 121–128.

Loague, K., & VanderKwaak, J. E. (2001). Simulating hydrologic-response for the R-5 catchment: comparison of two models and the impact of the roads. *Hydrological Processes, 16*, 1015–1032.

Martinez, E. (2000). Evento Meteorologico sobre el Litoral Central en Diciembre 1999. Informe inedito

Mehrabian, A., & Abousleiman, Y. N. (2015). Gassmann equations and the constitutive relations for multiple-porosity and multiple-permeability poroelasticity with applications to oil and gas shale. *International Journal for Numerical Analytical Methods in Geomechanics*. doi: 10.1002/nag.2399.

Montgomery, D. R., Dietrich, W. E., Torres, R., Anderson, S. P., Heffner, J. T., & Loague, K. (1997). Hydrologic response of a steep, unchanneled valley to natural applied rainfall. *Water Resources Research, 33*, 91–109.

Nieto, A. S., & Barany, I. (1988). Interim report on catastrophic rain-induced landslides in Rio de Janeiro and Petropolis, Brazil. In *Report to Committee on Natural Disasters, National Resource Council*, Urbana, University of Illinois, Department of Geology

Ogura, A. T., & Filho, O. A. (1991). The Morin debris-flow at Petropolis City, Rio de Janeiro State, Brazil. *Landslide News, 5*, 22–24.

Passman, S. L., Nunziato, J. W., & Walsh, E. K. (1984). A theory of multiphase mixtures. In C. Truesdell (Ed.), *Rational Thermodynamics* (pp. 286–325). Springer, New York.

Schuster, R. L., Salcedo, D. A., & Valenzuela, L. (2002). Overview of catastrophic landslides of South America in the twentieth century. In S. G. Evans, & J. V. DeGraff (Eds.), *Catastrophic Landslides: Effects, Occurrence, and Mechanisms* (pp. 1–33). Reviews in Engineering Geology, Boulder. Colorado: Geological Society of America.

Shlemon, R. J., Wright, R. H., & Montgomery, D. R. (1987). Anatomy of a debris flow, Pacifica, California. In *Debris Flows/Avalanches: Process, Recognition, and Mitigation* (pp. 181–199). Reviews in Engineering Geology, Vol 7. Colorado: Geological Society of America.

Smith, T. C., & Hart, E. W. (1982). Landslides and related storm damage, January 1982, San Francisco Bay region. *California Geology, 35*(7), 139–152.

Song, X., & Borja, R. I. (2014). Mathematical framework for unsaturated flow in the finite deformation range. *International Journal for Numerical Methods in Engineering, 97*, 658–792.

Song, X., & Borja, R. I. (2014). Finite deformation and fluid flow in unsaturated soils with random heterogeneity. *Vadose Zone Journal, 13*(5). doi:10.2136/vzj2013.07.0131.

Stadler, L., Hinkelmann, R., & Helmig, R. (2012). Modeling macroporous soils with a two-phase dual-permeability model. *Transport in Porous Media, 95*, 585–601.

Torres, R., Dietrich, W. E., Montgomery, D. R., Anderson, S. P., & Loague, K. (1998). Unsaturated zone processes and the hydrologic response of a steep, unchanneled catchment. *Water Resources Research, 34*, 1865–1879.

USAID. (2000). Venezuela factsheet, February, 2000. USAID-Office of Foreign Disaster Assistance.

Van Genuchten, M. T. (1980). A closed-form equation for predicting the hydraulic conductivity of unsaturated soils. *Soil Science Society of America Journal, 44*, 892–898.

Van Sint Jan, M., & Talloni, P. (1993). Flujo de sedimentos del 18 de junio de 1991 en Antofagosta: La Serena, Chile. *Tercer Congreso Chileno de Ingenieria Geotécnica, 1*, 247–265.

Warren, J., & Root, P. (1963). The behavior of naturally fractured reservoirs. *SPE Journal, 3*(3), 245–255.

White, J. A., & Borja, R. I. (2008). Stabilized low-order finite elements for coupled solid-deformation/fluid-diffusion and their application to fault zone transients. *Computer Methods in Applied Mechanics and Engineering, 197*(49–50), 4353–4366.

Chapter 21
Innovation in Instrumentation, Monitoring, and Condition Assessment of Infrastructure

Kenichi Soga

Abstract Future-proofing our existing and new infrastructure has become a constant theme in government and industry discussions. To achieve this, we need to understand the performance of our infrastructure better than before both during its construction and throughout its design life. Using recent advances in sensor systems and new technologies from the Internet of Things (IoT), it is hypothesized in this chapter that the rich information obtained from embedded sensors within infrastructure will allow us to develop a new paradigm in designing, constructing, operating, and maintaining our infrastructure systems. The chapter discusses the potential value of sensing not only to understand the behavior of monitored structures but also to make better decisions for management of infrastructure at the asset and city scales with a whole lifecycle-based thinking. New generations of sensor systems that satisfy these new requirements are introduced.

21.1 Introduction

Future-proofing our existing and new infrastructure has become a constant theme in government and industry discussions. We face with the fact that our old infrastructure continues to age and require monitoring and remedial interventions. Existing infrastructure is challenged by the need to increase load and usage and the requirement to maintain the existing infrastructure while operating at current capacity. The high cost involved in upgrading will lead to a desire to extend their life. For this reason, a need for better quality asset data to improve decision-making and reduce cost has been highlighted (e.g., HM Treasury 2014).

Active monitoring of the construction and operational processes of infrastructure is essential. That is, structures are instrumented to assess their performance against engineering design parameters or predictive models. The current monitoring systems do not fit well with this demand, especially for the long-term behavior.

K. Soga (✉)
University of Cambridge, Cambridge, UK

University of California-Berkeley, Berkeley, CA, USA
e-mail: ks207@cam.ac.uk; soga@berkeley.edu

P. Gardoni, J.M. LaFave (eds.), *Multi-hazard Approaches to Civil Infrastructure Engineering*, DOI 10.1007/978-3-319-29713-2_21

In recent years, sensing is rapidly becoming part of everyday life; the estimated global market in the sensor sector exceeds $300 billion. More advanced sensors and appropriate data analysis ensure better product quality, process safety, asset management, and shorter downtimes. Using such innovations, there are opportunities to advance our understanding in terms of the performance of civil engineering assets. If they can be used for long time, it can transform the industry with a whole-life cycle approach.

It is proposed here that the future of infrastructure relies on smarter information; that is, the information obtained from various sensors embedded within infrastructure or deployed around infrastructure provides an opportunity to create a new paradigm in design, construction, operation, and maintenance of our infrastructure systems, which is linked directly with user behavior patterns. This paper discusses the requirements for infrastructure sensing if it is designed, constructed, and maintained based on a whole-life cycle approach. The potential value of whole-life cycle-based sensing is not only to understand the behavior of monitored structures but also to make decisions for better management of infrastructure at the asset and city scale. New generations of sensor systems that satisfy these new requirements are introduced.

21.2 Infrastructure Sensing: Requirements

Figure 21.1 illustrates how "Infrastructure Sensing" can contribute to understanding the performance of our infrastructure at different scales and layers. This fully integrated approach starts from designing sensor systems, providing better information for decision makers and finally enabling more effective management of assets, cities, and infrastructure systems throughout their life.

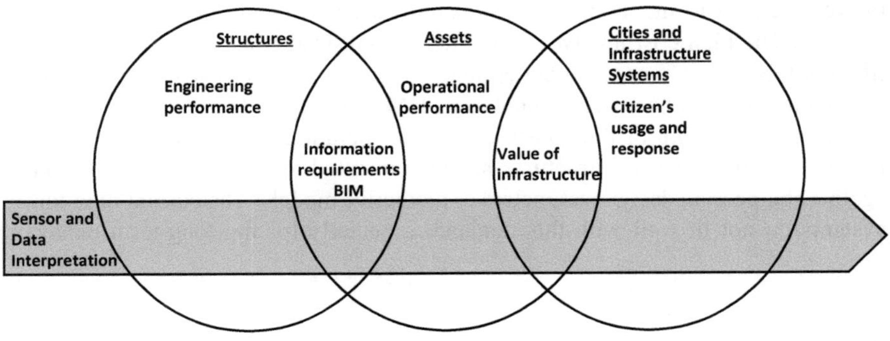

Fig. 21.1 Infrastructure sensing

21.2.1 Sensor and Data Interpretation Layer

Recent advances in 3D printing, printed electronics, ultra-low power-high performance microprocessors, fast speed electronics, digital signal processing, energy harvesting, and low-power wireless communication can radically change the quality and quantity of information we can get from infrastructure. These technologies allow any modification to be done at a much earlier stage of sensor system manufacturing specifically for civil engineering application. There is a great opportunity to create truly civil engineering-oriented sensor systems.

Possible questions related to sensor and data interpretation for infrastructure sensing are the following:

- What sensors are needed to measure the performance of our structures, assets, and cities?
- How can we make them robust and reliable?
- Should data be collected in automatic manner for the remainder of their lifecycle?
- How do we analyze the data to give reliable, meaningful results?
- What value should be measured directly? Should the value be discovered through data mining?
- Can citizens-as-sensors (see Sect. 21.4.4) replace some of the conventional sensor systems?
- How can we build confidence in using them?

21.2.2 Structures Layer

By considering a structure to be composed of "smart" building blocks with embedded sensors and information management systems, it is intended that a "living structure" can be created and continually adapted over time.

Possible questions for this layer using sensor data are the following:

- How to develop of new engineering design, construction, and maintenance processes for an integrated infrastructure system?
- How do we best design, construct, and monitor our structures to deliver the performance we need?
- Can our structures be more flexible and adaptable to synchronized growth?
- Can the shape and materials of structure be modified to provide the maximal future-proofing through re-configurability?
- What data do we need to do this, and how do we interpret it?
- How can real-time information about physical assets and human behavior inform utilization strategies and future design?

21.2.3 Asset Layer

ISO 55000 standards on asset management was published in 2014 (BSI 2014). They highlight the importance of the through-life management of physical assets and emphasize on realizing value rather than minimizing cost. The value of assets needs to be determined from a multi-stakeholder perspective. Therefore, asset owners face this multi-perspective challenge and the challenge of balancing cost and risk originating from decreasing funding and increasing regulation. BIM (Building Information Modelling) techniques that manage design and construction information in a transparent manner can aid to tackle these challenges. In 2012, the UK government mandated that all built assets procured by central government would by 2016 be delivered using BIM Level 2 techniques, in which Level 2 means new-build (CAPEX) delivery (http://www.bimtaskgroup.org/). The aim is to help develop a digitally enabled construction and facilities/asset sector, leading to the development of "BIM Level 3," which includes asset lifetime areas (CAPEX + OPEX). This "asset" layer is very much linked to this BIM vision.

Infrastructure sensing can support asset managers to secure the best value for money throughout an asset's lifetime and to make value-based asset management decisions. Possible questions for this layer using sensor data are the following:

- How do we future-proof our assets against changing requirements and against shocks?
- Can human behavior with data on user cost, human psychology, and daily/weekly activity patterns at both the macro and micro levels be used to determine the value of asset?
- How do we operate, manage, and maintain our assets to deliver best whole life value?
- What decisions do we need to take to do so? What information do we need to make those decisions?
- What kind of institutional objective utility/optimization is needed for asset owners?

21.2.4 Cities and Infrastructure System Layer

Changes in physical infrastructure, transportation, utilities, and communications cause citizens to adapt and evolve. By aiming to make infrastructure more adaptable and resilient to changes (demand, climate, etc.), the goal of this layer is to find ways to remove inefficiencies incurred when providing services to citizens whilst maintaining the integrity of city infrastructure systems. The vision is human behavior and infrastructure evolve hand-in-hand in order to enable enhanced living quality, vibrant business, trouble-free transportation and efficient, sustainable use of resources.

Possible questions for this layer using sensor data are the following:

- What economic value does our infrastructure create?
- How does our infrastructure best serve our communities?
- Where, how, and who will benefit from tapping the potential of infrastructure growth in urban master-planning projects across the city?
- How will it change the current spatial-temporal pattern of transportation networks, energy consumption, commercial activities, living lifestyles, and environmental quality of the city?
- What will be the knock-on effects?
- What kind of policies and planning procedures best provide incentives for change in infrastructure usage?
- How can cities and authorities coordinate policies/institutions for maximal delivery of change?

To answer these questions, inputs from experts in humanities and social sciences are necessary. Possible developments include novel frameworks identifying the relationships between human behavioral functions and infrastructure, and ways to accelerate the co-evolution.

21.3 The Value of Sensing

21.3.1 Why Measure

Any monitoring and sensing task should start from identifying variables to be measured and developing a framework to quantify the value of sensing by converting the individual sensor measurements into something usable. Hence there is a need to clarify end users/domains.

Monitoring systems may be applied to new or existing structures. For new structures the emphasis will often be on giving an early warning of significant deterioration taking place (for example, rebar corrosion, concrete spalling, crack development, strain and stress changes, and movements). For existing structures the emphasis will usually be on monitoring the rate of an already active deterioration process. The two cases may be treated with the same basic approach: monitoring to assess the current condition and to estimate the residual life of the structure.

For the quantification of performance, infrastructure models and material models are essential. They can be empirical, analytical, or computational and probabilistic, if possible. The models need to utilize specific sensing data and potentially can do an inverse analysis to find the value of sensing such as economical sensing requirements and sensor development for specific applications. New models can be developed based on the data previously not easily accessible is now available, or existing models can be modified based on data that new sensors can now measure. The models are used to optimize sensor networks to provide a balance between extent of data capture and confidence in condition assessment for residual life or retrofit. That is, for any infrastructure owner to invest in a new and potentially transformative sensing system, a quantitative evaluation of sensing value at monitoring points (such as for embedded sensors, moving sensors, and citizen-based sensors) is needed.

Ideally this integration of sensing and models should be done at scales across structures, asset, and city as shown in Fig. 21.1. The next question can be the following:

- Would existing sensors and monitoring systems be able to provide the data required at different scales?
- What is a new generation of sensors and monitoring systems that provide data for integration and multiple modalities?

21.3.2 Life Cycle of Sensor Systems

The rate of technical development of a monitoring system can be faster than the rate of infrastructure degradation. Some data can be from older sensors, and other sensors may be embedded now but the data will be used later. The data quality from sensors and monitoring system changes with time and their possible error propagation due to ageing needs to be assess so that system level decisions made by whole-life cycle models are appropriate.

The following questions related to the life cycle of sensors and monitoring systems need to be addressed:

- Are the currently available systems acceptable for the requirements given from the performance models in order to assess the whole-life cycle of buildings and infrastructure?
- What steps must be taken to ensure that the data to be harvested by embedded sensors are relevant, readable, and useful throughout the life cycle?
- How long sensors can last?
- Can they be replaced as newer ones become available?

These questions lead to a demand of a whole-life cycle model of monitoring and measurement systems. The model can include consideration of (1) data quality and its degradation with time, (2) survival rates of hardware and software components and the associated error propagation, and (3) costing of management and maintenance. Any gaps among the formats need to be identified and good data transfer links are essential. This is because such links may produce errors, which need to be quantified for accurate modelling and assessment of infrastructure performance.

21.3.3 Quantifying the Value of Sensing

The essential driver for monitoring and measurement is to create a value for business and/or public. The contrast between the immediate costs (e.g., sensor purchase) and the gains to be made downstream if the sensors are put in place needs to be articulated clearly to infrastructure owners and the public. It is important to justify the capital cost involved in upfront investment in data capture/sensing. The cost of managing and maintaining the measurement and monitoring systems in the long term needs to be evaluated. The possibility to share cost of acquiring data across government and private sectors can also be addressed.

It is also necessary to evaluate the data quality of the sensor systems adopted and to assess possible error propagation when the data is used for modelling, communication, and visualization. That is, the level of confidence in the data needs to be quantified, so that appropriate decisions can be made based on sensor data. The framework developed to quantify the value of sensing can be tested using "retrospective" value case (for example, if we could have measured and confidently communicated X, then this bridge would not have collapsed). It should be recognized that a quantification framework to retrofit sensing into existing buildings and infrastructure may be different from that for new buildings and infrastructure.

A knowledge database of monitoring systems and measured data is needed so that it is possible to conduct in-depth assessment of data quality and error propagation for measurement systems specifically designed to meet the performance requirements. The database should include immediate capital costs (e.g., sensor purchase) as well as cost of managing and maintaining the measurement and monitoring systems in the long term. It can also include data quality as well as error propagation when the data is used for modelling and communication. This will provide information to quantify the gains to be made downstream if the sensors are put in place.

21.4 Next Generation of Infrastructure Sensing

21.4.1 Intelligence for Life

Timescales in infrastructure and construction are long, with construction timescales alone stretching from 2 to 10 years, and asset lifetimes in the range of 60–100 years or even more. Civil engineering structures are very fixed in space and time (e.g., 120-year design life) and provide independent services for transportation, energy supply, water, sewage, and communication without any appreciable linkage. Each of these elements is operated with different business models, is guided by different performance metrics and deals with systems that involve different degrees of interconnectedness and timescales in terms of ageing and the requirements for repair and maintenance. There is a need to investigate the inter-relationship between different systems and different sensing data.

Due to different rates of technical developments between monitoring system versus infrastructure usage, some data we will be using may be from older sensors, and some of the sensors may be embedded now but data will be used in 10, 20, or 50 years later. At present, there is a mismatch between the life span of infrastructure and that of sensor systems, which makes the concept of whole-life cycle-based asset management difficult to achieve. Sensor systems to fulfill this concept ("intelligence for life") need to be either long-life or adaptable for replacement.

There are sensing systems potentially available for monitoring a variety of lifecycle attributes. These include (1) fiber optic sensors for movements and temperatures, (2) digital images for temperatures, damage, congestion, and occupancy, and (3) robust environmental sensors for temperature, humidity, and chemical changes. Data requirements for long-term management of infrastructure need to be identified and then a new generation of lifelong sensor systems can be developed. Information required include data quality and its degradation with time, survival rates of hardware and software components and the associated error propagation, and costing of management and maintenance.

21.4.2 Intelligence in Extremes

One of the challenges for civil engineers and infrastructure owners is to make decisions when a catastrophic event occurs. Infrastructure systems and projects face a multitude of hazards that must be assessed, communicated, and managed appropriately. The Fukushima Power plant event in 2011 in Japan and major flooding and extreme weather events in recent years around the world are good examples of this. In particular, sensing the performance of linear infrastructure systems such as power supply, buried pipelines, railways, and flood defense embankments is challenging because any break in a system can potentially disrupt the operation of the whole infrastructure as well as influences the other neighboring infrastructure.

To perform these assessments, there is a need to develop and deploy sensor networks for rapid condition assessment of critical structures in order to operate optimally and to ensure system safety and resilience. There are opportunities to extend the working conditions of the existing sensing devices to the range of extreme civil conditions, such as large deformation, high pressure, extremely high and low temperatures, and moisture. These sensor systems provide "intelligence in extremes." This effort can be coupled with high-performance computer simulation and advanced visualization tools to conduct risk assessments at the citywide scale, considering multiple hazards such as earthquakes, tsunamis, flooding, and fires.

Distributed fiber optic systems and embedded low-power sensor network are good candidates for reliable condition evaluation of local damage so that the residual safety margin of the structures can be checked after an extreme event. Also satellite images, high resolution LIDAR, and computer vision images are also effective in assessing the external damages. Robotic technologies are useful for effective data capture of difficult-to-access sites. Other potential candidates include mobile and wearable sensors. However, there is a need to design such systems to be used "in conjunction" with other in-placed sensor systems during emergency events to ensure a wide coverage of sensing.

21.4.3 Intelligence on Demand

It is often argued that every infrastructure asset has unique structural characteristics and any sensor systems need to be custom made or adapted so that they can provide satisfactory data for decision making. However, at present it takes time to create a sensor system that is robust and safe enough to be used for specific applications. By the time it is created, key decisions have often been made and hence there is "intelligence delay." However, recent advances in 3D printing, printed electronics, ultra-low power-high performance microprocessors, and wireless communication can radically change this situation. For example, inkjet printers can be used for prototyping and production of both individual sensors and printed circuit boards, whereas 3D printers will allow the production of bespoke sensor packaging capable tailored for particular applications. These will enable the concept of "intelligence on demand," in which sensors can rapidly be developed to produce the data required for decision making.

Recent developments in small-scale manufacturing techniques can be used for the rapid production of novel bespoke sensors. Techniques such as 3D printing, inkjet printing of electronic components and circuit boards are good candidates. It will focus on optimizing the time required to produce new sensor systems tailored for specific infrastructure applications, whilst also enabling the production of the civil engineering-orientated sensors. The development requires: (1) novel sensor device design and fabrication through fast prototyping, (2) device characterization, calibration, and operation for board-level system integration, (3) environmental testing for life time and extreme conditions, and (4) new production techniques to enable the creation of sensing devices with an extremely long life times in excess of 50 years.

21.4.4 Citizen as Sensors

Our understanding of mobility in a city can be improved by better tracking of where and how people move in space and time. Integrating such infrastructure information will lead to better management, operation, and allow cutting-edge technologies to be tested. The rich information provided will act as a catalyst for new design, construction, and maintenance processes for integrated underground service systems linked directly with user behavior patterns.

As an addition to conventional monitoring and measurement systems, it is now possible to take advantage of the "citizens as sensors", which are crowdsourcing aspect of citizen science. We can consider monitoring and measuring of real-time data using smart phones/watches, from occupants and customers that provide data on how they feel about the building, the environment, the products, and the management, for example. The applicability of such "citizen" sensors will be assessed in terms of privacy, data quality, and error propagation.

There are several technologies to gather data about space usage. These techniques will include instrumentation of users' mobile devices through (a) novel energy-aware software applications capable of gathering data from the phone's local sensors; (b) backend techniques to understand movement of crowds through WiFi and cellular logging; (c) use of data from geo-social networks and transportation cards; and (d) embedding qualitative techniques of measurement of user preferences (e.g., voice strengths) in order to measure their willingness to change and adapt. The data collected can often be of a sensitive nature and privacy protection methods are needed for users, including anonymization techniques and ongoing development of "privacy by design" principles.

The large quantities of data can be collected for real-time understanding of human activities. The outputs of the models can inform users about their movement and use of space (through their mobile devices and surrounding displays). This will allow information systems to be built that will inform retailers and other contractors about the current usage and provide guidance on future usage and efficiency of the area. The data will feed parameters to describe current behavior of existing users, and their willingness to change in response to new environments. The data will also allow development of models that consider the interplay of new technology and underground spaces with users' reaction. The models will guide the prediction of space usage.

21.5 Examples of Emerging Sensing Technologies for Infrastructure Sensing

21.5.1 Distributed Fiber Optics Sensing for Lifelong Monitoring

Distributed fiber sensing technique takes advantage of the sensitivity of the optical fiber with respect to ambient parameters like temperature, strain, vibration, and noise (acoustic). The ambient parameters, to which the fiber is subjected, influence the properties of a laser light signal travelling throughout the glass material in an optical fiber. When a light travels through a transparent media, the majority of it travels through, but a small fraction is backscattered. Scattering occurs due to inhomogeneity of the refractive index of the glass medium and allows coupling to acoustic waves known as phonons. The scattered lights can either propagate in the

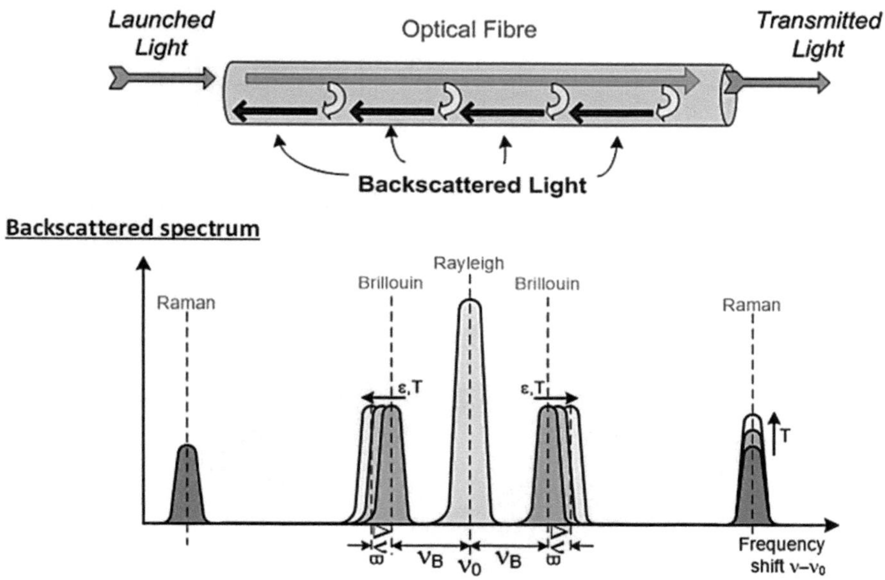

Fig. 21.2 Backscattering processes

same direction as the incident light or travel in an opposite direction to the incident light where the latter is called backscattering light (see Fig. 21.2). There are three notable scattering processes: (1) Rayleigh, (2) Brillouin, and (3) Raman scattering. Rayleigh scattering re-radiates at the same frequency as the incident light. It can be used to measure the loss distribution or attenuation along the length of the fiber by analyzing the Rayleigh scattered light power. A decrease in the scattered light power corresponds to the loss of light along the optical fiber. Brillouin scattering light is temperature and strain dependent, in which the frequency shift of the Brillouin spectrum varies with longitudinal strain and temperature in a fiber. The Raman scattering, on the other hand, has spectrum power levels that vary according to temperature changes.

As shown in Fig. 21.3, a typical highly distributed fiber optic sensing system includes two major components: optical fiber cables and an optical fiber analyzer that performs a number of tasks (including data acquisition, data processing, transmission, and storage). Different types of analyzer detect different scattering signals in a different way (e.g., injecting a broadband frequency laser or an extremely narrow linewidth laser with short pulse, analysis in time domain or frequency domain, using high power pulse to take advantage of non-linear scattering). For example, Phase-OTDR is capable of quantifying both acoustically induced and dynamic multiple strain perturbations (e.g., Lu et al. 2010). The technique is based on measuring the phase between the Rayleigh scattered light from two sections of the fiber which defines the gauge length. It can perform long distance dynamic

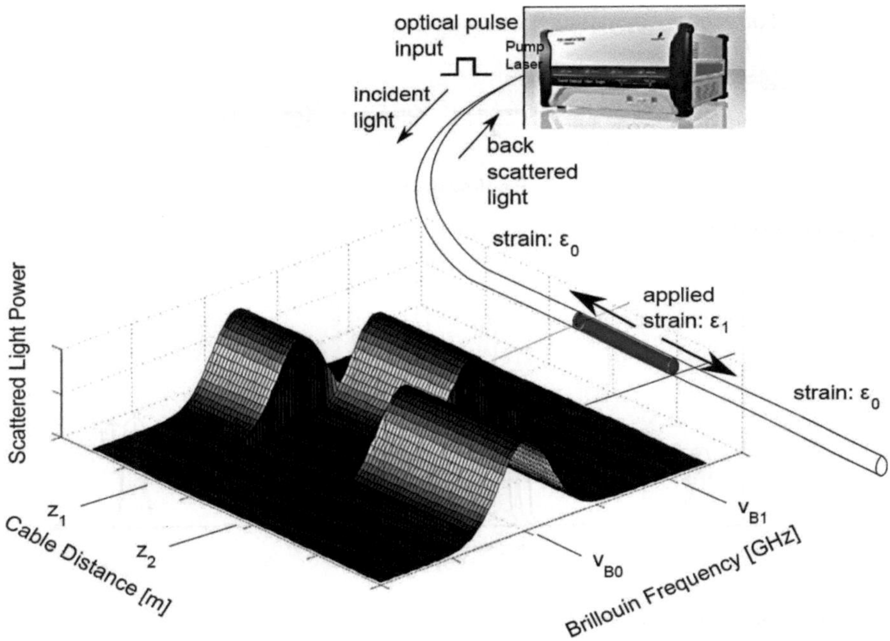

Fig. 21.3 Fiber optics sensing cable and an analyzer (BOTDR system in this case)

sensing for vibration and acoustics. The Raman optical time-domain reflectometry (ROTDR) is a classical technology to measure temperature at many points along the length of an optical fiber (e.g., Hartog et al. 1985).

Distributed temperature sensing systems (DTS) is a Raman scattering-based system, which can be used for power cable and transmission line monitoring, fire detection, and leakage detection at dikes, dams, and sewers. It is widely used for downhole temperature monitoring in oil and gas wells. Phase-OTDR can be used for security monitoring for long borders and perimeters at high-value facilities and high-level security locations, or real-time position and speed monitoring of trains.

Brillouin scattering-based techniques such as time domain techniques called "Brillouin optical time-domain reflectometry (BOTDR)" and "Brillouin optical time domain analysis (BOTDA)" are well-established for distributed strain measurement (Horiguchi et al. 1989; Kurashima et al. 1990; Niklès et al. 1996). BOTDR relies on the spontaneous Brillouin scattering, which uses the incident light to generate the Brillouin signal. BOTDA relies on stimulated Brillouin scattering, which uses an external excitation signal to amplify the Brillouin signal. As shown in Fig. 21.3, the frequency of this backscattered light is shifted from the original input frequency by an amount linearly proportional to the temperature and strain applied at the scattering location. The backscattered signal in time and frequency is resolved to obtain a complete strain profile along the full length of the fiber. There are other strain sensing techniques such as BOFDA (Garus et al. 1997) and BOCDA (Hotate and Hasegawa 2000). Further details can be found in Bao et al. (2012).

Engineering design limits are often based on strain and/or stress developing in the structure. Many structures interact with distributed load. In the context of monitoring strain in piled foundations, tunnels, pipelines, slopes, or embankments, capturing the continuous strain profile is often invaluable to pinpoint non-uniformly distributed soil-structure interaction loads and localized problem areas such as joint rotations and deformations. The current state of the art distributed fiber optic strain measurement systems provide data in the micro-strain range with a spatial resolution (strain is averaged over a specified gauge length) of 0.2 m or less. This means that it is possible to have thousands of "strain gauges" along a single cable connected to structures, embedded in civil engineering infrastructure. Some field trials and demonstrations of this technology for different civil engineering applications are shown in Fig. 21.4. Further details can be found in Soga (2014), Soga et al. (2015), and Kechavarzi et al. (2016).

Standard optical fiber becomes sensor and tens of kilometers of fiber can be sensed at once for continuous distributed measurement of the conditions around the optical fiber such as temperature, strain, acoustic noise, etc. Because of its simple and quick installation, distributed optical fiber sensing can be as equally practical as other conventional measurements. The cost of a standard optical fiber is potentially very low compared to other point measurement sensors. The material itself (silca) is relatively inert and can be ideal for long-term monitoring by embedding the fiber in structures. This implies that the quality of the data is expected to increase with time as the capability of analyzers should improve with time. Such features can potentially provide a relatively cheap but highly effective monitoring system for both short and long term. Most of the capital investment relates to the analyzer, which can be connected to a number of fibers or be shared at different sites. It is expected in the near future that more choice of analyzers from more manufacturers would give a reduction in price with time.

21.5.2 Low-Power Sensor and Energy Harvesting for Lifelong Monitoring

Micro-electro-mechanical systems (MEMS) are small integrated devices or systems that combine electrical and mechanical components varied in size from micrometers to millimeters, which can merge the function of computation and communication with sensing and actuation to produce a system of miniature dimensions. MEMS extend the fabrication techniques for semiconductor industry to include mechanical elements and the inherently small size of MEMS enables high-level integration of micro-machined components or structures to realize multiple functions or capabilities on the same silicon chip for greater utility.

A range of commercial MEMS sensors are now available in civil applications. For example, tilt measurement is common in civil engineering. MEMS sensors can a

b

c

Fig. 21.4 Examples of FO strain measurement in geotechnical structures by Cambridge University and CSIC. (**a**) Piles. (**b**) Tunnels. (**c**) Diaphragm walls

measure both static and dynamic accelerations and therefore they can be utilized to measure inclinations that are typically static accelerations. The inclinometer includes uniaxial or biaxial accelerometers which measure the gravity. The commercial MEMS inclinometer commonly incorporates an onboard microprocessor to automatically compensate the temperature effect of the tilt data. For instance, SCA103T by Murata-VTI Technologies is a 3D-MEMS-based single axis inclinometer family that uses the differential measurement principle. It has a resolution of 0.001°, with a range of 15°.

Strain sensing is highly critical for civil infrastructure applications. The conventional metal film strain gauge and vibrating wire strain sensors are not very well suited for wireless sensing civil applications, in which numbers of strain sensors are required to be deployed within large-scale infrastructure. Thus, high resolution, lower power, and small size MEMS strain sensors are in demand to replace the conventional strain gauges. A novel MEMS strain sensor has been prototyped at Cambridge University. The fabrication process of the strain sensor has been reported in Ferri et al. (2008, 2009, 2011), Belsito et al. (2013, 2016), and Do et al. (2016). Double-ended tuning fork (DETF) parallel-plate resonators with reduced coupling gaps (<1 μm) have been fabricated as shown in Fig. 21.5 (Ferri et al. 2010, 2011). A prototype MEMS crackmeter was manufactured with a thin steel bar fixed across a crack on the tunnel wall, onto which a multi-directional MEMS strain sensor was soldered (Fig. 21.5). The prototype MEM crackmeter was deployed in Prague Metro.

The conventional method to power sensor system in a remote location is to use a battery, which is limited to its life cycle. MEMS sensors offer major advantages in terms of low-power consumption. The powering of MEMS devices by capturing and storing energy from external sources present in the environment offers an opportunity to replace or augment batteries.

For example, an innovative vibration energy harvesting (VEH) utilizing the concept of parametric resonance has been developed at Cambridge University (Jia et al. 2013, 2014a, b). The technology works by harnessing vibration energy from ambient vibration sources such as freight, rail, tunnels, and traffic on bridges. The harvesters convert the mechanical energy into electricity in a decentralized manner at the device-level, essentially equipping the wireless systems to generate their own power.

By coupling with low-power MEM sensors, energy harvesting technologies provide long-term condition monitoring of assets in a range of remote and/or inaccessible locations. They also complement or replace existing battery solutions, reducing extensive battery replacement maintenance costs and the burden on hazardous waste disposal while.

21.5.3 Computer Vision for Extreme Event Assessment

The vast majority of buildings and infrastructure do not have usable as-built geometry information. New buildings have a design BIM model that may not been updated throughout construction or afterwards as changes and errors are common throughout the construction process. Older buildings, which constitute the vast majority in all developed nations, almost never have digital design information.

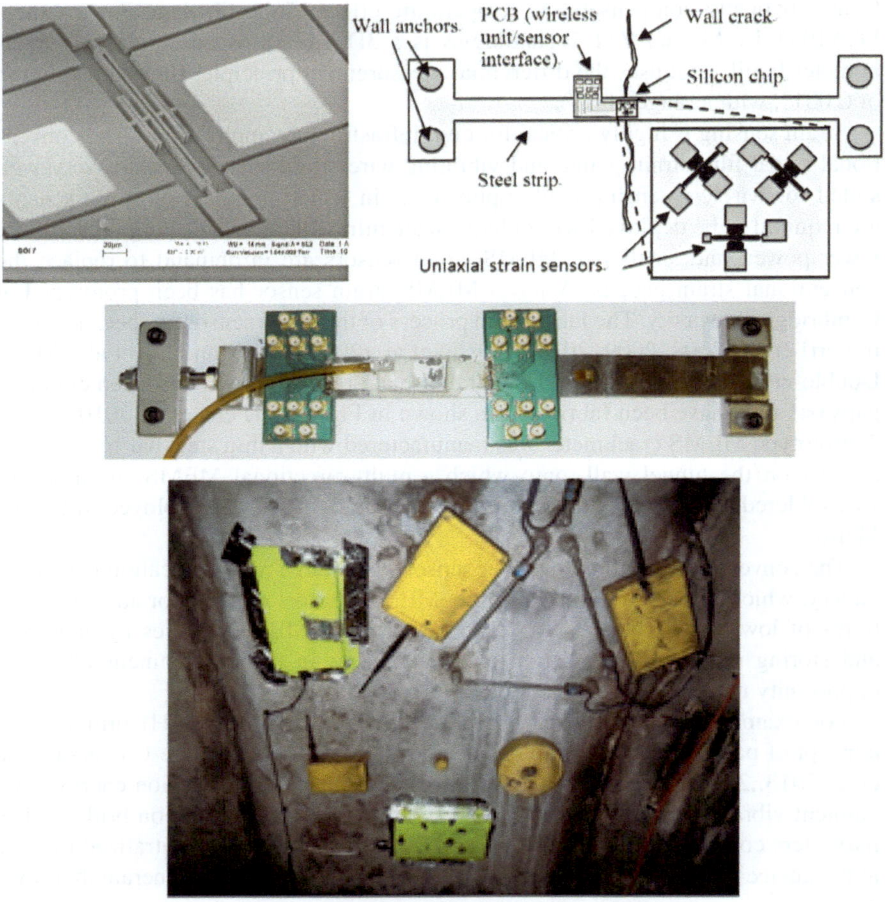

Fig. 21.5 Optical micrograph of MEMS DETF strain sensor and a prototype MEM crackmeter deployed in Prague Metro

Their current geometry and condition can be different from the initial design information. Configurations change or extensions may have been added by demand change. Damaged/deteriorated areas may have been fixed.

Geometric data is the key for Building Information Modelling (BIM) systems. Its current challenges are (a) acquiring raw spatial and visual data at a fraction of the cost/time needed today and (b) building geometric features from the raw data with minimum manual data handling. There are now a variety of laser- and optical-based systems that collect raw spatial and visual data (typically point clouds and images). At present, these data are used to manually create as-built geometrical models. This manual crafting process is extremely tedious. Bringing the modelling cost down by automating it makes such models reasonable for all asset owners.

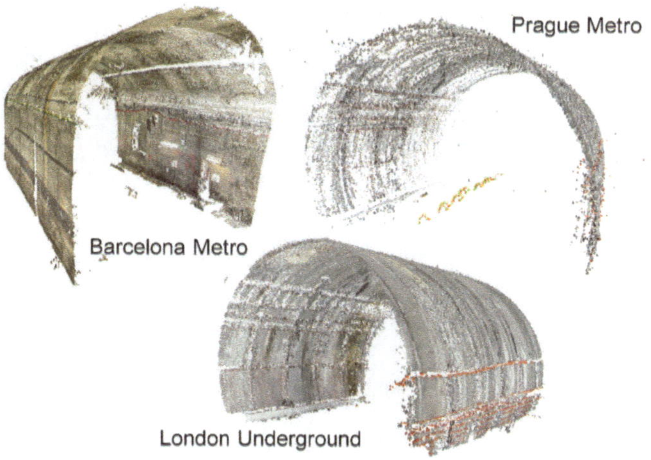

Fig. 21.6 3D reconstruction of tunnels from digital photos

The advances in digital photography have come to the stage that such technology can potentially replace more traditional and expensive surveying technologies. Structure from Motion (SFM) is a system that is able to simultaneously recover a 3D point cloud model and camera positions using only images. Some SFM systems have recently been commercialized. The software enables users to create a 3D point cloud model from uploaded photographs and allows users to browse and navigate through the photographs. It can simultaneously cope with free camera motion and more complex geometry of the scene. Figure 21.6 shows a 3D reconstruction of tunnels using digital images (Chaiyasarn 2011; Chaiyasarn et al. 2015). The accuracy is becoming close to that of a traditional laser scanning method. Further work is needed to capture material information when geometric data is acquired. Materials should be identified non-destructively and there is need to create a database of materials so that semantics of a captured scene can be harvested via various computer learning techniques (e.g., deep learning).

Visual inspection is commonly used for detecting anomalies such as cracks, spalling, and staining. Photographs and videos are used to record anomalies over years. As image collections often become large and difficult to organize and browse, improving the ways to access old images can lead to substantial progress in the effectiveness of monitoring. This is particularly true for monitoring locations like a deep shaft, where inspectors cannot easily access the inspection site.

Inspectors visually compare the images to determine whether any anomalies have arisen between inspections. The goal of change detection is to identify the regions of changes between multiple images of the same scene taken at different times. Figure 21.7 illustrates typical images taken from a tunnel site. An automatic tool that combines a large number of images into a single high-quality wide-angle

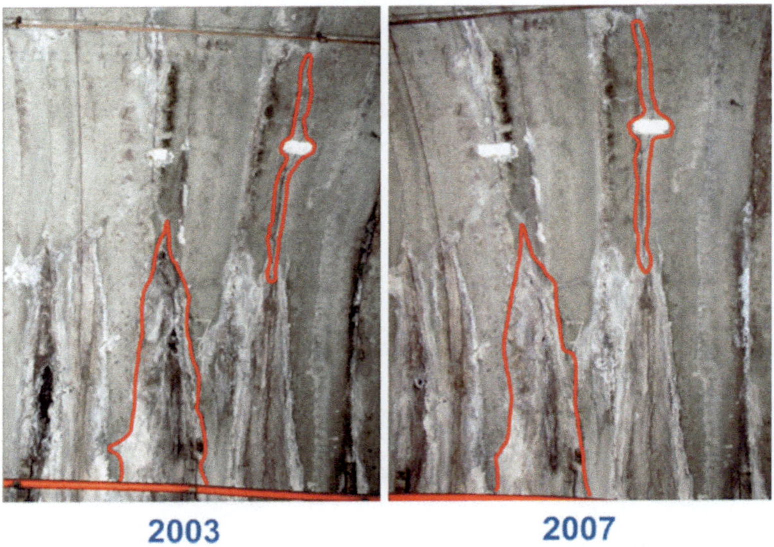

Fig. 21.7 Two photos taken at different time in Prague Metro. Note the difference in view angles (Chaiyasarn 2011)

composite view (see Fig. 21.8) is useful to organize a large collection of images for inspection. Images are registered by transforming image sets from varying and unknown coordinate systems into one single coordinate system. By assigning spatial coordinates to each image, it is possible to identify regions of change in images of the same scene taken at different times. Such system can reduce the workload for visual inspection significantly versus competing techniques, facilitating high frequency, effective tunnel inspections.

A prototype change detection system for tunnel inspection was developed by Chaiyasarn (2011) and was applied to a tunnel dataset from Prague Metro. The query image is shown in Fig. 21.9a, taken in 2003. Figure 21.9b shows four chosen reference images (see Fig. 21.9c) that have been pre-processed and overlaid on top of each other. All reference images are obtained in 2008. All four images are used in comparison with the query image so that a change mask will cover an entire area of the query image. Several changes such as a change in the color of water patches, new holes being created, and a new cable being added are labelled in Fig. 21.9a.

Stent et al. (2016) investigated the applicability of various change detection algorithms to detect visual changes on concrete tunnel linings as shown in Fig. 21.10. It includes results of various distance functions, the geometric prior mask, and a final change detection image for three sample queries. The changes shown in the first and second rows of the figure included leaking, fine chalk markings, discoloration, and new items attached to the surface. The geometric prior in all three cases correctly identified and removed most of the nuisance change caused by off-surface features. The bottom row gives a failure case, which had no change. The change detection

Fig. 21.8 Cylinder model and the flatten version (Stent 2014, personal communication)

algorithm detected unusual presence of some threads on the normally featureless red cable.

21.5.4 Wireless Sensor Network for Rapid Deployment

Installing traditional wired monitoring systems on large-scale civil infrastructure assets is time consuming and expensive. An alternative is to use wireless sensor networks (WSNs) that are faster to install and potentially cheaper on the sensor-by-sensor cost basis. The use of wireless sensor technology, which transmits the sensor data using radio, allows a rapid deployment due to elimination of some of the cabling and thus has significant potential benefits for infrastructure monitoring. Combined with MEMS sensors, there is the opportunity for significant overall cost savings for large-scale monitoring. The advantages of WSN for monitoring the behavior of civil engineering infrastructure are (1) large number of sensors can be deployed without needing to install cabling, (2) sensing is possible at difficult to

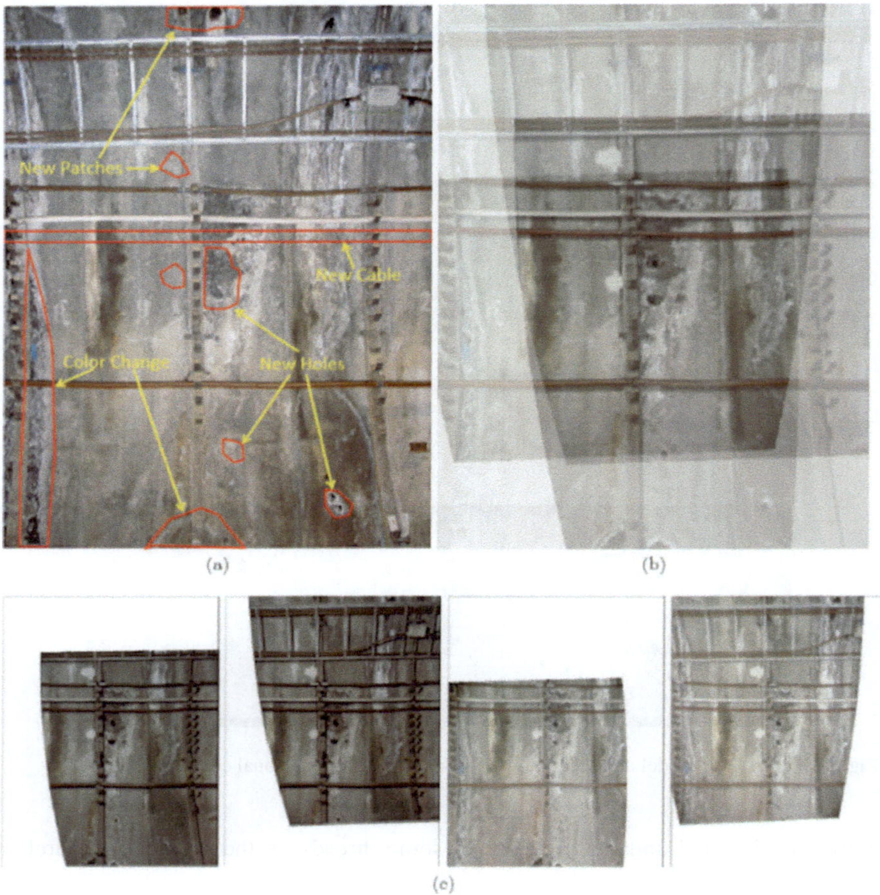

Fig. 21.9 (**a**) A query image, (**b**) four reference images after pre-processing and overlaying on top of each other, and (**c**) individual reference images after the geometric adjustment step (Chaiyasarn 2011)

access sites, and (3) quick deployment allows contractors and infrastructure owners to make better engineering decisions.

Although advantages of WSNs for conditioning monitoring of infrastructure have been identified, deployment of WSNs in real environment remains to be challenging (e.g., Stajano et al. 2010). There is also a perception that a wireless system may not be as reliable and robust as a wired system. The current limitations of WSN for civil engineering infrastructure monitoring are (1) sensor installation and establishment of a working wireless network currently requires expert knowledge, making it difficult for the clients to fully appreciate the reliability and risk involved, (2) interoperability of different WSN systems is very limited, and (3) no standards

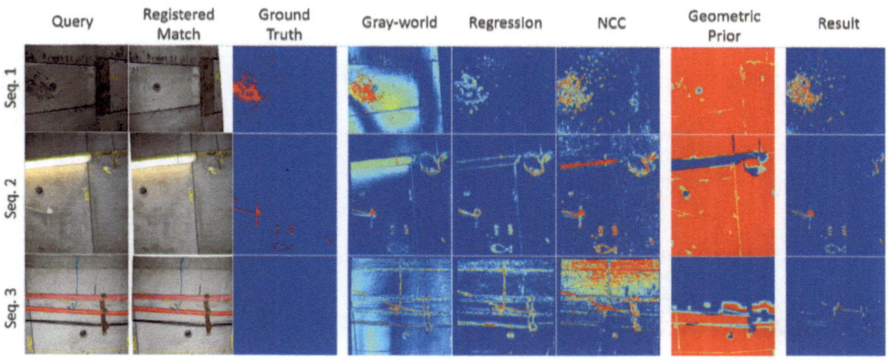

Fig. 21.10 Illustrative results for three cases (one per rom). By column: (*1*) Query image. (*2*) Registered matching image. (*3*) Ground truth. (*4*) Gray-world. (*5*) MSR. (*6*) NCC. (*7*) Mean-shift segmentation-based geometric prior. (*8*) Final probabilistic change mask (Stent et al. 2014)

Fig. 21.11 How to "trust" WSN data

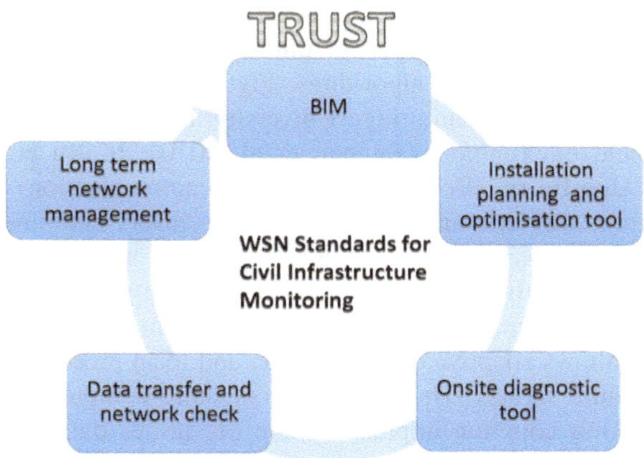

and/or guidance are available for clients to specify WSN systems that would allow better communication between WSN provider and clients.

One of the main focuses of WSN for civil engineering application should be to develop an integrated framework for planning, deployment, and management of large-scale wireless sensor networks so that users can trust the data coming from a wireless sensor network as shown in Fig. 21.11 (Rodenas-Herráiz et al. 2016). Although the wireless sensor network research community has looked into wireless communication, energy efficiency, limited network scalability, and several other problems, an integrated system for planning, deploying, and managing large-scale sensor networks is still missing. The current methods used when installing a wireless sensor network for monitoring civil infrastructure involve significant trial and error at the deployment site, and significant dependence on personal experience. This makes it extremely difficult to cost effectively deploy and manage large-scale wireless sensor networks that the end user can trust.

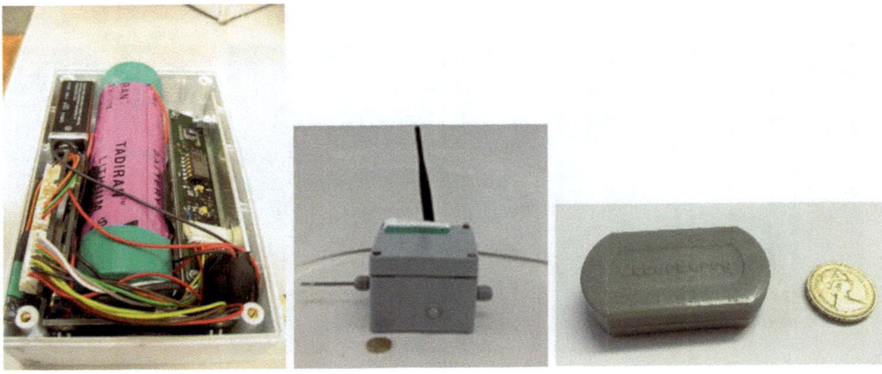

Fig. 21.12 Evolution of Cambridge WSN motes

A new generation of wireless sensor network hardware has sufficient processing capabilities to permit the development of computationally demanding signal-processing algorithms. Figure 21.12 shows the evolution of WSN motes developed at Cambridge University and they have been successfully deployed in field environment (Bennett et al. 2010a, b; Xu et al. 2015). The latest one is "UtterBerry" (Bevan 2014, personal communication). UtterBerry sensors are miniature, wireless, and ultra-low power sensors combined with artificial intelligence, specifically designed for infrastructure monitoring (Fig. 21.12, right photo). They use an artificial intelligence algorithm to perform onboard calculations to derive acceleration, inclination, and displacement in real-time without human intervention. It allows adaptive data collection and local processing of data for the extraction of failure signals. The availability of more computationally powerful platforms also allows common implementation of various data collection scenarios, in-network processing and compression algorithms.

Because of its small size and lightweight, it can be installed easily and placed in potentially unsafe or difficult-to-access sites. The sensors are self-calibrating after deployment and optimize their data communications within the sensor network according to environmental conditions. The sensors collect, process, and interpret data, reporting it to users remotely on any internet-enabled device and analyze trends in readings so that engineers can predict future events. An example of its field implementation is shown in Fig. 21.13.

21.6 Conclusions

In recent years, sensor and communications research has been undergoing a quiet revolution, promising to have significant impacts on new generation of monitoring technologies for civil engineering. In this chapter, the concept of whole-life cycle

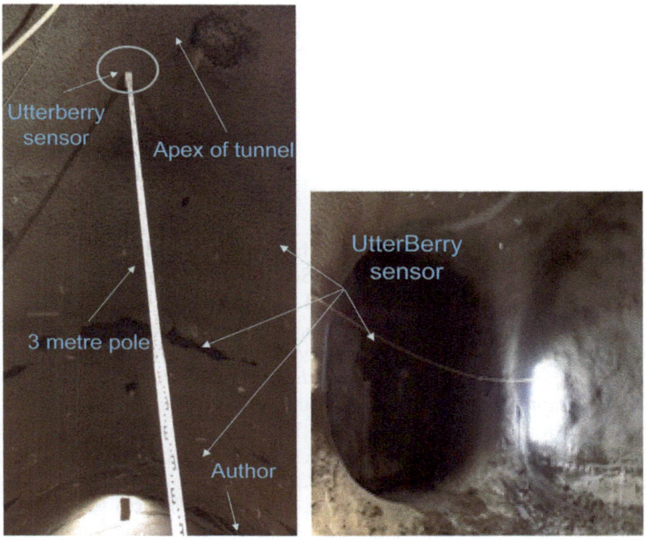

Fig. 21.13 Field installation of UtterBerry

monitoring has been introduced and discussed. Potential technical challenges have been identified to realize this. New emerging sensor technologies such as distributed fiber optic sensing, MEMS, WSN, and computer vision are introduced.

To accelerate the usage of such emerging technologies, field demonstrations are essential so that confidence within the community can be built. Infrastructure owners need to provide the market "pull" for smart technologies in response to the challenging targets set by the public. That is, the new infrastructure needs to be constructed and maintained economically and safely more than before. Our existing aging infrastructure requires better understanding of its extent of ageing and the consequent remaining design life. There is a need to assess the safety levels in extreme events such as flooding and earthquakes. Ultimately the development of "smart" infrastructure means true realization of performance-based design and maintenance.

Acknowledgements The author thanks the folowing past and current researchers at Cambridge University and the Centre for Smart Infrastructure and Construction (CSIC) who shared the research ideas and projects described in this chapter: Keita Abe, Sinan Açıkgöz, Mehdi Alhaddad, Mohamad Alserdare, Binod Amatya, Peter Bennett, Heba Bevan, Krisada Chaiyasarn, Xuesong Cheng, Loretta Cheung, Roberto Cipolla, Nicky de Battista, Vanessa Di Murro, Mohammed Elshafie, Tao Feng, Paul Fidler, Saleta Gil-Lorenzo, C. Y. Gue, Akio Hada, Chikara Hirai, Neil Hoult, Cedric Kechavarizi, Assaf Klar, Yusuke Kobayashi, Peter Knott, Vivien Kwan, Richard Laver, Katie Liu, Bo Li, Zili Li, Linqing Luo, Koson Janmonta, Yu Jia, Andy Leung, Robert Mair, Ying Mei, Tsukasa Mizutani, Hisham Mohamad, Masanari Nakashima, Sarfraz Nawaz, Tatsuya Nihei, Fuminano Okumura, Echo Ouyang, Loizos Pelecanos, David Rodnas-Harraiz, Satoko Ryuou, Hyungjoon Seo, Seda Torisu, Kiwamu Tsuno, Jennifer Schooling, Tina Schwamb,

Ashwin Seshia, Muneo Shibata, Simon Stent, Taro Uchimura, Eduard Vorster, Fei Wang, Ian Wassell, Matthew Wilcock, Yan Wu, Xiaomin Xu, Jize Yan, Seiji Yamada, Kyosuke Yasuda, Guoliang Ye, Yifei Yu, and Dan Zhang.

References

Bao, X., Li, W. and Chen, L. (2012). Distributed Fibre Sensors Based on Light Scattering in Optical Fibres (Chapter 19), In Handbook of Optical Sensors, V.1: Optical Sensors and Measurement Techniques; Section III: Fibre Optic Sensors; Taylor & Francis: Oxford, UK

Belsito, L., Ferri, M., Mancarella, F., Roncaglia, A., Yan, J., Seshia, A., et al. (2013). *High-resolution strain sensing on steel by silicon-on-insulator flexural resonators fabricated with chip-level vacuum packaging: Proceedings of the 17th International Conference on Solid-State Sensors, Actuators and Microsystems*. Barcelona.

Belsito, L., Ferri, M., Mancarella, F., Masini, L., Yan, J., Seshia, A. A., Soga, K., & Roncaglia, A. (2016). Fabrication of high-resolution strain sensors based on wafer-level vacuum packaged MEMS resonators. *Sensors and Actuators A, 239*, 90–101.

Bennett, P. J., Kobayashi, Y., Soga, K., & Wright, P. (2010a). Wireless sensor network for monitoring of underground tunnel. *Proceedings of the Institution of Civil Engineers: Geotechnical Engineering, 163*(GE3), 147–156.

Bennett, P. J., Soga, K., Wassell, I. J., Fidler, P., Abe, K., Kobayashi, Y., & Vanicek, M. (2010b). Wireless sensor networks for underground railway applications: case studies in Prague and London. *Smart Structures and Systems, 6*(5–6), 619–639.

British Standards Institution (2014). ISO 55000 Series: Asset Management.

Chaiyasarn, K. (2011). *Detection and monitoring of damage for tunnel inspection based on Computer Vision*. Ph.D. dissertation, University of Cambridge.

Chaiyasarn, K., Kim, T. K., Viola, F., Cipolla, R., & Soga, K. (2015). Towards distortion-free image mosaicing for tunnel inspection based on robust cylindrical surface estimation via structure from motion. *ASCE Journal of Computing in Civil Engineering*, 04015045.

Do, C. D., Erbes, A., Yan, J., Soga, K., & Seshia, A. A. (2016). Vacuum packaged low-power resonant MEMS strain gauge. *Journal of Microelectromechanical Systems*, accepted for publication.

Ferri, M., Mancarella, F., Roncaglia, A., Ransley, J., Yan, J., & Seshia, A. (2008). Fabrication of DETF sensors in SOI technology with submicron air gaps using a maskless line narrowing technique. *Proceedings of IEEE Sensors* (pp. 1131–1134). Lecce: IEEE.

Ferri, M., Mancarella, F., Seshia, A., Ransley, J., Soga, K., Zalesky, J. and Roncaglia A., (2009). Development of MEMS strain sensors for crack monitoring in ageing civil infra-structures, *Smart Structures and Systems, 6*(4), pp. 225–238.

Ferri, M., Mancarella, F., Seshia, A., Ransley, J., Soga, K., Zalesky, J., et al. (2010). Development of MEMS strain sensors for crack monitoring in ageing civil infrastructures. *Smart Structures and Systems, 6*(3), 225–238.

Ferri, M., Belsito, L., Mancarella, F., Masini, L., Roncaglia, A., Yan, J., et al. (2011). Fabrication and testing of a high resolution extensometer based on resonant MEMS strain sensors. *Proceedings of 16th International Solid-State Sensors, Actuators and Microsystems Conference, TRANSDUCERS'11* (pp. 1056–1059). Beijing: IEEE.

Garus, D., Golgolla, T., Krebber, K., & Schliep, F. (1997). Brillouin optical frequency-domain analysis for distributed temperature and strain measurements. *Journal of Lightwave Technology, 15*, 654–662.

Hartog, A. H., Leach, A. P., & Gold, M. P. (1985). Distributed temperature sensing in solid-core fiber. *Electronics Letters, 21*, 1061–1062.

Horiguchi, T., Kurashima, T., & Tateda, M. (1989). Tensile strain dependence of Brillouin frequency shift in silica optical fibers. *IEEE Photonics Technology Letters, 1*, 107–108.

Hotate, K., & Hasegawa, T. (2000). Measurement of Brillouin gain spectrum distribution along an optical fiber with a high spatial resolution using a correlation-based technique—proposal, experiment and simulation. *IEICE Transactions on Electronics, E83-C*, 405–411.

HM Treasury (2014). National Infrastructure Plan, UK. ISBN 978-1-910337-41-7, PU1726, p. 140.

Jia, Y., Yan, J., Soga, K., & Seshia, A. A. (2013). Parametrically excited MEMS vibration energy harvesters with design approaches to overcome initiation threshold amplitude. *Journal of Micromechanics and Microengineering, 23*(11), 114007.

Jia, Y., Yan, J., Soga, K., & Seshia, A. A. (2014a). A parametrically excited vibration energy harvester. *Journal of Intelligent Material Systems and Structures, 25*(3), 278–289.

Jia, Y., Yan, J., Soga, K., & Seshia, A. A. (2014b). Parametric resonance for vibration energy harvesting with design techniques to passively reduce the initiation threshold amplitude. *Smart Materials and Structures, 23*(60), 065011.

Kechavarzi, C., Soga, K., de Battista, N., Pelecanos, L., Elshafie, M., & Mair, R. J. (2016). *Distributed optical fibre sensing for monitoring geotechnical infrastructure—a practical guide*. London: ICE Publishing.

Kurashima, T., Horiguchi, T., & Tateda, M. (1990). Distributed-temperature sensing using stimulated Brillouin scattering in optical silica fibers. *Optics Letters, 5*, 1038–1040.

Lu, Y., Zhu, T., Chen, L., & Bao, X. (2010). Distributed vibration sensor based on coherent detection of phase-OTDR. *Journal of Lightwave Technology, 28*, 3243–3249.

Niklès, M., Thévenaz, L., & Robert, P. A. (1996). Simple distributed fiber sensor based on Brillouin gain spectrum analysis. *Optics Letters, 21*, 758–760.

Rodenas-Herráiz, D., Soga, K., Fidler, P., & de Battista, N. (2016). *Wireless sensor networks for civil infrastructure monitoring—a best practice guide*. London: ICE Publishing.

Soga, K. (2014). Understanding the real performance of geotechnical structures using an innovative fibre optic distributed strain measurement technology. *Rivista Italiana di Geotecnica, 48*(4), 7–48.

Soga, K., Kwan, V., Pelecanos, L., Rui, Y., Schwamb, T., Seo, H., et al. (2015). The role of distributed sensing in understanding the engineering performance of geotechnical structures. *Geotechnical Engineering for Infrastructure and Development, XVI ECSMGE* (pp. 13–48). London: ICE Publishing.

Stajano, F., Hoult, N., Wassell, I., Bennett, P., Middleton, C., & Soga, K. (2010). Smart bridges, smart tunnels: transforming wireless sensor networks from research proto-types into robust engineering infrastructure. *Ad Hoc Networks, 8*, 872–888.

Stent, S., Gherardi, R., Stenger, B., Soga, K., & Cipolla, R. (2016). Visual change detection on tunnel linings. *Machine Vision and Applications, 27*(3), 319–330.

Xu, X., Soga, K., Nawaz, S., Moss, N., Bowers, K., & Gajia, M. (2015). Performance monitoring of timber structures in underground construction using wireless SmartPlank. *Smart Structures and Systems, 15*, 769–785.

Part VII
Societal Impact of Extreme Events

Chapter 22
Theories of Risk Management and Multiple Hazards: Thoughts for Engineers from Regulatory Policy

Arden Rowell

Abstract Regulatory policy can offer a rich resource for engineers who are interested in managing multiple risks. Like engineers, regulatory policymakers routinely face decisions about risks from multiple hazards. The chapter begins by presenting two observations about risk that regulatory theorists have come to recognize as posing core challenges to multi-risk management: countervailing risk, which is ancillary risk that arises when (in)action is taken to reduce a targeted risk; and reciprocal risk, or risks that develop from interactions between multiple sources and multiple behaviors. The chapter then provides an overview of three approaches used in regulatory risk management to respond to the challenges posed by multiple risks: the precautionary principle, cost-benefit analysis, and the capability approach. These different approaches interact differently with the challenges posed by countervailing and reciprocal risks, have different strengths and drawbacks, and can pull decision-makers in different directions when managing multiple risks. Exploring the way that policymakers manage multiple risks—and the challenges they face in doing so—may thus provide valuable insights to engineers facing multiple hazards.

22.1 An Introduction to Regulatory Policy Regarding Multiple Risks

Regulatory policymakers must often manage risks from multiple hazards. In many ways, the challenges they face are comparable to the challenges engineers face in trying to make decisions to reduce, mitigate, and manage risks, including those that come from natural hazards. Understanding how regulators make these decisions, and the theoretical underpinnings on which those decisions rely, may prove fruitful for engineers who seek to manage risks to industry as well as to the public.

This chapter identifies two foundational concepts in current regulatory policy that bear on the risk management of multiple hazards: countervailing risk and reciprocal

A. Rowell (✉)
University of Illinois College of Law, Champaign, IL, USA
e-mail: karowell@illinois.edu

P. Gardoni, J.M. LaFave (eds.), *Multi-hazard Approaches to Civil Infrastructure Engineering*, DOI 10.1007/978-3-319-29713-2_22

risks. The chapter then describes three theoretical methods for managing multiple risks that have gained traction in regulatory literature and practice. Where possible, the chapter notes where these approaches are complementary to the decision-making methods already being used in engineering. It concludes with reflections on how engineers may be able to further leverage the theoretical work in risk management that has grown out of risk regulation.

22.2 Key Concepts in Regulatory Risk Management

Regulatory risk management typically seeks to manage risks to the public prospectively, before any injury actually occurs. This differentiates it from legal liability regimes like the civil tort system, which generally seeks to remedy harms retrospectively, after injury has accrued. In the USA, regulatory agencies now promulgate more legal rules than Congress, and many risk contexts that were once entrusted to retrospective judicial liability—such as health and safety, environmental risks, and transportation concerns—are now primarily managed through prospective regulatory policy. In recent decades, the regulatory portion of the USA legal system has grown so much that, by some measures, it exceeds the size of all of the three Constitutional branches of government combined.

As Sect. 22.3 discusses in more detail below, methods of regulatory risk management vary: regulators use different techniques for risk management for different types of issues and across different jurisdictions. For example, while the USA is highly precautionary about carcinogenic food additives, it typically relies on quantitative cost-benefit analysis to manage the multiple risks inherent in environmental policymaking. Other jurisdictions use different approaches for different topics. Regardless, theories of risk management generally grapple with at least two phenomena: (1) how to manage trade-offs among targeted and countervailing risks and (2) how to address reciprocal risks. The remainder of this section outlines these issues and describes how they pose puzzles to decision-makers—including engineers—who seek to manage multiple hazards.

22.2.1 Countervailing Risk

Countervailing risk is ancillary risk that arises when an action is taken to reduce a targeted risk (Graham and Wiener 1997). When a builder forgoes additional windows in a building to reduce seismic hazards, she increases the countervailing risks of blocked emergency egress. When a builder improves emergency egress options by adding additional points of exit, she increases the countervailing risks of unauthorized access to a building by terrorists or other dangerous persons. Budget allocated toward improved personnel access structures cannot be allocated toward flood risk mitigation. And so on.

In the regulatory risk literature, the recognition of countervailing risk has led some analysts to argue that it is impossible to reduce all risks (Sunstein 2005) and to general acceptance of the notion that countervailing risks pose fundamental challenges for decision-makers who attempt to be precautionary about multiple risks (Wiener 2011). As this chapter will discuss in Sect. 22.3, key approaches to risk management take different approaches to managing the challenges that arise from the existence of risks on all sides of a decision. In particular, they vary in how they approach the critical questions of (1) which targeted and countervailing risks are implicated by any decision-making structure, (2) which risks might be tolerated even if not targeted, and (3) how much to target or tolerate each of them.

Countervailing risks pose distinctive analytical challenges for planning reflective policies and effective projects. These challenges are particularly puissant in the context of multi-hazard engineering, where decision-makers face risks from multiple sources.

Regulatory theorists have developed typologies of countervailing risk that can help analysts identify and manage these risks. Different types of countervailing risk may have different implications for the analyst. One influential account distinguishes between countervailing risks by looking at two questions: the type of risk created and the population that is affected (Graham and Wiener 1997). When a different population is affected by a countervailing risk than a targeted risk, the project or policy may have distributional implications, putting additional pressure on ethical questions about whom it is appropriate to put at risk. When different types of risk are impacted by a countervailing risk than a targeted risk, it puts additional pressure on the technical analysis of those other risks and also creates a potential analytical tension between comparing two different kinds of risk. Thus, reflection on the types of risk implicated by a project, as well as the countervailing risks that arise from the attempt to address targeted risk(s), can help show analysts what features they need to look for in a decision procedure to manage the full suite of risks.

22.2.2 Reciprocal Risk

The concept of reciprocal risk is particularly valuable for understanding the social aspects of risk management. Here, the idea is that risks do not come from a single source: they come from an interaction of natural processes and human behavior. This idea can be traced to Ronald Coase, a critical figure in the history of law and economics, who won the Nobel Prize in part for this idea (Coase 1960).

Coase made his point in a series of examples. In one example, he encouraged us to imagine a community with some cattle ranchers and some farmers. The cattle periodically trample the farmer's fields and destroy and eat their crops. What is the source of the risk of the crops being trampled? Most people want to say the cows. But Coase said it is equally the crops. Both are necessary before it is possible for the cows to trample the crops. Furthermore, it's not just the farmers who experience risk. The farmers experience the risk of having their crops trampled, but the ranchers experience the risk of having their cows excluded from the land they need to access to eat. There are risks on both sides—countervailing risks—and those risks are reciprocal: they are being imposed by both the cattle ranchers and the farmers, through the integration of natural processes and human behavior.

Another of Coase's influential examples is a polluting factory. A factory emits pollution that makes nearby townspeople sick. What is the source of the risk of people becoming sick? Most people want to say the factory. But Coase said it is equally the people living near the factory. Both are necessary before it is possible for people to become sick from the pollution. Furthermore, it's not just the townspeople who experience risk. The factory also experiences a risk: the risk of being shut down by the townspeople. This doesn't mean that both are equally to blame from a social perspective or that legally or as a society we might not choose to attach liability to the polluter rather than the townspeople. But the example suggests that, to understand the source of a risk, we need to question our preconceptions and understand the multiple sources that lead to any risk being created.

Risk theorists outside of law have emphasized the theoretical importance of identifying the source of risks to fully develop strategies for managing those risks (Murphy and Gardoni 2011, 2012; Gardoni and Murphy 2013). The concept of reciprocal risk highlights the importance of interrogating those sources to consider both sides of the reciprocal risk coin: the ranchers and the farmers, the factory and the townspeople, and natural processes and human behavior.

22.3 Key Approaches to Risk Management

Regulators across the world, including in the USA, are constantly faced with the same kinds of questions that are addressed in this volume: trying to reduce or mitigate risks of various kinds by making decisions *before* a bad event occurs. Decision-makers have adopted different approaches to managing these risks, and to the challenges posed by countervailing and reciprocal risks. The discussion below addresses three key approaches to managing these concerns: the precautionary principle, cost-benefit analysis, and the capability approach. The remainder of this chapter describes these decision procedures for regulatory risk management, highlighting where possible their interaction with the reality of multiple hazards and the way they engage with the realities of countervailing and reciprocal risk. The discussion is summarized in Table 22.1.

For decision-makers interested in incorporating insights from risk regulation theory into engineering contexts, the main takeaway point should be that different approaches to risk management have foreseeably different strengths and challenges and that the selection of decision procedure should be matched to the goals of a project.

Table 22.1 Key risk management decision procedures

	Basic principle	Key strengths	Key challenges
Precautionary principle	"Better safe than sorry."	• simplicity • low information burden	• focuses attention on a single risk • limited guidance on "how far" to go
Cost-benefit analysis	"Weigh the costs and benefits."	• clear guidance on "how far" to go • allows apples-to-apples comparisons across risks	• high information burden • reliant on monetization
Capability approach	"Help individuals achieve what they value."	• significant guidance on prioritizing risks • can be tailored to different contexts and value structures	• challenging to compare across disparate persons and risks • requires lots of tailored analysis

22.3.1 Precautionary Principle

The simplest approach to managing risks is to identify a risk and attempt to reduce it. This approach is a form of the precautionary principle, a principle embodied in the familiar maxim that "it is better to be safe than sorry." It continues to enjoy significant popularity as a regulatory mechanism in the European Union; forms of it have been incorporated into a number of international agreements, including the Rio Declaration and the Kyoto Protocol; and it has also been selectively invoked in a number of regulatory contexts in the USA (Wiener 2011). One influential version of the principle is the so-called Wingspread Statement on the Precautionary Principle, which was created by a group of academics in 1998. That statement provided one formulation of the precautionary principle, which it described in this way: "When an activity raises threats of harm to human health or the environment, precautionary measures should be taken even if some cause and effect relationships are not fully established scientifically."

Like the other approaches discussed below—cost-benefit analysis and the capability approach—the precautionary principle exists in a number of forms. At a basic level, however, analysts must identify two things to use a precautionary principle: (1) the risk(s) against which they would like to be precautionary (such as "harm to human health or the environment") and (2) the strength of that preference (such as requiring action even when cause and effect have not been scientifically established). Nothing within the precautionary principle itself is designed to help with these decisions of risk prioritization and stringency: these must be added by the decision-maker according to the particular goals and concerns of a project. That said, once the targeted risk(s) and the strength of prioritization have been determined, the principle can act as a valuable explicit statement of risk priorities and preferences.

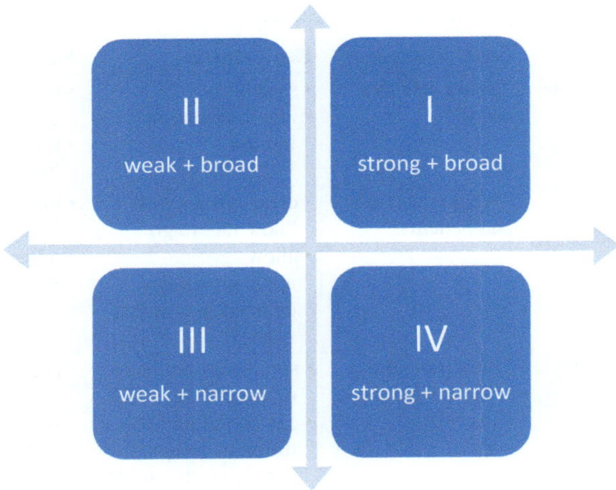

Fig. 22.1 Types of precautionary principles

As Fig. 22.1 portrays below, the characterization of the risk(s) to be targeted and the strength of the protection against them can be thought of as creating two spectra along which precautionary principles can be categorized. Principles in different quadrants then face predictable challenges and can be expected to interact differently well with policy and project contexts where there are multiple risks or multiple hazards in play.

The *Y*-axis here, or the "narrow vs. broad" spectrum, represents the generality with which the risk(s) to be targeted are identified. Principles that attempt to be precautionary against "all risks," "risks to the public," "harms to human health or the environment," or even "all engineering risks" are relatively broad—and therefore more likely to run into countervailing risk—than principles that attempt to be precautionary against more narrowly tailored categories of risk, such as "private bankruptcy," "building collapse in earthquake over a magnitude of 8.0," or "overfishing of specified species in specified waters."

The *X* axis, by contrast, represents the relative strength by which the targeted risk(s) are protected. The strongest principles would require action wherever there was any possible risk of the targeted type, regardless of cost and regardless of the weakness of supporting evidence. Weaker principles would establish weaker presumptions of action, for example the presumption that an action or policy might be warranted against the targeted risk even in the face of scientific uncertainty (Sunstein 2005; Mandel and Gathii 2006).

Broader precautions require less initial reflection on the part of the decision-maker, but tend to increase the chance that a principle will run into problems of countervailing risk. This is potentially a challenge for the precautionary framework, as when the same action would increase one type of targeted risk while decreasing another, the precautionary framework risks "paralysis by analysis:" preclusion of both action and inaction (Sunstein 2005). As the breadth of the principle increases, the number of types of risk targeted by the principle will increase as well, making paralysis by analysis increasingly likely. Some commentators have also suggested that this is more likely with strong precautions (Sunstein 2005). Strong precautionary principles have also been criticized for creating serious problems for decision-makers who face potential resource constraints, since they tend to require action regardless of cost (Sunstein 2005). Thus, quadrant I—or the establishment of a strong, broad principle—is likely to create distinctive challenges for decision-makers who face and must manage multiple hazards. Most likely they will have to choose some additional guide for which risk to reduce and that guidance will have to be external to the precautionary principle. Because a strong + broad principle itself will provide very little guidance in contexts of multiple risks, decision-makers should be particularly awake to the possibility that their choice among targeted risks could be affected by unconscious biases and by other nonreflective impacts (Sunstein 2005, 2009).

Weak precautionary principles are less likely to create both countervailing risk concerns and resource constraint problems. As narrower precautions are also less likely to sweep in additional risks, they too reduce the chance of analytical paralysis. Thus, Quadrant III, or weak + narrow precautions, are the precautions that are least likely to create trade-off issues across multiple values. Such principles may be particularly helpful as against risks for which there is significant uncertainty and/or when the decision-maker knows the specific types of risk against which she/he would like to protect. When principles are narrowly targeted enough, they can work as explicit risk prioritizations that help engineers as well as other stakeholders understand the reflective decision-making that went in to the relevant decisions.

In sum, then, the precautionary principle has the appeal of reduced difficulty in initial application, although subsequent invocation of the principle may be more complex than it appears at first glance, in light of the pervasiveness of countervailing and reciprocal risks. Different forms of the principle have different benefits and drawbacks, and decision-makers should be aware of those concerns before relying on the precautionary principle as a primary method of risk management. That said, the principle has the advantage of ease of communication, and where a narrow risk or set of risks can truly be prioritized, it can present a useful touchstone, including in contexts where existing information is incomplete or uncertain.

22.3.2 Cost-Benefit Analysis

If the decision procedure of cost-benefit analysis had a motto, it would be to "weigh the costs and benefits." But what exactly does that mean? As it happens, it can mean different things to different people: cost-benefit analysis comes in a variety of flavors that fall along a spectrum of formality (Rowell 2011). That said, one particular version of cost-benefit analysis has come to form a critical foundation for most major US regulatory decisions (Adler and Posner 2006; Sunstein 2002).

For this version of cost-benefit analysis, weighing the costs and benefits means counting up all the expected impacts of a policy (quantifying) and translating all those impacts into their monetary equivalents (monetizing), before comparing the monetized amounts in the "benefits" column to the monetized amounts in the "costs" column (Revesz and Livermore 2008; Rowell 2012). In the USA, most federal agencies are legally required to perform a quantitative and monetized cost-benefit analysis before they promulgate most major regulatory rules (EO 12,866; EO 13,563).

The key strengths and challenges of this approach are both related to quantification and to the conversion process between money and nonmonetary impacts. On the one hand, the creation of a single metric simplifies comparisons across multiple hazards and benefits. This is a particular benefit when a policy or project can be expected to have highly diverse impacts, or impacts over time, that are otherwise hard to compare to one another. In such cases, translating multiple risks and benefits into money can allow for an "apples to apples" comparison. On the other hand, dependence on the common metric of money requires complex and sometimes controversial "translations" of impacts like life saving, ecosystem protection, and health improvements into dollar figures. Unless cost-benefit analysis is put in context of the full range of project or policy goals, monetization may also draw attention away from ethical questions like how risks and benefits should be allocated across a population.

Monetization is therefore central to the process of cost-benefit analysis on which US agencies routinely rely. But how *do* regulators translate nonmonetary impacts—like reduced mortality risk, increased air quality, or infrastructure failure—into the common metric of money? For the most part, these conversations are performed using market- and preference-based studies of people's willingness to pay money to acquire nonmonetary goods (like environmental quality) or to avoid nonmonetary harms (like mortality risk) (Robinson and Hammitt 2013; Rowell 2010). To manage the complexities of monetization, agencies typically retain staff economists. Economists work with other agency officials to develop the inputs into the cost-benefit analyses and then monetize those inputs. When costs and benefits are spread out through time, they use the process of discounting to make the monetary amounts equivalent to one another, so that all of the costs and benefits can be expressed in money of the same dollar year. The cost-benefit analyses are then incorporated into federally published Regulatory Impact Analyses. Analyses are then reviewed (and sometimes rejected) by the US Office of Management and Budget. This process is required for most regulatory rules with an expected impact on the economy of at least $100 million, although agencies often perform cost-benefit analyses even on smaller proposed rules.

Individual agencies tend to specialize in monetizing the types of risks that they regulate and, in some cases, even provide training and software for stakeholders interested in performing cost-benefit analyses on their own. The Environmental Protection Agency (EPA), for example, publishes Guidelines for Preparing Economic Analyses (EPA 2014) for its own analysts, but these guidelines of best practices for environmental cost-benefit analysis are also publicly available and can be a starting point for information about how to monetize a variety of different environmental impacts. Other agencies provide other useful resources. The Federal Emergency Management Agency (FEMA), for example, administers the Stafford Act, which allows them to provide technical and financial assistance to state and local governments for purposes of developing hazard mitigation measures. FEMA requires a cost-benefit analysis prior to approval of any project under this program, but it provides guidelines, software, and methodologies for performing those analyses for a range of natural hazards, including flood, hurricane wind, earthquake, and wildfire. These resources can be valuable starting points for cost-benefit analysis by private actors, as well as provide useful information about how to engage with the agency through cost-benefit analysis.

That said, knowledge about how to monetize remains spread across agencies, and there is no single clearing-house for public education about monetization practices. Thus a stakeholder applying for aid to FEMA through the Stafford Act might do well to look to other agencies as well, for example to EPA for supplemental information about the monetizable benefits of stormwater drainage, as there is no guarantee that FEMA's software incorporates recent (or even any) research on any particular flavor of impact. Identifying the full range of risks and benefits posed by a project thus remains a critical foundation for a thorough cost-benefit analysis.

Furthermore, while various agencies can provide significant resources to educated analysts, the variety of agency practice can be bewildering on its own. In some cases, different agencies have come to different conclusions about how best to monetize the same risks. Perhaps most notoriously, agencies have historically used different monetary values for the amount they would allocate toward preventing a single human death or, in the technical lingo, different "valuations of a statistical life" or "VSLs" (Sunstein 2004). Nowadays, some agencies still use slightly different numbers, although the band has narrowed to around \$7–9 million (CRS 2010).

Although there are resources that can help with the monetization process, the dependence of cost-benefit analysis on monetization makes the decision procedure a potentially poor fit for certain types of projects, such as those that implicate important but difficult-to-quantify values like human dignity or equity (EO 13,563), those where information about probabilities or outcomes are particularly poor, or those where there are systematic market failures that make monetized valuations of a good into a poor proxy for human welfare or utility (Anderson 1995). Furthermore, even advocates of cost-benefit analysis recognize that it contains no intrinsic tool for addressing distributional inequities (Adler and Posner 2006; Adler 2011). Regulatory analysts attempt to address this incompleteness through distributional analyses that are performed separately from cost-benefit analyses (Rowell 2012). This is also an option for private decision-makers considering cost-benefit analysis as a tool for managing multiple risks: they can choose to use both cost-benefit analysis and other decision procedures for managing risk. This may be particularly appealing to analysts who wish to engage deeply with the ethical implications of their risk management decisions—implications that may not best explored through the distributionally-insensitive cost-benefit analysis—while still managing the countervailing risk trade-offs for which cost-benefit analysis is particularly helpful.

22.3.3 Capability Approach

Another theoretical approach to risk regulation, and one directly concerned with the ethics of risk management, is embodied in the capability approach to risk analysis. The capability approach to policymaking is focused on the promotion of human flourishing (Sen 1989, 1999; Nussbaum 2000, 2006). A capability approach to risk compares policies, risks, and outcomes based upon how the capabilities of individuals would be impacted, rather than the likely impact on resources or welfare (Murphy and Gardoni 2007, 2012).

At a very general level, a capability approach tries to identify what it is that allows people to live good lives and helps decision-makers focus on making lives better by improving people's ability to access those things. In the language of the discipline, the things that add value to a life—being educated, feeling safe from physical harm, building relationships—are called "functionings." (Sen 1989, 1992, 1999; Nussbaum 2000). A person's opportunity to achieve one of these functionings is a capability. A person's general capability is a function of all of the functionings she/he has an opportunity to achieve (Sen 1992). Under this approach, more desirable projects and outcomes will lead to higher general capabilities.

Different theorists differ regarding what capabilities are important and in determining the appropriate way to characterize those capabilities. In his initial formulation of the approach, Amartya Sen deliberately left capabilities largely unspecified, to promote context-specific analysis and deliberation (Sen 1992, 1993). Of course, creating project- or policy-specific lists of capabilities may fall outside of the realm of the expertise that most engineers feel themselves to have. But in that case, there are also existing specifications that can be helpful starting points. Perhaps most influential of these is Martha Nussbaum's list of capabilities, which includes life, bodily health, bodily integrity, emotions, practical reason, affiliation, being able to live with concern for other species, play, and control over one's own environment (both politically and materially) (Nussbaum 2006). Analysts interested in a capability approach can choose either to use an existing list, such as Nussbaum's, or could develop their own account of the types of human capabilities that might be impacted by a policy or project.

In addition to requiring an explicit analysis of the particular functionings by which capabilities should be measured, the capability approach differs from cost-benefit analysis and the precautionary principle because it explicitly imbeds the idea of minimal tolerable thresholds. In the context of risk analysis, this means that decision-makers must make judgments about the acceptability of risks to various functionings (Murphy and Gardoni 2008). In general, this tie between capability definition and risk thresholds is a double-edged sword, much as monetization is for cost-benefit analysis, and provides at the same time strengths and challenges for the capability framework.

The identification of important capabilities, the tracking of relationships between those capabilities and risks, and the selection of tolerable thresholds are all critical to the successful implementation of a capability approach to risk management. These are complex and controversial issues and create a significant analytical burden on analysts seeking to manage multiple hazards. At the same time, the capability approach forces explicit engagement with the relationship between human flourishing and certain levels of risk and with the selection of threshold quantities of risk, below which policies and projects should not fall. This means that the decision about how much risk is "too much" is an explicit one, providing the capability approach with some protection against the challenges of unconscious bias that have been leveled against application of the precautionary principle.

The existence of explicit minimal thresholds creates another set of puzzles in the context of trade-off analysis, reciprocal risk, and countervailing risk. Capability theorists recognize that there may be connections between different functionings—that decisions about promoting education may be connected to other functionings and that "[a]n individual might have a genuine opportunity to have a rewarding career and a large family, but not have an opportunity to achieve both at the same time" (Murphy and Gardoni 2012). But, at least within the framework of a capability approach, there is no intrinsic tool for choosing between projects or policies that would satisfy one functioning while allowing another functioning to fall below its minimal threshold. In this sense, a capability approach provides a mechanism for prioritizing between projects based upon how those projects can be expected to affect human flourishing, but it does not (necessarily) help the analyst decide what to do when multiple aspects of human flourishing are implicated by multiple hazards, and it does not (necessarily) provide guidance on how far to continue with the project once minimal standards are met.

These problems are not impossible to address, and one example of how they have been managed—and a valuable resource for analysts interested in applying a capability approach to risk management—comes from the most celebrated use of the capability approach in public policy: its incorporation into the United Nations' Human Development Report (HDR). The report has been published annually since 1990, which presents a Human Development Index (HDI) to assess development. The HDI uses a complex and reflective set of indicators to quantify each of the functionings on which the index is based, which include the ability to have a decent standard of living and to live a long and healthy life.

In sum, a capability approach to risk offers significant promise where the goal of an analysis is best characterized as being to help individuals achieve what they value. Particularly when the individuals in question have localized preferences or values that are not well represented in traditional marketplaces, the tailorability of this approach gives it an advantage over monetized cost-benefit analysis. Where the analytical burden of defining functionings and capabilities seems justified, this approach also offers significantly more (although not perfect) guidance than the precautionary principle on the question of how strenuously to act, although it does struggle, at least without supplement, to provide meaningful guidance when a decision-maker must choose between multiple bad options or where there are multiple hazards that will force some trade-off between different values.

22.4 Conclusion

Regulators, like engineers, routinely face decisions that pose risks from multiple hazards. There are legal and social science literatures on managing multiple risks. Literature on multiple risk management in the regulatory context builds on two important observations about managing multiple risks:

1. Countervailing risks are pervasive and to some extent unavoidable; we need risk management systems that recognize these trade-offs and that provide some way to manage them.
2. Reciprocal risks arise routinely; it is worth thinking about the directionality of risk and interrogating what all of the necessary conditions are for a risk to occur.
3. A key portion of any risk management exercise is the determination of how to value various risks and benefits and particularly whether the decision-maker will try to express diverse risks and benefits in terms of a single metric, like money. Three key approaches to risk management—the precautionary principle, the cost-benefit analysis, and the capability approach—balance these concerns differently and can lead decision-makers in different directions.

References

Adler, M. (2011). *Well-being and fair distribution: Beyond cost-benefit analysis* Oxford, Oxford University Press.

Adler, M., & Posner, E. (2006). *New foundations of cost-benefit analysis*. Boston: Harvard.

Anderson, E. (1995). *Value in ethics and economics*. Cambridge, Harvard University Press.

Coase, R. H. (1960). The problem of social cost. *Journal of Law and Economics, 3*, 1.

Copeland, C. (2010). *How agencies monetize statistical lives expected to be saved by regulations*. Congressional Research Service.

Environmental Protection Agency. (2014). *Guidelines for preparing economic analyses*.

Exec. Order No. 12,866 § 11, 3 C.F.R. 638, 649 (1994), reprinted as amended in 5 U.S.C. § 601 app. at 802–06 (2012).

Exec. Order No. 13,563, 3 C.F.R. 215 (2012), reprinted in 5 U.S.C. § 601 app. at 816–17 (2012).

Gardoni, P., & Murphy, C. (2013). A scale of risk. *Risk Analysis, 34*(7), 1208–1227.

Graham, G., & Wiener, J. (Eds.). (1997). *Risk vs. risk: Tradeoffs in protecting health and the environment*. Cambridge, Harvard University Press.

Mandel, G., & Gathii, J. (2006). Cost benefit analysis versus the precautionary principle: Beyond Cass Sunstein's laws of fear. *University of Illinois Law Review*, 1037.

Murphy, C., & Gardoni, P. (2007). Determining public policy and resource allocation priorities for mitigating natural hazards: A capabilities-based approach. *Science and Engineering Ethics, 13*(4), 489–504.

Murphy, C., & Gardoni, P. (2008). The acceptability and the tolerability of societal risks: A capabilities-based approach. *Science and Engineering Ethics, 14*(1), 77–92.
Murphy, C., & Gardoni, P. (2011). Evaluating the source of the risks associated with natural events. *Res Publica, 17*(2), 125–140.
Murphy, C., & Gardoni, P. (2012). The capability approach in risk analysis. In S. Roeser, R. Hillerbrand, P. Sandin, & M. Peterson (Eds.), *Handbook on risk theory* (pp. 979–997). Dordrecht: Springer.
Nussbaum, M. (1997). Capabilities and human rights. *Fordham Law Review, 66,* 273–300.
Nussbaum, M. (2000). *Women and human development: The capabilities approach,* Cambridge University Press.
Nussbaum, M. (2006). *Frontiers of justice: Disability, nationality, and species membership.*
Revesz, R., & Livermore, M. (2008). *Retaking rationality: How cost-benefit analysis can better protect the environment and our health.* Oxford, Oxford University Press.
Robinson, L., & Hammitt, J. K. (2013). Skills of the trade: Valuing health risk reductions in benefit-cost analysis. *Journal of Benefit-Cost Analysis, 4,* 107.
Rowell, A. (2010). The cost of time: Haphazard discounting and the undervaluation of regulatory benefits. *Notre Dame Law Journal, 85,* 1505–1542.
Rowell, A. (2011). Cost-benefit analysis. In *Measurements, indicators and research methods for sustainability, encyclopedia of sustainability* (Vol. 6). Berkshire.
Rowell, A. (2012). Allocating pollution. *University of Chicago Law Review, 79,* 985–1049.
Sen, A. (1989). Development as capabilities expansion. *Journal of Development and Planning, 19,* 41–58.
Sen, A. (1992). *Inequality reexamined,* Cambridge, Harvard University Press.
Sen, A. (1993). Capability and well-being, 9–29. In M. Nussbaum & A. Sen (Eds.), *The quality of life.* Oxford, Oxford University Press.
Sunstein, C. (2002). *Risk and reason: Safety, law, and the environment.* Cambridge: Cambridge University Press.
Sen, A. (1999). Development as Freedom. New York: Anchor Books.
Sunstein, C. (2004). Valuing life: A plea for disaggregation. *Duke Law Journal, 54,* 385.
Sunstein, C. (2005). *Laws of fear: Beyond the precautionary principle,* Cambridge: Cambridge University Press.
Sunstein, C. (2009). *Worst-case scenarios,* Cambridge, Harvard University Press.
Wiener, J. (2011). *The reality of precaution: Comparing the U.S. and Europe.* Washington, DC: Resources for the Future.
Wingspread Conference on the Precautionary Principle (1998).

Chapter 23
Disaster Risk Reduction Strategies in Earthquake-Prone Cities

Lori Peek, Stacia S. Ryder, Justin Moresco, and Brian Tucker

Abstract This chapter provides an overview of disaster risk reduction (DRR) activities underway in 11 earthquake-prone cities including Antakya and Istanbul, Turkey; Bandung and Padang, Indonesia; Chincha and Lima, Peru; Christchurch, New Zealand; Delhi and Guwahati, India; San Francisco, USA; and Thimphu, Bhutan. The goal is to provide information about the tools and resources that practitioners and organizations in these 11 cities have access to, in order to provide a more comprehensive understanding of DRR strategies and a better sense of the contexts in which future potential products may be created, evaluated, and ultimately adopted. Drawing on both survey and in-depth interview data with earthquake safety practitioners from government, business, health care, education, and grassroots groups, the chapter describes earthquake and disaster risk reduction programs and initiatives in these communities, explains what spurred the creation of those programs and initiatives, details the technical tools and resources that practitioners in these cities currently use to assess and mitigate their risk, and analyzes the communication channels that disaster and risk professionals now have access to and find most useful in their work. The chapter concludes with a discussion of overarching motivations for adopting new disaster and seismic risk reduction (SRR) practices and offers practical advice to help guide the development of risk reduction tools for use in earthquake-prone cities around the world.

L. Peek (✉) • S.S. Ryder
Department of Sociology, Center for Disaster and Risk Analysis, Colorado State University, Fort Collins, CO, USA
e-mail: Lori.Peek@colostate.edu; Stacia.Sydoriak@colostate.edu; stacia.s.ryder@gmail.com

J. Moresco
Model Product Management, Global Earthquake Models, Risk Management Solutions, Newark, CA, USA
e-mail: Justin.Moresco@rms.com

B. Tucker
GeoHazards International, Menlo Park, CA, USA
e-mail: Tucker@GeoHaz.org

P. Gardoni, J.M. LaFave (eds.), *Multi-hazard Approaches to Civil Infrastructure Engineering*, DOI 10.1007/978-3-319-29713-2_23

23.1 Introduction

Recent devastating earthquakes in the Indian Ocean, Pakistan, China, Haiti, New Zealand, Japan, and Nepal serve as painful reminders of the urgent need to disseminate both scientific and technical information to earthquake safety practitioners working in the world's most seismically vulnerable communities. Yet before this information can be successfully passed on, it is essential to establish an understanding of the development and implementation of disaster risk reduction (DRR) and seismic risk reduction (SRR) practices that are currently in place worldwide. Developing a knowledge base of these practices, which may be structural or nonstructural in nature, may help identify the possibility of reproducing, modifying, and integrating model risk reduction practices in different cities and communities around the world.

In order to continue to evaluate the feasibility and success of different practices, it is necessary to first uncover what current DRR and SRR programs and initiatives look like for earthquake-prone cities on a global scale. The purpose of this chapter is to identify the extent to which these practices are relevant, accessible, and relied upon in various higher- and lower-income nations. Specifically, this chapter examines catalysts for incorporating DRR and SRR strategies into community planning and building efforts in 11 earthquake-prone communities. It also identifies which of these strategies are considered most useful across the different contexts. Finally, we discuss the types of strategies that will likely best enhance future efforts of integrating risk reduction efforts on a global scale.

23.2 Theoretical Background

23.2.1 Disaster Risk and Risk Reduction

The recording of disaster events began in 1900 with the Centre for Research on the Epidemiology of Disasters (CRED). CRED's Emergency Events Database (EM-DAT) is an international database established in 1988 which provides disaster information on human mortality and injury, economic damages, and aid contributions and can be used for vulnerability assessments (CRED 2015). From 1979 to 2009, the EM-DAT shows a decrease in natural hazard mortalities (Smith 2013). Across the second half of the twentieth century (1950–1999), however, Munich Reinsurance's 2005 report demonstrated that the number of catastrophic events had increased 4.5 times from those reported in the 1950s, and financial losses increased from US$48 billion in the 1950s to US$575 billion in the 1990s. Auld (2008) recorded similar findings suggesting that since the 1950s, annual direct losses tied to natural catastrophes have continued to rise, as have the number of individuals who are affected by disasters. A recent assessment suggests that approximately

3.4 billion people worldwide reside in locations where they are at a high risk for experiencing at least one type of natural disaster (Dilley et al. 2005).

We know that geographic areas experience disasters at different rates and that increasing disaster frequency is not uniform across the continents (Guha-Sapir et al. 2013). Smith (2013, p. 38) notes that "disaster risks and disaster losses do not occupy the same geographic space." Furthermore, spatial differences in disaster trends also demonstrate how disaster risk and impact manifest differently depending on whether we are interested in discussing vulnerable populations or material losses or if we consider disaster risk reduction strategies (Smith 2013). Typically disasters within the "Global South" suffer more human loss, while the "Global North" experiences more extensive material and financial loss (Smith 2013).

Vulnerability, understood in the context of disasters as the likeliness a person or group will be affected by a hazard (Twigg 2004), helps to explain the disparity in the distribution of disaster risk and loss (Bankoff et al. 2004; Solomon et al. 2007). Vulnerability is largely dependent on "social, economic, cultural and political conditions which are up to internal changes and outside influences" (Rottach 2010, p. 6). The literature on disaster vulnerability has established that the level of vulnerability of individuals can depend on a variety of factors, including age, race, ethnicity, gender, land ownership, income, and beyond, impacting a person's disaster risk and experience across all phases of a disaster event (see Enarson and Morrow 1998; Haas et al. 1977; Hewitt 1983; Morrow 1999). Increasingly, researchers have demonstrated the way in which disasters reflect and often amplify preexisting inequalities and vulnerabilities within an impacted community (see Morrow 1999; Pielke and Pielke 1997). Essentially, vulnerability to disaster is "determined by every day patterns of social interaction and organization, particularly the resulting stratification paradigms which determine access to resources" (Morrow 1999, p. 2; see also Oliver-Smith 1986; Blaikie et al. 1994).

While ultimately it is unlikely we could ever eliminate disasters, through the institutionalization of DRR strategies, it is possible to drastically reduce the vulnerability of individuals and entire communities to disaster. Mitchell (2003, p. 5) suggests that in order to do so, it is first essential to "establish universal agreement over what constitutes DRR." To do this, it is crucial to establish a working definition of the concept. In the context of this chapter and the larger research project, we adopt the United Nations International Strategy for Disaster Risk Reduction definition of DRR, which is understood as:

> The concept and practice of reducing disaster risks through systematic efforts to analyse and manage the causal factors of disasters, including through reduced exposure to hazards, lessened vulnerability of people and property, wise management of land and the environment, and improved preparedness for adverse events. (2009, p. 10)

In simpler terms, DRR research focuses on what can be done to prevent or reduce disaster impacts, primarily prior to a disaster occurring. Disaster preparedness, which is a major component of DRR, refers to taking measures prior to a disaster to warn and inform about potential disasters, informing populations about how to be ready for a disaster, and teaching individuals how to respond appropriately to a potential disaster (Twigg 2004). Preparedness is one approach to mitigation, while mitigation efforts consist more broadly of actions taken across any phase of a disaster to minimize the impacts of a disaster (Twigg 2004). Both preparedness and mitigation efforts help to reduce disaster risks at the individual and community level.

Yet how do we move from an understanding of DRR into preparedness and mitigation efforts that ignite systemic change in disaster planning and preparedness? Schipper and Pelling (2006, p. 26) underscore the importance of understanding and addressing disaster vulnerability, suggesting that we must first change how we view and think about disasters:

> When disasters are seen as the outcome of accumulated risk produced by years of vulnerability and underlying hazard, the case for preventative action can be made more plainly. Contributing to such problems of perspective is a lack of data on vulnerability, hazard, risk, and disaster losses. Many international initiatives are pushing to overturn this barrier (IFRC 2003), but it remains in place, particularly where tools are needed to make an economic case for risk reduction. (Benson and Twigg 2004; Pelling 2005)

These increasing efforts to reframe our view on disasters have been accompanied by a focal shift in disaster research from "resistance" to "resilience" (Tierney and Bruneau 2007). In terms of practical application, this has led to explorations of "integrated disaster risk reduction concepts," such as "building-in resilience," as a DRR strategy that encourages building planning phases to incorporate strategies to resist and rapidly recover from an earthquake (Bosher and Dainty 2011, p. 2). Unfortunately, industry fragmentation between professionals working in the context of built environments often acts as a barrier to including DRR strategies in early building phases, and it remains unclear the extent to which these resilience efforts have been or are being incorporated in preparedness planning worldwide (Bosher and Dainty 2011; Wamsler 2006).

23.2.2 Seismic Risk and Risk Reduction

Seismic risk reduction (SRR) is a disaster-specific risk reduction strategy that stems from the larger DRR literature. Seismic risk is defined by Dowrick (2003, p. 1) as "the probability that social or economic consequences of earthquakes will equal or exceed specified values at a site, at several sites, or in an area, during a specified exposure time." Drawing on the definitions of DRR and seismic risk, SRR can be understood as *any element that can minimize the possibilities or probabilities of social or economic consequences that develop as a result of exposure to seismic events*.

Of the 3.4 billion people at high risk to disaster, 328 million individuals are specifically at risk to earthquakes (Dilley et al. 2005). Each year, there are approximately 60,000 persons who die in a natural disaster, the majority of whom are killed when buildings collapse during an earthquake (Kenny 2009). The 2008 Sichuan Earthquake in China, for example, killed over 70,000 people and the 2011

Haiti earthquake led to over 250,000 deaths. In addition to significant loss of life, earthquakes can cause direct economic losses, like the 2005 Marmara Earthquake that resulted in over $5 billion in losses (Trohanis et al. 2010). Given the proportion of earthquake-induced death and destruction in comparison to other disasters, seismic risk reduction is key to achieving a drastic decrease in damage caused by natural disasters on an annual basis. In fact, the 1994 Northridge earthquake demonstrates the benefits of SRR in terms of reducing mortality: only 72 lives were lost—which was in large part due to pre-disaster mitigation and preparedness efforts aimed at the built, natural, and social environments (see Trohanis et al. 2010). (It is worth noting, however, that the Northridge earthquake remains one of the costliest disasters in US history. This is a sharp reminder that disaster risk reduction may occur in one sphere (e.g., reduction in injuries and deaths) but not necessarily in another (e.g., reduction in dollars lost and economic downtime). These are issues worthy of further consideration.)

While SRR strategies are integral to diminishing some or all of the negative impacts earthquakes may have on communities and societies, less has been written about what these strategies actually look like in diverse geographic and social contexts. On the one hand, we know that disaster vulnerability differs, as persons in developing countries experience a significantly higher level of hardship in the face of natural disasters (see Auld 2008; Hao et al. n.d.; Solomon et al. 2007; McEntire 2001; Rottach 2010; Schipper and Pelling 2006). On the other hand, the literature has yet to account for the strategies that are currently being used to reduce these vulnerabilities worldwide—and the relative ability to transform these practices to be effective in different local contexts. Even less clear is an understanding of what factors serve to motivate these same communities and their leaders to begin thinking about and implementing DRR and SRR strategies in the most earthquake-prone cities.

To reduce disaster risk, Crompton and McAneney (2008, p. 44) call for the development of "reliable and valid information on hazards, society, infrastructure, and the environment." The purpose of this chapter, therefore, is to examine the circumstances which fostered the development of DRR and/or SRR strategies at selected research sites. We identify the SRR strategies that key stakeholders in communities are using in high-earthquake-risk cities and discuss if and how those strategies could be communicated, received, and implemented in other locations. This information is intended to contribute to the SRR and broader DRR literature as well.

23.3 Methodology

23.3.1 Methodological Approach

To best assess the SRR practices, initiatives, and programs being utilized in earthquake-prone cities, we gathered both *qualitative* and *quantitative* data. Given the moderate size of the study (surveys, $N = 119$; interviews, $N = 133$), employing a mixed method approach was the most appropriate way to conduct this research project. In doing so, we are able to treat each case as distinct for qualitative purposes and to treat the sample size as large enough for a simple, complementary quantitative analysis (Ragin et al. 2004).

The purpose of qualitative research methods is to study real people in natural settings, rather than in artificial experimental laboratories (Marshall 1996). The trend among disaster researchers to use qualitative approaches in data collection is increasing, with in-depth interviews and case studies being the most commonly used qualitative methods (Phillips 2014; Stallings 2002). In this study, we chose to conduct in-depth interviews with all participants (Rubin and Rubin 2012). This responsive interviewing style involves asking open-ended questions and directed follow-up questions, which are designed to evoke nuance and detail in participant narratives (Rubin and Rubin 2012). Indeed, the overarching goal of using qualitative methods is to gain an in-depth understanding of the experiences of a small group of research participants—including gathering rich contextual details and exploring related and contradictory themes (Rubin and Rubin 2012)—rather than a breadth of knowledge that can only be attained through surveying large, representative samples (Ambert et al. 1995). In our study, the goal of conducting responsive interviews was to garner in-depth understandings of the resources that key actors across several institutions relied on for developing DRR and SRR strategies in their own city.

The purpose of quantitative research methods is to provide numerical data about a study population. The numerical data are used to test hypotheses, evaluate relationships between phenomena, and generate counts or rates of incidence about a topic from a large number of people (Fowler 2009). Survey research is the most common quantitative method to gather numerical data about individuals' opinions or perspectives on a topic. We used surveys to gather opinions about access to risk reduction resources and information sources used by the participants. The information provided in the surveys allowed the team to perceive general patterns in the responses to these topics.

23.3.2 Research Sites

Choosing an appropriate research site is fundamental to the design of any qualitative study, given that researchers can never study *all* places or people (Marshall and Rossman 2011). Qualitative research design experts emphasize the importance of being critical when choosing research sites, in order to increase the quality of the data that is gathered. Scholars working in this area recommend choosing research locations where (a) entry is possible; (b) there is a high probability that a rich mix of the processes, people, programs, interactions, and structures of interest is present; (c) the researcher is likely to be able to build trusting relationships with the participants in the study; (d) the study can be conducted and reported ethically, and (e) data quality and credibility of the study are reasonably assured (Marshall and Rossman 2011). This research literature provided a logical framework with which to begin the city sampling criteria process. To further ensure the quality of our data collection, we developed 12 additional criteria for selecting our research sites[1]:

1. Representative of a range of World Bank Regions
2. Representative of a range of national incomes
3. Representative of a range of population sizes
4. Representative of recent and distant earthquake events
5. Regarded as at high earthquake risk
6. Varied in earthquake mitigation experience[2]
7. A secure working and political environment, open to SRR promotion
8. Generally within operational boundaries of the following grassroots organizations, where we had preexisting contacts:
    a. HelpAge International
    b. Huairou Commission
    c. Plan International
9. Representation of major international development organization offices in region/nation
10. Ability to identify/contact local partners
11. Research team members' previous experience in region/nation
12. Research team staffing availability to travel to country of interest

After identifying the target cities for this research, we drafted a 15–20 page "City Document" for each location. This document included relevant information for each city that team members would visit, including maps, sociodemographic and economic indicators, earthquake information on recent events and levels of seismic risk, earthquake hazard legislation and building codes standards for the city and country, and information on hazard-related public education efforts. These city documents provided important background information and helped to prepare our team for data collection in each city. Our final sample included 11 cities in seven countries, spanning five continents: Antakya, Bandung, Chincha, Christchurch, Delhi, Guwahati, Istanbul, Lima, Padang, San Francisco, and Thimphu.

23.3.3 Sample Population

For this study, we implemented a purposeful sampling procedure that requires a flexible and pragmatic approach (Marshall 1996). Participants were not chosen

[1]To be clear, not every site met each of the 12 criteria; rather, sites were chosen that met as many of the criteria as possible.

[2]Although no actual mitigation scale exists, the project team used rankings from the Human Development Index (HDI) as a proxy for risk reduction activity and experience. Those with lower, or worse, HDI scores were assumed to also be laggards in terms of mitigation activity.

Table 23.1 Study respondents by sector

Sector	Survey frequency	Percent	Interview frequency	Percent
Government	29	24	34	26
Business	20	17	22	16
Health care	22	18	23	17
Education	25	21	28	23
Grassroots	23	19	26	19
Total	119	100	133	100

randomly; instead, they were selected based on their knowledge of the research area, their ability to speak on behalf of the sector they represented, and their level of DRR and/or SRR experience in the field. Although choosing a random sample would have made the study more statistically generalizable, it is not the most effective way of developing an understanding of complex issues relating to human behavior (Marshall 1996). Instead, the sampling strategy was chosen based on the research questions and conceptual framework. This approach generated rich information about SRR activities.

We identified five key sectors to study: (1) government, (2) business, (3) health care, (4) education, and (5) grassroots or community-based organizations. The team focused on these sectors of society because, as UNISDR emphasizes, each plays a crucial role in reducing disaster risk and establishing a culture of safety in communities.[3,4] Data collection resulted in a total of 133 interview participants; 119 of these interviewees completed surveys from the key sectors. Table 23.1 shows the number of respondents by sector and survey and interview completion.

23.3.4 Data Collection and Research Instruments

Data were collected by members of our research team between June and November 2011. All interviews were audio recorded, so that they could be transcribed verbatim. Each interview lasted approximately 1 h, with 50 min of open-ended questions and 10 min to complete the survey and remaining follow-up questions. Each interviewer followed a semi-structured open-ended interview guide, allowed for consistency across interviews, while providing enough flexibility so that the

[3]See United Nations International Strategy for Disaster Reduction (UNISDR): http://www.unisdr.org/.

[4]Any given community may have other sectors—such as the media, criminal justice institutions, public- and private-sector housing organizations, and development associations—with important roles to play in risk reduction. However, the five sectors we studied are present in most communities, and they represent critical institutions for the promotion and implementation of risk reduction.

participant could frame and structure the responses from his or her own perspective (Marshall and Rossman 2011). After each set of interviews was completed, we created an interview matrix that included the name, date, time, and length of the interview. This document was sent with all of the audio files to the transcriptionist, who transcribed and returned the textual data for analysis.

In preparing for the responsive interviews, we drafted a two-page interview guide to be used in the 11 target communities. The guide included the open-ended questions to be asked during interviews. One of the hallmarks of qualitative research is that it allows for flexibility in ordering the questions asked, so often questions were asked in a different order, based on the flow of the conversation.

In addition to the interview guide, we drafted and revised a survey questionnaire and then sent it to outside research assistants from Colorado State University (CSU). These assistants checked the survey to ensure that the questions were clear, grammatically correct, and appropriate in relation to the research goals. After integrating these revisions, the final two-page survey included a total of 47 questions which in large part focused on asking participants to identify whether a particular SRR tool was something their organization "already had," "would like to have," or "did not need."

The survey was administered to participants following the interviews. The survey took approximately 5 min for participants to complete. After the participants finished the survey, we asked a final set of open-ended questions, in order to clarify responses and to help evaluate the resource needs of participants. At the end of the interview, we asked interviewees to fill out a demographic information form. Table 23.2 includes a breakdown of survey and interview respondents by city.

Table 23.2 Study respondents by city

City	Survey frequency	Percent	Interview frequency	Percent
Antakya	9	8	14	11
Bandung	11	9	12	9
Chincha	10	8	12	9
Christchurch	16	14	17	13
Delhi	11	9	11	8
Guwahati	8	7	11	8
Istanbul	10	8	10	8
Lima	14	12	13	10
Padang	10	8	11	8
San Francisco	11	9	11	8
Thimphu	9	8	11	8
Total	119	100	133	100

The study had a larger sample of interviewees ($N = 133$) than of survey respondents ($N = 119$), because additional practitioners often agreed to participate in the interviews but then had to leave before the survey was distributed toward the close of the interview.

The interview guide, survey questionnaire, and demographic information form were all created in English and, once finalized, were translated into Indonesian, Spanish, and Turkish to meet the language needs of the participants.[5] One of the responsibilities of the local partner[6] was to translate[7] the interview guide and survey questionnaire and, where necessary, to act as an interpreter during the interview. Local partners also "debriefed" with our team members immediately after each interview. This approach was especially useful, as it allowed the local partner to act as an interpretive guide and a coresearcher; this practice is highly recommended by social science scholars, to "strengthen the rigor and trustworthiness of qualitative cross-language research" (Berman and Tyyska 2011, p. 181).

In order to maintain consistency and accuracy across all of the project team members, meetings were held via Skype after each set of interviews for a particular focal city had been completed. This allowed the person leaving the field to debrief and share the successes, challenges, and lessons learned in each city, thereby strengthening the approach of the next team member.

23.3.5 Data Analysis

Once the interviews were transcribed, they were uploaded into Atlas.ti, a qualitative data analysis software program. Before beginning the analysis process, researchers at CSU created a codebook. The codebook included a list of potential major themes and initial codes that were likely to emerge from the data, given the research questions and design (Marshall and Rossman 2011). Once the coding process began, many more codes were added to the codebook, and new categories were created as new themes emerged.

Three CSU team members used Atlas.ti to code and analyze the interview transcripts. The analysis occurred in two major stages: (1) first cycle, or initial coding (i.e., searching for the most general themes and patterns that emerge in the

[5]The interview guide, survey instrument, and demographic information form—in English, Indonesian, Spanish, and Turkish language versions—are available upon request.

[6]We hired one to two people in each city (except San Francisco and Delhi, where members of our team already resided) to act as a "local partner." The local partners served as associates on the ground, with intimate local knowledge of the target city; this helped the project team to gain access to the field setting and to key interview participants. Before being hired, the team screened the potential local partner to ensure that he or she met our rigorous criteria.

[7]We used a dual translation strategy, which is encouraged in social science research to ensure the accurate translation research instruments. In the case of this project, a team member and a local partner worked together to translate the instruments from English to Indonesian, Spanish, or Turkish, respectively. The original document was then compared to the translated document, so the team could identify inconsistencies, mistranslations, issues with meaning, cultural gaps, and/or lost words of phrases (McGorry 2000). Then together, the translators worked through their differences, agreeing on the best translation for each discrepancy. The translated instrument was then proofread and finalized before being used in the field.

data), and (2) second cycle, or focused coding (i.e., searching for more generalizable thematic patterns). Once the coding process was complete, the team then began a process of representative coding (i.e., selecting interview quotes for inclusion in the written report that are representative of relevant findings) (for further information on coding, see: Marshall and Rossman 2011; Saldaña 2009).

The close-ended survey data were cleaned and entered into an Excel file. They were then uploaded and analyzed using Stata/IC 12.1, a quantitative data analysis and statistical software program.

23.4 Results and Discussion of Empirical Findings

The qualitative interviews allowed us to identify the primary factors that spurred the creation of many DRR and SRR strategies, programs, and initiatives. We found they fell primarily into five categories: (1) occurrence of an earthquake or other major disaster, (2) new risk reduction-oriented legislation and regulations, (3) available local/state/national funding, (4) external support and international guidance, and (5) hazard vulnerability concerns, the making of mitigation champions, and strong leadership. Below, we discuss each of these catalysts for developing risk reduction strategies and then describe the types of DRR and/or SRR strategies that were commonly relied upon within and across our research sites.

23.4.1 Primary Factors Spurring Risk Reduction Strategies

23.4.1.1 Occurrence of a Disaster

When the team asked practitioners what led to the creation of the programs that they help to coordinate, the most common response given was "a disaster." Most often, the disaster had occurred in the respondents' home city or nation, and these events were described as a "turning point" or "watershed moment" in their professional careers and organizational histories:

> These activities began with the 1999 Izmit earthquake, because it was a turning point for every sector, and after that, people started to do something to overcome these problems.—Health Care Respondent, Antakya

> In 1970, the history of natural disasters would never be the same again: 67,000 dead, 180,000 wounded, 60,000 homes destroyed... It's the black curtain that divides life from misfortune. This is the worst disaster that Peru had ever seen. Two years after that tragedy, in 1972, the Civil Defense system was created.—Government Respondent, Lima

> The [2001] Gujarat earthquake was a hallmark, a watershed moment in the preparedness of our own disaster management plans. Several surveys were conducted and Delhi was found to have an equal level of risk.—Government Respondent, Delhi

Predictably, earthquakes were the most frequently mentioned type of disaster that led to the creation of new SRR programs and initiatives. Other events were also influential, however, in many of the cities that the team visited. For example, in Bandung and Padang, respondents spoke of recurring flood losses, volcanic eruptions, and tsunamis as having motivated their program creation and activities. In Christchurch, interviewees discussed how the Severe Acute Respiratory Syndrome (SARS) epidemic, and all of the associated emergency response planning that occurred in New Zealand, also influenced their earthquake preparedness and response planning. In San Francisco, several respondents referenced Hurricane Katrina in 2005 and the September 11, 2001 terrorist attacks as having shaped their professional activities.

Respondents in all of the cities also referred to change that occurred in their organizations and earthquake preparedness activities, as a result of lessons learned from catastrophic events in other countries. Disasters mentioned several times in interviews included the 1988 Armenia earthquake, 1995 Kobe (Japan) earthquake, 2004 Indian Ocean earthquake and tsunami, 2008 Sichuan (China) earthquake, 2009 L'Aquila (Italy) earthquake, 2010 Haiti earthquake, 2010 Chile earthquake, and 2011 Tōhoku (Japan) earthquake. This finding supports previous literature that suggests that a disaster may lead to a renewed focus on preparedness efforts (see Birkmann and von Teichman 2010; Dikau and Weichselgartner 2005; Pelling 2006).

23.4.1.2 New Risk Reduction-Oriented Legislation and Regulations

Birkland (1996) refers to disasters as "focusing events"—sudden calamities that cause both citizens and policymakers to pay more attention to a public problem and to press for solutions. Such events can lead to the development of new programs as we mentioned above, but can also lead to policy revision and innovation as well as new or enhanced regulatory enforcement. Participants in this study similarly described how the earthquakes and other disasters that caused so much destruction in their home cities and surrounding regions led to stricter regulations and, in some cases, more oversight authority for the organizations where they work:

> Because the 1999 earthquake was a turning point for all of our country, and after that, the buildings... When a person wants to build new buildings, he or she has to obey the regulations.—Government Respondent, Antakya

In Christchurch, some interviewees indicated that they were engaged in earthquake risk assessment activities long before the September 4, 2010 and February 22, 2011 events, which caused so much damage. However, after those earthquakes and associated aftershocks, the work that they were doing became required by law. A Christchurch respondent described this shift:

> Part of that work has also been looking at what's at risk—people, property, infrastructure, things like that. I'm not sure whether you're aware, but we've carried out an engineering lifelines study. That's been going on in Christchurch now for—I think it started about 15

> years ago. It was a little bit like what's being done in California and other places as well, the local authorities, the university, local consultants all getting together pretty much on a voluntary basis and initially preparing a report on the risk to infrastructure from a whole range of natural hazards, earthquake obviously being a significant one.
>
> Interviewer: Were they mandated to do that?
>
> No, but we are now, under the new Civil Defense Emergency Management legislation, there is a requirement that structure providers carry out their works. It's obviously based on stuff that came out in the United States earlier, but it's caught on really well here.—Government Respondent, Christchurch

This respondent from New Zealand and other participants in the study demonstrate the two-way street of practice and regulation. Following a disaster, regulations may stimulate changes in risk reduction efforts, while conversely, success in risk reduction practices may also stimulate the development of new regulations.

23.4.1.3 Available Local/State/National Funding

Disasters can also lead to an influx of funding[8] dedicated to recovery, reconstruction, and risk reduction activities. In some of the cities that the team visited, this funding was used either to create new programs or to fund existing initiatives more fully:

> The Gujarat earthquake was the watershed development. That was when disasters of such magnitude were taken into the planning and budgeting of state governments and the government equally funded all such initiatives.—Government Respondent, Delhi
>
> This was two days after the [September 4, 2010] earthquake. The Deputy Prime Minister rang me. "How many businesses are affected?" So I said, "2,500." He said, "How much money do you want?" I said, "$15 million." He said, "Okay." So within two days, we had money being pumped into companies, subsidies going to employees ... After the [February 22, 2011] earthquake, he rang me again. The employment support subsidy, they made it much more generous because it was a much bigger disaster, February 22nd, it was ten times September 4th ... We pumped $200 million into companies over that six-week period. This was an enormous initiative.—Business Respondent, Christchurch

In the two wealthiest cities in the study sample—San Francisco and Christchurch—funding for disaster preparedness and mitigation actions was available on an annual basis, usually through competitive grant competitions sponsored by government agencies or the private sector. This funding, if received, was then used to develop new programs or activities within the respondents' organization, as described by this respondent:

> We applied for grants. I've got a grant in right now that we're hoping will come through. And the nice thing about FEMA [the U.S. Federal Emergency Management Agency] is that they're really starting to get it and to prioritize the disaster funding for people with disabilities. Each year they seem to have a theme with the grant cycles, and this year it's about including people with disabilities in the planning process. So hopefully this most

[8]Funding amounts referenced in this and other sections of the report refer to the respondents' national currency (e.g., New Zealand dollars, US dollars, etc.).

recent grant, it's for about $55,000, will come through. If it does, it will allow us to expand our installation activities into an affordable housing project for formerly homeless and disabled veterans.—Government Respondent, San Francisco

In some cases, the grants that these respondents received have been substantial: for instance, in San Francisco, one respondent described a $1 million, ten-year, city-funded effort to translate engineering standards into public policy; another respondent was responsible for a $500,000 grant dedicated to promoting emergency preparedness and to purchasing emergency supplies for schools in the San Francisco Unified School District. But even those with small budgets managed to stretch their funds, in order to develop new programmatic efforts. One faith-based leader elaborated on what he did with the first grant he received:

I think it was about $1,500. And what I was able to do was train my staff, and that's so very important, because I want them to come to work... But if they're worried about being prepared, worried about where their families are at, they may not come to work and you can't get anything done. So we bought wind-up flashlights for them, we gave them disaster kits to put in their car—so no matter where they are, [if a disaster happens], we made sure that they had all their numbers together, made sure that they knew where everybody was going to be. So doing all those different things, we were able to come up with a real constructive plan.—Grassroots Respondent, San Francisco

This suggests that while access to large sums of money does not automatically translate into successful DRR or SRR endeavors, it can indeed play an important role in spurring preparedness initiatives and providing support for developing community and/or organizational capabilities to enact successful risk reduction strategies.

23.4.1.4 External Support and International Guidance

Funding for preparedness and risk reduction activities came not only from local and national governments; several respondents from cities in developing countries indicated that they had also received vital monetary support from organizations such as the United Nations Development Programme (UNDP), the United Nations Children's Fund (UNICEF), and the World Bank. This funding allowed the practitioners to create new disaster preparedness and mitigation programs and to implement hazard education efforts. For instance, an interview participant from Lima described the benefit of UNICEF support:

We have an agreement with UNICEF, which is an agreement between the government and UNICEF. Before, those agreements focused on many themes, such as multiculturalism or basic education. But they never had the theme of risk management. So we created an opportunity in that area, so that with UNICEF we started working for the first time, as part of that cooperative plan that UNICEF has every year, to introduce risk management education, which was launched with some extremely low budgets. But since 2009, 2010, 2011, we now have a larger budget for educational materials on the topic of risk management, and it's been growing. Why? Because UNICEF understood that if they're going to address the situation of boys and girls, and the rights that boys and girls have, one of the rights they have is safety, the right to go to school, the right to a secure environment, and also the guarantee of the right to keep studying and going to school, even in an event of an emergency.—Education Respondent, Lima

In addition to providing monetary support, the abovenamed international organizations and other external nongovernmental organizations assisted community leaders with creating programs concentrated on capacity development. Save the Children, Plan International, and other major international organizations were especially focused on helping communities to create sustainable, culturally relevant risk reduction programs.

In terms of international guidance, numerous respondents in developing countries cited the 2005 Hyogo Framework for Action as critical to conveying the need to undertake new risk reduction actions in the cities where they worked. A respondent from Lima explained:

> We began to learn that there was this Hyogo Framework for Action agreement that had been signed in 2005. And [it said] that by 2015, the risk of disaster has to be reduced, minimized, in the areas of human, economic, and social losses. Never before could we have prioritized the reduction of risk without the Hyogo Framework for Action because we didn't have the knowledge, no one had given us the skills, no one had trained us. No one had told us that reduction of risk is a priority topic, just like nutrition and health. Because imagine when there's an earthquake, how much money does the central government lose? It loses a lot of money. But they never had the topic of prevention in mind. Because if you start to provide skills and train and prevent, the problem will be less severe as will the risk because at least people will be trained and things won't be as difficult as usual.—Grassroots Respondent, Lima

These responses from interviewees indicate that international support—whether monetary, educational, or otherwise—may also have played an important role in the growth of SRR strategy implementation.

23.4.1.5 Hazard Vulnerability Concerns, the Making of Mitigation Champions, and Strong Leadership

A final theme that emerged in the data provides insight into why new DRR and SRR programs were created in the cities we studied. In all cases, there were three vital components preceding the development of a new program or initiative: (1) individual level awareness and concern regarding earthquake risk, (2) action on the part of the concern, and (3) external support for the efforts of the concerned.

We found that the first step in the making of so-called mitigation champions was that an individual or small group of individuals became concerned about the vulnerability of the community to earthquake hazards. Yet concern alone was not enough for program implementation. The same individual(s) acted on their concern and became an actual mitigation champion. But what was it exactly that motivated citizens to become concerned to the point of transforming into mitigation champions? It turns out the root of their concern varied widely, including personal and professional experiences:

> Disaster planning has always been close to my heart, because I have been through so many earthquake disasters. I actually lived in the epicenter zone for the 1971 earthquake in Los Angeles. I was living here in San Francisco for '89 and responded [to Loma Prieta] as a city employee as well. So earthquake awareness and preparedness is very close to my heart.—Government Respondent, San Francisco

> We were planning for SARS and what that might mean, and what we recognized was that in any natural event, be it an infectious event like pandemic flu or whatever, or a natural disaster, the hospital in Christchurch only has about 500 beds, but in a big event like an earthquake, we realized we might need several thousand beds.—Health Care Respondent, Christchurch

In addition to actual first-hand experiences, sometimes individuals became concerned because their awareness was raised regarding earthquake risk. This then translated into a respondents' deepened commitment to disaster preparedness and mitigation:

> Turkey is a very disaster-prone country. Every few years, a flood occurs in certain places and really causes serious damage. And every other few decades, we have serious mass destruction disasters through earthquakes. So once we understood that in this country earthquakes and floods and natural disasters have a serious history, we sat down again and said, "Okay, we are going to form a voluntary search and rescue team that will work without anything in return, fully voluntary, as a charity, and we want to do this for mountaineering accidents, wilderness accidents, outdoor sports accidents. But then we said, "Why keep it only limited to the wilderness?" If we are going to organize such a voluntary team, let's use it for floods and big earthquakes and like that, if needs come up.—Grassroots Respondent, Istanbul

Despite the variation in experiences that led respondents to act as mitigation champions, the underlying commonality was that the practitioners in this study often progressed from an initial place of individual- or community-level concern to one of action. In other words, awareness and concern were prerequisites for action. Indeed, another respondent described how seeing the number of dead in consecutive natural disasters—and knowing that the losses could have been averted—prompted him and his colleagues to return to Delhi and increase their efforts to provide safe and sustainable housing that could withstand a disaster:

> We are trained as architects, planners, as you know. Our initial experience was in carrying out humanitarian activities, response and relief activities, for people who were getting affected by these disasters. There was the earthquake in 1999 in Turkey. Similar such incidents included a cyclone in Gujarat in 1998. So we were like volunteers, we were just going through to distribute relief materials, to help out other big organizations who were actually carrying out such operations. And when we were doing that, we realized that so many lives were lost because of simple things that people could have done. And a lot of that has to do with the level of technical knowledge. Somewhere there was this burning flame . . . I think our mission should be to bridge that knowledge gap. So we have at one end the best knowledge institutions in the world, but at the other end people are dying because they don't know simple things to do that can save their lives. So that was one major spark that led to the creation of this organization.—Grassroots Respondent, Delhi

Still, it was essential that concerned actors in vulnerable communities had additional support for their action. Support by an authoritative leader was the final core commonality in the development of SRR programs. That is, SRR programs were established in cities where mitigation champions' ideas and efforts were supported by someone in a leadership position with the authority to help enact change. With this support, initiatives to produce tangible, lasting change could be implemented and funded. This point was evident in the narrative offered by a respondent from Christchurch:

> On September 6th, the Monday after the September 4th earthquake, I went on national radio. The city was closed down. The central business district was closed off. I said, "We have to do something to protect the businesses in our community, because there is no cash flow . . . " The Deputy Prime Minister heard me on the radio and he rang me and he said, "I've just heard you on the radio. I agree with every word that you've said. What do you want us to do?"—Business Respondent, Christchurch

From these experiences it seems that while organizational or institutional action is vitally important, so too are the *individuals within* the organizations or institutions who have relevant experiences and skills and will take initiative in moving forward with collaborative SRR efforts. Just what do these SRR efforts look like? Below we examine the type of SRR strategies respondents frequently discussed using in their communities.

23.4.2 *Risk Reduction Strategies*

In addition to expanding our understanding of *why* communities initiate or implement SRR strategies, we relied on information gathered from the open-ended interviews and the surveys to determine the types of programs, initiatives, and communication channels that interviewees identified as integral in their local DRR and SRR efforts.

23.4.2.1 Programs and Initiatives

Discussions of strategies for reducing risk in the event of an earthquake are often dominated by structural engineering measures (see Kenny 2009; Mulargia and Geller 2003; Thomalla et al. 2006). While structural mitigation efforts are indeed invaluable in terms of SRR, they by no means are the only way in which these initiatives take shape in a community. Respondents in this study expressed a wide variety of efforts aimed at enhancing SRR practices, including, but not limited to, structural mitigation efforts. Table 23.3[9] displays the risk reduction practices that

[9]Three additional features of the table are worth noting: First, when respondents from different target cities reported engaging in the same general activity, then a generic name was used for that activity (e.g., building retrofit programs). Second, when respondents reported engaging in locally developed programs, then the formal names of those programs were included (e.g., "Map Your Block" preparedness program). Third, when respondents reported that they and their organizations were engaged in regional or national programs and initiatives, then the formal names of those

Table 23.3 Earthquake mitigation and preparedness programs

Mitigation	Preparedness
Structural mitigation (building retrofit programs; structural strengthening program; unreinforced masonry removal)	Map Your Block, neighborhood resource mapping programs
Nonstructural mitigation (fastening contents in buildings)	Neighborhood Empowerment Network
Enhanced building design	Staff preparedness training
Identification of collapse hazards	CPR training; emergency medical care training; psychological first aid training
Identification of high-priority buildings for retrofitting	Public risk education; disaster awareness trainings
Structural assessment program	School-based hazard education
Lifeline protection	School-based emergency drills
Earthquake and Megacities Initiative	Public emergency drills
Environmental conservation programs	Distribution of emergency kits and emergency supplies
Slope stability efforts	Neighborhood capacity, skill, resource mapping
Hazard risk mapping	Disaster preparedness teams
Investment planning for mitigation actions	Community organizing for disaster preparedness
Disaster mitigation awareness programs (promoting an understanding of the importance of mitigation)	Safe School Initiatives (school disaster management and evacuation plans)
Population relocation programs (moving persons and businesses out of vulnerable structures and areas)	Emergency training exercises
Micro-zonation maps	Business continuity planning; tabletop exercises and disaster simulations
Planning regulations or policies that incentivize mitigating actions	Disaster volunteer recruitment and training programs
Improved building codes	Parent–child reunification programs
Inventory contents of buildings and homes for insurance purposes, should a disaster occur	First 72 h: Are You Prepared? program

respondents described within their community. The table should be read *vertically*. Each cell in the Mitigation or Preparedness column lists a program or activity described by respondents during the in-depth interviews. There is no relationship across rows in the table.

Table 23.3 illustrates the wide variety of SRR strategies that are being implemented across the 11 communities included in this study. The mitigation and

programs and initiatives were included (e.g., "Earthquake and Megacities Initiative" mitigation program), as they clearly influenced local action.

preparedness efforts outlined above were designed to assist with risk assessment, to encourage physical protection, or to increase response and recovery capacity through pre-event planning.

The types of programs described by the respondents were aimed at assisting and/or engaging one or more levels of social organization (Petal 2007), ranging from:

- The *micro*-level, which includes individuals and households
- The *meso*-level, which comprises schools, hospitals, businesses, local government, community- and faith-based organizations, neighborhoods, and communities
- The *macro*-level, which covers regional, national, and international policymaking bodies

Table 23.4, which organizes the programs by these levels, should be read *vertically*. Each cell in the micro-, meso-, or macro-level column lists a program target described by respondents during the in-depth interviews. There is no relationship across rows in the table.

We also learned that these risk reduction practices that develop at a macro-, meso-, or micro-level may work their way down—from an international or national program to a local level—or up, from a local program to a national or international best practice. In addition, the diversity in approaches may indicate that some strategies might require access to many resources (i.e., structural mitigation), while others may be put into place with little to no need for specialized resources (i.e., inventorying building contents). In fact, the practitioners whom we interviewed often had limited budgets, but still managed to serve many residents, groups, and organizations in their local communities.

One of the primary ways in which respondents were able to achieve successful programmatic outcomes was through partnering with and/or learning from trusted organizations and individuals, both within their cities and from outside regions. Understanding the wide array of individuals, groups, and institutions involved, as well as the variety of groups targeted in preparedness programming, can lead to a clearer sense of the collaborative abilities and responsibilities related to DRR and SRR actions and programs. Effective collaboration (both externally and internally) with representatives from the government, business, health care, education, and grassroots sectors could mean that SRR practices will be able to reach a significantly larger portion of people within a given community.

23.4.2.2 Technical Resources

In order to understand the risk that cities faced, practitioners in this study relied upon a variety of technical resources amassed from many sources. The number and types of resources used varied by sector and by city. The qualitative data regarding technical resources suggest three centrally important patterns. First, the overwhelming majority of practitioners interviewed in this study, across all cities

Table 23.4 Earthquake mitigation and preparedness program targets

Micro level	Meso level	Macro level
Children	Elementary and secondary schools	Policymakers: advocacy to move toward a "culture of disaster prevention"
Elderly	Colleges and universities	Policymakers: change building code standards
Women; pregnant women	Hospitals	Policymakers: change land-use planning regulations
Adults with disabilities; children with disabilities	Elder care facilities	Policymakers: make preparedness guidelines more socially inclusive
Low-income individuals	Businesses	Policymakers: include the public in mitigation planning decisions
Drug-addicted individuals	Government	
Homeless	Churches, mosques, and other faith-based organizations	
Incarcerated populations	Prisons and jails	
War veterans	Nonprofits	
Renters	Media	
Homeowners	Lifelines	
Small business owners		
Nonprofit volunteers and staff		
Faith-based leaders; faith-based congregations		
School administrators; teachers		
Government workers		
Health-care staff, including doctors, nurses, home health aides, emergency medical technicians, and ambulance drivers		

and sectors, acquired technical information regarding earthquake risk (including information on hazard exposure and physical and social vulnerability) from secondary sources. Practitioners found this information online or in technical reports that their organizations (or other organizations in their city) commissioned.

Second, only a small number (i.e., <10) of more technically sophisticated practitioners collected primary data from various sources (e.g., Hazus™, US Geological Survey, city-level building inventories, etc.), in order to generate their own "risk profiles" for their organizations and the cities that they serve. This was a technically difficult and time-consuming process to undertake. Moreover, many of the respondents lack access to even the most basic information that they would need to be able to generate such a risk profile.

The most consistent finding across all 11 cities was that respondents lacked one central tool or piece of technology that could provide a comprehensive portrait of earthquake risk in their cities. Instead, practitioners attempted to draw together technical resources from different sources to get some sense of the potential impacts of an earthquake on their city's lifelines, critical infrastructure, and population groups. For example, one participant described difficulty in developing a contingency plan as the health department lacked the technical resources to determine the effects of a disaster across a variety of sectors in the community. The information was not previously compiled or easily accessible, and as such, the participant had to collect all of this information in a piecemeal fashion:

> All of the technical data that [is] in this contingency plan, [I] get together with every single department that is related to this plan, such as the number of schools they have in Padang, I will go to the education department, and find out about the number of people, I go to the statistics department. And I go to the health department to find out about the number of health facilities. Making the scenarios involves working together with other departments.—Health Care Respondent, Padang City

This suggests that in many cities, despite engaging in a variety of SRR strategies that in part increase information sharing, a major gap still exists in terms of a central knowledge source for understanding earthquake risks and risk-reducing strategies within a particular local context.

23.4.2.3 Communication Channels and Technologies

We asked respondents a series of closed-ended survey questions about preferences for receiving and sharing information for professional purposes, with the goal of uncovering which communication channels respondents in the target cities find to be the most useful. The final survey included 13 questions that asked respondents to specify whether the following items or activities were of "low," "medium," or "high" usefulness for professional purposes; "not available"; or "available, but not useful":

- Newspapers
- Radio
- Television
- Social media (such as Facebook, Twitter)
- Scientific publications (such as books, journal articles, trade magazines)
- Email
- Telephone
- Talking in person with community members
- Talking in person with scientific experts
- General news websites
- Government websites
- Earthquake- or disaster-focused websites
- Earthquake hazard maps

Table 23.5 Average usefulness of communication channels

Source	Average usefulness	Standard deviation
Talking in person with community members	3.51	0.65
Earthquake- or disaster-focused websites	3.46	0.76
Talking in person with scientific experts	3.43	0.75
Earthquake hazard maps	3.36	0.85
Television	3.36	0.8
Email	3.34	0.8
Scientific publications (such as books, journal articles, etc.)	3.22	0.78
General news websites	3.19	0.79
Telephone	3.09	0.91
Newspapers	3.09	0.83
Government websites	2.96	0.81
Radio	2.92	0.95
Social media (such as Facebook, Twitter, etc.)	2.87	0.87

In our analyses, we summarized response counts and percentages across all 119 survey respondents for each of these 13 communication-related items. Few respondents rated any of the communication channels as "not useful" or "not available." That result shows that earthquake safety practitioners in various geographic regions are likely reachable through a variety of communication channels, and even the least useful communication channels are valuable to some extent. In addition, this suggests that perhaps certain communication outlets, such as government websites or social media, might be more effective if evaluated and revised.

To summarize and more fully compare the usefulness of the 13 communication channels that the survey assessed, the team created a rank-order scale and assigned the following numbers to the relevant survey responses: "not useful" (1), "low" (2), "medium" (3), and "high" (4).

Table 23.5 shows the average score and standard deviation on the scale for each communication item. Higher scores indicate that the resource is, on average, perceived as more useful than the other sources of professional information. Because the scale only ranges from 1 to 4, the averages appear rather tightly clustered. However, they indicate important differences in the perceived usefulness of different communication channels. For example, talking with community members, perceived as the most useful, has an average score of 3.51, which falls between medium- and high-perceived usefulness on the scale. In comparison, the 2.87 average score for social media, perceived as the least useful from this group, indicates that respondents perceive it as having between low and medium usefulness.

23.4.2.4 Current Communication and Outreach Activities

The practitioners who participated in this study communicated with and conducted outreach to the various community members and groups that they served in numerous ways, including through running disaster simulations, workshops, trainings, educational classes, and more. The interviews revealed a wide range of activities that the practitioners are engaged in, as well as underscored the ongoing need for new and innovative ways to communicate with a variety of "publics" that these practitioners serve.

Across the spectrum of city and sector, responses demonstrate the importance of having face-to-face interactions with others—through collaborative projects, conferences and workshops, giving and attending presentations, and engaging the broader community which they serve. Given the emphasis on face-to-face interactions at a time when communications increasingly occur across other forms of technology, the value of this approach in successfully employing SRR strategies may be of significant interest for the future direction of risk reduction research.

23.4.3 Limitations and Challenges

As with every study, there are challenges and limitations that should not be overlooked. First, a key limitation of this study is that it is not generalizable. Because this project was exploratory, the goal was not to draw a random, representative sample. Indeed, it would have been impossible to do so, as in order for a true random sample to be selected, the characteristics under study of the entire population should be known. This would have required a complete listing of all earthquake safety practitioners in all 11 cities; such lists are simply not available. While results cannot be generalized within or across communities, they also cannot be viewed as developing a "one-size-fits-all" approach to SRR strategy. Indeed, each city is unique and as stated previously, understanding and incorporating the local context into SRR planning are essential for effective action and programming. Still, it is important to consider which practices were commonly relied upon across the research sites as a way to gauge what plans and programs may be adaptable across different local contexts.

A second issue related to the importance of local context is that language and cultural barriers were unavoidable in this research. When working across multiple sites as we did in this study, it is likely that participants will apply different meaning and value systems to interview questions. Similarly, the analysis will be unable to capture all of the variation between participants. Although the research team relied on local partners and used rigorous methods when translating the research instruments, concepts may still be lost in translation between the interviewer, the translator/transcriber, and the participant.

23.5 Conclusion

While there are some forces, such as the occurrence of a disaster or earthquake, beyond our control that act as catalysts which spur SRR practices, our research also demonstrates that disaster practitioners and community leaders can also undertake SRR initiatives when able to garner sufficient internal and external economic, regulatory, and social support. Hence, these findings suggest that collaboration between the actors and institutions involved in these roles is a key component of launching a successful risk reduction campaign.

Additionally, the value of this study is that SRR methods uncovered here can be compared, contrasted, modified, and potentially transferred from one city's local risk reduction strategy toolbox to another. While we have strived to distinguish locally effective SRR strategies in each of the 11 communities we visited, in a broader sense, we have also identified the way in which "context-sensitive resilience frameworks" or risk reduction strategies might be adapted and developed in the future (Bosher and Dainty 2011, p. 7).

Here, we have attempted to uncover and synthesize information that paints a clearer picture of seismic risk reduction initiatives around the world. By identifying the underlying drivers to SRR creation, development, and implementation, as well as the SRR programs and communication strategies, we have also attempted to strengthen the foundational pillars of SRR in practice. In doing so, we have created an initiative which engages SRR practitioners directly in discussions of SRR strategy development and needs, something which is often not assessed prior to the development of scientific or technical tools. That lack of needs assessment has contributed to a gap between what decision-makers and end users say that they want from science and technology and what science and technology offer to decision-makers and end users.

The nature of this project is one that applies academic rigor to disaster risk reduction actions. As a final contribution, this chapter helps to bridge the current gap between scientific communities and policymaker and practitioner communities. Engaging in dialogue and practice assessments with key institutions and actors—such as we have done here through a collaborative, interdisciplinary endeavor—helps to build understanding across these communities and is a positive direction for future disaster and seismic risk reduction research and successful practices.

Acknowledgments This work was funded by the Global Earthquake Model. Research was conducted by teams at GeoHazards International and the Center for Disaster and Risk Analysis at Colorado State University. In addition to the authors of this chapter, Veronica Cedillos, Hari Kumar, Michelle Meyer, Liesel Schilperoort, and Jennifer Tobin-Gurley contributed to research design, data collection, and data analysis efforts.

References

Ambert, A.-M., Adler, P. A., Adler, P., & Detzner, D. (1995). Understanding and evaluating qualitative research. *Journal of Marriage and the Family, 57*(4), 879–893.

Auld, H. (2008). Disaster risk reduction under current and changing climate conditions. *WMO Bulletin, 57*(2), 118–125.

Bankoff, G., Frerks, G., & Hilhorst, D. (Eds.). (2004). *Mapping vulnerability: Disasters, development, and people*. London: Earthscan.

Benson, C., & Twigg, J. (2004). *Measuring mitigation: Methodologies for assessing natural hazard risks and the net benefits of mitigation—A scoping study*. Phase 1 Report. ProVention Consortium Secretariat, Geneva.

Berman, R., & Tyyska, V. (2011). A critical reflection on the use of translators/interpreters in a qualitative cross-language research project. *International Journal of Qualitative Methods, 10*(1), 178–190.

Birkland, T. (1996). Natural disasters as focusing events: Policy communities and political response. *International Journal of Mass Emergencies and Disasters, 14*, 221–243.

Birkmann, J. & von Teichman, K. (2010). Integrating disaster risk reduction and climate change adapation: *Key challenges–scales, knowledge, and norms. Sustainability Science, 5*(2), 171–184.

Blaikie, P. T., Cannon, I. D., & Wisner, B. (1994). *At risk: Natural hazards, people's vulnerability, and disasters*. London: Routledge.

Bosher, L., & Dainty, A. (2011). Disaster risk reduction and 'built-in' resilience: Towards overarching principles for construction practice. *Disasters, 35*(1), 1–18.

Centre for Research on the Epidemiology of Disasters. (2015). EM-DAT The International Disaster Database. About. Accessed at http://www.emdat.be/about on 25 June 2015.

Crompton, R., & McAneney, J. (2008). The cost of natural disasters in Australia: The case for disaster risk reduction. *Australian Journal of Emergency Management, 23*(4), 43–46.

Dikau, R., & Weichselgartner, J. (2005). *Der unruhige Planet: Der mensch und die maturgewalten*. Darmstadt: Wissenschaftliche Buchgesellschaft.

Dilley, M., Chen, R. S., Deichmann, U., Lerner-Lam, A. L., Arnold, M., Agwe, J., Buys, P., Kjekstad, O., Lyon, B., & Yetman, G. (2005). *Natural Disaster Hotspots: A Global Risk Analysis*. Disaster Risk Management Series, Issue No. 5. The World Bank, Washington, D.C.

Dowrick, D. (2003). *Earthquake risk reduction*. West Sussex, UK: Wiley.

Enarson, E., & Morrow, B. (1998). *The gendered terrain of disaster*. New York: Praeger.

Fowler, F. (2009). *Survey research methods* (4th ed.). Thousand Oaks, CA: Sage.

Guha-Sapir, D., Santos, I., & Borde, A. (2013). *The economic impacts of natural disasters*. Oxford: Oxford University Press.

Haas, J. E., Kates, R. W., & Bowden, M. J. (Eds.). (1977). *Reconstruction following disaster*. Cambridge, MA: MIT.

Hao, X., Lai, A., & Norton, E. (n.d.). *Implementing a disaster risk reduction framework in China: Addressing the 'top-down culture' to reduce disaster related risk*. Working Paper. George Warren Brown School of Social Work, Washington University, St. Louis, Missouri.

Hewitt, K. (1983). The idea of calamity in a technocratic age. In K. Hewitt (Ed.), *Interpretations of calamity: From the viewpoint of human ecology*. London: Allen and Unwin.

International Federation of Red Cross Red Crescent Societies (IFRC). (2003). *World disasters report*. Geneva: IFRC.

Kenny, C. (2009). *Why do people die in earthquakes?: The costs, benefits and institutions of disaster risk reduction in developing countries*. World Bank Policy Research Working Paper. The World Bank Sustainable Development Network, Washington, DC.

Marshall, C., & Rossman, G. (2011). *Designing qualitative research* (5th ed.). Thousand Oaks, CA: Sage.

Marshall, M. (1996). Sampling for qualitative research. *Family Practice, 13*(6), 522–525.

McEntire, D. (2001). Triggering agents, vulnerabilities and disaster reduction: Towards a holistic paradigm. *Disaster Prevention and Management: An International Journal, 10*(3), 189–196.

McGorry, S. (2000). Measurement in a cross-cultural environment: Survey translation issues. *Qualitative Market Research: An International Journal, 3*(2), 74–81.

Mitchell, T. (2003). *An operational framework for mainstreaming disaster risk reduction*. Disaster Studies Working Paper 8. Benfield Hazard Research Centre, University College of London, London.

Morrow, B. H. (1999). Identifying and mapping community vulnerability. *Disasters, 23*, 1–18.

Mulargia, F., & Geller, R. (Eds.). (2003). *Earthquake science and seismic risk reduction*. Dordrecht: Kluwer Academic.

Oliver-Smith, A. (Ed.). (1986). *Natural disasters and cultural responses*. Studies in Third World Societies No. 36. College of William and Mary, Williamsburg, VA.

Pelling, M. (2005). *International data on disaster risk* (pp. 172–181). World Disasters Report 2005. Geneva: IFRC.

Pelling, M. (2006). Incentives for reducing risk. A reflection on key themes, issues and ideas on risk reduction raised at the 2006 ProVention Forum. Provention Consortium, Geneva.

Petal, M. (2007). Disaster risk reduction education: Material development, organization, and evaluation. *Regional Development Dialogue Journal, 28*(2), 1–25.

Phillips, B. D. (2014). *Qualitative disaster research: Understanding qualitative research*. New York: Oxford University Press.

Pielke, R. A., Jr. & Pielke, R. A., Sr. (1997). Hurricanes: Their Nature and Impacts on Society. New York: John Wiley and Sons.

Ragin, C., Nagel, J., & White, P. (2004). *Workshop on scientific foundations of qualitative research*. Washington, DC: The National Science Foundation.

Rottach, P. (2010). *Background and components of disaster risk reduction*. Berlin: ACT Alliance (ACT), Diakonie Katastrophenhilfe.

Rubin, H., & Rubin, I. (2012). *Qualitative interviewing: The art of hearing data* (3rd ed.). Thousand Oaks, CA: Sage.

Saldaña, J. (2009). *The coding manual for qualitative researchers*. Los Angeles, CA: Sage.

Schipper, L., & Pelling, M. (2006). Disaster risk, climate change and international development: Scope for, and challenges to, integration. *Disasters, 30*(1), 19–38.

Smith, K. (2013). *Environmental hazards: Assessing risk and reducing disaster*. London: Routledge.

Solomon, S., Qin, D., Manning, M., Chen, Z., Marquis, M., Averyt, K. B., Tignor, M. & Miller, H. L. (Eds.). (2007). Contribution of working group I to the Fourth Assessment Report of the Intergovernmental Panel on Climate Change. IPCC 2007. Cambridge, UK: Cambridge University Press.

Stallings, R. (2002). Methods of disaster research: Unique or not? In R. Stallings (Ed.), *Methods of disaster research* (pp. 21–44). Philadelphia: Xlibris.

Thomalla, F., Downing, T., Spanger-Siegfried, E., Han, G., & RockstrÖm, J. (2006). Reducing hazard vulnerability: Towards a common approach between disaster risk reduction and climate adaptation. *Disasters, 30*(1), 39–48.

Tierney, K., & Bruneau, M. (2007). Conceptualizing and measuring resilience: A key to disaster loss reduction. *TR News, 250*, 14–17.

Trohanis, Z., Dastur, A., Xu, T., & Cira, D. (2010). *Disaster risk management in East Asia and the Pacific*. Working Paper Series No. 13, EAP DRM Knowledge Notes. Washington, DC: World Bank.

Twigg, J. (2004). Good practice review. Disaster risk reduction. Mitigation and preparedness in development and emergency programming, ODI/HPN.

United Nations International Strategy for Disaster Risk Reduction. (2009). UNISDR Terminology on Disaster Risk Reduction, Geneva: UNISDR.

United Nations International Strategy for Disaster Risk Reduction. (2015). Accessed at www.unisdr.org on 24 June 2015.

Wamsler, C. (2006). Mainstreaming risk reduction in urban planning and housing: A challenge for international aid organisations. *Disasters, 30*(2), 151–177.

Chapter 24
Community Resilience: The Role of the Built Environment

Therese P. McAllister

Abstract Buildings and infrastructure systems play a key role in communities by supporting social needs and institutions, including housing, business, government, industry, and other vital services. The concept of community resilience addresses the way that communities prepare for and recover from disruptive events. This chapter focuses on the role that buildings and infrastructure systems play in developing community resilience. The needs of citizens and institutions in a community, including public safety, define the performance requirements for buildings and infrastructure systems. However, current practice does not adequately address interdependencies between buildings and infrastructure systems or the role they play in recovery following a hazard event.

Recent examples of how the built environment performs during hazard events, such as Hurricane Katrina or Superstorm Sandy, are used to illustrate the uneven performance and interdependence of infrastructure systems, as well as cascading events, that dramatically affect recovery of the community. This chapter presents some recommendations on how to address these deficiencies by improving guidance, standards, and tools that supports community resilience planning. The research plan starts with the development of guidance documents, with stakeholder input across multiple disciplines, to identify best practices for achieving community resilience as well as research needs. The longer-term research includes development of performance goals and metrics for buildings and infrastructure systems, development of modeling tools at a community systems level, and a scientific foundation for developing improved standards, codes, guidelines, and tools to enhance community resilience.

T.P. McAllister (✉)
National Institute of Standards and Technology, Gaithersburg, MD, USA
e-mail: therese.mcallister@nist.gov

P. Gardoni, J.M. LaFave (eds.), *Multi-hazard Approaches to Civil Infrastructure Engineering*, DOI 10.1007/978-3-319-29713-2_24

24.1 Introduction

Natural, technological, and man-made hazards cause significant damage to communities, particularly to buildings and infrastructure systems. Buildings[1] and infrastructure systems,[2] also referred to as the built environment, play critical roles in community resilience. Resilience of the built environment depends upon the capacity of each facility and infrastructure system to maintain acceptable levels of functionality during and after a disruptive event and to recover full functionality within a specified period of time. Other aspects of a resilient community—security, protection, emergency response, business continuity, and social issues related to human health, safety, and general welfare—are also important and inform performance goals for the built environment. Despite substantial progress in science and technology toward improved performance of the built environment during disasters, natural and man-made hazards in the United States are responsible for loss of life, disruption of commerce and financial networks, damaged property, and loss of business continuity and essential services.

This chapter examines the current performance of the built environment in communities—primarily buildings and infrastructure systems—and how the implementation of resilience concepts and methods can improve that performance.

24.2 What Is Resilience?

There are many definitions for the term "resilience" depending on the intending scope or application (NAP 2012). With regard to research at NIST and activities at other federal agencies, resilience is broadly defined as "the ability to adapt to changing conditions and withstand and rapidly recover from disruption due to emergencies" (PPD-8 2011) and "the ability to prepare for and adapt to changing conditions and withstand and recover rapidly from disruptions" (PPD-21 2013). The term "community" is also defined in a variety of ways. When considering resilience of the built environment, and supported social functions, a community is an area with defined boundaries under the jurisdiction of a local government, such as a town, city, or county (NIST 2015).

Community resilience depends on the capacity of the built environment to maintain acceptable levels of functionality during and after disruptive events and to recover full functionality within a reasonable period of time specified by the

[1]The term building includes all the systems necessary for its functional operation, including architectural, structural, life safety, mechanical, electrical, plumbing, security, communication, and IT systems.

[2]The terms infrastructure systems include the physical plants, transmission, and distribution networks for transportation facilities (e.g., roads, bridges, airports, tunnels, ports, rail) and utilities (e.g., electric power, water and wastewater, fuels, and communication).

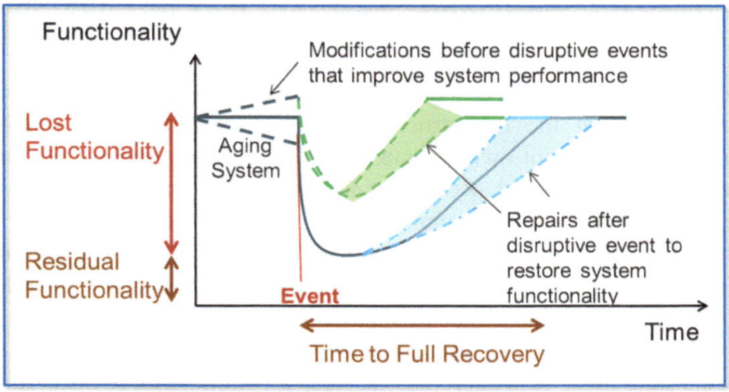

Fig. 24.1 Resilience can be simply expressed in terms of system functionality and the time to recovery of functionality following a disruptive hazard event (adapted from Bruneau et al. (2003) and McDaniels et al. (2008))

community, based on the role of each facility or system within the community. Functionality is the ability of the building or system to support its intended purpose, such as providing healthcare or delivering potable water. Loss of functionality may occur suddenly during a hazard event, but recovery may take anywhere from hours to years.

Figure 24.1 illustrates the concept of functionality versus recovery time. Loss of functionality typically occurs suddenly due to physical damage to either the buildings or supporting infrastructure systems that is sustained during a discrete hazard event of minutes to days, whereas recovery of functionality may take anywhere from hours to years. There is uncertainty in the condition of the built system prior to the event, depending on the design, age, and maintenance of the system and whether any improvements have been made. The degree of lost functionality following a hazard event depends on the system capacity at the time of the event and the magnitude of the hazard event. Typically, a lesser degree of damage will reduce the time to full recovery. Thus, mitigation of hazard effects prior to the event can greatly reduce the level of damage and time and cost of recovery to full functionality. However, repair costs and recovery time after a disruptive event may have large uncertainties, depending on the degree of interdependency among building and infrastructure systems and the availability of resources after a hazard event. At present, many communities do not plan for recovery of their physical infrastructure following disruptive events, and both the time to full recovery and the accompanying costs are highly uncertain.

24.2.1 How Can Community Resilience Be Achieved?

Resilience of the built environment can be assessed at local, regional, or national scales, depending on the infrastructure systems under consideration. While a community likely does not own or manage all of the infrastructure systems within their boundaries, public and private owners and operators should coordinate in planning for the integrated performance and recovery of these systems.

Communities can reduce damage caused by hazard events and compounded by time-dependent effects of degradation and climate change (e.g., sea level rise) through coordinated actions that include developing community-level goals that guide long-term planning that addresses implementation of mitigation, response, and recovery activities. This approach provides a comprehensive basis for decision making and prioritization, where the relative costs of alternative mitigation, performance, and recovery plans are considered together. This comprehensive approach is a key concept for community resilience and can provide tangible resilience benefits to communities.

To achieve compatible levels of performance and recovery within defined groups of buildings and systems (e.g., those that support essential services), a risk-informed methodology is needed that accounts for the community performance goals for desired levels of functionality and recovery, the costs and consequences of actions (or lack of actions) before and after a disruptive event, and associated uncertainties.

24.2.2 What Is the Problem?

Many communities address hazards that may threaten their safety and well-being by reducing vulnerabilities through preparedness, mitigation, and design efforts to minimize the risk of damage and losses. However, across the nation, communities have continued to experience significant damage and losses, even those with robust code adoption and enforcement.

Current best practices, regulations, codes, and standards primarily address life safety issues for buildings, reliability of utility service during normal operation, and aspects of community preparedness and immediate response to disruptive events. These guidance documents and requirements are largely developed independently for each infrastructure system. Accordingly, the integrated performance and reliability of buildings and infrastructure systems at the community level, the interdependencies between social and physical systems, or the plan for recovery of function across the community are not addressed (Rinaldi et al. 2001; Pederson et al. 2006). For instance, delivery of emergency services depends on fuel, power, communication, and transportation systems. Identifying and understanding the dependencies between and among systems, and the potential cascading effects from damage in one system to other systems and services, provides an informed basis for setting performance goals for community response and recovery.

Model codes and standards primarily ensure life safety in buildings and other structures, and regulations address reliability of service for utilities, but these documents generally do not address resilience or dependency issues. Codes, standards, and regulations for the built environment are often developed independently through varying public or private processes. Stove-piped development can lead to variable hazard and performance criteria for buildings and infrastructure systems. For instance, many infrastructure systems are not designed for the same hazard levels or service life, leading to varying reliability across the built environment of the community. Buildings often use a 50-year service life for design (ASCE 2010), and transportation structures often use a 75-year service life (FHWA 2012).

Communities in the United States (US) may not adopt or enforce current building codes and standards or may exempt critical sections (such as seismic requirements). Many states and communities have begun to recognize the value of codes and standards and have begun adopting current versions over the last 5–10 years (ICC 2015; LSUCCC 2014). Additionally, older facilities and infrastructure systems often do not meet current design standards or may have degraded performance due to aging effects or inadequate maintenance. However, it will take decades before a substantial portion of the building stock and infrastructure are replaced given the nominal 50-year design life (or greater) of many buildings and infrastructure systems. Thus, while some immediate actions can be taken, community resilience requires long-term planning to achieve community goals.

Resilience is more than adopting and enforcing the current codes and standards. In communities that adopt and enforce the latest codes, there is still uncertainty about the expected performance of the built environment when subjected to hazard events. There are additional potential sources of uncertainty in expected performance, including hazard and performance design requirements in codes and standards that change over time, or hazards that are not addressed, such as tornados or sea level rise. Many coastal communities are now addressing sea level rise as a time-varying hazard condition. The current condition of the built environment is also a source of uncertainty in expected performance, as it depends on maintenance, improvements or retrofits, and (for some) use beyond their intended service life or original function. Such sources of uncertainty add significant challenges to assessing the performance of the existing built environment for each of the three hazard levels.

Presidential disaster declarations spanning the period from January 2000 to January 2011 (FEMA 2015) provide an indicator of the performance of the existing built environment for a variety of hazards. In the United States, a governor seeks a presidential declaration by submitting a written request to the president through the Federal Emergency Management Agency (FEMA). The governor certifies that the combined local, county, and state resources are insufficient and that the situation is beyond their recovery capabilities. Following a FEMA review of the request and the findings of a preliminary damage assessment, FEMA provides the president an analysis of the situation and a recommended course of action. FEMA considers a number of factors in developing a recommendation, such as extent of damage, impact on infrastructure and critical facilities, threats to public health and safety, level of insurance in place, available assistance from other sources, and frequency of disaster events. Between 45 and 81 declarations were made every year between 2000 and 2011 for floods, hurricanes, tornadoes, earthquakes, fire events, and severe storms. Many of the disaster declarations were based on economic recovery costs, even though the hazard intensities experienced during the events fell below current design thresholds. During this period, resilience activities largely focused on hardening critical infrastructure as well as hazard mitigation for individual facilities and emergency response activities to meet social needs (McAllister 2013). This approach has resulted in a piecemeal protection and risk reduction and typically did not significantly improve the resilience of the community.

24.3 Observed Community Performance

The risk across the United States for substantial damage due to hazard events continues to increase, due to the combined effects of urban development and population growth (NOAA 2005; NRC 2006). Much of the physical infrastructure is susceptible to natural hazards (e.g., along coastlines, in the wildland-urban interface, in tornado alley, and in earthquake-prone regions). Additionally, much of the United States infrastructure is reaching the end of its useful service life or operating in a degraded state. The American Society of Civil Engineers' (ASCE) 2013 report card for America's infrastructure evaluated aviation, bridges, dams, drinking water, energy, hazardous waste, inland waterways, levees, ports, public parks and recreation, rail, roads, schools, solid waste, transit, and wastewater systems at a national level. In their report, "A" is exceptional, "B" is good, "C" is mediocre, "D" is poor, and "F" is failing. For these 16 categories of infrastructure systems, the highest grade was a B for solid waste management and 11 systems received a D. While this deteriorated state is a cause for significant concern, it is also an opportunity to develop and implement a new paradigm—resilience—as we plan and envision our future communities.

Recent disaster events demonstrate the need for improved resilience planning in communities. Additionally, the following events significantly influenced the development of resilience concepts in the United States: Hurricane Andrew in 1992, the World Trade Center (WTC) and Pentagon terrorist attacks in 2001, Hurricane Katrina in 2005, and Superstorm Sandy in 2012. The following summaries highlight community-level impacts and the subsequent evolution of resilience concepts after each disaster event.

Hurricane Andrew, 1992. Hurricane Andrew struck Dade County on August 24, 1992, as a Category 5 hurricane. The storm caused an estimated $25 billion in damage. Approximately 49,000 homes were destroyed, and an additional 108,000 damaged (NWS 2012). The widespread structural damage from Hurricane Andrew led to improved building codes and practices in South Florida (FBC 2004). Some of the important changes included mitigation measures that are widely used today, including adoption of science-based wind provisions from a national standard (ASCE 1990), requirements for impact-resistant glazing through testing to reduce debris damage, and enforcement of positive ties for a continuous load path to resist uplift forces (Tsikoudakis 2012). These mitigation measures have greatly reduced damage and recovery times for communities following hurricane events.

WTC Attacks, 2001. On September 11, 2001, large aircraft were flown into the World Trade Center (WTC) 1 and 2 buildings and the Pentagon by terrorists. The fires following the aircraft impact caused WTC 1 and WTC 2 to collapse (NIST 2005). The WTC 7 building also collapsed due to uncontrolled fires (NIST 2008). The collapse of these buildings damaged surrounding buildings and power, communication, and water systems in lower Manhattan and interrupted financial markets, raising issues about the cascading effects of building collapse on the surrounding community. Damaged power, communication, and water systems were rebuilt to improve their redundancy and minimize the probability of system failure across the Manhattan area. However, these efforts were conducted without the benefit of standardized assessment methods or quantitative metrics for resilience.

Hurricane Katrina, 2005. Hurricane Katrina struck the Gulf Coast region on August 29, 2005. The storm surge reached 8.5 m (28 ft) at locations along the Mississippi Gulf Coast and made landfall with maximum sustained winds of 56 m/s (125 mph) (NIST 2006). Along coastal areas and in New Orleans, storm surge was the dominant cause of damage. Storm surge and wave action breached the flood protection system and approximately 75 % of New Orleans flooded. Bridges, seaports, and petrochemical facilities were damaged by wave uplift and lateral loads. Almost one million electric power distribution poles were damaged, as well as a number of high-voltage transmission towers (NIST 2006). Away from coastal areas, wind and wind-borne debris caused substantial damage in many locations where the winds were lower than the design wind speeds specified in codes and standards—suggesting that the structures did not perform as expected (NIST 2006). Two of three bridges across Lake Pontchartrain were damaged by storm surge uplift and wave impact during Hurricane Katrina. These bridges provided major transport routes to New Orleans and affected rebuilding and recovery efforts. The lack of support tools to help communities evaluate the performance of their social and built systems and to make rational planning decisions for the future clearly showed the need for future research. Communities need science-based tools that allow consideration of alternative plans and associated risks for integrated systems. The extent of damage across several states reminded the nation of the vulnerability of our communities to natural hazards and the role of the built environment in community recovery from such devastation.

After Katrina, the following reports identified the need for risk-informed strategies to help communities achieve disaster resilience, as well as standardized tools and metrics. The "Grand Challenges for Disaster Reduction" report (OSTP 2008) identified needs for (a) predictive technologies and mitigation strategies to improve the performance of buildings and infrastructure systems and (b) standard methods to assess the resilience of buildings and infrastructure. The National Infrastructure Protection Plan (DHS 2009) identified needs for (a) analytical tools to quantify interdependencies and cascading consequences across infrastructure sectors, (b) decision-support systems, and (c) rapid mitigation and recovery technologies.

The National Academies "Disaster Resilience: A National Imperative" report (NAP 2012) recommended developing a risk-informed management strategy that addresses structural and nonstructural measures. Most recently, FEMA issued Comprehensive Preparedness Guidance (CPG) 201 (DHS 2012) to help communities conduct a risk assessment process that identifies hazards, vulnerabilities, and consequences to provide a basis for assessing preparedness, mitigation, response, and recovery plans.

Superstorm Sandy, 2012. Superstorm Sandy was one of the largest hurricanes recorded in the Atlantic Ocean, with a radius of maximum winds over 100 nautical miles (185 km, NHC 2013). Following its landfall in New Jersey, a surge level of 12.6 ft (3.8 m) was recorded at King's Point at Long Island Sound and extended across Long Island and Manhattan to New Jersey. This region had not previously experienced significant coastal inundation. Three key issues were identified for essential facilities affected by the storm surge during Superstorm Sandy (McAllister 2014). Hospitals, wastewater treatment plants, transit facilities, and data centers experienced significant damage and durations of recovery due to loss of equipment and utilities as well as loss of public power, communication, and transportation services. Many of these facilities are recovering in stages, but full recovery may take years for some of these facilities as they require unique equipment with long lead times. The loss of these community services illustrates the need to consider higher levels of performance and/or improved recovery rates to support community recovery.

Critical infrastructure in the region, such as bridges or emergency facilities, was hardened against blast and other hazards at a number of discrete locations after the WTC Event in 2001. However, when Hurricane Sandy occurred in 2012, many buildings and infrastructure systems in the New York and New Jersey metropolitan area were damaged that had not been hardened. Thus, while hardened facilities were functional, many other buildings and infrastructure systems were not. Further, mitigation activities for one hazard may not improve performance for another hazard. The communities in the area had isolated functional facilities, but the community was not resilient.

The local and national response to these events recognized the need for not only rapid recovery, but recovery that led to improved performance and community resilience. Nationally, the Sandy Recovery Improvement Act (FEMA 2013) improved FEMA public assistance delivery, coordination for emergency support of transit system recovery, and required the development of a national strategy to reduce costs on future disasters. Locally, many municipalities have embraced recovery that integrates long-term community resilience, rather than restoring the same type of construction (NYC 2015; Hoboken 2015 are examples).

24.4 Risk-Informed Resilience Assessments

Risk assessments consider the threat or hazard, the vulnerability (or probability of exceeding a limit state) of a facility or system, and the resulting consequences. Risk assessment for community resilience would ideally include all consequences, particularly direct and indirect losses and recovery costs due to damage caused by a given hazard event. However, at present, such a comprehensive assessment of consequences, especially prior to an event, is virtually impossible as there are many dependencies and indirect effects that need to be considered but for which there is little available guidance or data. A risk-informed structured methodology with stakeholder input is needed to identify the most important potential consequences.

There is a considerable body of work in reliability and risk assessments for buildings and infrastructure systems that provide a foundation for risk-informed resilience assessments. Structural reliability accounts for the probability of exceeding a limit state during the service life of a structure for a given hazard. Risk assessments consider the structural reliability (e.g., probability of failure or exceeding a limit state during a specified service life) and the associated consequences of that failure or damage. Reliability analysis and risk assessment methods are well established for addressing hazard events, expected performance, life cycle costs, and estimated damage costs. To support resilience assessments, these approaches need to be integrated into a comprehensive assessment that also considers recovery of functionality. For community-level assessments, methods are needed for estimating the aggregated performance and recovery of a "system of systems" and dependencies between and among systems.

Resilience is a concept that focuses on timely recovery of function. Methodologies that address anticipated damage levels and associated recovery goals are needed to expand the current level of risk assessments, which often evaluate the probability of failure and the consequences of damage in terms of costs, but do not address recovery of function. The following standards and guidance provide an initial basis for evaluating community resilience of the built environment:

- The SPUR (2009) community planning process establishes performance goals and measures for the built environment for a range of hazard intensities.
- The RAMCAP standard (ANSI/ASME-ITI/AWWA 2010) provides a risk-based process for evaluating threats, vulnerabilities, and consequences, with either qualitative or quantitative measures.
- The RR/SAP (ASME-ITI 2011) process uses a "system of systems" approach, based on RAMCAP, to evaluate infrastructure dependencies and interdependencies within a regional system.
- FEMA CPG 201 (DHS 2012) supports risk-informed decision making with a process where hazards, vulnerabilities, and consequences provide a basis for evaluating and improving existing plans and capabilities.

Risk-informed community resilience assessments can be used to evaluate alternative mitigation or design options or to compare the performance of a given alternative option for multiple levels of a selected hazard. With this application in mind, the probability of losses can be expressed as conditional on a given hazard, H, where

$$P\left[L > x \middle| H\right] = \sum\nolimits_{D} P\left[L \middle| D\right]\, P\left[D \middle| H\right]\, P[H] \tag{24.1}$$

and the probability of recovery costs can be similarly expressed as

$$P\left[R > y \middle| H\right] = \sum\nolimits_{D} P\left[R \middle| D\right]\, P\left[D \middle| H\right]\, P[H] \tag{24.2}$$

where $P[H]$ is the probability of a hazard event, H, $P[D|H]$ is the conditional probability of damage, D, given the hazard event, $P[L>x|H]$ is the conditional probability of losses L exceeding a level of x for a given hazard, and $P[R>y|H]$ is the conditional probability of recovery costs R exceeding a level of y for a given hazard. This formulation also supports analyses of hazards that may be expressed as scenarios, such as wildfires or tornadoes, as similarly shown in Chapter 3 of NIST TN 1681 (2010).

At present, available cost data after events for direct and indirect costs of recovery are incomplete. Readily available data include insured losses which, depending on the degree of insurance coverage, only partially address recovery costs expended by communities. The lack of reliable data and methods to estimate recovery costs leads to predictions that underestimate the risk associated with damage. The total costs of mitigation, losses, and repair or replacement need to be considered for each design alternative and set of performance criteria.

24.5 Community Resilience Research for the Built Environment

While resilience can be addressed at different scales (e.g., local, regional), the community scale provides a rational basis and sufficient capacity for bringing together stakeholders to develop long-term community resilience plans and governance to support implementation. Furthermore, having a plan in place will enable prepared communities to quickly recover and be better prepared for future events.

The performance of the built environment in supporting social needs and institutions during and after hazard events is critical to achieving community resilience goals. However, the methodology and tools for establishing community performance goals for recovery of functionality of social and physical systems is not well defined or understood in current practice. Research is needed to meet the immediate short-term need to establish methodologies and tools to provide guidance to communities, based primarily on currently available tools, information, and experience. Research is also needed to provide a risk-informed basis for validated, science-based tools and metrics for community resilience to improve guidance and develop quantitative methodologies.

24.5.1 Short-Term Research Needs

Communities need guidelines based on a risk-informed methodology for resilience planning and decision making that can be tailored for their individual needs, from small, rural to large, urban communities. Needs for community resilience guidance include planning based on long-term community goals, social needs supported by the built environment, and system-level performance goals and recovery metrics for system and community functionality.

These topics are not addressed through application of current codes, standards, regulations, and best practices. System dependencies and cascading effects are also not addressed in current practice, but need to be accounted for in recovery planning. Current metrics for recovery are based on time to recovery of function, as shown in Fig. 24.1, and associated direct and indirect costs. Other metrics already in use by communities for social and economic systems may also be useful for community recovery measures.

There are a number of resilience initiatives and activities at regional, national, and international levels, including assessment methodologies that engage stakeholders in a variety of ways. Resilience activities in the United States include the NIST Community Resilience Planning Guide (NIST 2015), SPUR (2009) Framework, Baseline Resilience Indicators for Communities (BRIC) (Cutter et al. 2014), the Community and Regional Resilience Institute's (CARRI) Community Resilience System (2013), the Oregon Resilience Plan (2013), NOAA's Coastal Resilience Index (2010), and the Communities Advancing Resilience Toolkit (CART) (Pfefferbaum et al. 2013). International initiatives include the United Nations International Strategy for Disaster Reduction (UNISDR 2014) Resilience Scorecard and Rockefeller Foundation's 100 Resilient Cities initiative (Rockefeller 2014). Some methods use qualitative indicators while others use quantitative approaches. Often, the outcomes are presented in the form of scorecards or dashboards. Such visual representations provide a simple but limited way of conveying information both for experts or decision makers. In general, most of these methodologies emphasize social issues, and in some cases, the focus is on one particular social service or system.

The listed initiatives provide a set of categories that contribute to community resilience and, in many cases, include a list of indicators or variables for each category. In cases where the methodologies involve the engagement of community stakeholders, process-oriented guidelines for implementation are included. For methodologies that are heavily quantitative—typically involving readily available data—details are provided about strategies for data analysis and modeling. Most of these resilience initiatives have minimal integration of physical, social, and economic systems and do not address dependencies.

24.5.2 Long-Term Research Needs

Quantitative science-based assessment tools and metrics for community resilience based on an integrated "system of systems" model with incorporated reliability and risk principles will require long-term research. Research needs include development of systems modeling approaches, resilience assessment and economic analysis tools, and validation data.

Systems modeling of integrated infrastructure systems will provide the science basis for community resilience assessment and decision-support tools. Systems models need to account for interdependencies among buildings and infrastructure systems, and the social systems that they support, and address uncertainty in data and knowledge through reliability and risk-informed methods. As robust models for studies of community resilience are developed, a scientific basis for tools and metrics will be established to help communities with their decision making and resilience assessments.

Significant research tasks are associated with integrated models for infrastructure systems. The models, data, and performance criteria can be quite different for each infrastructure system or social representation. For instance, social impact models may be agent-based or system dynamics, economic models may be based on empirical/statistical data, power systems may use network models, and buildings may have fragility representations. Methods for simulating dependencies or cascading effects between these models are not readily available.

A key aspect of any model is the data needed for constructing the models, input data to the model, and output data used for other tools and metrics. The tremendous range of systems and data types encountered in conducting community resilience analyses and assessments requires a broad range of data types and formats. The development of common data architectures and data management tools to enable disaster resilience planning for emergency and decision-making officials, code and standard professionals, engineering design experts, and researchers is critical to the viability of such a large-scale, integrated modeling approach.

Application of risk-informed methods, as indicated in Sect. 24.4, requires the development of data and models for community-level assessments that include the following:

- Community-level performance goals for each physical infrastructure systems based on social needs and dependencies in the integrated community system
- A rational basis for selecting multiple hazard levels to support sensitivity studies of community resilience over a range of demands
- Quantitative reliability-based representations of infrastructure systems and components that include aging and deterioration effects on existing infrastructure performance
- Models that capture the physical performance and service provided by each infrastructure system in a community, including uncertainties in input data, system conditions, and modeling methods
- Models that address social and economic factors to facilitate decision making

- Models that address recovery based on damage levels and available resources
- Integrated models of system dependencies and cascading effects of one system failure on other systems, including uncertainties
- Data for validation of models, particularly recovery of data
- Criteria for determining when model results have adequate confidence intervals for decision making

Resilience assessment tools are needed to assess resilience at the community scale, particularly for physical infrastructure systems. The tools need to address both physical and social systems, including their attributes, functions, and interdependencies. Identification of assessment parameters will be challenging and may benefit from input by communities and coordination with systems modeling methods.

Economic analysis tools are needed to facilitate cost-effective resource allocations that achieve the constrained optimization of short-/long-term costs and benefits to the community. Economic analysis tools, combined with the resilience assessment tools, should provide decision makers at the community/regional level a means to evaluate alternate investment decisions.

No model or tool is ready until it has been validated against experimental or event data. Historical disaster event data can be mined to support model validation and to develop case studies that illustrate community resilience issues, such as the benefits (or lack of) community performance goals, recovery plans, impacts on functionality, and resource allocations and benefits.

To begin addressing these short and long term needs, NIST established a Center of Excellence that is led by Colorado State University (CSU 2016). The Center has three main objectives: to develop an integrated, multi-scale computational environment with systems-level models, develop data architectures and management tools to enable use of multi-disciplinary data, and conduct studies to validate models and data tools for a variety of hazards.

24.6 Summary

Community resilience is a process that takes place over time, with prioritized implementation tasks undertaken as funds and opportunities are available. If the resilience planning process takes place at the individual owner level, without a comprehensive understanding of the building or infrastructure system's role in the community, community resilience will likely not be attained. Planning for resilience and performance goals at the community level provides a rational, integrated basis for determining the resilience and performance goals of individual building and infrastructure systems.

Planning for the integrated performance of community infrastructure and social systems has the added benefit of reducing vulnerabilities, reducing economic losses, and improving overall quality of life through community stability and productivity. However, standardized methods to determine compliance or to estimate expected performance are lacking. Quantitative metrics and validated tools for evaluating performance of the built environment at component (e.g., water facility) and system levels (e.g., water supply) need to be developed.

Current practice in risk analysis predicts the probability of failure or damage for buildings and infrastructure systems, but does not address integrated recovery of function at a community level. The addition of community performance goals and recovery of functionality to the traditional elements of hazard mitigation, system performance, and estimation of damage and losses will provide a more complete risk-based assessment of community resilience through a comprehensive evaluation of risk and consequences. To advance community resilience beyond current practice, evaluation of design, mitigation, and recovery alternatives within a risk-informed methodology is needed to support rational decision making under uncertainty. Prioritization of alternative solutions is necessary for achieving resilience, as community resources and opportunities are limited.

To address these gaps in tools and metrics, research is needed to meet the immediate short-term need to establish methodologies and tools to provide guidance to communities, based primarily on currently available tools, information, and experience. Research is also needed to provide a risk-informed basis for validated, science-based tools and metrics for community resilience to improve guidance and develop quantitative methodologies. A Center of Excellence for community resilience research has been established by NIST to begin addressing these gaps and research needs.

References

ANSI/ASME-ITI/AWWA. (2010). Standard J-100-10, Risk Analysis and Management for Critical Asset Protection (RAMCAP®) Standard for Risk and Resilience Management of Water and Wastewater Systems, American Water Works Association (AWWA). New York, NY: Denver, CO and ASME Innovative Technologies Institute (ASME-ITI).

ASCE. (1990). *Minimum design loads for buildings and other structures, ASCE Standard 7-88*. Reston, VA: American Society of Civil Engineers.

ASCE. (2010). *Minimum design loads for buildings and other structures, ASCE Standard 7-10*. Reston, VA: American Society of Civil Engineers.

ASCE. (2013). *A failure to act, the impact of current infrastructure investment on America's Economic Future, prepared by the Economic Development Research Group Inc*. Reston, VA: American Society of Civil Engineers.

ASME-ITI. (2011). *A regional resilience/security analysis process for the Nation's critical infrastructure systems, December 2011*. New York, NY: ASME Innovative Technologies Institute (ASME-ITI).

Bruneau, M., Chang, S. E., Eguchi, R. T., Lee, G. C., O'Rourke, T. D., Reinhorn, A. M., et al. (2003). A framework to quantitatively assess and enhance the seismic resilience of communities. *Earthquake Spectra, 19*(4), 733–752.

CARRI. (2013). Community and Regional Resilience Institute, Community Resilience System, http://www.resilientus.org/recent-work/community-resilience-system, http://www.resilientus.org/wp-content/uploads/2013/05/CRS-Final-Report.pdf.

Cutter, S. L., Ash, K. D., & Emrich, C. T. (2014). The geographies of community disaster resilience. *Global Environmental Change, 29*, 65–77.

CSU. (2016). Center for Risk-Based Community Resilience Planning, http://resilience.colostate.edu.

DHS. (2009). *National Infrastructure Protection Plan, partnering to enhance protection and resiliency*. Washington, DC: Department of Homeland Security. http://www.dhs.gov/xlibrary/assets/NIPP_Plan.pdf.

DHS. (2012). Threat and Hazard Identification and Risk Assessment Guide, Comprehensive Preparedness Guide (CPG) 201, First Edition, April 2012. Washington, DC: Department of Homeland Security, http://www.fema.gov/library/viewRecord.do?fromSearch=fromsearch&id=5823.

FEMA. (2013). *Sandy Recovery Improvement Act (SRIA) of 2013*. Washington, DC: Federal Emergency Management Agency. https://www.fema.gov/sandy-recovery-improvement-act-2013.

FEMA. (2015). *Presidential disaster declarations by year*. Washington, DC: Federal Emergency Management Agency. https://www.fema.gov/disasters/grid/year.

FBC. (2004). History of the Florida Building Commission, Florida Building Commission, Department of Community Affairs, Building Codes and Standards, Tallahassee, FL. http://www.floridabuilding.org/fbc/information/building_commission.htm.

FHWA. (2012). *Steel bridge design handbook: Limit states* (Publication No. FHWA-IF-12-052-Vol. 10). Washington, DC: US Department of Transportation, Federal Highway Administration.

Hoboken. (2015). Hoboken Recovery Resource Center, Hoboken, NJ, http://www1.nyc.gov/site/recovery/index.page.

ICC. (2015). *International codes-Adoption by Jurisdiction (January 2015)*. International Code Council, http://www.iccsafe.org/gr/Documents/jurisdictionadoptions.pdf.

LSUCCC. (2014). *Uniform Construction Codes and Amendments Effective 1-1-14*. Louisiana State Uniform Construction Code Council, Department of Public Safety, http://lsuccc.dps.louisiana.gov/codes.html.

McAllister, T. P. (2013). *Developing guidelines and standards for disaster resilience of the built environment: A research needs assessment, NIST TN 1795*. Gaithersburg, MD: National Institute of Standards and Technology.

McAllister, T. P. (2014). The performance of essential facilities in Superstorm Sandy. *Structures Congress 2014* (pp. 2269–2281). American Society of Civil Engineers.

McDaniels, T., Chang, S., Cole, D., Mikawoz, J., & Longstafff, H. (2008). Fostering resilience to extreme events within infrastructure systems: Characterizing decision contexts for mitigation and adaptation. *Global Environmental Change, 18*, 310–318.

NAP. (2012). *Disaster resilience, a national imperative*. Washington, DC: The National Academies Press. www.nap.edu.

NHC. (2013). Tropical Cyclone Report, Hurricane Sandy (AL182012) 22–29 October 2012. National Hurricane Center, http://www.nhc.noaa.gov/data/tcr/AL182012_Sandy.pdf.

NIST. (2005). Final Report of the National Construction Safety Team on the Collapses of the World Trade Center Towers (NIST NCSTAR 1). Gaithersburg, MD: National Institute of Standards and Technology, http://www.nist.gov/disaster-resilience/.

NIST. (2006). Performance of physical structures in Hurricane Katrina and Hurricane Rita: A Reconnaissance Report, NIST Technical Note 1476. Gaithersburg, MD, June.

NIST. (2008). Final Report of the National Construction Safety Team on the Collapses of the World Trade Center Building 7. Gaithersburg, MD: National Institute of Standards and Technology.

NIST. (2010). *Best practice guidelines for structural fire resistance design of concrete and steel buildings, NIST TN 1681*. Gaithersburg, MD: National Institute of Standards and Technology.

NIST. (2015). *Community resilience planning guide for buildings and infrastructure systems* (NIST Special Publication 1190). Gaithersburg, MD: National Institute of Standards and Technology.

NOAA. (2005). *Economic statistics for NOAA* (4th ed.). National Oceanic and Atmospheric Administration, U.S. Department of Commerce, May 2005. www.publicaffairs.noaa.gov/pdf/economic-statistics2005.pdf.

NOAA. (2010). *Coastal resilience index: A community self-assessment*. Washington, DC: National Oceanographic and Atmospheric Administration. http://masgc.org/assets/uploads/publications/662/coastalcommunity_resilience_index.pdf.

NRC. (2006). *Improved seismic monitoring – Improved decision making: Assessing the value of reduced uncertainty*. Washington, DC: Committee on the Economic Benefits of Improved Seismic Monitoring, Committee on Seismology and Geodynamics, National Research Council, The National Academies Press.

NWS. (2012). *National Weather Service*. Miami, South Florida: Weather Forecast Office. http://www.srh.noaa.gov/mfl/?n=andrew.

NYC. (2015). *NYC recovery*. New York City, NY, http://www1.nyc.gov/site/recovery/index.page.

Oregon. (2013). *The Oregon Resilience Plan: Reducing risk and improving recovery for the next Cascadia Earthquake and Tsunami*. Salem, OR, February, http://www.oregon.gov/OMD/OEM/osspac/docs/Oregon_Resilience_Plan_Final.pdf.

OSTP. (2008). Grand Challenges for Disaster Reduction, National Science and Technology Council, Committee on Environment and Natural Resources, A Report of the Subcommittee on Disaster Reduction, Jun 2005, Second Printing Jan 2008. Washington, DC: Executive Office of the President, Office of Science and Technology Policy, http://www.sdr.gov/SDRGrandChallengesforDisasterReduction.pdf.

Pfefferbaum, R., Pfefferbaum, B., Van Horn, R. L., Klomp, R. W., Norris, F. H., & Reissman, D. B. (2013). The Communities Advancing Resilience Toolkit (CART): An intervention to build community resilience to disasters. *Journal of Public Health Management & Practice, 19*(3), 250–258.

Pederson, P., Dudenhoeffer, D., Hartley, S., & Permann, M. (2006). *Critical infrastructure interdependency modeling: A survey of U.S. and International Research, INL/EXT-06-11464*. Idaho National Laboratory.

PPD-8. (2011). Presidential Policy Directive, PPD-8 – National Preparedness. The White House, March 30, 2011, http://www.dhs.gov/presidential-policy-directive-8-national-preparedness.

PPD-21. (2013). Presidential Policy Directive/PPD-21. The White House, March 30, 2011, http://www.whitehouse.gov/the-press-office/2013/02/12/presidential-policy-directive-critical-infrastructure-security-and-resil.

Rinaldi, S. M., Peerenboom, J. P., & Kelly, T. K. (2001). Identifying, understanding, and analyzing critical infrastructure interdependencies. *IEEE Control Systems Magazine*, December.

Rockefeller. (2014). *City resilience framework*. New York City, NY: The Rockefeller Foundation, http://www.rockefellerfoundation.org/uploads/files/0bb537c0-d872-467f-9470-b20f57c32488.pdf.

SPUR. (2009). *When is a building safe enough?* San Francisco, CA: San Francisco Urban Research Association, Issue 479, February 2009, http://www.spur.org, http://www.spur.org/sites/default/files/publications_pdfs/SPUR_Seismic_Mitigation_Policies.pdf.

Tsikoudakis, M. (2012). *Hurricane Andrew prompted better building code requirements*. Business Insurance, August 19, 2012, http://www.businessinsurance.com/article/20120819/NEWS06/308199985#full_story.

UNISDR. (2014). Disaster Resilience Scorecard for cities, based on the "Ten Essentials" defined by the United Nations International Strategy for Disaster Risk Reduction (UNISDR) for Making Cities Resilient, Developed for UNISDR by IBM and AECOM, Version 1.5, March 10, 2014, http://www.unisdr.org/2014/campaign-cities/Resilience%20Scorecard%20V1.5.pdf.

Chapter 25
Digital Technologies, Complex Systems, and Extreme Events: Measuring Change in Policy Networks

Louise K. Comfort

Abstract The increasing incidence and mounting costs of extreme events globally in developed and developing societies alike create a compelling need to design effective methods of anticipating and reducing risk in practice. Three basic issues confound systematic measurement of the likely impact of hazards upon global communities: (1) scale of operations, (2) degree of uncertainty, and (3) rate of change in the interaction between hazards and human communities. Each of these issues involves access to, and flow of, information among a complex set of participating organizations and jurisdictions and can be partially addressed by advances in information technology, if designed and used appropriately. We examine different methods of digital data collection, analysis, and modeling that have been used to assess the complex, dynamic interactions between known hazards and communities at risk. We present applications of three digital technologies in reference to three different types of hazards: 2012 Superstorm Sandy, 2012 Pittsburgh International Airport fuel leak scenario, and the 2013 Yarnell Hill, Arizona, wildland fire. We conclude with a preliminary design for a complexity index to measure the interactions among operations, uncertainty, and rate of change over time in a region of risk and calibrate these measures against the existing capacity of a region's sociotechnical system to manage risk.

25.1 The Cost of Global Risk

The increasing incidence and mounting costs of extreme events globally in developed and developing societies alike create a compelling need to design effective methods of anticipating and reducing risk in practice. The global risk assessment conducted by the United Nations International Strategy for Disaster Reduction (ISDR) in preparation for the World Conference on Disaster Risk Reduction

L.K. Comfort (✉)
Center for Disaster Management, Graduate School of Public and Inernational Affairs, University of Pittsburgh, Pittsburgh, PA 15260, USA
e-mail: comfort@gspia.pitt.edu

P. Gardoni, J.M. LaFave (eds.), *Multi-hazard Approaches to Civil Infrastructure Engineering*, DOI 10.1007/978-3-319-29713-2_25

held in Sendai, Japan, March 14–17, 2015, presents a sobering profile of the impact of extreme events on the collective global community. Average losses from earthquakes, typhoons, tornadoes, tsunamis, floods, and wildfires have risen to approximately US$250 to US$300 billion per year, while future losses to infrastructure alone are estimated at US$314 billion per year (UNISDR 2015). Not only do these extreme events destroy critical infrastructure and disrupt economic, social, and cultural activity in nations at risk, but repeated occurrences of such destructive events consume global resources that could otherwise be used to develop stable, sustainable communities for the world's burgeoning population.

The challenge is to anticipate when and where the next extreme event will occur and the extent to which damaging consequences can be mitigated, if not avoided. Anticipating risk is not easy and requires a different frame of analysis than most analytical techniques, based largely on past records, allow. First, it is essential to capture the direction and rate of change in existing physical, engineered, social, and cultural systems that characterize communities exposed to risk. Second, it is necessary to use that information to inform decisions that reduce exposure and enable communities to recognize risk and implement, collectively, informed strategies to build resilience to continuing hazards. Third, it is vital to support public decision making to adapt and adjust collective action dynamically as risk ebbs and flows, calibrating available resources to known risk in timely, informed ways, while acknowledging the potential of sudden, catastrophic events that exceed the capacity of communities to protect their residents.

Developing coherent public policy to address increasingly urgent risk is not trivial. It requires a reconceptualization of the interaction between hazards that occur in physical and meteorological environments beyond human control and the engineered, organizational, social, economic, and cultural environments that communities construct for their safety and development. Translating a reframed conception of this interaction into measurable units means acknowledging the multiple organizations, jurisdictions, and disciplines involved and identifying sources of data and methods of measurement that can be aggregated and integrated into credible models of change to support informed decision making.

Previous efforts to estimate the impact of hazards on built environments have largely focused on single hazards, e.g., earthquakes (FEMA HAZUS-MH Earthquake Model 2014a; OpenQuake 2015) or flooding (FEMA HAZUS-MH Flood Model 2014b), and have built loss scenarios largely on the basis of the engineered environment, excluding more qualitative measures of disruption to social, economic, and organizational performance. The difficult task is to assess the interaction of the likely occurrence of known hazards with the existing technological infrastructure for communications and the capacity that it brings to support rapid communication of risk and mobilization of collective action for communities at risk. Advances in technical communication and the wider availability of social media to community residents mean the possibility of engaging a more informed, active community in managing risk. This possibility, however, is not certain, and the assessment is especially critical for public safety organizations that have legal responsibility for protecting their communities from harm. Further, it is essential to base strategies for mitigation and risk reduction on the underlying science of the hazards to which the communities are exposed. The task of building a common knowledge base to support collective action in environments of risk becomes increasingly complex as interactions among disciplines, organizations, jurisdictions, and social groups influence the degree of shared understanding that is essential to mobilize collective action rapidly in response to an actual extreme event. This inquiry focuses not on a single "system," but on a "complex adaptive system of systems" (Glass et al. 2011) that represents more accurately the societies that confront recurring hazards.

Three basic issues confound systematic measurement of the likely impact of hazards upon global communities. These issues are (1) scale of operations, (2) degree of uncertainty, and (3) rate of change in the interaction between the occurrence of hazards and the actions of human communities seeking to reduce exposure to risk. Each of these issues involves access to, and flow of, information among a complex set of participating organizations and jurisdictions and can be partially addressed by advances in information technology, if designed and used appropriately. The dark side of technology is that it can escalate error as well as insight and create the opposite of what its proponents intend. Consequently, a major challenge for technologists who propose to use advances in information and communications technology to inform decisions in dynamic situations is to recognize the limits of these technologies and identify the threshold points of failure as well as successful performance under different conditions and at different levels of operation in complex societal systems.

This chapter examines different methods of data collection, analysis, and modeling that have been used to address the complex, dynamic interactions between known hazards and communities at risk. The research questions that drive this inquiry are the following: (1) What methods of modeling risk are valid in assessing the interaction between hazards and communities under which conditions; (2) how do these methods change under different social, economic, and cultural conditions; (3) how do these methods change at different scales of operation, local, state, national, and international; and (4) what are the limits of systematic measurement of risk at different thresholds of scale, uncertainty, and change? Addressing these questions offers a beginning means to understand the complex dynamics that drive systems of systems as they adapt (or fail) to extreme events.

25.2 Complex Adaptive Systems of Systems

Four streams of research from different disciplinary perspectives reflect the difficulty, complexity, and cross-scalar character of the problem of managing extreme events.

25.2.1 Complex Adaptive Systems of Systems

Interdisciplinary teams at Sandia National Laboratory (Glass et al. 2011) have developed a promising theoretical and analytical initiative to address *large-scale, eco-socio-economic-technical systems* (LESETs) such as the interaction between recurring hazards that threaten communities and the communities' response to known hazards by adapting their physical, engineered, social, and economic systems to reduce that risk. This theoretical framework, termed "Complex Adaptive Systems of Systems (CASoS) Engineering," includes three phases: (1) define measures of the complex system under study, define factors that influence the system, and create a conceptual model that connects the two; (2) design and test influences based on a mathematical representation of the conceptual model (discrete to continuous agents, networks, and dynamics) in uncertain conditions; and (3) implement a prototype solution in an actual test bed to assess its effectiveness (Glass et al. 2011, pp. 882–883). The process evolves continuously, as interactions among the subsystems adapt to one another and change the environment in which they are operating, precipitating fresh adaptations in the meta-system.

Figure 25.1 adapts the CASoS process of Glass et al. (2011) to the problem of assessing the impact of known hazards on existing communities. This framework offers a means of examining multiple subsystems operating at different scales of operation and monitoring the degree of uncertainty and rate of change within each subsystem. The interaction among these subsystem operations, in turn, drives the rate of change within the entire system of systems. The degree of uncertainty that runs throughout this framework in part reflects the task of defining the boundaries of the subsystems and the intensity of their interaction within the whole system of systems. This issue of uncertainty merits careful exploration, as setting the boundaries of the subsystems at different points affects the degree of interaction among them and consequently the power of the dynamics that drives the meta-system in either positive or negative directions. The framework is based on data from existing subsystems and serves as a starting point for iterative analysis, rather than forecasting potential outcomes.

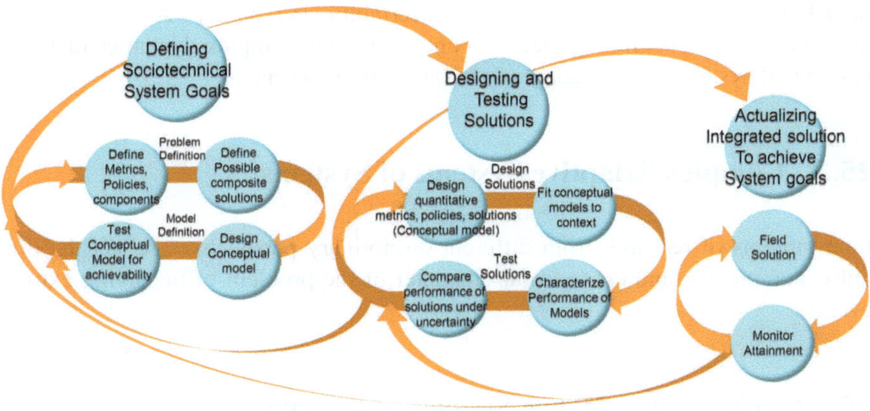

Fig. 25.1 CASoS engineering design process, adapted from Glass et al. (2011), p. 383

25.2.2 Cognitive Machines

Current advances in information technology offer a possible means to bridge the dynamics of natural and technical systems with the development of "cognitive machines" (Nobre et al. 2009) that include sophisticated, timely, and accurate decision support systems. Arguing from Simon's (1997) premise that individuals have limited problem-solving capacity, but large long-term memory space, Nobre et al. (2009) extend the concept of decision support systems for individuals and small groups to collective decision processes for organizations and communities by incorporating machines directly into the problem-solving arena for LESET problems. These researchers assert that organizing the presentation of information regarding complex problems through visual analytics increases the capacity of human decision makers operating in different locations and different disciplines and at different scales of time and responsibility to understand the dynamics of changing conditions in urgent situations. Designing decision support systems that can support the exchange and analysis of information in real time as an event is occurring supports coordinated action in systems of systems. Creating a large-scale, collective problem-solving space supported by cognitive machines extends the capacity of human communities to design innovative adaptations in the macro-system.

25.2.3 Organizational Adaptation to Support Technical Performance

Changes in the technical decision support infrastructure must be accompanied by changes in the organizational and policy processes that drive action. Perrow (2007) examines adaptations in the organizational and communication processes that characterize organizational efforts to introduce change in complex, sociotechnical environments. Each complex adaptive system consists of multiple agents or subsystems that are interacting to some degree with one another and their environment. The degree of interdependence or dependence among the organizational and computational agents engaged in operations to maintain a CASoS under uncertainty is a major factor that shapes the performance of the whole system and its vulnerability to failure in environments exposed to risk (Perrow 2007). The extent to which one system made up of multiple agents depends upon another, either for a physical resource, spatial proximity, or updated status of the changing situation, increases the vulnerability of the second system to failure of the first. If this dual dependency then triggers failure in a third system, a cascade of failure may ripple across all related systems in catastrophic failure of the whole system of interacting systems or the meta-system. The vulnerability of organizational systems that conduct basic response operations in a region exposed to risk cannot be calculated separately from the technical systems that support action, but rather must be based on careful estimates of the degree of interdependence or dependence across the entire sociotechnical system that provides urgently needed services to a stricken region.

25.2.4 Social Skills in Uncertain Environments

Patterns of social interaction influence organizational capacity to mobilize effective strategies of collective action. Fligstein and McAdams (2012) outline a theory of fields of strategic action, highlighting the influence of social skills in building collective capacity to act in uncertain environments. These concepts are central to understanding different organizational and cultural responses to similar threats. Understanding the roles of macro-, meso-, and microlevels of operation in mobilizing community action to mitigate hazards is fundamental to effective management of risk. Rethinking the technical requirements for supporting effective information flow among fields of action becomes a critical task in managing complex systems under stress.

The challenge is to identify and model dynamic, temporally changing, and adapting interactions among physical, engineered, and sociotechnical systems that occur during hazard emergence and response as a complex, adaptive system of systems, so that the CASoS in practice enhances resiliency and enables communities to manage the risk of hazards within existing resource and time constraints.

25.3 Modeling Risk: Methods and Metrics

Measuring risk presents a particular challenge for policy analysts. Risk is an assessment of a threatening event that has not yet happened, but might. The only data available to analysts is from past events and current conditions. Even more difficult is capturing change as it occurs and using that information to inform the next estimate of risk. Methodologists seeking to assess risk and estimate the rate of change in evolving conditions face two tasks: (1) identifying the baseline characteristics in a community for which they are seeking to measure change and (2) identifying the critical variables in that community that are likely to precipitate system-wide change, if triggered by an extreme event. The first task is relatively easily achieved by using widely available geographic information system (GIS) technologies to develop a profile of the existing status of a community at risk. Such a profile needs periodic updating to remain current, as communities undergo continuous change. Further, it is important to include in that profile scientific measures underlying the hazards to which the region is exposed. Without a clear understanding of the geological rock formations, soil structure, and seismology of a region, as well as the behavior and recurrence patterns of earthquakes, engineering codes for buildings may be seriously inadequate. Without comprehension of the meteorological conditions that generate hurricanes or tornadoes, policies for the construction of schools based primarily on cost will be unsafe for children in high wind-hazard regions. If the area is exposed to recurring seismic risk, it is useful to include records of prior earthquakes in the area, including magnitude, depth, duration of shaking, and peak ground acceleration (pga) in a regional risk profile. If the region is exposed to flooding risk, identifying the watersheds that drain into the area and the frequency and depth of flooding from past events is useful. But these data cannot be assumed to serve as predictors of extreme events, due to the influence of multiple other factors that may interact with existing conditions in novel and disruptive ways. To cope with such challenges, digital technologies, such as seismic networks that monitor ground motion or river gages that monitor the rising level of water in the riverbanks during heavy rains, offer opportunities for enhancing timely, valid assessment of risk in near real time.

25.3.1 Sociotechnical Transformation in Information Processing

The distinctive contribution of technical advances in information processes is the potential transformation that adoption of these technologies precipitates in organizational decision making and action. This transformation is not automatic, and it must necessarily be designed and implemented, but technical advances create the potential for more effective management of extreme events. By providing timely, valid access to the current status of a community at risk and rapid transmission of risk information, real-time communication alters the performance of organizations in anticipating and responding to hazards as they occur. This capacity creates the potential for more timely recognition of risk and rapid mobilization of response to extreme events, enhancing risk reduction strategies in practice. By monitoring patterns of change in physical and environmental conditions over time, responsible managers can adjust strategies of coping with complex interactions among technical structures and socioeconomic conditions. Importantly, technical advances now allow easier access to expert knowledge regarding specific hazards and changing contexts of action.

25.3.2 Innovative Methods of Data Collection, Analysis, and Simulation

Critical for researchers, digital technologies also create the potential for exploring more comprehensive and more timely means of data collection, aggregation, and analysis of the actual occurrence of hazards and their impact upon communities. These analyses permit the creation of a broader profile of the impact of a given hazard, such as a hurricane, not only on the stricken community but also the wider affected area. As decision makers view the potential impact of the hazard on the region at risk and share that information graphically, they can explore alternative strategies for action within the wider operational arena to achieve greater functionality within existing constraints and available resources. Creating a "common operating picture," a function supported by digital technologies, increases the potential for collaborative performance among organizations and jurisdictions in the region at risk. Further, this function allows responsible managers to identify possible points of failure in the overall system and estimate the likely consequences of actions taken as they interact in the dynamic process of a complex, adaptive system coping with risk. This common profile of risk would be shared, first, by the local emergency managers who are legally responsible for the safety of their communities but also by managers of nonprofit organizations such as hospitals or schools who are responsible for the welfare of dependent groups of patients or students and by managers of business organizations responsible for employees and clients. Using digital information technologies and peer-to-peer communications, this profile of risk could also be shared throughout the community via households and community groups.

25.3.3 Digital Technologies

Digital technologies provide greater opportunity to monitor performance of component subsystems in any complex, dynamic system. Using sensors to monitor physical processes such as seismic movement that affects electrical generation provides timely information about critical functions in a metropolitan community, such as water flow, traffic lights, and gasoline pumps. Digital technologies allow public managers to monitor interactions among subsystems as risk escalates or diminishes. Careful management of these information flows allows analysts to aggregate and integrate measures from different performance functions to create a common index of risk for the region. This potential, while promising substantial improvement in managing extreme events, can only be achieved through careful design, iterative review, and informed management to achieve optimum performance. Misused or poorly executed, these same technologies create the potential for error and distortion which also can cascade throughout a complex, dynamic system. Digital technologies, in sum, create the structure for sociotechnical systems, but the design, implementation, and analysis of the information they generate remain the responsibility of the organizational and jurisdictional managers who operate the CASoS or complex, adaptive system of systems. These managers could include the local fire chief, police chief, emergency coordinator, school superintendent, hospital administrator, and bank manager, as well as the next level of managers with responsibility for larger jurisdictions and greater numbers of people, at county, state, and national levels of operation.

25.4 Digital Technologies in Practice

Using digital technologies in the analysis of complex, adaptive systems requires a continual process of monitoring, updating, and aggregating incoming information to meet immediate needs. There are basically four phases to this process: data collection, data analysis, visual representation of the results to aid comprehension among different groups, and simulation of complex processes drawn from actual events to anticipate plausible strategies of action in future events. In each of these phases, digital technologies offer innovative approaches that, used appropriately, could aid decision making in complex regions responding to extreme events. Returning to the challenge that pits changing scales of operation against shifting degrees of uncertainty under dynamic rates of change over time, researchers have used digital technologies in creative ways to address and measure these intersecting tensions.

25.4.1 Data Collection in Multiscale Operations

Since systematic measurement requires "sufficient structure to hold and exchange information..." (Kauffmann 1993, p. 174), the first phase in the analysis of any system of action is to identify the structure of operations and the actors who are engaged in the system. Traditionally, data collection to document the actors and actions engaged in response to an extreme event has been a painstaking process, requiring months of detailed review of news accounts, situation reports, agency memoranda, survey of participants, coding, cleaning, and validating before decision makers had sufficient confidence in the data to use it as a basis for decision. In rapidly evolving events, such data collection practices are too late to have any practical contribution to the decision making process. Decision makers were left with limited options and often need to make decisions based on incomplete, inaccurate, or outdated assessments. One means of addressing this need is a technique called rapid ethnographic assessment (REA).

Rapid ethnographic assessment (REA) is an adaptation of the well-known anthropologists' technique of ethnographic assessment, used to characterize the context of action and identify the actors engaged in that context, the relationships among the actors, and the norms that govern those relationships (Mead 1954, 2002; Fairweather and Tornatsky 1977; Fairweather et al. 2013). In its classic form, ethnography is conducted by anthropologists who live in a community for a period of time, become participant observers to the daily routines of the members, and document the structure, content, and norms of the group's behavior. REA seeks the same goal of characterizing the context of action, identifying the actors and the premises that govern their relationships, but does so by using digital data broadly available for any extreme event. Given the rich array of digital data available for any extreme event, Pfeffer and Carley (2012) developed a method of rapid ethnographic assessment, which provides a first analysis of a rapidly emerging event within a short period of time. This method uses automatically prestructured text data from multiple news sources and creates a network of actors based on co-occurrence of words in a common article. This technique is a semantic analysis of words in relation to one another and is based on the assumption that if names of persons, organizations, locations, or events occur in a text within a specific span of words (e.g., window of five or seven words), a relationship is implied among them. This method makes no assumption about the type of relationship and direction or strength of that relationship but simply that co-occurrence in the same unit of text implies a relationship that can be further identified and studied. This technique, used as a rapid screening method, allows a quick review of a vast array of digital data as an initial step in characterizing the context and actors involved in an actual event in near real time.

25.4.1.1 REA in Application to Superstorm Sandy, 2012

To demonstrate the utility of the REA technique, we used it to screen a large collection of digital data in reference to Superstorm Sandy, from October 24, 2012, 5 days before landfall on October 29, 2012, until January 31, 2013. This period included the first warnings about the storm and the first 3 months of an extended recovery period. The event generated a very large volume of digital information—news reports, situation reports from public agencies, video clips, professional assessments, and Facebook and Twitter posts—so that any kind of traditional coding and classification would have taken months of work. Such a delay renders the results of traditional methods of analysis virtually useless for practicing decision makers who are coping with the event in near real time. Using the REA method, a team of graduate student researchers at the Center for Disaster Management (CDM), University of Pittsburgh, successfully screened this large volume of digital information in roughly 5 days, producing a preliminary network diagram of actors by organization and location who were involved in response operations to this event (Comfort et al. 2013). The data collection involved a two-stage process: data extraction and data mining. Following the steps outlined by Pfeffer and Carley (2012) for REA, J. Yeo, a CDM researcher, used the Lexis-Nexus Smart Indexing System to identify relevant articles for this event, using the keywords *Sandy*, *Superstorm Sandy*, and *Hurricane Sandy*. Using this search process, she extracted 1000 articles from the Lexis-Nexus news archive in reference to the storm. Next, she conducted an iterative cleaning and sorting process to remove duplicates from this set of articles, resulting in a total of 681 articles out of the initial 1000 articles, which then served as the basic knowledge base for further analysis (Yeo and Comfort 2015). The second step of the REA method involved data mining to screen the selected 681 articles for known missing items, that is, names of persons, organizations, and events that had been identified in other sources. This process involved iterative review and data cleaning in order to produce a clean dataset for further analysis.[1]

[1]Mark Voortman, postdoctoral fellow at CDM, assisted J. Yeo in this process.

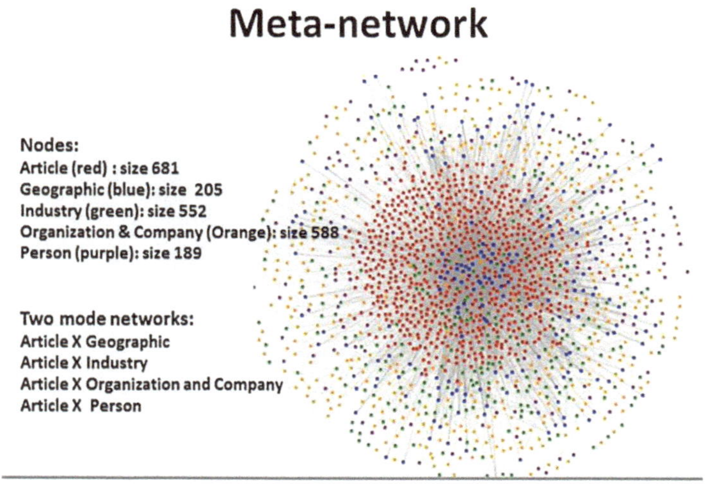

Fig. 25.2 Meta-network of Sandy response system. Lexis-Nexus Archives, 2012. Diagram by J. Yeo

25.4.1.2 Visual Products

The meta-network produced by this process is shown in Fig. 25.2, which identifies the nodes of the network by color according to the six categories used by Lexis-Nexus to classify the data included in 681 news articles by their initial categories. The red nodes represent the 681 articles that were identified through the REA screening process. The blue nodes represent the geographic locations that are being reported in the articles. The green nodes represent the types of industry that are identified in the articles. The orange nodes represent the organizations or companies that are identified in the news articles. The purple nodes represent the individual persons who are named as actors in the articles. The meta-network is interactive when it is launched on an Internet connection, such that a manager could click on any blue node, and it would display the geographic location of that node or any orange node and identify the organization or company that has been identified in the article. Such a method of digital display of information greatly aids a manager who is planning strategic actions to deliver supplies or identify obstructions in humanitarian assistance after disaster. The lines, or edges, show the connections among the nodes.

A second diagram documenting the geographic extent of the actors involved in the Sandy response is shown in Fig. 25.3. The diagram reveals the international and national reach of the response network, from Toronto, Canada, in the north, to Haiti in the Caribbean region, to California on the Pacific Coast, to the heavily affected cities of New York and New Jersey on the Atlantic Coast. The nodes are sized by betweenness centrality, that is, the size of the node indicates roughly the number of organizations that are connected through that specific node to other nodes.

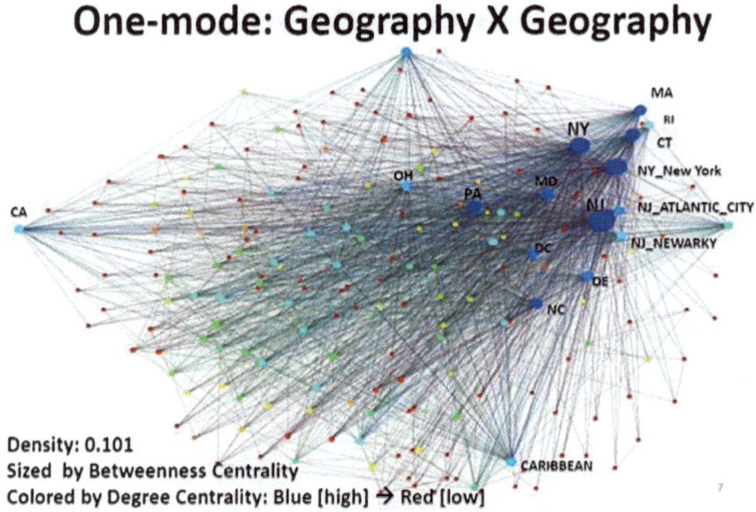

Fig. 25.3 Geographic extent of Sandy response system. Data: Lexis-Nexus Archive, 2012. Diagram by J. Yeo

The larger nodes represent key players in the response system. The graphic display of data from the news articles provides a vivid profile of the geographic impact of the storm, as well as the range of organizations and resources that responded to this event. The database created using the REA technique could then be used for other types of analysis or contrasted with a database created by manual coding from the same datasource (Yeo and Comfort 2015).

Figure 25.4 shows a network diagram of organizations engaged in the Sandy response system during the first week of response operations (October 28–November 3, 2012) by jurisdiction, type of organization (source of funding), and number of transactions performed in response operations. The network diagram, interestingly, reveals a sub-net of interconnected actors, with two actors that had substantial interactions with other organizations. The large red square indicates a national public organization, which is FEMA, and the large red up triangle indicates a national nonprofit organization, which is the Red Cross. The smaller red square indicates the office of President Obama, and the two yellow squares indicate public state organizations: the Governor's Office, New York, and the Governor's Office, New Jersey. The smaller red squares are national public organizations, such as the US Department of Energy, the US Army Corps of Engineers, and the US Small Business Administration. The blue squares indicate municipal public agencies such as the Office of the Mayor of New York City and New York Police Department. The diagram also shows a scattering of organizations that are operating in smaller clusters disconnected from the main cluster of organizations, including a distinct subcluster of international airlines and triads of news media, private organizations,

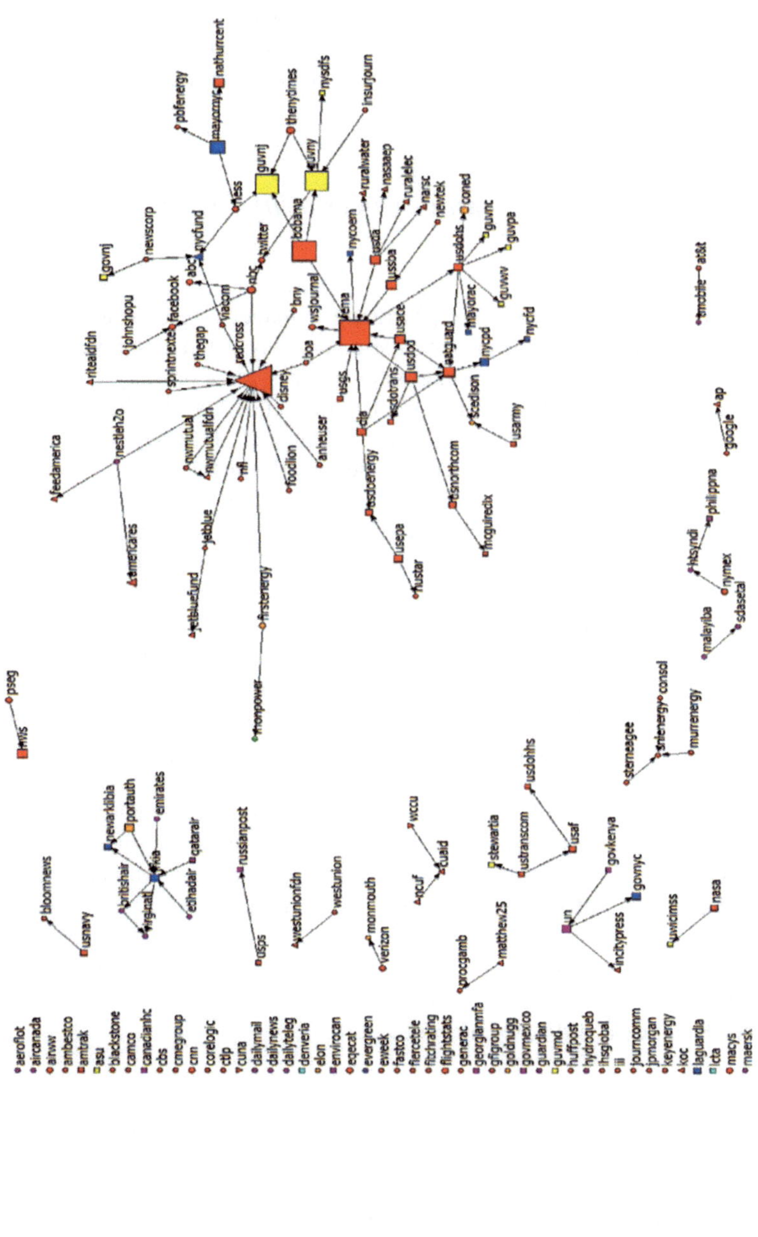

Fig. 25.4 Network diagram of the Sandy response system, October 28–November 3, 2012. Symbols sized according to the total number of transactions (more transactions = larger). Symbols shaped by funding source (*square* = public, *circle* = private, *up triangle* = nonprofit, *down triangle* = not-for-profit/credit union). Symbols colored by jurisdiction (*magenta* = international, *red* = national, *orange* = subnational, *yellow* = state, *green* = substate, *cyan* = county, *blue* = municipal, *purple* = local). Diagram by Jee Eun Song

and other national and international agencies. More striking is the long list of organizations on the left side of the diagram that are not interacting with any other organization. Centrality statistics for this network were calculated using UCINET software (Borgatti et al. 2002). The average degree centrality for the network is 0.44, that is, the average number of connections per node is 0.44, indicating lack of interaction among the responding organizations. This is further supported by the low degree of connectedness among all nodes in the network at .056 and a density measure of 0.056. Although it is the first week of response operations, the network is not strongly interconnected.

25.4.2 Data Analysis: Anticipating Uncertainty Using Bayesian Networks

One of the classic problems confronting practicing managers is decision making under the stress of actual extreme events. While emergency managers have recognized and characterized the need for "situation awareness" as a primary condition for effective decision making (Klein et al. 1993; Comfort et al. 2004), other scholars have noted the importance of "option awareness" (Drury et al. 2009; Klein et al. 2011; Comfort et al. 2013) as a critical factor in the effectiveness of decisions made in extreme events. Option awareness, as defined by G. L. Klein and his colleagues, is that brief moment between the recognition of the threat of an extreme event and the actual decision made to take action (Klein et al. 2011). In that moment, often only seconds, a decision maker reviews the options available and makes a choice for action. In well-structured or familiar environments, the choice is often obvious, and the choice is virtually automatic. But in ill-structured, complex, dynamic environments, there may be multiple factors interacting to produce strategies of action that lead to radically different or unexpected outcomes. If a decision maker follows the scripted rules for standard response strategies in a rapidly changing, uncertain situation, the outcome may be ineffective or, worse, dangerous. In these circumstances, common in complex, dynamic environments, creating awareness of the potential decision space before the hazard occurs can stimulate creative thinking among decision makers about options for action under existing constraints of time and resources (Klein et al. 2011). One technique for exploring the decision space available to practicing managers and increasing awareness of potential options for action in different contexts of threat and uncertainty is to create Bayesian networks to map the potential interdependencies among different possible strategies in an action situation (Klein et al. 1993; Comfort et al. 2013).

25.4.2.1 A Bayesian Network Approach: Pittsburgh Airport Scenario

One means of exploring the validity of option awareness in extreme events is to work with a set of practicing managers to map the decision space available to them in a probable extreme event as a Bayesian network, specifying the actual constraints of time, resources, and location. In a collaborative project with the Mitre Corporation, the CDM staff conducted a field study in conjunction with the Airport Aircraft Rescue and Firefighting (ARFF) facility at the Greater Pittsburgh International Airport to test this concept. The ARFF at Pittsburgh is operated by a small department of 52 highly trained firefighters, most with years of experience and many with advanced degrees in fire science and public management. The firefighters welcomed this opportunity to explore a range of decision strategies with interest and willingness to participate, an important factor in the conduct of the study. There were five stages to this field experiment: (1) mapping the decision space for action in response to a probable hazard in collaboration with working firefighters; (2) identifying all possible strategies for action in the situation, with their interdependencies and limits, given the constraints of physical location, resources, and time defined for the scenario; (3) weighting the identified options for action according to a specified set of priorities given by the firefighters; (4) programming the identified options for action to run in a prototype decision support system that would display the choices made in a comparative format; and (5) testing the validity of the Bayesian model with the practicing firefighters (Comfort et al. 2013).

The scenario used for this field study was a moderate event that might occur any day at the Pittsburgh International Airport. In it, a small tug collides with a plane loaded for takeoff. There are 137 passengers on board; the fuel tank is damaged, and fuel is leaking. The outside temperature is 35 °C, and there is a possible source of ignition. Given this scenario, the firefighters, tasked with responsibility for protecting life and property at the airport, were asked the following questions: (1) *What is the goal of the ARFF?* (2) *Who takes what action first?* (3) *What interdependencies exist among the actions?* (4) *What are the likely consequences of each action and of the cumulative set of actions?* (5) *What is the overall risk for the system?* Their responses were coded, weighted, and incorporated into a Bayesian network; see Fig. 25.5.

In the conduct of this field study, two technologies played key roles. The first was the development of a Bayesian influence diagram using GENIE, a software program developed by Marek Druzdzel at the Decision Sciences Laboratory, University of Pittsburgh, and the second was the software program to run the decision support model, developed by Scott Connelly, CDM. Both technologies contributed to identifying plausible strategies for action, given the constraints of real time, resources, and local conditions defined for the scenario, weighting those strategies in terms of professional priorities for action, and visualizing different outcomes to allow the decision makers to compare strategies for action. The outcome of the field study experiment was a favorable judgment from the ARFF crew regarding this method of exploring new options for decision that they had not considered for this scenario, a validation of the concept of option awareness.

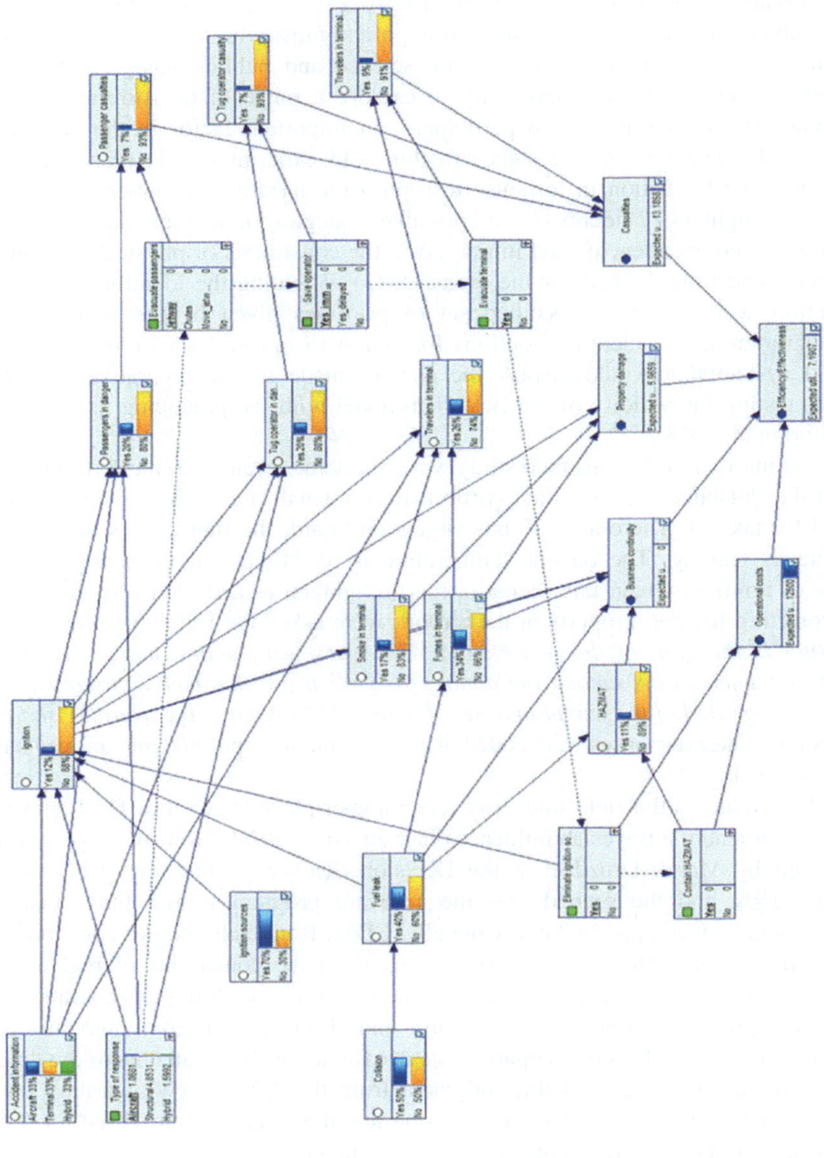

Fig. 25.5 Bayesian influence diagram. Airport Aircraft Rescue and Firefighting team, diagram by M. Voortman, Center for Disaster Management, 2012

25.4.3 *Measuring the Rate of Change in Extreme Events: Using Simulation to Explore Alternatives*

Simulation represents a third technology that is useful in exploring interactions among actors at different scales of operations in large-scale, sociotechnical systems that evolve in response to extreme events. It offers a practical means of estimating the rate of change in rapidly evolving contexts, and it is especially useful in studying situations that are too large, too dangerous, or too costly to replicate in reality. Simulation can be adapted to multiple contexts and offers a useful means of contrasting existing plans for mitigating risk to actual practice in response operations following extreme events. It can be used to replicate, computationally, models of information flow in extreme events and allows the identification of points of system failure in past events. Further, it serves as a means for testing concepts for improved practices in future events in a cost-effective way. Importantly, simulation creates a learning environment for managers of complex, sociotechnical systems, one in which managers can replay actual threats, change the underlying conditions, evaluate different outcomes, and use insights gained from the exercise to improve practice in future events.

Used to reconstruct information flow in actual extreme events, simulation facilitates the redesign of decision processes in programs, organizations, and social interaction in more constructive ways. Computational modeling offers both retrospective and prospective approaches to designing complex adaptive systems. The process is iterative, but it is based on experience and insight gained from practice. By reviewing the chain of decisions in an event that failed, managers can gain fresh insight into complex problems and apply that insight to prospective designs. Current analysts of complex interactions that lead to extreme events recognize the need for innovative approaches (Epstein 2006; Carley 2002), but those approaches need to be tested in actual contexts to discover both their limits and potential for effective action. Critical to improving collective performance in complex systems is identifying the threshold points of change that lead to dynamic models of adaptation and learning. This process is possible, given a digital record of communications gleaned from multiple sources. In an effort to analyze the decision processes that characterized the deadly Yarnell Hill, Arizona, wildfire, June 30, 2013, Karim Hardy, an experienced firefighter/researcher, and I compared professional standards specified for fire personnel against the actual record of communications and actions taken on that day, using the AnyLogic simulation software program (Hardy and Comfort 2015). The simulation revealed points of breakdown in the decision processes among actors in the complex multi-organizational system involved in responding to a rapidly escalating wildfire, despite extraordinary efforts by highly trained, skilled firefighters.

25.4.3.1 Simulation in Application to the Yarnell Hill, AZ, Wildfire, June 30, 2013

Wildfires cross all dimensions of scale, uncertainty, and rapid change. These are among the most difficult events to manage, as there are multiple interacting conditions that are beyond the control of the firefighters on the ground. Physical conditions, such as ground cover and topology of the region, vary, winds change, availability of equipment and personnel may be limited, and importantly, the demands of the event may exceed the capacity of the interorganizational structure within which the operations are being conducted. All of these variables were interacting in 3 days in late June 2013, culminating in a loss of system control and the deaths of 19 members of the Granite Mountain Hotshot Crew. A brief summary of critical decision points is listed below:

- **June 28**: Lightning ignites small fire in dry canyon near Yarnell, AZ.
- 5:30 p.m.: Fire noted by AZSF; no action taken.
- **June 29**: AZSF contacts US BLM; air tankers fly over the area and drop retardant; 37 fires burning in the state; fire expands, but one of many in the state.
- 4:00 p.m.: Wind shifts and fire escalates and expands rapidly; AZSF arrives.
- 11:30 p.m.: Plans response and requests support.
- **June 30**:
- 7:00 a.m.: Interagency planning meeting. Area under fire expanded overnight; Granite Mountain Interagency Hotshot Crew arrives, does not attend the meeting, and goes directly to fire field.
- 3:22: Wind shifts, velocity increases, and fire spreads rapidly.
- 4:00 p.m.: GMIHC radios that it is en route to safety.
- 4:22: Wind gusts escalate; fire rages; all crews ordered out.
- 4:37: Air tanker flies over, hears broken radio transmission, and assumes GMIHC is safe.
- 4:42: Radio transmission is silent.
- **6:35: All 19 members of GMIHC are found dead, overtaken by fire.**

Using the AnyLogic simulation program, we compared the professional standards for fire personnel as listed in the Incident Command System (FEMA 2013; National Wildfire Coordinating Group 2013) against the actual record of communications provided by the Arizona State Forestry Division (2013) to identify the critical points of breakdown in the complex, multiagency, multijurisdictional system that evolved in response to this fire. Figure 25.6 shows the general organizational relationships involved in fighting wildfires, based on documentation from the National Wildfire Coordinating Group's task book (2009) for position performance required for fire personnel. Figure 25.7 shows breakdown in the system.

These factors are more specific to the Yarnell Hill Fire incident and show how inadequate management leads to inadequate response and, unfortunately, to losses. "After-action reports from ineffective incident responses find that response problems are far more likely to result from inadequate management than from any other single cause" (FEMA 2013).

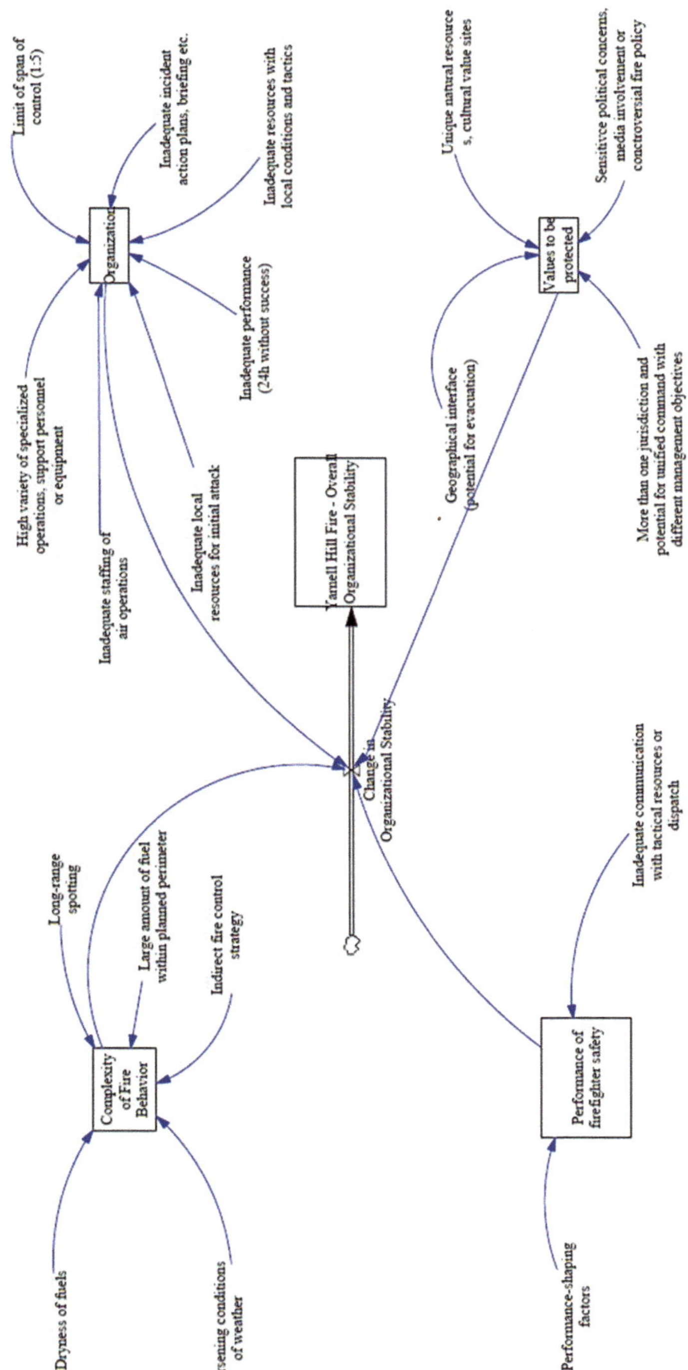

Fig. 25.6 General organizational factors in wildland firefighting adapted from operational documentation, National Wildfire Coordinating Group, 2009. Adapted from PMS 210, 2013. Figure created by K. Hardy

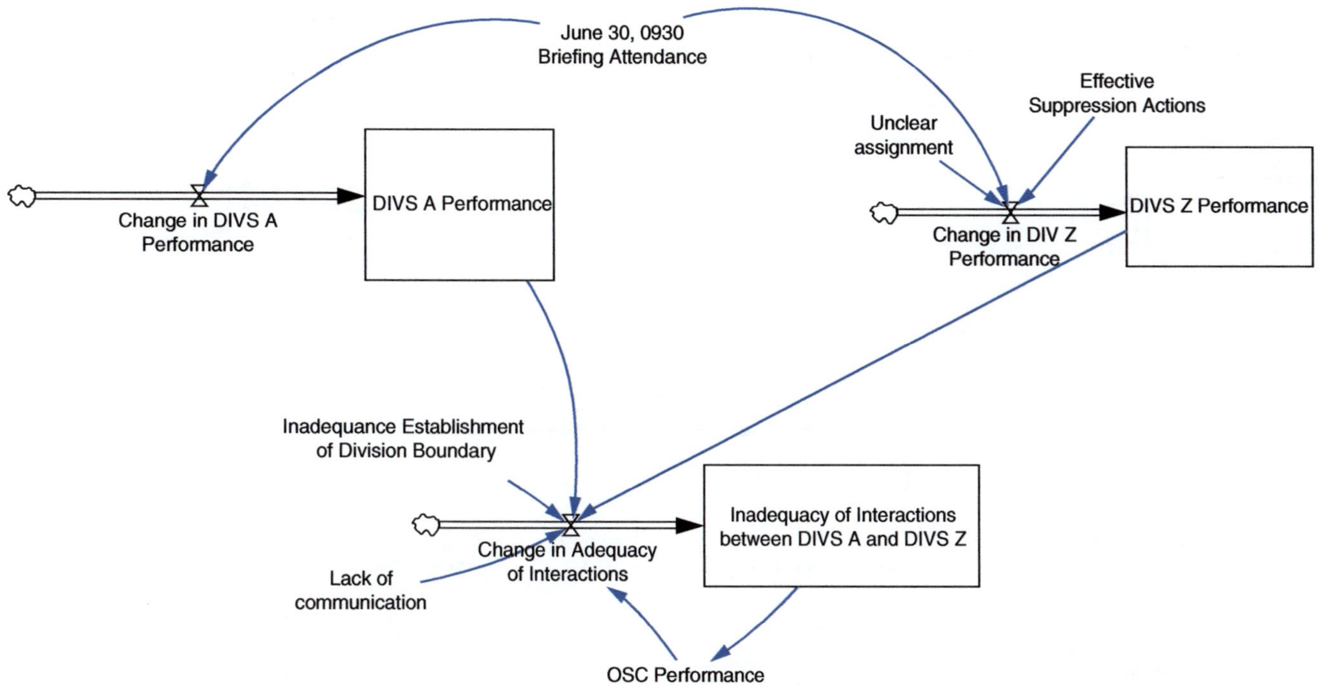

Fig. 25.7 Factors influencing interactions between DIVS A and DIVS Z (Blue Ridge IHC Superintendent) prior to the accident. Adapted from WFA (2013). Figure created by K. Hardy

The use of simulation in reference to the Yarnell Hill, AZ, fire provided a retrospective examination of a set of documented decisions and actions taken in comparison with the set of performance requirements specified for each position in a firefighting crew, according to professional documentation from the National Wildfire Coordinating Group. Such use of simulation is possible when the documented record of actual decisions is provided by real-time records of communications kept by the responsible agency, in this case, the Arizona State Forestry Division. As public agencies increasingly use digital technologies to record and manage extreme events, the data collected makes possible a reconstruction of interacting conditions that have exceeded the capacity of the existing system to manage the event.

25.5 Proposed Complexity Index

The three extreme events characterized briefly above—Superstorm Sandy, Pittsburgh Airport Rescue and Firefighting scenario, and the Yarnell Hill, AZ, wildfire, June 30, 2013—shared a common vulnerability in responding to actual extreme events. The complexity generated by the interacting dimensions of scale, uncertainty, and rate of change exceeded the capacity of the existing organizational systems to anticipate the risk, mobilize response actions at multiple scales of operation, and bring the extreme situations under control with minimal loss. The three dimensions—scale of operations, degree of uncertainty, and rate of change in performance—can be tracked and measured separately. Yet the interactions among them drive the dynamics of system performance—or failure—in extreme events. The critical issue in each case is to enable the affected organizations and communities to recognize the risk of extreme hazards, reallocate their resources, and adapt their actions in time to avoid severe consequences. Building the capacity for collective action in a community that is resilient to hazards requires fresh conceptual models, but sound analytical skills. In this conception, the flow of information through a community drives adaptation, but this dynamic flow is necessarily embedded in a larger context of informed interaction among attentive community participants. Building collective awareness and commitment to risk reduction develops over time with systematic guidance from experienced community and professional leaders. In such a resilient community, there is a shared knowledge of both risk and resources that is based on evidence. The steps to achieve this shared knowledge are similar to any systematic effort to analyze a system. They are:

- Identify key functions in operational system.
- Identify points of interdependence among functions in system operation.
- Identify range of adaptation within each function, among the set of interacting functions.
- Identify key points of potential failure in overall system performance.

- Identify the range of resources—internal and external—available to support action.
- Measure performance of interactions over time.

Yet these steps, none of which are surprising or esoteric, can best be achieved by using systematic methods of digital data collection, storage, analysis, and exchange, creating a "knowledge commons" within the community. A knowledge commons is a shared platform to collect, store, and exchange information that is systematically updated and made accessible to participating community members, preferably in visual format. It allows analysts to model risk in near real time. The goal is to reduce the surprise of unexpected, unanticipated, destructive events by increasing the general awareness of risk and building the collective capacity of individuals, organizations, and jurisdictions to hold and exchange information to support action. One possible approach to achieve this goal is to develop a complexity index that captures the dynamics of the three interacting dimensions of scale, uncertainty, and rate of change. In an initial effort to develop such an index, I offer the following set of propositions:

- As the scale of operations increases, the number of actors and organizations increases.
- As the number of actors and organizations increases, the number of novel interactions within the system also increases.
- As the number of novel interactions within the system increases, the degree of uncertainty regarding possible actions also increases.
- As the degree of uncertainty increases, the number of possible points of error increases.
- As the number of possible points of error increases, the rate of change in operating conditions increases.
- As the rate of change in operating conditions increases, the likelihood of loss of stability or system failure increases.
- As appropriate actions are taken to assess the status of system increase, potential error is likely to decrease.
- As undetected errors cumulate, performance strains and the system, unreinforced, loses stability.

The continuing challenge in communities exposed to increasing risk is to develop and maintain a "knowledge commons" in communities exposed to multiple hazards. Using digital technologies and advanced methods of analysis, analysts can assess the scale of risk for each level of operation, acknowledging the degree of uncertainty and rate of change in interactions among levels of operation. This is a complex set of tasks, but it offers a means of measuring risk in a dynamic society and estimating the cumulative effect of risk generated by interacting physical conditions, actors, organizations, and jurisdictions over time on the wider society.

25.6 Conclusions

Reviewing the questions posed for the use of digital methods in studying extreme events, we offer the following conclusions. First, we explored digital technologies for data collection, data analysis, and modeling and found, not unexpectedly, that the utility of the methods depends in large part on the time and resources available for relevant staff to use them in the context of extreme events. Regarding data collection, the rapid ethnographic analysis method allows screening of a vast amount of digital data in a very short time period, hours or days in comparison with weeks and months for manual methods of data review. Timely coverage of extreme events is more extensive than other methods since agencies and organizations are increasingly making digital data available online. Nonetheless, the data collected represents only a first-level effort at identifying key actors, conditions, and characteristics of an extreme event. Further steps are essential for cleaning the data of duplicate references, coding, and integration of data from multiple sources for coherent analysis.

Second, the availability of digital data is highest at the national level, when federal agencies, national news media, and national organizations post their data online. This means that special efforts need to be taken to ensure adequate coverage at the local level, such as cross-checking the findings from digital data with expert interviews or professional reports. Increasingly, local governments and local organizations are using digital data, but building a valid profile of a region at risk requires ensuring adequate coverage from multiple sources and multiple disciplines.

Third, identifying the limits of measuring risk at different thresholds of scale, uncertainty, and change is an ongoing process. There is no magic solution to this task, and it requires a continuous process of monitoring the dimensions of performance that we do know, while seeking to identify from that baseline, what we do not know. The most effective strategy is likely a continuous cycle of defining, designing, and testing in practice, followed by rigorous review and redesign

Finally, by using digital data and technologies, it is now possible to model conceptual designs to test their strengths and weaknesses *before* we implement them in practice. By using methods such as the development of a Bayesian network of decision choices in an airport fuel leak or a simulation of rules for firefighting against the actual capacity of organizations, personnel, and equipment in the dangerous and rapidly changing environment of a wildfire, professional managers have tools that will enable them to make more timely, effective decisions in extreme events.

Acknowledgments I acknowledge, with thanks and appreciation, the talented young researchers at the Center for Disaster Management, University of Pittsburgh, who assisted with the application of these technologies in the analysis and modeling of Superstorm Sandy and the Pittsburgh International Airport scenario. They are Brian Chalfant, Jee Eun Song, Mengyao Chen, Jungwon Yeo, Mark Voortman, and Brian Colella. I warmly thank Karim Hardy for his modeling of organizational interactions in the 2013 Yarnell Hill, Arizona, wildfire. We also acknowledge, with gratitude, the National Association of Workforce Boards for its support of the Sandy study and the Mitre Corporation for its support of the Pittsburgh International Airport study.

References

Anylogic Software version 6. (2013). Imported and exported Vensim PLE models. Chicago, IL.

Arizona State Forestry Division. (2013). Yarnell Hill fire serious accident investigation report, September 2013. Tucson, AZ.

Borgatti, S. P., Everett, M. G., & Freeman, L. C. (2002). *Ucinet 6 for windows: Software for social network analysis*. Harvard, MA: Analytic Technologies.

Carley, K. M. (2002). Smart agents and organizations of the future. In L. Lievrouw & S. Livingstone (Eds.), *The handbook of new media* (pp. 206–220). California: Sage.

Comfort, L. K., Chalfant, B., Song, J. E., Yeo, J., Chen, M., & Colella, B. (2013). *The impact of extreme events on business organizations: Innovative approaches to managing risk for the national association of workforce boards*. Available via http://www.nawb.org/publications.asp.

Comfort, L. K., Colella, B., Voortman, M. Connelly, S., Drury, J. L., & Klein, G. L. (2013). Real-time Decision Making in Urgent Events: Modeling Options for Action. In Comes, T., Fiedrich, F., Fortner, S., Gerldermann, J. and Mueller, T. eds. Proceedings of the 10th International ISCRAM Conference, Baden Baden, Germany, May, 2013.

Comfort, L. K., Dunn, M., & Johnson, D., Zagorecki, A., & Skertich, R. (2004). Coordination in complex systems: Increasing efficiency in disaster mitigation and response. *International Journal of Emergency Management, 2*(1–2), 62–80.

Drury, J. L., Klein, G. L., Pfaff, M., & More, L. (2009). Dynamic decision support for emergency responders. In: *2009 IEEE technologies for homeland security conference*. Waltham, MA, May 2009.

Epstein, J. (2006). *Generative social science: Studies in agent-based computational modeling (Princeton studies in complexity)*. Princeton: Princeton University Press.

Fairweather, G. W., & Tornatsky, L. G. (1977). *Experimental methods for social policy research*. New York: Pergamon.

Fairweather, G. W., Tornatsky, L. G., & Goldstein, A. P. (2013). *Experimental methods for social policy research*. New York: Pergamon.

FEMA. (2013). *ICS-100.a – Introduction to incident command system*. Washington, DC

FEMA. (2014a). *HAZUS-MH earthquake model*. https://www.fema.gov/hazus-mh-earthquake-model#. Accessed 26 September 2015.

FEMA. (2014b). *HAZUS-MH flood model*. https://www.fema.gov/hazus-mh-flood-model. Accessed 26 September 2015.

Fligstein, N., & McAdams, D. (2012). *A theory of fields*. Oxford: Oxford University Press.

GEM. (2015). *Openquake: Global Earthquake Model*. www.globalquakemodel.org/openquake/. Accessed 26 September 2015.

Glass, R. J., Ames, A. L., Brown, T. J., Maffitt, S. L., Beyeler, W., & Finkey, D. (2011). *Complex adaptive systems of systems (CASoS) engineering: Mapping aspirations to problem solutions*. Paper presented at 8th International Conference on Complex Systems and 6th IEEE International Conference on Systems of Systems Engineering (SoSE), June (SAND 2011-3354 C). Also available via http://www.sandia.gov/CasosEngineering/docs/ICCS_Mapping_Aspirations_2011-3354.pdf.

Hardy, K., & Comfort, L. K. (2015). Dynamic decision processes in complex, high-risk operations: The Yarnell Hill fire, June 30, 2013. *Safety Science, 71*, 39–47.

Kauffmann, S. (1993). Origins of Order. Oxford: Oxford University Press.

Klein, G. A., Orasanu, J., Calderwood, R., & Zsambok, C. E. (1993). *Decision making in action: Models and methods*. Norwood: Ablex Publishing.

Klein, G. L., Drury, J. L., & Pfaff, M. S. (2011). Providing an option awareness basis for naturalistic decision making. *Journal of Cognitive Technology, 16*(2), 10–19.

Mead, M. (1954). *2002 methods of research of contemporary cultures. Reissued as understanding ourselves: Theory and method in anthropologies of contemporary western cultures*. New York: Berghahn Books.

National Wildfire Coordinating Group. (2013). Wildland fire incident management field guide, PMS 210, April 2013.

Nobre, F. S., Tobias, A. M., & Walker, D. S. (2009). Organizational technological implications of cognitive machines: Designing future information management systems. IGI Global 15

Perrow, C. (2007). *The next catastrophe: Reducing our vulnerabilities to natural, industrial, and terrorist disasters*. Princeton: Princeton University Press.

Pfeffer, J., & Carley, K. M. (2012). Rapid modeling and analyzing networks extracted from pre-structured news articles. *Computational and Mathematical Organization Theory, 18*(3), 280–299.

Simon, H. A. (1997). *Administrative behavior: A study of decision-making processes in administrative organizations*. New York: The Free Press.

UNISDR. (2015). *Global risk assessment report on disaster risk reduction*. Geneva: United Nations International Strategy for Disaster Reduction.

WFA. (2013). Granite mountain IHC entrapment and burnover investigation, Yarnell Hill fire – June 30, 2013. Wild Fire Associates Report. Prepared for Arizona Division of Occupational Safety and Health, November 2013.

Yeo, J., & Comfort, L. K. (2015). *Toward rapid and valid information for disaster and emergency rescue and response operations: Rapid network assessment method vs. manual content analysis method*. Working Paper, University of Pittsburgh, Center for Disaster Management, Pittsburgh.

MIX
Papier aus verantwortungsvollen Quellen
Paper from responsible sources
FSC® C105338

If you have any concerns about our products,
you can contact us on
ProductSafety@springernature.com

In case Publisher is established outside the EU,
the EU authorized representative is:
Springer Nature Customer Service Center GmbH
Europaplatz 3, 69115 Heidelberg, Germany

Printed by Libri Plureos GmbH
in Hamburg, Germany